PROPERTY OF FUNK SEEDS

NAME _____ W9-BVH-720

ANNUAL REVIEW OF PHYTOPATHOLOGY

EDITORIAL COMMITTEE (1979)

R. D. BERGER

J. A. BROWNING

R. J. COOK

E. B. COWLING

R. G. GROGAN

R. M. LISTER

R. L. MILLAR

G. A. ZENTMYER

Responsible for the organization of Volume 17
(Editorial Committee, 1977)

K. F. BAKER

J. A. BROWNING

E. B. COWLING

R. G. GROGAN

R. M. LISTER

J. TAMMEN

P. H. WILLIAMS

G. A. ZENTMYER

R. D. DURBIN (Guest)

D. P. MAXWELL (Guest)

G. L. WORF (Guest)

Production Editor   T. HASKELL

Indexing Coordinator   M. A. GLASS

Subject Indexer   D. C. HILDEBRAND

ANNUAL REVIEW OF PHYTOPATHOLOGY

RAYMOND G. GROGAN, *Editor*
The University of California, Davis

GEORGE A. ZENTMYER, *Associate Editor*
The University of California, Riverside

ELLIS B. COWLING, *Associate Editor*
North Carolina State University

VOLUME 17

1979

ANNUAL REVIEWS INC.   4139 EL CAMINO WAY   PALO ALTO, CALIFORNIA 94306

ANNUAL REVIEWS INC.
Palo Alto, California, USA

COPYRIGHT © 1979 BY ANNUAL REVIEWS INC., PALO ALTO, CALIFORNIA, USA. ALL RIGHTS RESERVED. The appearance of the code at the bottom of the first page of an article in this serial indicates the copyright owner's consent that copies of the article may be made for personal or internal use, or for the personal or internal use of specific clients. This consent is given on the condition, however, that the copier pay the stated per-copy fee of $1.00 per article through the Copyright Clearance Center, Inc. (P. O. Box 765, Schenectady, NY 12301) for copying beyond that permitted by Sections 107 or 108 of the US Copyright Law. The per-copy fee of $1.00 per article also applies to the copying, under the stated conditions, of articles published in any Annual Review serial before January 1, 1978. Individual readers, and nonprofit libraries acting for them, are permitted to make a single copy of an article without charge for use in research or teaching. This consent does not extend to other kinds of copying, such as copying for general distribution, for advertising or promotional purposes, for creating new collective works, or for resale.

REPRINTS The conspicuous number aligned in the margin with the title of each article in this volume is a key for use in ordering reprints. Available reprints are priced at the uniform rate of $1.00 each postpaid. The minimum acceptable reprint order is 5 reprints and/or $5.00 prepaid. A quantity discount is available.

International Standard Serial Number: 0066–4286
International Standard Book Number: 0–8243–1317–8
Library of Congress Catalog Card Number: 63–8847

Annual Reviews Inc. and the Editors of its publications assume no responsibility for the statements expressed by the contributors to this Review.

PRINTED AND BOUND IN THE UNITED STATES OF AMERICA

PREFACE

The topics and authors for Volume 17 of the Annual Review of Phytopathology were selected by the editorial committee at a meeting in Madison, Wisconsin, during August of 1977. In common with the committees who planned previous volumes, we were faced with the difficult task of choosing topics that would be most timely and still would be in the forefront of significant research progress in phytopathology two years later. But of even greater importance was choosing prospective authors who, because of outstanding comprehension and leadership in the subject areas, could be expected to produce up-to-date, authoritative, instructive, and stimulating reviews for our diverse readers.

The accomplishment of our objectives is due in large part to the suggestions for topics and authors that have been made to members of the editorial board. We solicit your comments on this volume and welcome your suggestions for topics and authors for subsequent volumes.

We welcome R. D. Berger as the new member of the Editorial Committee, replacing P. H. Williams who concludes his five-year tenure, and we thank R. D. Durbin, D. P. Maxwell, and G. L. Worf, guest committeemen who helped plan this volume.

THE EDITORIAL COMMITTEE

CONTENTS

SOME RELATED ARTICLES IN OTHER ANNUAL REVIEWS

From the *Annual Review of Biochemistry,* Volume 47, 1978

Explorations of Bacterial Metabolism, H. A. Barker
*The Outer Membrane Proteins of Gram-Negative Bacteria:
Biosynthesis, Assembly, and Functions,* Joseph M. DiRienzo, Kenzo
Nakamura, and Masayori Inouye
Amino Acid Biosynthesis and Its Regulation, H. E. Umbarger
The Plasma Lipoproteins: Structure and Metabolism, Louis C. Smith,
Henry J. Pownall, and Antonio M. Gotto, Jr.

From the *Annual Review of Ecology and Systematics,* Volume 9, 1978

The Statistical Prediction of Population Fluctuations, Robert W. Poole

From the *Annual Review of Entomology,* Volume 24, 1979

Integrated Pest Control in the Developing World, L. Brader
Pest Management in Rice, Keizi Kiritani

From the *Annual Review of Genetics,* Volume 12, 1978

Some Features of Genetic Recombination in Procaryotes, Maurice S.
Fox
Modes of Gene Transfer and Recombination in Bacteria, K. Brooks
Low and Ronald D. Porter
Transmission Genetics of Mitochondria and Chloroplasts, C. William
Birky, Jr.

From the *Annual Review of Microbiology,* Volume 32, 1978

Phagocytosis as a Surface Phenomenon, C. J. van Oss
Formation, Properties, and Germination of Actinomycete Spores,
Jerald C. Ensign
Regulation of Bacterial Growth, RNA, and Protein Synthesis, Donald
P. Nierlich
Plasmid-Determined Resistance to Antimicrobial Agents, Julian Davies
and David I. Smith

From the *Annual Review of Plant Physiology,* Volume 30, 1979

Enzymic Controls in the Biosynthesis of Lignin and Flavonoids,
Klaus Hahlbrock and Hans Grisebach
DNA Plant Viruses, Robert J. Shepherd
Biosynthesis and Action of Ethylene, Morris Lieberman
Physiological Aspects of Desiccation Tolerance, J. Derek Bewley

ANNUAL REVIEWS INC. is a nonprofit corporation established to promote the advancement of the sciences. Beginning in 1932 with the *Annual Review of Biochemistry*, the Company has pursued as its principal function the publication of high quality, reasonably priced Annual Review volumes. The volumes are organized by Editors and Editorial Committees who invite qualified authors to contribute critical articles reviewing significant developments within each major discipline.

Annual Reviews Inc. is administered by a Board of Directors whose members serve without compensation. The Board for 1979 is constituted as follows:

Dr. J. Murray Luck, Founder and Director Emeritus of Annual Reviews Inc.
Professor Emeritus of Chemistry, Stanford University
Dr. Joshua Lederberg, President of Annual Reviews Inc.
President, The Rockefeller University
Dr. James E. Howell, Vice President of Annual Reviews Inc.
Professor of Economics, Stanford University
Dr. William O. Baker, *President, Bell Telephone Laboratories*
Dr. Robert W. Berliner, *Dean, Yale University School of Medicine*
Dr. Sidney D. Drell, *Deputy Director, Stanford Linear Accelerator Center*
Dr. Eugene Garfield, *President, Institute for Scientific Information*
Dr. William D. McElroy, *Chancellor, University of California, San Diego*
Dr. William F. Miller, *Professor of Computer Science and Business, Stanford University*
Dr. Colin S. Pittendrigh, *Director, Hopkins Marine Station*
Dr. Esmond E. Snell, *Professor of Microbiology and Chemistry, University of Texas, Austin*
Dr. Harriet Zuckerman, *Professor of Sociology, Columbia University*

The management of Annual Reviews Inc. is constituted as follows:

John S. McNeil, Chief Executive Officer and Secretary-Treasurer
William Kaufmann, Editor-in-Chief
Lawrence C. Russell, Business Manager
Sharon E. Hawkes, Production Manager
Ruth E. Severance, Promotion Manager

Annual Reviews are published in the following sciences: Anthropology, Astronomy and Astrophysics, Biochemistry, Biophysics and Bioengineering, Earth and Planetary Sciences, Ecology and Systematics, Energy, Entomology, Fluid Mechanics, Genetics, Materials Science, Medicine, Microbiology, Neuroscience, Nuclear and Particle Science, Pharmacology and Toxicology, Physical Chemistry, Physiology, Phytopathology, Plant Physiology, Psychology, and Sociology. The *Annual Review of Public Health* will begin publication in 1980. In addition, four special volumes have been published by Annual Reviews Inc.: *History of Entomology* (1973), *The Excitement and Fascination of Science* (1965), *The Excitement and Fascination of Science, Volume Two* (1978), and *Annual Reviews Reprints: Cell Membranes, 1975–1977* (published 1978). For the convenience of readers, a detachable order form/envelope is bound into the back of this volume.

William B. Hewitt

Ann. Rev. Phytopathol. 1979. 17:1–12
Copyright © 1979 by Annual Reviews Inc. All rights reserved

CONCEPTUALIZING IN PLANT PATHOLOGY

❖3696

William B. Hewitt[1]

Professor of Plant Pathology Emeritus, University of California, Davis 95616, and San Joaquin Valley Agricultural Research and Extension Center, Parlier, California 93648

> Tell me not in mournful numbers,
> Life is but an empty dream!
> For thy soul is dead that slumbers,
> and things are not what they seem.
>
> Henry Wadsworth Longfellow,
> *Psalm of Life*

It is an honor beyond dreams to have been invited to write this chapter and to choose the topic.

About the quotation from Longfellow, a reviewer of an earlier version of this chapter commented: "How is this quotation related to your ideas about concepts?" I might reply in paraphrases: "Let not your soul slumber, for things are not what they seem." If we merely follow the beaten path or just sit in our canoe and float with the stream, relying on the imagination of others, accepting results of scientific endeavor without question, and/or doing more of what has been done, then "Life is but an empty dream." Whereas, if we take a new path or paddle our canoe across the current and into the byways of the stream, are inquisitive about scientific works, and are imaginative in exploring things, then life is real and things are more like what they seem.

OUR GOAL AS PLANT PATHOLOGISTS

Our goal as plant pathologists is to keep plants in good productive health. That goal is important because healthy and productive plants not only are

[1]Present address: 2325 Glacier Lane, Santa Maria, California 93454.

1

0066-4286/79/0901-0001$01.00

essential but are the very essence of humankind—environment, food, fiber, energy, and general well-being. To reach our goal we plant pathologists must continue to grow not only in knowledge but also in breadth and depth of mind. We need to excel in performance, increase our knowledge, enlarge our scope of concepts in our science, and improve the science and art of plant pathology.

CHAPTER OBJECTIVE AND CONTENTS

My objective in this chapter is to encourage creative thinking in solving our problems and, in so doing, to take the unbeaten path; it is to encourage conceptualizing in terms as broad as the mind can comprehend, to be inquisitive about things in terms of knowledge in all biological and impinging sciences, and to perform to the maximum of our potential.

This chapter discusses some of my experiences in thinking, forming concepts, and solving plant-pathological problems. I also discuss some of my ideas about departments of plant pathology and the performance of individuals.

The examples given stress my own experiences in exploring, solving problems, and forming concepts in plant pathology. Naturally, my own work is familiar to me and I feel free to discuss it; so much of this chapter is in the first person—which is not to imply that I am an authority. I am simply relating some of my experiences in my career in plant pathology, as the editor suggested.

ABOUT THINKING AND CONCEPTS

It is conventional to believe that the mind is improved by acquiring knowledge, and it is remarkable the amount of information that one can acquire and recall. However, there is more to improving the mind than acquiring knowledge. Thinking is a vital and powerful function. Furthermore, creative thinking is very important to us in plant pathology. I believe that creative thinking can be cultivated—or it can be suppressed—that thinking processes can be exercised, and that, with reasonable health, one is never so old that these functions cannot be improved. Although the thinking processes appear not to be fully understood, I like to believe that I think by forming mental images of things and processes. At times, however, I have had difficulty with my imagination and in thinking clearly. In my efforts to put an end to such episodes, I discovered that many forces have influenced my thinking—and not always the same forces.

Before continuing with this discourse, permit me to digress. The title of this chapter is "Conceptualizing in Plant Pathology." My dictionary defines concept as "the product of the faculty of conception; an idea of a class of objects, a general notion." Conception is defined as "the action or the faculty of conceiving in the mind," and conceptualize is "to make or form a concept or concepts." We are discussing concepts, the power of the faculty of the mind to think, and some of the forces that influence creative thinking in solving plant disease problems. Let us now discuss some of the forces that influence thinking.

FORCES THAT INFLUENCE THINKING

In retrospect, I find it important that I like what I am doing, that it is a challenge, and that it is constructive. Forces that have influenced my thinking and at times still do so are my environment, emotions, ability to concentrate, perception and judgment, acquired knowledge and ability to recall information, my background, and/or my concepts of and about things. Let us discuss each of those forces briefly.

ENVIRONMENT   I prefer a zealous atmosphere; one that encourages and fosters exploration. Some other factors of the environment that have influenced my thinking include the attitude of the administration in support of my department, and cooperation and rapport among my colleagues.

EMOTIONS   Aside from an occasional eruption, my emotions are usually under reasonable control. Emotions can disrupt or block out thinking processes. I have observed that fear has a great influence on some of my colleagues. For example, fear about insecurity, of not being promoted, of making a mistake, and of the "publish or perish" concept. These fears are interrelated and can usually be overcome when one considers that published papers are reports on work done; they are our principal mode of communication. There should be no fear in reporting accomplishments, only pleasure in completing a job.

CONCENTRATION   I have always been subject to distractions and to the drifting of the mind. Consequently, I had to train myself to follow a subject and now I can tune out almost all sorts of distractions.

PERCEPTION AND JUDGMENT   An artist has good powers of perception. In plant pathology, it is important that we see a problem and evaluate what we see (use our best judgment). In viewing problems, one might consider such questions as: Do I see the real problem? Do I see all of the problem? Have I looked at the problem from all angles or do I see only that which I expect or wish to see? And can I define or delimit the problem? A satisfactory answer to such questions has helped me solve problems.

ACQUISITION AND RECALL OF KNOWLEDGE Since all of us have devoted much time to learning and recalling information and it appears we do very well at it, I shall only call attention to the fact that it is important for us to gain and use as much knowledge as possible in all biological and other sciences impinging on plant pathology. Such information broadens our vista in solving problems and forming concepts.

BACKGROUND AND CONCEPTS My background—early family life, traditions, taboos, schooling, play, work, etc—has influenced and at times continues to emerge in my thinking. My thinking is influenced also by the books I read, the radio and television programs I hear and view, and current events, experiences, etc.

We all have our concepts about this, that, and other things. Concepts of long standing and traditions become "part of us." Concepts of our behavior tend to control our emotions and actions and permit us to function subconsciously on many routine matters. Furthermore, every group of individuals, be it a family, a trade, or a profession, has its own concepts and traditions, and so do we as plant pathologists. We have learned to think and to talk like plant pathologists, and we even have our own jargon. We were trained that way and that is the way we train our students. We tend to follow tested and accepted lines; we become conventional; we take the "beaten path." As a group we have kindred feelings. One can sense this at our meetings; at times we may even feel a bit out of place at a meeting of another profession. A certain amount of subjective control is good, especially to the degree that it supports a healthy self-esteem and promotes self-discipline, though not when it restricts or blocks creative thinking. Concepts, tradition, and custom, all influence what we do as plant pathologists. Given a plant disease problem, the usual approach is to delimit it within our preconceived notions. In essence, we observe the problem in retrospect and tend to analyze it from our most active and/or recent experiences. I do not question the values of conventional procedures and/or approaches in resolving and solving plant-disease problems and forming concepts. My point is that they may or may not complete the job. The following are some examples from my own experiences.

SOLVING PROBLEMS AND FORMING CONCEPTS: SOME EXAMPLES

Nematode Vector of Soilborne Fanleaf Virus of Grapevines

The mode of transmission of soilborne virus diseases was a subject of much research activity during the 1950s. Under various names, grapevine fanleaf disease occurred in many viticultural countries of the world (6). The soilborne nature of the disease was known in Europe before the turn of the century (6). It was also known that some ten years of fallow were required before grapevines could be replanted in the soil without the disease developing in the new planting (6). In California, the relative short life of new vineyards and replanted vineyards as a result of degeneration problems prompted our program of studies on virus and virus-like diseases. The control concept—plant clean materials in clean soil—became the basis of our program. We had developed some disease-free planting materials for new vineyard soils, but, because fanleaf, the dominant degenerative vine disease, was soilborne, old vineyard soils could not be replanted safely. Although it had been demonstrated in Europe that soil fumigation would destroy the soilborne nature of fanleaf (6), the mode of retention and transmission in soil was not known. Development of that knowledge became our immediate objective. Although the literature at that time indicated others were exploring virus-soil relationships, we reasoned that since many viruses were spread by biological vectors the soilborne viruses could also have biological vectors and soils contained many organisms that fed on or were parasitic on roots of plants.

Our approach was conventional and simple. Three-gallon containers were filled with soil from the root zone of fanleaf-diseased grapevines (fanleaf-soil). Containers with sterilized and unsterilized fanleaf-soil were planted with disease-free grape rootings. Other containers were planted with fanleaf-diseased rootings. All containers were placed in a screen enclosure. In the spring following planting, the vines in most containers had fanleaf disease, indicating that the fanleaf soil in them had the soilborne virus and the transmitting agent (vector) (6). In further work, we used fanleaf-soil from these containers to obtain mites, root aphids, and different species of nematodes that were tested for capability to transmit fanleaf virus.

Nematodes were screened from the soil. Bulk lots of mixed nematodes from these screenings and handpicked lots of specific species known to feed on roots were placed about the roots of small healthy grape plants. Samples of bulked nematode species transmitted fanleaf virus to grape rootings and so did samples of *Xiphinema index* alone whereas other species did not (6). Since lots of other species of nematodes tested were also very likely to carry any other possible microvector but did not transmit fanleaf, it was apparent that *X. index* was a vector of grapevine fanleaf virus. Furthermore, comparative examinations of vineyard soils showed a correlation of *X. index* with fanleaf disease in grapevines. Of course, the tests were repeated and enlarged upon to be certain of our results (6).

The important points for this discussion are that the conventional approach solved the problem. The demonstration that nematodes were vectors of a soilborne virus was a new concept and stimulated much work on the relationship of nematodes to soilborne plant viruses and on nematodes as disease-inducing pathogens.

Big-Vein of Lettuce: Another Soilborne Disease Problem

This work is a good example of a step-by-step exploration in developing facts and analyzing a disease problem. As early as 1930 the big-vein disease of lettuce was reported to be soilborne. The cause and mode of transmission, however, had not been demonstrated (3). R. G. Grogan was exploring this soilborne disease of lettuce at the same time that work was in progress on soilborne fanleaf disease of grapevines and he and I frequently compared developments in our work and shared in some experiments.

The association of big-vein with *Olpidium brassicae* was determined through a series of experiments which, in turn, demonstrated that the chemical fumigation of big-vein soil destroyed the ability of soil to transmit the disease to lettuce (infectivity); chemical surface sterilization of whole plants, roots, root pieces, and scrapings from roots of lettuce plants destroyed infectivity; and infectivity was associated with scrapings from the surface of roots but not with inner tissues of roots. Furthermore, the infectivity accompanied juice extracted from roots and it passed through a fritted-glass filter with 40 μm pores. After determination of the minimum size of lettuce rootlet portion that usually resulted in infection when used as soil inoculum, a comparative microscopic examination of roots of about the same size showed that the roots of lettuce with big-vein were infected with *O. brassicae* but roots of healthy lettuce were not. Roots from big-vein lettuce were placed in one side of a shallow pond of sterile water and roots of healthy, intact lettuce seedlings were positioned 5 inches distant. The seedlings were exposed for different time periods and then transplanted and examined after about four weeks for big-vein symptoms and root infection by *O. brassicae*. The results showed that both the big-vein agent and *O. brassicae* had been released and transferred to the seedlings during about 4 hr of exposure to BV-infected roots. In other experiments, suspensions of swarm spores of *O. brassicae* were used as inoculum and lettuce seedlings as bait plants to determine that big-vein survived 10 min at 50°C and *O. brassicae* 55°C, and that big-vein was transmitted by spore suspensions diluted 10^{-2} but some infectivity of *O. brassicae* was retained in spore dilutions of 10^{-3}. In some root-treatment tests with chemicals both big-vein and *O. brassicae* survived whereas when roots were immersed in a 10^{-3} water solution of $HgCl_2$, big-vein survived for 2 min and *O. brassicae* for 10 min. In this series of tests, some bait plants that were negative for both big-vein and *O. brassicae* and all noninoculated control plants remained healthy; 57 bait plants had both big-vein and *O. brassicae,* and 20 were infected only with *O. brassicae*. Thus big-vein and *O. brassicae* appeared to be separable. However, the numbers of plants in each test were small, and the significance of the separation was to be evaluated later (1, 2).

From these tests it was concluded that big vein of lettuce was associated with *O. brassicae* (3) but no conclusion was reached concerning the nature of the big-vein pathogen. Later experiments, however, demonstrated that big-vein was graft-transmissible in the absence of *O. brassicae* (1, 2).

Although big-vein has been called a virus disease, the infectious BV agent has been neither isolated nor observed with the electron microscope. Thus, I believe that this disease is better termed *virus-* or *viroid-like* until properly described.

Summer Bunch Rot

Summer bunch rot (SBR) of Thompson Seedless grapes became epidemic in the San Joaquin and Coachella valleys of California in the 1960s. Annual loss in some vineyards exceeded 30% (4). The disease is characterized by a sour vinegar odor, juice dripping from clusters of grapes, the presence of numerous fruit flies (*Drosophila melanogaster*), and the rotting out of the center of clusters which contain dried fruit beetles (*Carpophilus hemipterus*), numerous maggots, and spore masses of *Aspergillus niger* and other fungi. Our problem was to determine the cause and nature of the disease and to work out a control. Some 35 microorganisms, mostly fungi and an *Acetobacter* sp., were isolated from rotting grapes. However, when inoculated into grape clusters, none of them reproduced SBR. Furthermore, chemical sprays did not control the disease.

Since conventional methods failed to identify a causal pathogen of SBR, we changed our approach. Comparative epidemiological studies of vineyards with and without a history of SBR provided useful information that led to discovery of the primary pathogen responsible for SBR and to resolution of the disease complex. The comparative study included visual observations at intervals from spring to harvest, and a periodic survey of the ecology of the developing grape cluster from before bloom until SBR first appeared. Differences between vineyards with and without SBR were few but contrasting: (*a*) all vineyards with SBR had a tip blight of canes caused by *Diplodia natalensis* (with blight always starting at the tips of canes in contact with the soil), and also numerous fruit flies and dried fruit beetles; (*b*) all vineyards without SBR had only a few fruit flies and dried fruit beetles, and no tip blight of canes; and (*c*) bunch rot caused by *Botrytis cinerea* and other organisms was common to all vineyards. The ecological study was in two parts: (*a*) a culturing of the microflora from the surface of grape flowers and developing fruit, and a culturing of flowers and fruits after surface-sterilization, each at intervals from the time of prebloom until SBR began to develop in vineyards; and (*b*) the enclosing of clusters within paper bags to include and/or exclude factors of the environment.

Let me sum up the differences between vineyards with and without SBR by listing the factors found only in vineyards with SBR: (*a*) SBR vineyards had a cane tip blight caused by *Diplodia natalensis;* (*b*) this same fungus was isolated from the styles and styler ends of grape ovaries; (*c*) *D. natalensis* caused rot of grapes only in clusters bagged after bloom, and not in other clusters that developed from bagged blossoms; and (*d*) infection of grapes by *D. natalensis* occurred during bloom. Further, inoculations demonstrated that *D. natalensis* would rot grapes after they were nearly mature and contained some 10% soluble solids. Additional information that we had at that time was that many organisms including fruit flies, dried fruit beetles, maggots, *Acetobacter* sp., yeasts, etc. were a part of the SBR complex; SBR first appeared when grapes had some 10–12% soluble solids; and *D. natalensis* was the only pathogen isolated from grapes in SBR vineyards that did not occur in other vineyards also.

Experimental tests to determine the cause of SBR consisted of enclosing clusters of grapes in chambers infested with fruit flies and inoculating with various organisms. The results demonstrated that SBR developed if the grapes in the upper shoulder of a cluster were inoculated first with *D. natalensis,* and then, after localized rot developed, the clusters were infested with fruit flies, yeast, *Acetobactor roseum,* and spores of *Aspergillus niger* or *Rhizopus arrhizus* (4). Thus it was concluded and demonstrated in vineyard tests that *D. natalensis* initiated the disease; infection occurred at bloom time through the stigma and style; the fungus became latent in the stylet end of the young grape, but resumed growth and produced rot after the grapes matured (9); cracks formed in the skin of rotting grapes; fruit flies and dried fruit beetles fed at cracks, laid eggs, left spores of various microorganisms (yeasts and *A. roseum,* etc.); activity of these organisms resulted in dripping of juices onto other grapes that caused their skins to crack and the rot spread, *D. natalensis* died out; and the complex became SBR (4).

Control of SBR

Chemical sprays, we found, would not control SBR or cane blight. In searching for another approach to control, blighted canes were viewed as an inoculum source and as a possible control point. However, in winter pruning, blighted canes, along with many healthy canes, were cut off the vines, shredded, and disked into the soil.

In tests to determine times of inoculum dispersal, we observed that spores of *D. natalensis* were recovered on traps during bloom, but only among numerous soil particles, indicating that dispersal occurred at times when there was vineyard dust. Many pieces of blighted canes found in the surface layers of vineyard soil had dark masses of dried spores of *D. natalensis.* Using pieces of green cane as bait in soil samples, we found that the fungus occurred generally in soils of vineyards with SBR and at depths up to 30 in. or more. Pieces of mature canes did not become infected with the fungus when put in infested soil. Machines were often operated in vineyards and created dust during bloom time. It then was postulated, and later proved, that keeping the tips of canes out of the soil by cutting them off would control cane blight, and by reducing inoculum would also control SBR. In vineyard plots, cane blight was controlled in the first year; SBR in the second year; and *D. natalensis* had mostly died out of the vineyard soil by the third year (4). The fungus is likely not new in vineyards. No doubt, cane blight and some SBR had occurred at low levels for a long time. Earlier practices, before the SBR epidemic, were to burn prunings. Shredding and disking prunings into soil, a labor-saving practice, permitted the fungus to build up in vineyards and the disease to become epidemic. Thus, the change in cultural practice from burning prunings to disking them into the soil led to the disease epidemic. Another change, cutting off tips of canes to keep them out of the soil, controlled the disease.

This investigation into the cause, nature, and control of SBR demonstrates the importance of perception, exploring, and gathering as much information as possible on a disease problem before designing experiments. Had perception been exercised more fully in the beginning, it is likely that the experimentation period could have been shortened by some two years.

Early Botrytis Rot of Grapes

In our explorations on summer bunch rot (4), we discovered, as with *D. natalensis,* that *Botrytis cinerea* infected grapes at bloom time through the stigma-style, became latent and, when the grapes began to mature, resumed growth and caused grape berry and cluster rot. We named this disease early botrytis rot (7). It was controlled by applying a fungicide spray to clusters in the early bloom stage (8). I point this out because in exploring any plant disease problem one can often learn much that relates to other problems.

A BROAD CONCEPT OF PLANT PATHOGENS

Concepts of plant pathology can be limited in scope by definition and thus restrict our thinking and willingness to study a problem, or they can be unlimited, open into infinity, and thus encourage investigation over a broader field. I envision plant disease to include any malfunction in the growth of the healthy productive plant. In other words, any thing, event, structure, function, nutrition, even frost, hail, galls, that may engender malfunction of plant tissues or systems is a pathological problem. The cause or causes of disease thus become broader in concept, which permits thinking and exploring into any facet that may be involved in disease whereas more restricted concentration on the disease and the diseased plant in terms of finite plant pathogens, such as viruses, bacteria, and fungi, tends to limit the scope of thinking. A good example of the need for a broad concept on plant pathogens is demonstrated by my work in the 1960s on a flower-abortion problem of grapevines in some California north-coast wine-grape vineyards, and by later work (in the 1970s) on a similar disease in the south-central coast. An example follows.

The Exanthema Disease in Grapevines

In the early 1960s I worked on a flower-cluster abortion problem on Pinot Noir, Pinot St. George, and Sauvignon vert grapevine cultivars in some vineyards in north coast California. Conventional procedures of isolation, inoculation, and spraying with fungicides led nowhere. Later, a similar disease problem was shown to me in some young vineyards of Pinot Noir in the newly developing wine-grape region of the southcentral coast (5). In addition to flower and flower-cluster abortion, the disease here expressed symptoms in leaves, shoots, canes, and overall vine growth. The symptoms could have been induced by fungi, bacteria, or viruses, none of which were recovered; nor did chemical sprays control the disease. Leaf-petiole analysis showed very high nitrates, some 3000 to 6000 μg/g dry weight, yet many of the vineyards had not been fertilized. Furthermore, irrigation water and soil water were low in nitrates. At this stage, attention was turned from the diseased vine to exploring many vineyards, agricultural crops in general, pasture lands, and cattle feeding. It was this exploration that provided clues to the nature of the disease. In certain areas the chick pea (*Cicer arietinum*) required copper to set and mature a crop, and cattle in some pastures needed a copper and/or molybdenum supplement. In another analysis, copper in grape petioles was low—from 1.5 to 8 μg/g dry weight and lower copper content of tissues correlated fairly well with high nitrates.

Copper deficiency was tested by spraying foliage with a chelated-copper compound. This copper treatment corrected leaf symptoms, improved bloom, and increased yield. Copper in petioles from sprayed vines increased (from 12 to 40 μg/g), and nitrates decreased [from 700 to 2000 μg/g (5)]. In our 1978 test plots, grapevines treated with copper-chelate spray yielded significantly more grapes and had significantly more petiole copper than did control vines. However, petiole nitrate nitrogen was high in all plots and had no apparent relationship to petiole copper (W. B. Hewitt, unpublished).

In review, the flower-abortion problem in north-coast vineyards occurred in areas where the exanthema disease (copper deficiency) was known to have existed in apples. The diseases in the north coast and in the south-central coast apparently have a similar etiology inasmuch as many symptoms are the same and copper deficiency is known in other plants in both areas. All the same, the true nature of the disease, the cause of the symptom syndrome, has not been resolved because the apparent negative relationship of copper to nitrates in petioles as found in 1977 did not recur in 1978.

The function of copper and probable role of nitrates in the exanthema disease are not known. Furthermore in 1978, some vineyards with mild exanthema were fertilized with ammonia and produced nearly normal crops. This indicates the vines used the ammonia nitrogen and implies that molybdeum, known to function in nitrate reduction, may be deficient and also involved in the disease. However, in these studies the molybdenum content of grapevine petioles has not yet been determined. Thus, the problem has not been solved completely.

Some feel that such problems are for experts in plant nutrition, and I agree that they should work on nutrition and functions of nutrients. However, it is my opinion that exanthema, which appears to be related to copper deficiency, is a plant-pathological problem.

PERFORMANCE

Let us now consider performance from the standpoint of the Department of Plant Pathology and from that of the individual plant pathologist. Departments of plant pathology are what we make them. Departments serve not only as administrative units but also as centers for teaching, study, and research. They include a group of scientists with common interests and objectives, a grouping designed to foster or stimulate new ideas. Departments of plant pathology have images in their respective institutions and elsewhere. The images were formed over time and have a basis in performance.

Departments are composed of scientists, housing, equipment, and supporting staff. Functions common to departments are research, teaching, and practice. The breadth of a department may be viewed in the research interests and the depth in the composite capabilities of the constituent scientists. Performance is measured in research reported in published papers. Excellence is evident in quality of work, adventure, discovery, and formation of new ideas. For example, research that repeats what has been done, such as finding a known virus or enzyme system in a different host or pathogen, adds to the store of knowledge but does little to broaden or increase concepts in plant pathology. In contrast, research that creates new concepts and engenders further exploration adds to understanding of the nature of diseases. Eminence is gained through continued and/or repeated excellence in performance. Thus, it is performance that distinguishes the scientist; and the composite performance of the scientists distinguishes the department.

Performance in teaching is more difficult to measure but it can be assessed to a degree on the basis of the composition of the courses taught, approaches to teaching, rapport with students, staff and student evaluations, and student performance in the long term.

IN CONCLUSION

We are what we think we are; we can do what we think we can do; we are our own best critic. I envision that I know myself. I know my accomplishments and what I can do.

If my accomplishments reflect the best of my abilities, then my contributions are satisfying, worthy, and significant. If, however, I have not taxed my capacity for work and thought, I have merely done a job and my work falls short of my potential. Put in another way, if my concept of plant pathology is limited, the results of my work in plant pathology will also be limited; whereas, if my concept of plant pathology is unlimited, the potential of my work becomes unlimited, and more stimulating, significant, and rewarding.

> Let us, then, be up and doing,
> With a heart for any fate;
> Still achieving, still pursuing,
> Learn to labor and to wait.

Henry Wadsworth Longfellow
Psalm of Life

ACKNOWLEDGMENTS

I am grateful to Dr. and Mrs. Charles J. Delp, Dr. Gary A. Strobel, and Richard N. Hewitt for commenting on an early draft of this manuscript. Their remarks were of great value in preparing this chapter. Appreciation and thanks I give to Maybelle, my wife, for her patience, help, and continued interest, and to the editors of this volume for their constructive assistance.

Literature Cited

1. Campbell, R. N., Grogan, R. G. 1964. Acquisition and transmission of lettuce big-vein virus by *Olpidium brassicae*. *Phytopathology* 54:681–90
2. Campbell, R. N., Grogan, R. G., Purcifull, D. E. 1962. Studies on the transmission of the virus causing big-vein of lettuce. *Phytopathology* 52:5 (Abstr.)
3. Grogan, R. G., Zink, F. W., Hewitt, W. B., Kimble, K. A. 1958. The association of *Olpidium* with the big-vein disease of lettuce. *Phytopathology* 48: 292–97
4. Hewitt, W. B. 1974. Rots and bunch rots of grapes. *Univ. Calif. Agric. Exp. Stn. Bull. 868.* 52 pp.
5. Hewitt, W. B., Coburn, J. 1978. Exanthema disease of grapevines in California. *Grape Grower,* Feb., pp. 4, 5, 33
6. Hewitt, W. B., Raski, D. J., Goheen, A. C. 1958. Nematode vector of soilborne fanleaf virus of grapevines. *Phytopathology* 48:586–95
7. McClellan, W. D., Hewitt, W. B. 1973. Early botrytis rot of grapes: Time of infection and latency of *Botrytis cinerea* Pers. in *Vitis vinifera* L. *Phytopathology* 63:1151–57
8. McClellan, W. D., Hewitt, W. B., La Vine, P., Kessler, J. 1973. Early botrytis rot of grapes and its control. *Am. J. Enol. Vitic.* 24:27–30
9. Strobel, G. A., Hewitt, W. B. 1964. Time of infection and latency of *Diplodia viticola* in *Vitis vinifera* var Thompson Seedless. *Phytopathology* 54:637–39

Ann. Rev. Phytopathol. 1979. 17:13–20
Copyright © 1979 by Annual Reviews Inc. All rights reserved

LEADERS IN PLANT PATHOLOGY: L. R. JONES

♦3697

J. C. Walker

Department of Plant Pathology, University of Wisconsin,
Madison, Wisconsin 53706

Born in the closing days of the Civil War in a farming community in Fond du Lac County, Wisconsin, Lewis Ralph Jones had the usual boyhood of a pioneer family. His mother, a native of Vermont, was a country school teacher. His father, a native of Wales, migrated to America as a young man with his family to southern Wisconsin and later moved to Fond du Lac County.

There was a high school in Jones' home village of Brandon, where he was encouraged by the principal, Kirk Spoor, to go to college. Ripon College, a Congregational school, was not far away. After two years there he was encouraged by C. Dwight Marsh, Professor of Chemistry and Biology, to go to the University of Michigan, where he matriculated in 1886. He majored in botany under Professor Volney M. Spalding. After one year he withdrew to teach three terms at Mt. Morris Academy, Mt. Morris, Illinois, in 1887–1888. Returning to the University of Michigan in 1888, he completed his undergraduate education in 1889.

During his final year at Michigan he was invited by Professor Spalding to attend a final PhD examination. The candidate was Erwin F. Smith, who had returned from his position with the United States Department of Agriculture (USDA) to present his thesis, which consisted of a report of his pioneering work during the previous three years on peach yellows. This experience influenced Jones' early interest in plant pathology. It was only four years after the discovery of Bordeaux mixture by Millardet in France and three years after the establishment of a section of mycology in the USDA devoted primarily to research on plant diseases. With the passage

13

0066-4286/79/0901-0013$01.00

of the Hatch Act in 1887, which provided annual federal grants for agricultural research to state experiment stations, state botanists were being appointed in many states. In 1889, Jones accepted the new position of Botanist at the Vermont Agricultural Experiment Station. The appointment also carried an instructorship in natural history. He advanced to Professor of Botany in 1893 and retained this joint position until 1910.

During his twenty years at Vermont, Jones rose in stature as a leader in botany and plant pathology in the United States. His lovable disposition suited him admirably for his dual role as teacher and investigator. He told me that soon after going to Vermont he heard of a group of lay botanists who went on frequent forays and realized that many of them knew the local flora much better than he did. He was instrumental in organizing them into the Vermont Botanical Society, of which he served as secretary for several

Figure 1   L. R. Jones.

years. He was also active in study of the physiology of sap flow in sugar maple. He initiated the organization of the Vermont Forestry Association, served for a time as its president, and maintained a vital interest in this field during his entire period in Vermont. The L. R. Jones State Forest is a perpetual tribute to his interest in and service to forestry.

Potato was then a major crop in Vermont and late blight found there an ideal climate in which to prosper. I remember his telling me how he took a knapsack sprayer and a cask of Bordeaux mixture in a wagon, and sprayed certain patches in farmers' fields to demonstrate the need to adopt this new and effective control measure.

Carrot was another important food crop adapted to winter storage. He showed that the principal storage decay was a bacterial soft rot, of which he described the causal organism, *Erwinia carotovora* (Jones) Holland. During this time he was given a semester leave which he spent in E. F. Smith's laboratory in the USDA at Washington DC. Here he associated not only with Smith but with other pioneers in plant research, such as A. F. Woods, B. T. Galloway, M. B. Waite, M. A. Carleton, and Theodore Holm. He was not content to describe the causal organism during this notable period when Smith was engaged in a debate with certain German scientists who maintained that bacteria could not act as plant pathogens. He went on to show how it acted as a pathogen—by secretion of pectolytic enzymes which dissolved the middle lamella of carrot cells. This was the first report of this fundamental process in bacterial plant pathology. He used the report on this research as the basis of his PhD thesis which he presented at the University of Michigan in 1904.

Mrs. Jones once told me that the year 1904 was a very notable one for Ralph, for in that year he took his doctorate, built them a house, and went to Europe. His preeminence in potato disease research was responsible for his being assigned by the US Department of Agriculture the task of traveling to Europe to study, among other things, the possibility of controlling late blight of potato through disease resistance. He found European varieties, as a class, somewhat more tolerant than American varieties and brought back an extensive collection, which he turned over to William Stuart, his colleague in horticulture at Vermont, who was at that time transferring to the USDA in Washington where he spent the rest of his active life in potato improvement through breeding.

By the middle of the first decade in the twentieth century, plant pathology in the United States was about to take off as a distinct science. In 1908 the American Phytopathological Society was formed not without some hard feeling on the part of some prominent mycologists who tried their best to keep it under their wing. Jones told me that he took no active part in this controversy, but he did consent to become the first president of the new

Society, and some two years later as he moved to Wisconsin he became the first editor in chief of *Phytopathology.*

His call to found the new department of plant pathology in the University of Wisconsin came at a time when similar action was taking place in several states including California, Minnesota, and New York.

Already well grounded in botany in the broad sense and fully appreciative of the scientific needs of agriculture, Jones was well fitted to meet the peculiar needs immediately ahead in plant pathology. He realized that future growth in this area rested upon the best possible postgraduate education along with high-quality research. His cadre of graduate students grew steadily as he attracted students from most states and many foreign countries. His success in development of an outstanding graduate department rested upon several fundamental principles. The first of these was an innate interest in young people and their development. He realized, however, that there was little to be gained in expending his efforts on individuals who were temperamentally or inherently unfit to become good scientists. He felt that a successful teacher of science must be continually on the alert for students whose talents fitted them peculiarly for a career in science. This conviction is well illustrated in one of his favorite anecdotes. During their Vermont days together, Cyrus Pringle made frequent collecting trips in Mexico and often tried, without success, to get Jones to accompany him. On this particular occasion, as an inducement, Pringle, in his enthusiasm, exclaimed: "Jones, if you will only come along with me, you may even have the thrill of discovering a new plant." Jones' quiet answer was: "Yes, Pringle, and so I might, but while you are in Mexico, I may have the thrill of discovering a new scientist right here at home."

Despite this faculty of sizing up and selecting his students carefully, there never was a teacher or adviser who was more patient and sympathetic with his students, nor one who strove harder and with more success in bringing out the best talents of the individual. Many a time he deftly guided a temperamental student or colleague away from personal or selfish ideas and induced him to respond to the fundamental challenge of scientific achievement—a challenge that could be accepted best with teamwork. He once remarked to me concerning his great and good friend, Liberty Hyde Bailey, that the latter had the uncanny faculty of "bringing out the Bailey" in those with whom he was associated. Perhaps he little realized his own great talent of "bringing out the Jones" in his own students and associates.

Jones insisted that successful education in plant pathology must be built on a thorough foundation in botany. Thus, most of his students spent more graduate course hours in botany than in plant pathology. He insisted on having an up-to-date departmental library and directed his courses in such a way that students became familiar with the literature of their special field.

His seminars laid strong emphasis on the history of plant pathology, the nature of parasitism and disease resistance, the relation of climatic factors to disease development, and other fundamental topics. An important feature of his educational method was the principle of learning by doing. He insisted on holding formal course instruction to a minimum in order that each student would have more time for original research.

He believed that the best way to develop the research instincts of a student was to have plenty of research going on about him. Thus, in spite of increasingly heavy administrative duties, he set a pace for his staff and students by always having a number of research problems under way. His long list of scientific papers attests to his continuous productivity as a scientist.[1] At the time of his retirement after 25 years at Wisconsin, nearly 150 doctors' degrees had been granted to students majoring in his department. However, his world-wide reputation in the field of plant pathology rests as much, if not more, upon the continuous flow of original research papers from his laboratory.

Before he left Vermont he was fully aware that research in plant pathology during the previous fifty years was concerned predominantly with the description of diseases and their causative pathogens. As he arrived at Wisconsin he was determined to shift emphasis to fundamental aspects of the nature of disease. He chose as his first direction of emphasis the relation of environmental factors to disease development. Possibly because some of the diseases needing immediate attention in Wisconsin were induced by soil-inhabiting pathogens, e.g. cabbage yellows, pea wilt, tobacco root rot, seedling blights of corn and wheat, his first emphasis was on soil factors. With his associates he perfected thermostatically controlled equipment in which plants could be grown over a range of constant soil temperatures in infested soil. This equipment soon became widely known as the "Wisconsin soil-temperature tank." With this equipment and with supplementary devices, many plant diseases were studied and the focus of attention on phytopathological research was influenced remarkably under his leadership.

Early in this century with the rediscovery of Mendel's laws of inheritance and the founding of the science of genetics, an increasing amount of attention was being directed toward disease resistance as a means of control. Successes reported by Bolley with flax wilt in North Dakota, by Orton with cotton wilt in South Carolina, and by Bain and Essary with tomato wilt in Tennessee were attracting attention. The most immediately pressing prob-

[1] A complete list of Jones' publications is included in a biographical memoir by J. C. Walker and A. J. Riker published in *Biographical Memoirs of National Academy of Sciences* 31:156–79 (1958).

lem presented to Jones by Dean H. L. Russell when he came to Wisconsin in 1910 was disease in the cabbage crop in the southeastern part of the state. Russell had described the black rot disease of cabbage, induced by *Xanthomonas campestris* (Pam.) Dows, in this area some fifteen years earlier, but Jones found that the practically unknown disease he named "yellows" to be the major trouble. It was a fusarium vascular disease similar in many respects to the three wilt diseases mentioned above. Moreover he noticed at once that in most severly devastated fields a few plants remained healthy. From these plants he was able to develop a resistant variety which he released only six years later.

He stimulated others among his students and staff to keep on the lookout for possibilities of controlling specific diseases by means of developing resistant varieties. The work he initiated on disease resistance in cabbage has continued up to the present time. Resistance to five different diseases of cabbage have been incorporated into a single variety. In his laboratory, success in development of varieties was extended to pea wilt, cucumber scab and mosaic, bean mosaic, onion pink root, and wilt of China aster.

One of the keys to success in this area is the development of rapid techniques for accurate screening of resistant individuals in segregating populations. By the use of controlled environment, this was done in several instances, e.g. cabbage yellows, onion pink root, cucumber scab, cucumber mosaic, pea wilt. Professor Jones was continually reminding his students and colleagues to watch out for leads by which to enlighten our science as to the *nature of disease resistance in plants*. While skepticism prevailed generally over the possibility of tying the resistant character to specific genes, this was done in due time in the cases of cabbage yellows, pea wilt, cucumber mosaic, cucumber scab, bean mosaic, onion smudge and others. The isolation of the chemical which formed the basis of resistance in colored-bulb onions to smudge and neck rot was accomplished with his encouragement by the author and Karl Paul Link in the Department of Biochemistry.

It is natural that a man of such outstanding ability and leadership should be drafted into many extracurricular activities. While he shunned many such calls within and without the university, he seldom refused to give support to what he considered a worthy cause for which he could wisely spare the time.

He was elected to the National Academy of Sciences in 1920. He was one of the organizers of the Division of Biology and Agriculture of the National Research Council, on which he served as vice-chairman from 1919 to 1921 and as chairman in 1922. During his period of service on the council, *Biological Abstracts* was launched on its successful career. He was president of the Tropical Research Foundation from 1924 to 1943. He was on the original board of trustees which organized the Boyce Thompson Institute for Plant Research and continued to serve on that board for several years. In 1934 he was appointed by President Franklin D. Roosevelt to the President's Science Advisory Board.

Professor Jones was a regular attendant at meetings of plant science societies. In addition to being the first president of the American Phytopathological Society in 1908, he was president of the Botanical Society of America in 1913. He was vice-president of Section O (Agriculture) of the American Association for the Advancement of Science in 1924, chairman of the Section of Mycology and Plant Pathology of the Fifth International Botanical Congress at Cambridge, England, in 1930, and an honorary president of the Third International Congress of Microbiology in New York City in 1939.

He received honorary degrees from University of Vermont (1910), Cambridge University (1930), University of Michigan (1935), and University of Wisconsin (1936). He was an honorary member of the following foreign scientific societies: British Association of Applied Biologists, Phytopathological Society of Japan, Société de Pathologie Végétale et Entomologie Agricole de France, and Verein für Angewandte Botanik in Germany.

Professor Jones had many nonprofessional interests. He was a lover of the out-of-doors and enjoyed long hikes in summer and winter. He frequently organized parties of his students to accompany him. In his fifties he took up golf and became an enthusiastic devotee. He was a member and regular attendant at the First Congregational Church of Madison and served a term as deacon. He was a regular and later an honorary member of the Madison Rotary Club, an association which he greatly enjoyed and which he continued after retirement.

In 1890, Jones married May Bennett, a classmate at Ripon College. They had no children of their own but were very fond of young people and made a practice of inviting graduate students into their home. These were memorable occasions for all of us.

Mrs. Jones passed away on September 26, 1926. They had only recently returned from a trip to Hawaii and had attended a gathering of his former students at the International Congress of Plant Science at Cornell University. At this gathering his students gave him an oil portrait that hangs today in his beloved departmental library at the University of Wisconsin.

On July 27, 1929 he married Anna Clark, a former student at Vermont and professor of biology at Hunter College, New York City. The following year he asked to be relieved of administrative responsibilities at Madison. He and Mrs. Jones spent several months of that year in the British Isles, western Europe, and Russia. In 1931 they traveled extensively in Japan, Korea, and China. After retirement in 1935 they continued to visit colleagues and friends making their headquarters in the summer at Mrs. Jones' home at Brookfield, Vermont, and during the winter at Orlando, Florida, where he passed away peaceably in his eighty-first year on March 31, 1945.

It is appropriate in closing this treatise to quote a tribute to L. R. Jones written by Gardner L. Green of the University of Vermont.

> A man of science to the manner born,
> Who labored hand in hand with the Divine.
> All prejudice he taught himself to scorn
> As deadly sin. Truth was his holy shrine.
> The plastic minds of youth he looked upon
> As gardens to be nurtured for the gods.
> His spirit, gentle as the breaking dawn,
> Revealed a kindness that the world applauds.
> His cup of joy was brimful just to know
> The secrets of God's children of the soil.
> And nothing that was ornament or show
> Intrigued him like experiment and toil.
> He needs no shaft to tell us when he went—
> His work will always be his monument.

Ann. Rev. Phytopathol. 1979. 17:21–28
Copyright © 1979 by Annual Reviews Inc. All rights reserved

IMPORTANT LITTLE-KNOWN CONTRIBUTORS TO PLANT PATHOLOGY: MASON BLANCHARD THOMAS

❖3698

G. C. Kent

Professor Emeritus, Department of Plant Pathology, New York State College of Agriculture and Life Sciences, Cornell University, Ithaca, New York 14853

> A great teacher is far more than a purveyor of information; he must have that peculiar insight and inerrant touch which reveal to young men their possibilities, and arouse within them the forces of thought and action which send them on to realization. Professor Thomas had this gift in an unusual degree. (5)

At the turn of the century, one of the outstanding teachers of botany in the United States was Mason Blanchard Thomas at Wabash College. Other great teachers of the time were known primarily for their research and their graduate programs. Thomas' reputation rests entirely on his undergraduate teaching and his stimulation of students to become botanists and to continue graduate work elsewhere.

Thomas taught the entire gamut of the field of botany. He taught botany as a basis for the applied fields of forestry, plant pathology, plant physiology, etc. A number of his students went into medicine and biochemistry. He is reputed to have sent over 50 students on to graduate work during his 20 years of teaching. There remains little evidence of any particular characteristic which made him such a productive teacher. Was it the man? Was it the particular institution at that time? Or, was it a group of intelligent, energetic, inquiring students stimulating each other while being directed by an unusual teacher? Perhaps it was all three.

Thomas interviewed nearly every student who took his courses, or who sought advice of a known counselor. He was able to inspire nearly any student, to get close to him, and constantly to convey his confidence in and respect for him. He rapidly developed a reputation as an excellent judge of

21

men and their potential. To those whom he judged capable he transferred his love of nature, his unbounded energy, his enthusiasm for the botanical sciences, and his philosophy of learning (3).

Mason Blanchard Thomas was born December 16, 1866, at New Woodstock, New York, about 25 miles southeast of Syracuse. Little is known of his youth. He prepared for college at Cazenovia Academy, 6 miles north of New Woodstock, and, having won a competitive scholarship, entered the College of Arts and Sciences at Cornell University in the fall of 1886 (2, 3).

It is unclear why he chose to study botany. He early began to work with W. R. Dudley who was at that time establishing his section of histology. Dudley was an experienced teacher at Cornell, and a half-time investigator

Figure 1   Mason Blanchard Thomas.

of plant disease in the Cornell University Experiment Station. Dudley was a proponent of the understanding of disease through laboratory analysis, particularly histological examination. Through Dudley, Thomas knew and worked on microscopy with Simon H. Gage (2, 3).

Thomas became a student fellow in botany and as an undergraduate began independent study. He refined the then new European technique for the use of collodion for embedding plant tissues. He read a paper on his research at the thirteenth Annual Meeting of the American Microscopical Society in 1890, and published a separate report in *The Botanical Gazette.* His work on sectioning fern prothallia and other delicate materials became a part of Professor G. F. Atkinson's book on ferns, published in 1894 (4).

Thomas obtained his BA with a major in botany, in 1890. Because of his outstanding academic record, and research, he was awarded a graduate fellowship for the following year. But that fellowship apparently expired after nine months. In the fall of 1891, with no other means of supporting himself during continued graduate studies, Thomas accepted a position as Rose Professor of Biology and Geology at Wabash College in Crawfordsville, Indiana. In this position he replaced John M. Coulter who left Wabash to become president of Indiana University (2, 3).

Thus, with only one year of graduate work behind him, Thomas replaced an experienced researcher and teacher who was considered one of the outstanding faculty members of Wabash College. This was at a time when Wabash College, Indiana University, and Purdue University were more or less on a par. On the basis of his undergraduate research he had been elected to Sigma Xi at Cornell. Shortly after arriving at Wabash, he was elected one of the charter members of the Wabash Chapter of Phi Beta Kappa.

With his former teacher, W. R. Dudley, Thomas published a manual of plant histology (7). It is clear in the book that the manuscript was developed by Thomas for use by students in his classes. From the Preface, signed by Dudley alone, it seems that Thomas obtained some of his concepts of teaching and student direction from Dudley (7):

> Almost every thoughtful teacher of botany in the colleges and universities of our country is confronted by two problems in connection with his laboratory instruction. He is forced to provide a course which shall give the general student a fair knowledge of what the teacher deems the most important phase of plant life; and on the other hand, if a conscientious instructor, he will encourage students to advanced work, inaugurating courses which are intended not only to inform the mind, but to train the powers of observation, comparison and scientific judgment, and finally produce the investigator capable of pursuing problems of science without aid or admonition, if not without suggestions, from his professor.

It seems that Thomas, with his forceful personality and ability to influence students solved these problems by persuading his students to major in botany and then go on to graduate work in botany, plant pathology, plant physiology, mycology, and forestry.

In describing the program of the Department of Botany at Wabash in the "Wabash College Souvenir" (8), Thomas, still then a one-man department, wrote (8):

> The student is taught to see things around him, to become familiar with the story of nature's plans and methods of operation. But more important than this in a college the work aims to accomplish its share of what the Royal Commission says of science training: 'It quickens and cultivates directly the faculty of observation which in very many persons lies dormant almost through life, the power of rapid and accurate generalization and the mental habit, method and arrangement. It accustoms young persons to trace the sequence of cause and effect. It familiarizes them with a kind of reasoning which interests them and which they can promptly comprehend, and it is perhaps the best corrective for that indolence which is the vice of half-awakened minds and which shrinks from any exertion that is not, like an effort of the memory, merely mechanical.'

In 1893, Thomas married Anne Davidson, a young lady from Crawfordsville, who contributed to his warm personal relationships with students. She retained the affection of many Wabash students after Thomas' death. They had no children.

In 1895 biology at Wabash was divided. Thomas became professor of botany and Dr. Donaldson Bodine joined the faculty as professor of biology and geology. Opposites in many ways, the two men complimented each other, and remained the best of friends and colleagues. Bodine directed his program to the general education of all students, whereas Thomas concentrated on selecting and promoting specialists (6).

In 1907 Wabash conferred on Thomas an honorary doctorate degree and made him Dean of the Faculty. For the next five years he served as Dean in addition to handling the Department of Botany and continuing on many committees. About that time he also obtained his first official, paid assistant in the department.

First as professor of biology and geology, then as professor of botany, and after 1907 as dean, Thomas interviewed all students with an interest in the science curriculum. J. R. Schramm wrote Paul W. Cook, Jr., President of Wabash, that Thomas once told him that when he found Wabash had no library he had to decide whether to make books or men.

When Dr. Coulter left Wabash College he left Thomas an active program of classes and students, mostly in the area of biology. One of these students was E. W. Olive who became Thomas' first student assistant and in 1893 became his first "graduate student" at Harvard. Thomas apparently did not immediately begin to attract students to majors in biology, but during the second decade of his teaching he rarely sent less than one away for graduate work in any year. Even this partial list of "his boys," who went on to graduate work in biology is impressive:

E. W. Olive	'93	E. E. Davis	'10
G. T. Moore, Jr.	'94	W. H. Rankin	'10
R. B. Miller	'94	H. L. Rees	'10
B. R. Hoobler	'01	J. R. Schramm	'10
H. H. Whetzel	'02	C. E. Taylor	'10
S. J. Record	'03	L. R. Hesler	'11
W. D. Funkhouser	'05	H. M. Jennison	'11
J. D. Reddick	'05	G. A. Osner	'11
S. S. Stewart	'06	D. C. Babcock	'12
W. A. Ruth	'06	L. M. Massey	'12
H. W. Anderson	'07	J. H. Muncie	'12
M. F. Barrus	'07	C. C. Thomas	'12
H. F. Dorner	'08	J. L. Weimer	'12
H. M. Fitzpatrick	'09 ex.	W. H. Burkholder	'13
E. C. Pegg	'09	C. D. Chupp	'13
Earl Price	'09	C. B. Gibson	'14
V. B. Stewart	'09	W. E. Pickler	'14
P. J. Anderson	'10	F. B. Wann	'14

Most of these students were from the midwestern United States and came to Wabash with no intention of studying biology. Many of Thomas' students reported how he advised them to take his general course, then encouraged them to major with him, and, finally, directed them to graduate school. Many students indicated their indebtedness to him for his stimulation, his teaching ability, and his interest in them as individuals.

Throughout his teaching, Thomas taught three full-year courses in botany each year. The first course was required of all students. It was given in the sophomore year. The second- and third-year courses were elective and required completion of previous courses. The content varied constantly, keeping up with current advances in botanical sciences. In 1908–1909 he first entitled the fall term of the third course "Mycology and Plant Pathology."

During the years when Thomas was active (1890–1912) Wabash was a small institution struggling with several others in the Indiana-Illinois area for support and recognition. There were seldom over 200 students in the collegiate program. Because of the requirement of Latin and Greek for entrance to the collegiate program there was an equally strong college-preparatory program. The 15 to 20 faculty members were outstanding and standards were high. Students received a great deal of individual attention in small classes (6).

The faculty consisted of dedicated men who knew their fields and believed in the value of education. It was considered a great faculty. The graduates of the college went to graduate schools all over the country to enter professions or become scientists. No graduate came to feel his preparation had been superficial or inadequate except because of his own indolence.

A classical philosophy of education dominated Wabash during the entire period of Thomas' career. Until about 1900, admission to the collegiate program at Wabash required prior preparation in both Latin and Greek. Continued study of both languages also was required throughout the college course. After a change in these requirements, about 1901, Latin and one modern language were required for entrance and either Latin or Greek and a foreign language were required for continuing study in the collegiate program.

These requirements served as a screen of those admitted and as a solid base for the graduate and research programs of the period. But they also severely limited enrollment because most of the high schools of the surrounding states offered no instruction in these languages.

For the reasons outlined above, the student body of Wabash College consisted of serious students who knew their objectives and wanted an education. In 1891 when one or two teachers did not meet their standards of classroom discipline, the student magazine, *The Wabash,* declared: "We believe there is nothing next to competence that contributes to the success of an instructor or so commands the respect of students as earnestness, strictness, and order in the recitation room. . . . Let the professors themselves show that they mean business, and not only expect but require the same spirit from the students."

Most of the students who came from Thomas' program graduated after 1901. Was this the result of his growing maturity and experience as a teacher, or the development of better high schools in the Midwest? Or, was it the result of liberalization of entrance requirements and the change in the curriculum to allow 84 credit hours of elective courses (out of 188 required for graduation) instead of the two to three elective courses permitted in earlier years? This change, which "liberalized" the liberal arts education, and the move away from the classics, was bitterly contested among the faculty. Thomas' leadership on the faculty committee which recommended substitution of Greek by a modern language brought him the unsparing criticism of half the faculty for the rest of his life. It also, apparently, helped to bring him, in 1907, the honorary doctorate and deanship of the faculty.

In the first year of his teaching at Wabash, Thomas was recognized as an outstanding instructor with an exceptionally able mind. From the first he exhibited an unsurpassed power to stimulate his students' curiosity, enthusiasm, and desire to understand the plan and functions of plants and plant communities. His published syllabus for general botany reveals that he used the microscope, living materials, and research techniques in his courses. His students remarked that although not an active investigator he always was up-to-date on new discoveries and procedures.

Evidence from students and papers published in the Indiana Academy of Science indicates that Thomas encouraged his students to do independent research work. Many of these students served as assistants in summer work with Thomas at the Marine Biology Laboratory at Woods Hole, Massachusetts or with Whetzel and others at Cornell University. Even as freshmen, some of his students were recognized as doing research comparable in quality with most current graduate schools.

Thomas was always a perfect gentleman, quiet, even in his words of encouragement, but always exuding his enthusiasm for knowledge and understanding, and for his students. Students were always addressed as "Mister." He gave them direction, embued them with the devotion and determination to achieve, kept in contact with them, continued to counsel them, and relished in their achievements. The one word for his relation with students was "inspiration" (3).

About 1908 his students founded the Botanical Society of Wabash College, which continued as an active botanical club for 20 years. He appeared often as adviser and friend, counseling them on their activities and their philosophies.

Evidence tendered in the memorial edition of the Wabash College Record (3) indicates the degree to which Thomas "won the high esteem and respect of all his colleagues by his quiet, refined and dignified manner, by his fair consideration of opinions of others, and by his rare good sense, sound judgment, and thoroughness of knowledge." He has been described variously as an able administrator, as "one of the ablest men ever to be associated with Wabash in any capacity" (6), as "The Man Who Does Things," where the man was not named but apparently all knew it referred to Thomas (1).

In his first year at Wabash, Thomas' remarkable energy and his interest in students led him to become coach of the new football team. During a practice session he received a knee injury which ended his coaching career, caused him continual pain, and necessitated use of a cane for the rest of his life. However, he continued to assist the football team as manager and in 1900 he became chairman of the Committee on Athletics, and was recognized as responsible for the revival of athletics at Wabash.

When Thomas arrived on the Wabash campus, it contained many large trees 35 to 45 meters in height. The grassy areas were seldom cut, although previously neighborhood cows had grazed it; the paths were indifferently covered with cinders or gravel. Thomas became interested in the appearance of the campus and was soon charged by the president with general oversight of college buildings and grounds; finally in 1897, Thomas persuaded the Executive Committee of the Trustees to bring in a landscape architect to develop a landscaping plan for the campus. It was his first excursion into administration. Later he chaired several campaigns that resulted in the erection of monuments and plaques on the campus honoring Indiana veterans and Wabash notables, and generally adding to the beauty of the campus.

His executive abilities were used fully by President Kane after 1900. He was called on to head or sit on committees handling many difficult problems of the college involving long-range decisions. He was responsible for reviving athletics; maintaining and managing the supply of athletic equipment; organizing and reorganizing several student groups, societies, and fraternities; changing admission requirements and the curriculum; and organizing many memorials to former students and contributors to the college and state. He was as much as anyone responsible for the resurgence of the college after 1905, and wholly responsible for the unprecedented popularity of his department, of which he was the only staff member until 1908.

Thomas was recognized for his unbounded energy and his keen mind. He did not confine his activities to Wabash. He became a fellow of the Indiana Academy of Science and from 1900 to 1901 served as its president. He was a fellow of the AAAS and served on its council. He was secretary of the Indiana Forestry Association and chairman of its Educational Committee. He served as a member of the Corporation of the Marine Biological Laboratory at Woods Hole, where he and his students spent several summers. He was a member of the American Microscopical Society, the American Forestry Association, the American Phytopathological Society, the Botanical Society of America, and the Society of Western Naturalists. He spent a good deal of time with and served as Secretary of the Board of Trustees of the Boys School at Plainfield, Indiana.

In February 1912 Thomas became ill and was confined to bed with what was diagnosed as severe pleurisy. For two weeks he continued to conduct classes at his bedside. As he continued to decline, his former student, Dr. B. R. Hoobler, came to Crawfordsville to assist in his treatment. Thomas died March 6, 1912 at age 45. He lived up to his theory of life: It is better to wear out than to rust out.

Literature Cited

1. Anonymous 1906. The man who does things. *The Wabash* June, p. 6
2. Anonymous 1912. Mason Blanchard Thomas. *Proc. Indiana Acad. Sci.* 1911:4–5
3. Anonymous 1912. Mason Blanchard Thomas. *Wabash Coll. Rec.* 10: 1–54
4. Atkinson, G. F. 1894. *The Study of the Biology of Ferns, by the Collodion Method.* Part II, Chapter 1. New York: Macmillan
5. Millis, W. A. 1912. Memorial issue. *Wabash Coll. Rec.* 10:49
6. Osborne, J. I., Gronert, T. G. 1932. Wabash College—*The First Hundred Years.* Crawfordsville, Ind: Banta. 395 pp.
7. Thomas, M. B., Dudley, W. R. 1894. *A Laboratory Manual of Plant Histology.* Crawfordsville, Ind: The Journal Co. 115 pp.
8. Thomas, M. B., Bodine, D. ca. 1900. *Wabash College Souvenir.* 25 pp.

Ann. Rev. Phytopathol. 1979. 19:29–35
Copyright © 1979 by Annual Reviews Inc. All rights reserved

ROLAND THAXTER ❖3699

James G. Horsfall

The Connecticut Agricultural Experiment Station, Box 1106, New Haven, Connecticut 06504

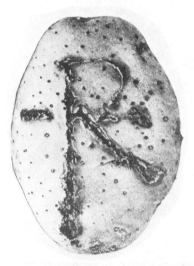

Figure 1 RT: A monogram etched in scab by a Boston Brahmin on a lowly Connecticut potato.

I can think of no better logo than this to signify Roland Thaxter. It is all there, his ego, his showmanship, his extraordinary imagination, his mycological bent, his basic research, and his grudging assent to the need for useful research.

He published this photograph in his second report (2). Look at the exquisite experimental design. You scrabble away the soil from a single young potato, dip a nail into your test culture, and scratch your logo onto the tuber; no statistics, no sweat, no error, just unequivocal results. Look at the basic research. He establishes the etiology of a disease. Look at the practical research. He provides a route to control of the disease.

With one thrust this deceptively simple experiment slices through all the accumulated guff that scab on a potato tuber is caused by mechanical

29

0066-4286/79/0901-0029$01.00

irritation from stones, by barnyard manure, or alkaline soils, by excessive soil moixture, by iron oxide, or by drought. David slays Goliath! Thaxter [(2) p. 82] in his inimitable style, says "It is needless to examine the different theories [about scab] in detail, since in all cases they lack the confirmation of crucial experiments and are involved and involve each other in a mass of contradictions, with a result of nothing definite in any case."

I once shook Thaxter's hand. In so doing I had shaken the hand of the man who had shaken the hand of Farlow, who had shaken the hand of deBary. I record here my respect and admiration for my distinguished predecessor, the first plant pathologist at The Connecticut Agricultural Experiment Station.

Figure 2   Roland Thaxter, Squirt Gun Botanist par excellence.

HIS INTELLECTUAL BACKGROUND

Thaxter came from a long line of intellectuals. W. H. Weston (3) writes:

> His father, Levi Lincoln Thaxter (Harvard, 1843), was by nature and inclination a scholar, an authority on the life and works of Browning [in fact Browning wrote a poem for his gravestone] . . . a respected member of a literary and artistic group comprising such men as James Russell Lowell, Henry D. Thoreau . . . and Nathaniel Hawthorne. His mother, Celia Layton Thaxter, will always be remembered for her poetry, in which her love of nature and her deep religious feeling were manifest.

I assume that it was his biologically minded mother who interested him in biology. Entomology was his first love, and his first six papers were on insects. He must have decided that he could not earn a living thus, and so he entered Harvard Medical School to become a doctor of medicine.

In the meantime he came under the influence of W. G. Farlow (whose first paper was on the potato rot, by the way). Farlow persuaded him to enter cryptogamic botany and his seventh paper was on *Gymnosporangium.* After he fled from New Haven back to Harvard in 1891, he combined his mycology with his entomology by working on fungal parasites of insects. Mind you, if anyone had told him that this was pioneering research in biological control of insects, he probably would have dropped it.

HIS ENTRY INTO PLANT PATHOLOGY

It was 91 years ago on July 1, 1888, that he got off the horse car on Whitney Avenue in New Haven, Connecticut, presented himself to the Director of the Agricultural Experiment Station, and stalked into the pages of the history of plant pathology. He thus became the tenth plant pathologist in the Nation after W. G. Farlow, T. J. Burrill, J. C. Arthur, K. F. Kellerman, C. E. Bessey, S. M. Tracy, F. Lamson-Scribner, B. T. Galloway, and A. F. Woods.

He did not call himself a plant pathologist. Thaxter called himself a botanist, sometimes a mycologist. He called his new laboratory a "myco-theca" from two Greek words, *"mikos"* for fungus and *"theca"* for box. He grew his plants in what he called a "sass garden," now, of course converted to a parking lot.

CONFLICT BETWEEN BASIC AND USEFUL SCIENCE

Upon his arrival in New Haven, he was immediately dumped into the schizophrenic conflict still waged between basic and useful science in the

Nation in general and in plant pathology in particular. As an intellectual he yearned to do only basic research far removed from "the madding crowd's ignoble strife," but being in a tax-supported institution, he found intense pressures to be useful. Thaxter's thrashings in the boneyard of primitive plant pathology helped set it on its way to becoming the science it is today, however.

Let us look at the record.

In the first report of the Experiment Station in 1876, Director Atwater (1) had stated the conflict, 13 years ahead of Thaxter. He wrote:

> It has been felt from the first that the more abstract scientific investigations would afford not only the proper, but also the most widely and permanently useful field of labor [for an agricultural experiment station]. But the need of a fertilizer control system was so pressing and so vital to the interests of a considerable portion of the farmers of the state that it seemed absolutely necessary to turn the first efforts in that direction.

In his letters to Farlow, Thaxter has stated his position on the two sides of the case. "Original research," he wrote, "is a precious slow coach and is not wanted by the constituency for which the Stations are created. They do not want pure science; they want mud pies, the sloppier the better." He labored to explain it and later wrote to Farlow. "Mr. _____ of Greens Farms said to me the other day; Humph, this learning just for the sake of learning! I never took any stock in it. It's all I can do to get a living."

Of course, the Stations's constituency was not interested in subsidizing the study of fungi parasitizing the insects that crawl in the slime on the bottom of a pond (as Thaxter was). They undoubtedly felt that the parasites on an onion or a potato were just as interesting, but useful as well. Nevertheless, even a Brahmin has pangs of conscience and Thaxter wrote to Farlow saying that in writing his annual report he was "looking back at my year's work with a sickish feeling when I balance my practical accomplishment with my cash recompense."

His onion smut research showed, however, that he could carry water on both shoulders. He investigated the basic biology of the disease, showed that it is soilborne, not seedborne, showed that seedlings are susceptible for only a few days, applied potassium sulfide or sulfur dust in the row with the seed, and protected them until they could reach the immune stage. He then persuaded a machinist to make a gadget to sow fungicide and seed together, took it to a farmer's field, and supervised its trial.

This is an elegant case for the three stages of technological development: basic research to discover new knowledge, useful research to transduce it to technology, and work to transfer its technology to the farmer.

He went on to describe and name *Phytophthora phaseoli* on lima bean. I suspect that Thaxter would be pleased secretly but offended publicly to learn that in the 1960s the USDA released for use a resistant lima bean and named it, Thaxter.

SQUIRT GUN BOTANY

In public Thaxter viewed plant pathology with a jaundiced eye, sometimes calling it "pocketbook mycology," sometimes "squirt gun botany." I suspect that the latter term arose from his pioneering research to introduce fungicides, especially Bordeaux mixture to American agriculture. Nevertheless, he detested Bordeaux mixture and wrote with some feeling, and, I must say, with some justification, that "Bordeaux mixture is the vilest compound imaginable, but it would give me intense satisfaction to spray [some] Connecticut farmers with it until . . . the moss started from their backs."

In private, however, he was more affectionate toward plant pathologists. In 1909, 18 years after he left the "bucolic constituency" of The Connecticut Agricultural Experiment Station, he formalized his fellowship with the squirt gun botanists by signing up as a charter member of the American Phytopathological Society. And in 1917, 26 years after he had returned to the protecting folds in the skirts of his alma mater, he visited Clinton, then plant pathologist in Connecticut. He wrote, "I am sorry to make such a flying visit to my old and happy hunting grounds, but I enjoyed myself greatly."

To make sure of his continuing identity with plant pathology he turned out 12 students as follows: J. B. Rorer, J. R. Johnston, G. P. Clinton, H. P. Barss, H. S. Jackson, J. H. Faull, R. H. Colley, F. C. Stewart (note Stewart's wilt of corn), W. A. McCubbin, O. F. Burger, Charles Drechsler, and F. C. Meier. Three of these, Clinton, Stewart, and Barss, became presidents of the American Phytopathological Society.

A RACHEL CARSON OF 1891

The environmentalists pestered Thaxter, too. Here in Figure 3 is a letter from a farmer worrying about the possible poisonous properties of pesticides back in 1891. He didn't have Carson's elegant mastery of the King's English, but his case was just as real to him and he used Carson's bogie. *It might be.*

I can't find Thaxter's response. My guess is that he was offended by the spelling and irritated, that after all his labors to introduce Bordeaux mixture, he was suspected of poisoning the horses.

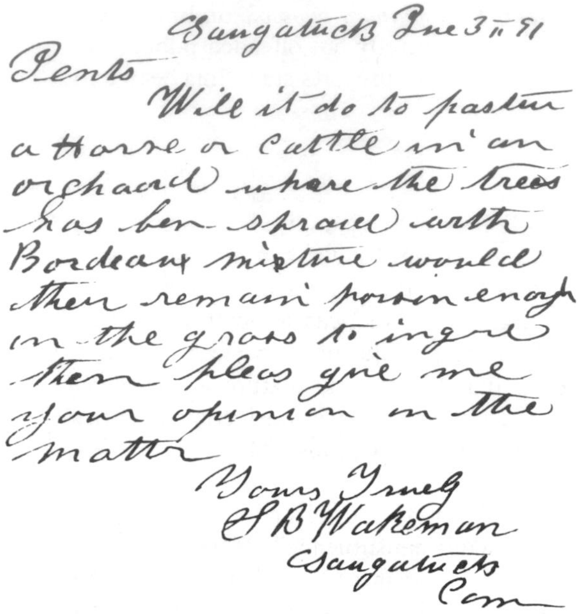

Saugatuck, June 3 II91

Jents

Will it do to pastur
a Horse or cattle in an
orchaord whare the trees
has been spraid with
Bordeaux mixture would
their remain poison enough
in the grass to ingre
them pleas give me
your opinion on the
mattr

Yours truely

S. B. Wakemen

Saugatuck

Conn.

Figure 3   A letter from an environmentally minded farmer.

THE BATTLE OF KITTERY POINT

And now we turn to a few anecdotes.

Thaxter, being well off, had a summer place on the coast of Maine at Kittery Point. He had a garden, [see photograph by Weston (3)], and in his garden he grew a few currants and gooseberries, anathema to the federal gumshoes charged with the ill-fated program to eradicate white pine blister rust by eradicating *Ribes* spp. When they came to root out Thaxter's currants, he stopped them at the gate with his shotgun—thus the famous Battle of Kittery Point. He won the battle, but, of course, he lost the war for the time being. Now we know that Thaxter really won after all. The Federals have quit trying to kill off all the *Ribes.*

Thaxter disliked J. C. Arthur intensely because in a meeting Arthur guffawed when Thaxter mentioned that Bordeaux mixture was not all powerful because he had seen molds growing on the surface of the liquid in a barrel of copper sulfate.

When his paper on the lima bean *Phytophthora* was done, he sent Farlow two reprints and got back a thank-you letter saying, "I am glad to have two copies of your papers . . . one copy . . . to feed the little pigs in the laboratory [graduate students, that is] and the other I keep for the capacious devouring of the old hog himself."

Thaxter was often afflicted by poor health. He made a typically astute observation about an attack of malaria in New Haven. "I do not see why it should be so prevalent here, for the bad drainage and night smells cannot account for its continued occurring back in the country." Dr. Thaxter, had you continued your entomology with an aim to make it useful, you could have discovered insect transmission of human disease ahead of Walter Reed.

HIS KUDOS

During his lifetime, Thaxter garnered many kudos, the presidencies of the Mycological Society, Botanical Society of America, and the New England Botanical Club. He was elected a member of the National Academy of Sciences and the American Academy of Arts and Sciences. He was made an honorary member of the Russian Mycological Society, the Linnean Societies of London and Lyons, the Royal Botanical Society of Belgium, the Royal Academies of Sweden and Denmark, the Botanical Society of Edinburgh, the Academy of Science of the Institute of France, the British Mycological Society, and the Deutsche Botanischen Gesellschaft—not bad for a squirt gun botanist!!

A FEW CONCLUSIONS

We wonder how he did so much in three short years after his doctorate. This was not three short years after any doctorate. This was three short years after one of the very first doctorates. I suggest that he did it by choosing the productive paths, wasted no time on trivia, no coffee breaks, no long weekends. His letters were often signed, "Hastily."

His major contribution was not that he proposed the first soil treatment, not that he named *Oospora scabies* or *Phytophthora phaseoli.* His major contribution, I think, was to resist the excruciating pressure by the locals to give them aspirin for their several pneumonias. Thaxter researched the biology of their pneumonias for significant answers and thereby set some very high goals for our science.

As we said at the outset, Thaxter had no experimental farm for his research, only a small "sass garden." Sass or garden sauce is said to be of two kinds, long-sass (as beets and carrots) and short-sass (as onions, potatoes, and pumpkins). Since Thaxter worked on onions and potatoes, we may assume that professionally Thaxter was a short-sass man.

Literature Cited

1. Atwater, W. O. 1876. Preliminary report. Agricultural Experiment Station. *Rept. Conn. Board Agric.* 1875:360–65
2. Thaxter, R. 1891. The potato scab. *Ann.* *Rept. Conn. Agric. Exp. Stn.* 1890: 81–95
3. Weston, W. H. Jr. 1933. Roland Thaxter. *Mycologia* 25:69–89

Ann. Rev. Phytopathol. 1979. 17:37–58
Copyright © 1979 by Annual Reviews Inc. All rights reserved

MYCOPLASMAS, SPIROPLASMAS, AND RICKETTSIA-LIKE ORGANISMS AS PLANT PATHOGENS

❖3700

F. Nienhaus and R. A. Sikora

Institut für Pflanzenkrankheiten, Rheinische Friedrich-Wilhelms-Universität
Bonn, Nussallee 9, 5300 Bonn, Federal Republic of Germany

INTRODUCTION

Mycoplasmas, spiroplasmas, and rickettsia-like organisms are considered to be the causal agents of yellows diseases, formally thought to be of viral origin. They belong to various groups of procaryotic microorganisms that are morphologically similar to or identical with members of the Mycoplasmatales and Rickettsiales (6). Their similarity to these two orders has given rise to the terms mycoplasma-like organisms (MLO) and rickettsia-like organisms (RLO). The genus *Spiroplasma* has been placed in the order Mycoplasmatales.

Mycoplasma-like organisms have been found in more than 200 plant species and in a large number of arthropods. Pathogenicity has not been proved in that Koch's postulates have not been fulfilled. The main reason for the failure to prove pathogenicity is that most attempts to isolate and cultivate these organisms have failed.

Spiroplasmas occur as internal parasites in plants, leafhoppers, *Drosophila,* honey bees, and ticks. They have also been recently isolated from the surface of several flowers (25). Spiroplasmas have been cultivated on artificial media and their pathogenicity to plants has been demonstrated. *Spiroplasma citri* is the only recognized species in the family Spiroplasmataceae; all other spiroplasmas lack binomial names. Although many isolates are

37

related serologically, they are not considered as one species (26a). Electrophoretic comparisons of protein and nucleic acid patterns and DNA hybridization tests are considered necessary for accurate classification (130).

Rickettsia-like organisms seem to be a heterogeneous group of parasites, most commonly found in the xylem or phloem of diseased plants. Pathogenicity has been demonstrated with RLO isolated from xylem tissue and cultivated on artificial media. They were later renamed rickettsia-like bacteria (21, 61). Koch's postulates have not been fulfilled with most of the other RLO forms. The results to date indicate that in some cases RLO are contaminants or parasites of minor importance which only in isolated cases are responsible for disease development. In this article all forms are discussed under the name rickettsia-like organisms (RLO).

Early discoveries concerning MLO and RLO in plant tissue and research advances in the field have been discussed in a number of reviews (24, 27, 33, 53, 55, 61, 65, 66, 77, 86, 87, 105, 108, 119, 124, 129, 130, 161, 169). In this review an attempt has been made to cover only the most recent advances in the field. The number of references cited has been restricted because of space limitations.

MORPHOLOGY AND ULTRASTRUCTURE

The MLO and RLO found in plants and insects are similar to true mycoplasmas and rickettsias in both morphology and ultrastructure. In the cell, MLO form elementary, small spherical, large globular, and filamentous bodies. The elementary bodies are 60 to 100 nm in diameter, electron dense, and slightly granular. Spherical bodies, which in some cases are less than 60 nm in diameter, are considered to be the smallest MLO forms thus far detected. The large globular bodies are 150 to 1100 nm in diameter and transparent. Filamentous bodies have been observed in sieve tubes, where they resemble migrating stages passing through the sieve pores. The MLO change structurally in form as the disease in the plant progresses. These changes, which parallel those found with cultured mycoplasmas, are believed to be stages in the reproductive cycle. The sensitivity of MLO to changes in environmental conditions, however, leaves this interpretation in doubt. "Inclusion bodies" or "inner bodies," for example, were formed by large MLO bodies that invaginated to form sacs in aster plants grown under reduced lighting conditions. These bodies were considered the end product of a degeneration process most probably resulting from nutritional deficiencies and drastic physiologic changes affecting the host. Comparable structural changes have been seen in old mycoplasma cultures (11). Mycoplasma-like organisms possess a triple-layered membrane that is almost identical with those found in mycoplasmas. The surface structures of several mycoplasmas have been studied by freeze etching and scanning electron microscopic techniques. *Acholeplasma laidlawii,* for example, has been shown to be covered with a tile-like membrane surface. These techniques, which are being applied to MLO, should produce interesting comparisons between true mycoplasmas and those found in plant tissue. The granulated contents of small spherical MLO bodies that are bounded by a membrane, and similar granules in the peripheral zone of the large globular bodies, have been described as ribosome-like osmiophilic particles. The central core of MLO contains fibrillar, presumably DNA, material. Vacuoles, which are only occasionally found in filamentous bodies, are frequently encountered in the large globular bodies.

Helical bodies, which have been observed with spiroplasmas, are 3 to 25 μm in length and 100 to 200 nm in diameter. They are often seen attached to spherical structures termed blebs. Spiroplasmas also change morphologically during cultivation. The spiral cells become increasingly distorted with time, collapsing in old cultures, until only round structures and the remains of nonhelical filaments can be seen. These cultures are still viable and begin forming new spiral cells when subcultured. It should be noted that nonhelical and therefore nonmotile forms of spiroplasmas exist which are closely related to *Spiroplasma citri* (154). Three-dimensional granular structures are seen on the surface of spiroplasmas when treated by freeze etching. These projections look similar to those of several mycoplasmas and T-strains of bacteria. Protein patterns of isolated and cultivated organisms obtained by gel electrophoresis have proven to be a valuable tool in the study of biochemical aspects and taxonomic relationships (19, 26a). Spiroplasmas from honeybees, for example, can be distinguished from *S. citri* by their cell protein patterns (31). Seventeen different polypeptides were separated from membrane extracts of spiroplasmas by the same method. The main fraction with an amino acid composition different from those of the other fractions was named spiraline (174). The protein patterns of the nonhelical isolate from citrus differs from those of *S. citri* in the absence of one protein. This protein may be associated with an actomyosin-like contractile complex presumably responsible for the maintenance of the spiral structure and motility of spiroplasmas (155).

Rickettsia-like organisms are comparable in both morphology and ultrastructure with true rickettsias. The cells are usually rod-shaped, 0.2 to 0.5 μm wide, and 1.0 to 3.0 μm long. They are usually surrounded by a well-defined cell wall and plasma membrane, both trilaminar in structure. The walls are generally ridged or rippled in a periodic manner. These ridges appear to spiral around the long axis of the cell. The cell wall is separated from the inner cytoplasmic membrane by an electron-lucent space (3, 84, 115). Several RLO associated with plant diseases appear to lack the layer of dense-staining material, often termed the R-layer, lying between the outer and inner membranes (3, 84, 90, 102). The R-layer has been seen in gram-negative bacteria and in some rickettsia-like bacteria, for example, in the Pierce's disease agent (103). Demonstration of the existence of the R-layer in RLO may be dependent on preparation techniques (3). Such a demonstration could be of taxonomic importance (102). The ridges or folds, which account for the rippled appearance of the outer wall of RLO bodies, are similar to those of *Rickettsia prowazekii* (115). Some RLO bodies are pleomorphic, spherical, or filamentous and have a thin cell wall (60, 63, 132). Internally, RLO consist of electron-dense areas and one or more areas of low density. The latter areas contain strands of material similar to the DNA strands seen in bacteria. Ribosome-like granules are also present.

REPRODUCTION

As already stated, MLO in infected plant cells take on a number of different forms that resemble stages in the replication processes of the Mycoplasmatales. Hypotheses on the mode of MLO reproduction refer to binary fission, budding, chain formation, release of inclusion bodies, and reproduction through elementary bodies. The presence of hantle-like forms led to the binary fission theory of reproduction, and the separation of dense globular regions from large MLO bodies suggested budding. These processes have been also interpreted as the beginning stages in the development of filamentous forms (87). Filaments have been shown to give rise to chains of coccoid bodies which were considered to be replicating forms if they formed segments able to continue further development. Similar processes have been observed in mycoplasma cultures. Spherical bodies predominate in very young cultures of spiroplasmas. Later, helical filaments are formed that increase and divide by constriction. The resulting short segments resemble those found with mycoplasmas (130).

Hantle-like structures and forms that seem to be in a state of binary fission have been frequently demonstrated with RLO (84, 109, 132). Pleomorphic bodies, sometimes bounded by only a membrane, were common in RLO suspensions taken from actively reproducing stages in chick embryos. They were also occasionally found in RLO-invaded root cells (111, 112, 132). These bodies were considered to be early RLO developmental stages. Small spherical particles, 28 to 72 nm in diameter, bounded by a single membrane, were seen in RLO-invaded cells of clover plants. When the work with RLO on grape is taken into consideration (132), these particles could be considered stages in the RLO reproductive cycle. Some of the pleomorphic structures found in plant cells may also be degenerative stages, especially when they appear after antibiotic treatments (90).

ISOLATION AND CULTIVATION

Most attempts to cultivate MLO taken from plant material have proven negative. This may be due to the fragile nature of the microorganisms, which makes isolation difficult, or to still unknown nutritional requirements. The requirements of different MLO isolates for essential nutrients, pH, and osmotic pressure in media derived from substrates favorable for mycoplasma cultivation have been studied (48). Anaerobic conditions were reported to aid in the cultivation of the agent responsible for aster yellows (28). Although typical "fried egg" colonies were obtained by some workers (17, 83, 107, 151) most attempts to reproduce the results failed or the results were unclear (9, 57).

Spiroplasmas seem to be the only plant pathogenic mycoplasma that form "fried egg" or similar colonies when cultivated on solid media. The first spiroplasma to be successfully cultured on artificial media was the agent of stubborn and little leaf diseased citrus, *Spiroplasma citri* (5, 14, 18, 44, 135, 136). Spiroplasmas were also isolated from stunted corn plants (12, 23, 118, 172), the cactus *Opuntia tuna monstrosa* (79), yellow dwarf diseased rice plants, and green leaf bug vector of yellow dwarf (149). They have been recently isolated from nonsurface sterilized flowers of several plant species. The latter did not form typical "fried egg" colonies on solid media (25, 26, 30). Three newly isolated spiroplasmas seem to be associated with aster yellows (Day-strain), western-X (peach yellow leaf roll strain), and pear decline (84a, 115a, 127a). Spiroplasmas have also been detected in *Oncometopia*-sharp shooter (94a).

Two spiroplasma forms pathogenic to vertebrates and carried by ticks have been isolated which show some serological cross-reactivity with *S. citri* (157).

The development of new culture techniques should stimulate research on the isolation and cultivation of MLO. A new and promising approach might be the use of phloem sap in culture media preparation. Good results have been obtained in studies utilizing the sap of palm trees, *Yucca,* and species of *Fraxinus* (33), although these substances are difficult to collect. Phloem sap of coconut and *Veitchia* palms has also been shown to be favorable growth media for both spiroplasmas and mycoplasmas. The resistance of 'Malayan Dwarf' coconut palm to lethal yellowing, therefore, is most probably not the result of toxic effects or nutritional deficiencies in the phloem sap, since the sap, instead of inhibiting, supports the growth of a broad range of mycoplasmas (94).

Many workers have had no success in their attempt to isolate and cultivate RLO on artificial media used in bacteriology (40, 49, 63, 100, 132). Techniques using new media for cultivating RLO of the xylem-inhabiting group that causes Pierce's disease of grapevine, almond leaf scorch, and alfalfa dwarf have been recently described. It should be noted that all three parasites are considered to be identical (22, 62, 152). Another method of culturing RLO that do not reproduce on artifical media involves the use of chick embryo culture, which is used with true rickettsias. These techniques have been recently applied to RLO isolated from plants (11, 112).

PATHOGENICITY

The detection of MLO and RLO in plants, their intracellular multiplication in diseased tissue, and their transmission by insect vectors from diseased to healthy plants is not proof of pathogenicity. Proof of pathogenicity requires the fulfillment of Koch's postulates. Thus far these postulates have been fulfilled in tests with isolates of spiroplasmas and some RLO. Leafhoppers injected with or fed spiroplasmas extracted from pure cultures have been shown to transmit the organisms to healthy plants which later developed the respective plant disease. The same spiroplasmas were then recovered from the diseased plant tissue (18, 91, 128, 144). *Spiroplasma citri* produces at least two toxins that may contribute to symptom production in infected plants. The toxins extracted from spiroplasma cultures broke down plant cells and inhibited seed germination (20). Cultures of xylem-inhabiting RLO have been transmitted by different mechanical inoculation techniques to their initial hosts where they produced typical disease symptoms (22, 62, 152). Phytotoxins produced by these organisms when grown in culture induced characteristic symptoms (81a).

Several points favoring the hypothesis that MLO and several RLO cause plant diseases are their association with diseased plants, their adverse effects on invaded and surrounding plant cells, disease development following vector and graft transmission, symptom remission and the decrease in their numbers after treatment with antibiotics, and the positive effect of thermotherapeutic treatment on both the disease and the organisms. The association of viruses with certain MLO and MLO-diseases has added support to the hypothesis that the MLO are merely virus vectors. There is, however, strong evidence to refute this assumption (53).

LOCALIZATION AND SPREAD IN PLANT TISSUE

Mycoplasma-like organisms are generally confined to the sieve elements of the phloem. They have been occasionally observed in the parenchyma and companion cells of the phloem, and in some cases in the cortical parenchymatous tissues. They are seldom found in xylem or mesophyll cells or in the pericycle of roots. In general, MLO are found in the intact cytoplasm of mature cells. On the other hand, they have been demonstrated in cells in early developmental stages (117), in callus tissue (158), and in meristematic tissue near the apical extremity of infected plants (15).

Rickettsia-like organisms can be divided into two groups depending on their location in the plant, the first present in the xylem and the second in the phloem tissue. Although some forms occur in both types of tissue, they normally show preference for one or the other. A number of RLO have also been reported to invade meristemic and young differentiating tissues. In some cases, however, the tissue preference is still unknown or unreported.

The number of MLO and RLO in ultrathin sections can reach densities of 100 per cell. In such cases the cells are completely filled with the organisms. Sieve pores have been shown to play a major role in MLO spread within the diseased plant. A number of authors have also suggested that the plasmodesmata may have some importance in MLO movement from cell to cell. The motile stages of spiroplasmas may be able to spread actively through the tissue. With RLO, spread is most probably by way of xylem vessels or through sieve tubes, since many agents have pleomorphic forms which could move through the sieve pores. The mechanism by which they invade meristematic or similar cells is still unknown.

MODE OF TRANSMISSION

The transmission of MLO and RLO from plant to plant is, in general, dependent on insect vectors. Transmission by grafting, mechanical transmission, transmission through the parasitic seed plant *Cuscuta,* and seed transmission have to varying degrees been demonstrated. Mechanical transmission has been in most cases unsuccessful, because of the sensitivity of MLO and RLO to plant extracts and to environmental conditions. Successful transmission has been obtained by dipping roots, cut stems, and leaves in *Mycoplasma* and *Acholeplasma* culture solutions and by injecting these suspensions into the plant tissue (1, 95, 139). *Spiroplasma* inoculated by root dipping invade citrus seedlings and are transported to the leaves where they can be reisolated (96). Rickettsia-like organisms have been mechanically transmitted with suspensions obtained from pure cultures (21). Although grafting is normally restricted to closely related plants, it has been shown that the implantation of diseased plant tissue onto unrelated indicator plants can result in transmission (35). A number of species of *Cuscuta* have been shown to act as plasmatic bridges for MLO and RLO. The MLO multiply in the *Cuscuta* tissue which may also develop disease symptoms. Rickettsia-like organisms were transmitted from rosette diseased sugar beets to healthy plants by *C. planiflora,* whose vascular bundles were completely filled with the agent (52). Seed transmission, which has been demonstrated with a number of diseases, seems to be absent in many cases (78). Arthropods are the most important vectors of MLO and RLO. They are transmitted by about 60 species of leafhoppers and several species of aphids and psyllids (53). Although in most cases phloem- and xylem-feeding leaf-hoppers transmit RLO agents (61), two species of *Psyllina* transmit the citrus greening agent (92, 101), and a leaf bug, *Piesma quadratum,* trans-mits the RLO of rosette disease of sugar beets (157). Soilborne transmission of MLO by vectors of Margarodinae (32) and of RLO by nematodes (133) has been presumed.

Most research on MLO transmission has been performed with leafhop-pers. The acquisition period in leafhoppers varies between 3 hr and 5 days. Long incubation periods have been reported and may be related to MLO migration in the vector or to the influence of environmental conditions. The multiplication of MLO in leafhoppers, their transovarial transmission, and their passage through a successive number of leafhopper generations has been described (87). The lack of efficient defense mechanisms in insects to mycoplasmas may be the reason for the acquisition of spiroplasmas in laboratory tests with *Drosophila,* species of cicadellids, and other insects (128, 172). Furthermore, the multiplication of saprophytic species of *Acholeplasma* in vector and nonvector insects has been observed (76, 170). The MLO were detected extra- and intracellularly in a number of organs in the vector. Mycoplasmas have been reported to have both beneficial and harmful effects on their vectors (88).

The presence of RLO in arthropod vectors collected from diseased plants is not valid evidence that they are plant pathogenic agents, since both nonpathogenic and animal pathogenic rickettsias are often found in insects and mites. The simultaneous occurrence of these different forms in artho-pods complicates research studies on the interaction between RLO and their vectors.

DISEASES

Diseases caused by, or associated with, MLO, spiroplasmas, and RLO are most prevalent and of greater economic importance in warm climatic re-gions where environmental conditions favor multiplication and vector ac-tivity (5, 8, 14, 16, 33, 61, 89, 114, 167).

Mycoplasma-like organisms disrupt normal translocation processes in the diseased plant. Cells not associated with the invaded tissue are also adversely affected. Phytotoxin production increases in the presence of MLO and may be involved in disease development. Substances toxic to plant tissue have also been detected in spiroplasma cultures (20, 148). Excessive callose formation and degeneration of phloem tissue in MLO diseased plants may lead to stunting, wilting, or leaf yellowing, the latter being less specific. The presence of MLO also disrupts normal growth regulator levels resulting in premature development of adventitious buds and altered apical dominance, leading to the proliferation of axillary shoots and the condition known as witches'-broom. This imbalance is also responsible for phyllody and antholysis in the flowering stage.

The diseases associated with MLO belong to the yellows type which have been divided into the following four types: aster yellows (elongation of internodes, leaf yellowing), stolbur (apical dwarfing, stunting, leaf roll, epinasty, wilting, virescence), witches'-broom (proliferation of axillary shoots), and decline (degeneration). This system is an oversimplification, since under natural conditions these organisms may produce symptoms belonging to a number of the different types depending on the host species affected, the time and mode of infection, environmental factors, and the agent involved. The inability to use symptoms to differentiate these disease agents and the lack of biochemical and serological diagnostic techniques make identification and grouping at this time unreliable. Several diseases, each belonging to a different disease type, may be caused by the same MLO or by different strains of the same agent. Based on the literature up until 1976, MLO diseases affect more than 200 different plant species in 59 families (53).

The widest host range has been described for the California and New York strains of the aster yellows agent (39, 81, 140, 141). The stolbur types mainly affect solanaceous plants. The disease affecting tomato was named "big bud" in Australia and North America (17, 137), "stolbur" in Europe (164), and "Mal Azul" in South America (4). Other disease types are parastolbur and metastolbur in potato (162, 163).

The corn stunt disease, first described in Texas and California in 1945, was differentiated into several types. The disease caused by the Rio Grande strain was found to be associated with spiroplasmas (29), whereas that incited by the Mesa Central strain seems to have a MLO origin (2).

The rice yellow dwarf disease, first reported in Japan in 1910, is widely distributed in the tropical areas of Asia (116). Spiroplasmas have been isolated from diseased plants and from the vector, the green leaf bug *Trigonotylus ruficornis* (149).

Among decline diseases caused by MLO or spiroplasmas, the western-X disease of peach has a fairly wide host range including other stone fruits and a number of herbaceous plants (69, 131). Pear decline is associated with graft-union phloem failure that affects trees with *Pyrus* and *Cydonia* rootstocks (59).

A number of leafhopper-transmissible, xylem- and phloem-limited RLO diseases were discussed in detail in a recent review article (61). Pierce's disease of grape, alfalfa dwarf, and almond leaf scorch, which are presumably caused by the same agent, seem to be restricted to the North American continent. Severe winter temperatures seem to be the major factor limiting the northern distribution of this RLO in the USA (126). Serologically identical RLO were found associated with plum leaf scald in South America (73) and with a similar disease in southwestern USA where it had the same geographic range as the phony disease of peach (41, 42).

Three phloem-inhabiting diseases of clover—club leaf of *Trifolium incarnatum* in the USA, club leaf-like disease of *T. repens* in England, and rugose leaf-curl disease of different *Trifolium* species and other legumes in Australia—may be caused by closely related RLO. Similar structures have been observed in the phloem tissue of trees affected with citrus greening, which occurs in South Africa and Asia and can be differentiated from citrus stubborn (*Spiroplasma citri*) by symptom expression. These organisms were also found in the vascular fluid of citrus trees affected with young tree decline growing on rough lemon rootstock in Florida (37). A new disease of sugar beet and other species of Chenopodiaceae named latent rosette disease, because symptoms develop only under warm greenhouse conditions, seems to be induced by a phloem-inhabiting RLO (52, 113). Other phloem-inhabiting RLO were found in dodder (46) which can serve as an efficient RLO vector (52). These organisms were also detected in the phloem of potato affected with leaflet stunt (75) and in *Sida cordifolia* with little leaf symptoms (60).

Diseases in which the RLO simultaneously inhabit the phloem and xylem tissue, and occasionally the parenchyma and meristematic cells, have been reported from a number of countries. They have been detected in infectious necrosis of grape in Czechoslovakia and in yellows disease of grape in Germany and Greece, the latter similar to bois noir in eastern France. In all three cases the RLO were mainly concentrated in the root system. A number of workers have suggested, therefore, that soilborne transmission by nematodes may be important (110, 132, 133, 159). The presence of RLO in the plant-parasitic nematode *Xiphinema index,* found associated with yellows diseased grapes, lends strong support to this theory (133). Soilborne transmission may also be a factor in a witches'-broom disease affecting larch trees (*Larix decidua*) reported from Germany (108, 109).

Another agent considered to be a RLO, even though it lacks a cell wall, may be the causal agent of yellow wilt disease of sugar beets (marchitéz amarilla) in Argentina and Chile (160). Crinkle disease thought to be incited by RLO is also an important limiting factor in hop production in eastern Europe (165). Rickettsia-like organisms were found associated with chlorosis and aspermy of wheat (125), and with pollen sterility of onion (*Allium sativum*), where they produced nutritional disturbances in the anthers during microsporogenesis (80). They were also found to be associated with a decline of coconut palms (147), and together with MLO, responsible for proliferation-disease in carrots (45). Experimental evidence has further shown that MLO are important in the etiology of apple proliferation, whereas the function of an associated RLO could not be clarified (120).

METHODS OF DIAGNOSIS

One of the most basic and simple methods of determining the presence or absence of MLO and RLO has been the application of tetracyclines to the diseased plants. Symptom remission following treatment is considered strong evidence that these agents cause plant diseases. The most important and accepted diagnostic tool in diagnosis is the electron microscope. Its main disadvantage lies in the small amount of plant tissue that can be studied in thin sections. Negative results are often misleading since the agent might not be uniformly distributed in the plant. Another disadvantage is the laborious techniques required for preparation of test material. Negative staining has been tried with sap extracted from plants and with suspensions from in vitro cultures. This method is not specific for MLO, since similar bodies observed in diseased and healthy plant tissue were considered to be artifacts (99, 173). The purification of MLO with Sepharose-gel filtration and density gradient centrifugation techniques was considered more difficult (143).

Vacuum extraction of vascular fluid and the concentration of the RLO in the extract by centrifugation before electron microscopic examination have also been used with some success (37). The detection of the helical structures of spiroplasmas requires thicker ultrathin sections or a series of sections. Three-dimensional reproductions of MLO have been produced by stereo-pair photographs taken from sets of serial photographs (38).

A digestion technique in which ultrathin sections fixed in glutaraldehyde were treated with pepsin has been shown to be helpful in differentiating MLO and RLO structures from morphologically similar cell parts (127).

Scanning electron microscopy has been successfully applied with mycoplasmas, and more recently with MLO and spiroplasmas (54, 93, 122, 153). Diagnostic work with this method could prove important, because the preparation of test material is easier, the amount of tissue examined larger, and the test for MLO and spiroplasma structures more dependable. In addition, the distribution of the organisms in the tissue can be readily determined.

Light microscopic examination used together with different staining techniques has been employed with yellows diseased plant tissue and could be of significant diagnostic value even though the techniques are nonspecific

(142). Ultrathin sections embedded in epoxy resin, for example, can be stained with toluidine blue. Distinct differences can be detected in the amount of stain absorbed by the vascular bundles of MLO invaded tissue as seen under the light microscope. This technique also aids in selecting areas of the tissue section for electron microscopic examination. Histochemical techniques, such as staining nucleic acids with Feulgen stain or methylene green-pyronine, have also been used to demonstrate the presence of MLO and RLO (15, 52, 72).

Phase contrast microscopy has been applied in tests for spiroplasmas and RLO. In club leaf diseased clover plants, for instance, RLO could be detected with this technique in a small sample taken from the leaf midrib, or from a single leafhopper vector (82). In addition, phase contrast microscopy can be used with fluids extracted from the vascular bundles and with culture suspensions (40).

Different fluorchromatic techniques in combination with the UV light microscope have also proven promising. There was a positive correlation between callose formation and degeneration processes in diseased phloem tissue when stained with aniline blue. More dependable results have been obtained by treating glutaraldehyde fixed tissue with polyethyleneglycol-4000 and staining with azur I, or fast green (50). The demonstration of DNA by staining with DAPI (4'-6-diamidino-2-phenylindole), DNA-binding benzimidole derivatives, and with other less specific cytochemical fluorescence techniques has been used recently to detect mycoplasmas, large viruses, and RLO in diseased tissue (132, 134, 138). It has been further suggested that a combination of DAPI and berberinesulfate can be used for detection of these organisms (121). The pretreatment of plant tissue with pectinase to induce tissue separation before the application of DAPI has also been recommended (10).

Successful transmission of MLO by injection of diseased plant sap into a vector has been reported (47, 145, 150). This has proven to be a difficult but valuable tool in diagnostic and pathogenicity tests.

Serological tests are dependent on antisera produced from pure MLO, spiroplasma, and RLO cultures. Growth inhibition tests and precipitin ring tests have been performed with spiroplasma antisera (156). In the presence of specific *S. citri* antibodies the organism takes on the spiral structure that can be seen by light microscopy (97). Other attempts to produce specific antibodies to partially purified plant MLO antigens have met with only limited success. It has been suggested that the poor results were related to inadequate periods of immunization, small amounts of immunizing antigen, or to insufficient concentrations of MLO (156). Tests utilizing the more sensitive ELISA techniques (13) have been started to determine the possible uses of specific serological tests in diagnostic work, which if positive, could be of immense value (34). Immunofluorescence techniques have been recently applied with MLO and RLO. In combination with other serological methods these techniques aided in differentiating between the two microorganisms (22, 34, 43, 62, 152).

The application of polyacrylamide gel electrophoresis in diagnostic work should be looked at closely. The protein patterns obtained from different isolates of *S. citri* grown on media were found to be similar. This was not the case when the *S. citri* patterns were compared with those of *Acholeplasma laidlawii* (5, 89). Furthermore, this technique aids in distinguishing *S. citri* from spiroplasma isolated from the honeybee and constant diseased plants (26a, 31). The technique has also been used to differentiate between the xylem-inhabiting RLO of grapevine, almond, and alfalfa. In this case it showed that one agent was responsible for different diseases in different hosts (22, 152).

CONTROL

The economic importance of many diseases presumed to have a MLO or RLO etiology has stimulated the search for acceptable control measures. Because of the similarity between viruses and these organisms, most of the prophylactic measures developed for virus control can be applied to MLO and RLO diseases. In addition to chemical control, therapeutic treatments with heat and the establishment of meristemic cultures used by virologists have become important.

The selection of propagating material from plants lacking disease symptoms is a common practice which has to be supplemented by more specific tests. Techniques that could prove to be of practical importance have been described in the section on methods of diagnosis. Their application in indexing programs will in most cases require further intensive research efforts.

The natural spread of diseases caused by MLO and RLO transmitted by insect vectors can be controlled in most cases with insecticides. In addition, the results indicate that organophosphorus and oximecarbamate insecticides besides controlling the vectors also act directly on mycoplasmas. Nematicides could also prove to be applicable in cases involving nematode-transmitted RLO.

Eradication of diseased plants has been recommended in many instances in order to reduce the source of new infections. The control of weeds and wild plants that may act as overwintering hosts for the pathogens affecting annual crops serves the same purpose. Breeding for resistance is another way to protect crops from heavy losses, and is probably the only effective long-term control method available (51, 56, 58, 64, 85, 104, 106, 116, 146).

The application of nutrients and soil improvement techniques, which stimulate the regeneration processes of the root system, aid in combating RLO infections (132, 165).

Treatment with tetracycline was shown to cause mulberry dwarf symptom remission (68). Similar treatments have been applied to a wide range of diseases for diagnostic and control purposes. Antibiotics which inhibit bacterial cell synthesis, for example penicillin, are ineffective against mycoplasmas and MLO, but are to a certain extent effective against RLO (61). The acquisition and transport of these antibiotics in plants have been thoroughly studied. The addition of $CaCl_2$ or $MnCl_2$ to tetracycline solutions, for example, increases the uptake and distribution of the antibiotic when applied to leaves (171). In addition to leaf sprays, antibiotics can be applied by root immersion, soil treatment, and by trunk injection. The latter technique allows a sufficient concentration of the antibiotic to reach the leaf area without causing simultaneous damage to the root system (67, 98, 166). The early findings of Stoddard that sulfa compounds influence disease development in yellows diseased aster plants have also been recently reconfirmed (74). Since such compounds have no effect on mycoplasmas in vitro (87), the reaction in plants seems to be indirect. According to other observations these compounds also act on rickettsias (27). New chemotherapeutic substances which in some cases seem to be more effective than tetracyclines (7) should be tested on a wider range of diseases. The treatment of stolbur-infected tomato with kinetin, for example, resulted in MLO degeneration and plant recovery (123).

Heat treatment and mist propagation have been shown to be useful in producing healthy plant material (36, 49, 70, 71, 73, 175). The importance of cold therapy, which is effective in controlling a number of diseases restricted to warm climates, has been discussed in other reviews (126, 131, 168). The possible use of meristem culture to control diseases produced by persistent types of MLO has also been examined. Tobacco plants reconstituted from callus tissue cultures containing 2,4-D, initially started from MLO-infected tobacco plants, were found to be free of MLO (158).

CONCLUSIONS

Since their discovery, mycoplasma-like organisms have been detected in more than 200, and rickettsia-like organisms in over 30, different plant species. The number of scientific reports dealing with these organisms in this period increased phenomenally. The discovery of spiroplasmas in plant and arthropod tissues as well as on plant surfaces was similarly followed by a rise in research activity. Proof of pathogenicity, which to date has been demonstrated in only a few cases, is the main task presently facing the scientific community. The recent breakthrough in culturing spiroplasmas and RLO on artificial media and in tissue culture should stimulate research on their physiology, biochemistry, and host-parasite interactions. The incorporation of new and improved serological, electrophoretic, and hybridization techniques should bolster research involving the use of protein and nucleic acid patterns as taxonomic criteria, and in clarifying the relationship between plant-inhabiting MLO and RLO and true forms of mycoplasmas and rickettsias. Although the classical diagnostic technique of testing ultrathin sections with the electron microscope still predominates, initial steps have been taken to develop reliable and simplified alternatives utilizing fluorescence, phase contrast, and scanning electron microscopy. The economic importance of many diseases associated with these organisms has stimulated research into their possible control. Rapid and simple serological techniques will be required, however, before efficient control measures can be developed. The environmental and economic restrictions on the use of antibiotics have increased the need for the development of new therapeutic compounds. The initiation of prophylactic control programs for MLO and RLO diseases of perennial crops, which are similar to those used in virus control, are an important step forward. These programs, which include propagation of healthy plant material, breeding resistant or tolerant varieties, and vector control, can be easily integrated into established plant protection programs.

Literature Cited

1. Babbar, O. P., Shukla, U. S., Agnihotri, V. P., Singh, K. 1973. Appearance of transmissible disease syndrome in maize (*Zea mays* L.) plants infected with avian mycoplasma. *Curr. Sci.* 42:190–92
2. Bascope, B., Galindo, J. 1978. Mycoplasmic nature of the Mesa Central corn stunt. *Int. Congr. Plant Pathol., 3rd, Munich* 1978:78 (Abstr.)
3. Behncken, G. M., Gowanlock, D. H. 1976. Association of a bacterium-like organism with rugose leaf curl disease of clovers. *Aust. J. Biol. Sci.* 29:137–46
4. Borges De Lourdes, M. V., David-Ferreira, J. F. 1968. Presence of mycoplasma in *Lycopersicon esculentum* Mill. with 'Mal Azul'. *Bol. Soc. Broteriana, Sér. 2* 42:321–33
5. Bové, J. M., Saglio, P., Tully, J. G., Freundt, A. E., Lund, Z., Pillot, J., Taylor-Robinson, D. 1973. Characterization of the mycoplasma-like organism associated with 'stubborn' disease of Citrus. *Ann. NY Acad. Sci.* 225:462–70
6. Buchanan, R. E., Gibbons, N. E., eds. 1974. *Bergey's Manual of Determinative Bacteriology.* Baltimore: Williams & Wilkins. 1268 pp. 8th ed.
7. Capoor, S. P., Thirumalachar, M. J. 1973. Cure of greening affected citrus plants by chemotherapeutic agents. *Plant Dis. Reptr.* 57:160–63
8. Caudwell, A., Kuszala, C., Bachelier, J. C., Larrue, J. 1970. Transmission de la flavescence dorée de la vigne aux plantes herbacées par l'allongement du temps d'utilisation de la cicadelle *Scaphoïdeus littoralis* Ball et l'étude de sa survie sur un grand nombre d'espèces végétales. *Ann. Phytopathol.* 2:415–28
9. Caudwell, A., Kuszala, C., Larrue, J. 1974. Sur la culture *in vitro* des agents infectieux responsables des jaunisses des plantes (MLO). *Ann. Phytopathol.* 6: 173–90
10. Cazelles, O. 1978. Mise en évidence, par fluorescence, des mycoplasmes dans les tubes criblés intacts isolés des plantes infectées. *Phytopathol. Z.* 91:314–19

11. Chen, M-h., Hiruki, C. 1977. Effects of dark treatment on the ultrastructure of the aster yellows agent in situ. *Phytopathology* 67:321–24

12. Chen, T. A., Liao, C. H. 1975. Corn stunt spiroplasma: Isolation, cultivation, and proof of pathogenicity. *Science* 188:1015–17

13. Clark, M. F., Adams, A. N. 1977. Characteristics of the microplate method of enzyme-linked immunosorbent assay for the detection of plant viruses. *J. Gen. Virol.* 34:475–83

14. Cole, R. M., Tully, J. G., Popkin, T. J., Bové, J. M. 1973. Morphology, ultrastructure, and bacteriophage infection of the helical mycoplasma-like organism (*Spiroplasma citri* gen. nov., sp. nov.) cultured from 'stubborn' disease of citrus. *J. Bacteriol.* 115:367–86

15. Cousin, M. T., Kartha, K. K. 1975. Electron microscopy and plant mycoplasma-like agents (MLA). *Proc. Indian Natl. Sci. Acad. B* 41:343–54

16. Dabek, A. J. 1977. Electron microscopy of Kainkopé and Cape St. Paul wilt diseased coconut tissue from West Africa. *Phytopathol. Z.* 88:341–46

17. Dana, B. F. 1940. Occurrence of big bud of tomato in the Pacific northwest. *Phytopathology* 30:866–69

18. Daniels, M. J., Markham, P. G., Meddins, B. M., Plaskitt, A. K., Townsend, R., Bar-Joseph, M. 1973. Axenic culture of a plant pathogenic spiroplasma. *Nature* 244:523–24

19. Daniels, M. J., Meddins, B. M. 1973. Polyacrylamide gel electrophoresis of mycoplasma proteins in sodium dodecyl sulphate. *J. Gen. Microbiol.* 76:239–42

20. Daniels, M. J., Stephens, M. A., Field, S. A. 1978. Toxin production by *Spiroplasma citri*. *Ann. Rep. John Innes Hortic. Inst., 68th* 1977:104–5

21. Davis, M. J., Purcell, A. H., Thomson, S. V. 1978. Pierce's disease of grapevines: Isolation of the causal bacterium. *Science* 199:75–77

22. Davis, M. J., Thomson, S. V., Purcell, A. H. 1978. Pathological and serological relationship of the bacterium causing Pierce's disease of grapevines and almond leaf scorch disease. *Int. Congr. Plant Pathol., 3rd, Munich* 1978:64 (Abstr.)

23. Davis, R. E. 1974. Spiroplasma in corn stunt-infected individuals of the vector leafhopper *Dalbulus maidis*. *Plant Dis. Reptr.* 58:1109–12

24. Davis, R. E. 1974. New approaches to the diagnosis and control of plant yellows diseases. *Proc. Int. Symp. Virus Dis. Ornamental Plants, 3rd,* ed. R. H. Lawson, M. K. Corbett, 1972:289–302. College Park, Md: Univ. Maryland Press.

25. Davis, R. E. 1978. Spiroplasmas from flowers of *Bidens pilosa* L. and honeybees in Florida: Relationship to honeybee spiroplasma AS 576 from Maryland. *Phytopathol. News* 12:78 (Abstr.)

26. Davis, R. E. 1978. Diverse cell wall-free prokaryotes isolated from surfaces of flowers and cultivated in vitro. *Int. Congr. Plant Pathol., 3rd, Munich* 1978:78 (Abstr.)

26a. Davis, R. E., Lee, I. M., Basciano, L. K. 1978. Spiroplasmas: Identification of serological groups and their analysis by polyacrylamide gel electrophoresis of cell proteins. *Phytopathol. News* 12:215 (Abstr.)

27. Davis, R. E., Whitcomb, R. F. 1971. Mycoplasmas, rickettsiae and chlamydiae: Possible relation to yellows diseases and other disorders of plants and insects. *Ann. Rev. Phytopathol.* 9:119–54

28. Davis, R. E., Whitcomb, R. F., Purcell, R. 1970. Viability of aster yellows agent in cell-free media. *Phytopathology* 60:573 (Abstr.)

29. Davis, R. E., Worley, J. F. 1973. Spiroplasma: Motile, helical microorganism associated with corn stunt disease. *Phytopathology* 63:403–8

30. Davis, R. E., Worley, J. F., Basciano, L. K. 1977. Association of spiroplasma and mycoplasma-like organisms with flowers of tulip tree (*Liriodendron tulipifera* L.). *Proc. Am. Phytopathol. Soc.* 4:185–86 (Abstr.)

31. Davis, R. E., Worley, J. F., Clark, T. B., Moseley, M. 1976. New spiroplasma in diseased honeybee (*Apis mellifera* L.): Isolation, pure culture, and partial characterization in vitro. *Proc. Am. Phytopathol. Soc.* 3:304 (Abstr.)

32. Delattre, R., Giannotti, J., Czarnecky, D. 1974. Maladies du cotonnier et de la vigne liées au sol et associées à des cochenilles endogées. Présence de mycoplasmes et étude comparative des souches in vitro. *C. R. Acad. Sci. Ser. D* 279:315–18

33. de Leeuw, G. T. N. 1977. Mycoplasma's in planten. *Nat. Tech.* 45:74–89

34. de Leeuw, G. T. N., Polak-Vogelzang, A. A. 1978. Serological characterization of plant mycoplasmas with the microplate method of the enzyme-linked immunosorbent assay (ELISA). *Int.*

Congr. Plant Pathol., 3rd, Munich 1978:61 (Abstr.)

35. Dimock, A. W., Geissinger, C. M., Horst, R. K. 1971. A new adaptation of tissue implantation for the study of virus and mycoplasma diseases. *Phytopathology* 61:429–30

36. Edison, S., Ramakrishnan, K. 1972. Aerated steam therapy for the control of grassy shoot disease (GSD) of sugarcane. *Mysore J. Agric. Sci.* 6:516–18

37. Feldman, A. W., Hanks, R. W., Good, G. E., Brown, G. E. 1977. Occurrence of a bacterium in YTD—affected as well as in some apparently healthy citrus trees. *Plant Dis. Reptr.* 61:546–50

38. Florance, E. R., Cameron, H. R. 1978. Three-dimensional structure and morphology of mycoplasmalike bodies associated with albino disease of *Prunus avium. Phytopathology* 68:75–80

39. Frazier, N. W., Severin, H. H. P. 1945. Weed-host range of California aster yellows. *Hilgardia* 16:621–50

40. French, W. J. 1974. A method for observing rickettsialike bacteria associated with phony peach disease. *Phytopathology* 64:260–61

41. French, W. J., Latham, A. J., Stassi, D. L. 1977. Phony peach bacterium associated with leaf scald of plum trees. *Proc. Am. Phytopathol. Soc.* 4:223 (Abstr.)

42. French, W. J., Stassi, D. L. 1978. Further observations on plum leaf scald and peach phony: Rickettsia-associated diseases of stone-fruits. *Int. Congr. Plant Pathol., 3rd, Munich* 1978:79 (Abstr.)

43. French, W. J., Stassi, D. L., Schaad, N. W. 1978. The use of immunofluorescence for the identification of phony peach bacterium. *Phytopathology* 68:1106–8

44. Fudl-Allah, A. E-S. A., Calavan, E. C., Igwegbe, E. C. K. 1972. Culture of a mycoplasmalike organism associated with stubborn disease of citrus. *Phytopathology* 62:729–31

45. Giannotti, J., Louis, C., Leclant, F., Marchoux, G., Vago, C. 1974. Infection à mycoplasmes et à micro-organismes d'allure rickettsienne chez une plante atteinte de prolifération et chez le psylle vecteur de la maladie. *C. R. Acad. Sci. Ser. D* 278:469–71

46. Giannotti, J., Marchoux, G., Devauchelle, G., Louis, C. 1974. Double infection cellulaire à mycoplasmes et à germes rickettsoïdes chez la plante parasite *Cuscuta subinclusa* L. *C. R. Acad. Sci. Ser. D* 278:751–53

47. Giannotti, J., Vago, C. 1971. Rôle des mycoplasmes dans l'étiologie de la phyllodie du Trèfle: Culture et transmission expérimentale de la maladie. *Physiol. Veg.* 9:541–53

48. Giannotti, J., Vago, C., Marchoux, G., Devauchelle, G., Czarnecky, D. 1972. Charactérisation par la culture in vitro des souches de mycoplasmes correspondant à huit maladies différentes des plantes. *C. R. Acad. Sci. Ser. D* 274:330–33

49. Goheen, A. C., Nyland, G., Lowe, S. K. 1973. Association of a rickettsialike organism with Pierce's disease of grapevines and alfalfa dwarf and heat therapy of the disease in grapevines. *Phytopathology* 63:341–45

50. Goszdziewski, M., Petzold, H. 1975. Versuche zum fluoreszenzmikroskopischen Nachweis mykoplasmaähnlicher Organismen in Pflanzen. *Phytopathol. Z.* 82:63–69

51. Gourley, C. O., Bishop, G. W., Craig, D. L. 1971. Susceptibility of some strawberry cultivars to green petal. *Can. Plant Dis. Surv.* 51:129–30

52. Green, S. K. 1978. Association of rickettsialike organisms with rosette disease of sugar beets. *Int. Congr. Plant Pathol., 3rd, Munich* 1978:79 (Abstr.)

53. Grunewaldt-Stöcker, G., Nienhaus, F. 1977. Mycoplasma-ähnliche Organismen als Krankheitserreger in Pflanzen. *Acta Phytomed.* 5:27–87

54. Haggis, G. H., Sinha, R. C. 1978. Scanning electron microscopy of mycoplasmalike organisms after freeze fracture of plant tissues affected with clover phyllody and aster yellows. *Phytopathology* 68:677–80

55. Hampton, R. O. 1972. Mycoplasmas as plant pathogens: Perspectives and principles. *Ann. Rev. Plant Physiol.* 23:389–418

56. Harries, H. C. 1973. Selection and breeding of coconuts for resistance to diseases such as lethal yellowing. *Oléagineux* 28:395–98

57. Hayflick, L., Arai, S. 1973. Failure to isolate mycoplasmas from aster yellows-diseased plants and leafhoppers. *Ann. NY Acad. Sci.* 225:494–502

58. Hewitt, W. B. 1958. The probable home of Pierce's disease virus. *Plant Dis. Reptr.* 42:211–15

59. Hibino, H., Kaloostian, G. H., Schneider, H. 1971. Mycoplasma-like bodies in the pear psylla vector of pear decline. *Virology* 43:34–40

60. Hirumi, H., Kimura, M., Maramorosch, K., Bird, J., Woodbury, R. 1974.

Rickettsialike organisms in the phloem of little leaf-diseased *Sida cordifolia*. *Phytopathology* 64:581–82 (Abstr.)

61. Hopkins, D. L. 1977. Diseases caused by leafhopper-borne, rickettsia-like bacteria. *Ann. Rev. Phytopathol.* 15:277–94

62. Hopkins, D. L. 1978. Comparisons of Florida isolates of the Pierce's disease bacterium. *Int. Congr. Plant Pathol. 3rd, Munich* 1978:62 (Abstr.)

63. Hopkins, D. L., Mollenhauer, H. H., French, W. J. 1973. Occurrence of rickettsia-like bacterium in the xylem of peach trees with phony disease. *Phytopathology* 63:1422–23

64. Hopkins, D. L., Mollenhauer, H. H., Mortensen, J. A. 1974. Tolerance to Pierce's disease and the associated rickettsia-like bacterium in Muscadine grape cultivars. *J. Am. Soc. Hortic. Sci.* 99:436–39

65. Hull, R. 1971. Mycoplasma-like organisms in plants. *Rev. Plant Pathol.* 50:121–30

66. Hull, R. 1972. Mycoplasma and plant diseases. *PANS* 18:154–64

67. Hunt, P., Dabek, A. J., Schuiling, M. 1974. Remission of symptoms following tetracycline treatment of lethal yellowing-infected coconut palms. *Phytopathology* 64:307–12

68. Ishiie, T., Doi, Y., Yora, K., Asuyama, H. 1967. Suppressive effects of antibiotics of tetracycline group on symptom development of mulberry dwarf disease (jap.). *Ann. Phytopathol. Soc. Jpn.* 33:267–75

69. Jensen, D. D. 1971. Herbaceous host plants of western X-disease agent. *Phytopathology* 61:1465–70

70. Jones, A. L., Hooper, G. R., Rosenberger, D. A. 1974. Association of mycoplasmalike bodies with little peach and X-disease. *Phytopathology* 64:755–56

71. Kahn, R. P., Lawson, R. H., Monroe, R. L., Hearon, S. 1972. Sweet potato little-leaf (witches'-broom) associated with a mycoplasmalike organism. *Phytopathology* 62:903–9

72. Kartha, K. K., Cousin, M. T., Ruegg, E. F. 1975. A light microscopic detection of plant mycoplasma infection by Feulgen staining procedure. *Indian Phytopathol.* 28:51–56

73. Kitajima, E. W., Bakarcic, M., Fernandez-Valiela, M. V. 1975. Association of rickettsialike bacteria with plum leaf scald disease. *Phytopathology* 65:476–79

74. Klein, M., Maramorosch, K. 1970.

Sulfa therapy of aster yellows. *Phytopathology* 60:1015 (Abstr.)

75. Klein, M., Zimmerman-Gries, S., Sneh, B. 1976. Association of bacterialike organisms with a new potato disease. *Phytopathology* 66:564–69

76. Kleinhempel, H., Karl, E., Lehmann, W., Proeseler, G., Spaar, D. 1974. Zur experimentellen Übertragung von Mykoplasmen auf Insekten. *Arch. Phytopathol. Pflanzenschutz* 10:351–52

77. Kleinhempel, H., Lehmann, W., Müller, H. M., Spaar, D. 1974. Gegenwärtiger Stand der Untersuchungen über mykoplasmaähnliche Organismen als mögliche Erreger von Pflanzenkrankheiten. *Int. Z. Landwirtsch.* 4:417–22

78. Kleinhempel, H., Müller, H. M., Spaar, D. 1975. Zur Frage der Übertragung mykoplasmaähnlicher Organismen durch das Saatgut infizierter Pflanzen. *Arch. Phytopathol. Pflanzenschutz* 11:89–99

79. Kondo, F., McIntosh, A. H., Padhi, S. B., Maramorosch, K. 1976. A spiroplasma isolated from the ornamental cactus *Opuntia tuna monstrosa*. *Proc. Soc. Gen. Microbiol.* 3:154 (Abstr.)

80. Konvička, O., Nienhaus, F., Fischbeck, G. 1978. Untersuchungen über die Ursachen der Pollensterilität bei *Allium sativum* L. *Z. Pflanzenzücht.* 80:265–76

81. Kunkel, L. O. 1926. Studies on aster yellows. *Am. J. Bot.* 13:646–705

81a. Lee, R. F., Raju, B. C., Nyland, G. 1978. Phytotoxin(s) produced by the Pierce's disease organism grown in culture. *Phytopathol. News* 12:218 (Abstr.)

82. Liu, H. Y., Black, L. M. 1974. Study of the rickettsia-like organism of clover club leaf by phase contrast microscopy. *Proc. Am. Phytopathol. Soc.* 1:97 (Abstr.)

83. Lombardo, G., Pignatelli, P. 1970. Cultivation in a cell-free medium of a mycoplasma-like organism from *Vinca rosea* with phyllody symptoms of the flowers. *Ann. Microbiol. Enzimol.* 20:84–89

84. Lowe, S. K., Nyland, G., Mircetich, S. M. 1976. The ultrastructure of the almond leaf scorch bacterium with special reference to topography of the cell wall. *Phytopathology* 66:147–51

84a. Lowe, S. K., Raju, B. C. 1978. The morphology of spiroplasmas associated with aster yellows and pear decline diseases: A comparative study by negative staining in vitro. *Phytopathol. News* 12:216 (Abstr.)

85. Maia, E., Bettachini, B., Beck, D., Vénard, P., Maia, N. 1973. Contribution a l'amélioration de l'état sanitaire du lavandin, clone 'Abrial'. *Ann. Phytopathol.* 5:115–24

86. Maramorosch, K. 1974. Mycoplasmas and rickettsiae in relation to plant diseases. *Ann. Rev. Microbiol.* 28:301–24

87. Maramorosch, K., Granados, R. R., Hirumi, H. 1970. Mycoplasma diseases of plants and insects. *Adv. Virus Res.* 16:135–93

88. Maramorosch, K., Jensen, D. D. 1963. Harmful and beneficial effects of plant viruses in insects. *Ann. Rev. Microbiol.* 17:495–530

89. Markham, P. G., Townsend, R., Bar-Joseph, M., Daniels, M. J., Plaskitt, A., Meddins, B. M. 1974. Spiroplasmas are the causal agents of citrus little-leaf disease. *Ann. Appl. Biol.* 78:49–57

90. Markham, P. G., Townsend, R., Plaskitt, K. A. 1975. A rickettsia-like organism associated with diseased white clover. *Ann. Appl. Biol.* 81:91–93

91. Markham, P. G., Townsend, R., Plaskitt, K., Saglio, P. 1977. Transmission of corn stunt to dicotyledonous plants. *Plant Dis. Reptr.* 61:342–45

92. Martinez, A. L., Nora, D. M., Armedilla, A. L. 1970. Suppression of symptoms of citrus greening disease in the Philippines by treatment with tetracycline antibiotics. *Plant Dis. Reptr.* 54:1007–9

93. Marwitz, R., Petzold, H. 1978. Examination of mycoplasma-like organisms in yellows diseased plants by scanning electron microscope. *Int. Congr. Plant Pathol., 3rd, Munich* 1978:78 (Abstr.)

94. McCoy, R. E. 1977. Growth of mycoplasmas in phloem sap from lethal yellowing resistant Malayan dwarf coconut palm. *Proc. Am. Phytopathol. Soc.* 4:108 (Abstr.)

94a. McCoy, R. E., Tsai, J. H., Thomas, D. L. 1978. Occurrence of a spiroplasma in natural populations of the sharpshooter *Oncometopia nigricans*. *Phytopathol. News* 12:217 (Abstr.)

95. McIntosh, A. H., Maramorosch, K. 1973. Mycoplasma and acholeplasma in plants. *Ann. NY Acad. Sci.* 225:330–33

96. McIntosh, A. H., Maramorosch, K. 1973. *Spiroplasma citri* in experimentally inoculated plants. *Int. Congr. Plant Pathol., 2nd, Minneapolis* 1973:0642 (Abstr.)

97. McIntosh, A. H., Skowronski, B. S., Maramorosch, K. 1974. Rapid identification of *Spiroplasma citri* and its rela-tion to other yellows agents. *Phytopathol. Z.* 80:153–56

98. McIntyre, J. L., Walton, G. S., Dodds, J. A., Lacy, G. H. 1977. Declining pear trees in Connecticut: Symptom remission by oxytetracycline treatments and associated mycoplasma-like-organisms. *Proc. Am. Phytopathol. Soc.* 4:193 (Abstr.)

99. Mendonça, A. de V. E., Sequeira, O. A. de, Mota, M. 1974. Mycoplasma-like structures extracted from *Vitis vinifera* L. and observed by the negative staining technique. *Agronomia Lusit.* 35:277–82

100. Mircetich, S. M., Lowe, S. K., Moller, W. J., Nyland, G. 1976. Etiology of almond leaf scorch disease and transmission of the causal agent. *Phytopathology* 66:17–24

101. Moll, J. N., Martin, M. M. 1973. Electron microscope evidence that citrus psylla (*Trioza erytreae*) is a vector of greening disease in South Africa. *Phytophylactica* 5:41–44

102. Moll, J. N., Martin, M. M. 1974. Comparison of the organism causing greening disease with several plant pathogenic Gram negative bacteria, rickettsia-like organisms and mycoplasma-like organisms. *Inserm* 33:89–96

103. Mollenhauer, H. H., Hopkins, D. L. 1974. Ultrastructural study of Pierce's disease bacterium in grape xylem tissue. *J. Bacteriol.* 119:612–18

104. Morvan, G., Castelain, C. 1972. Induction d'une résistance de l'abricotier à l'enroulement chlorotique (ECA) par le greffage sur *Prunus spinosa*. *Ann. Phytopathol.* 4:418

105. Müller, H. M., Surgučeva, N. A., Fedotina, V. L., Schmidt, H. B., Kleinhempel, H., Procenko, A. E., Spaar, D. 1974. In der UdSSR und der DDR durchgeführte Untersuchungen zum elektronenmikroskopischen Nachweis mykoplasmaähnlicher Organismen in Pflanzen. *Arch. Phytopathol. Pflanzenschutz* 10:15–23

106. Nariani, T. K., Raychaudhuri, S. P., Viswanath, S. M. 1973. Tolerance to greening disease in certain citrus species. *Curr. Sci.* 42:513–14

107. Nasu, S., Jensen, D. D., Richardson, J. 1974. Primary culturing of the western X mycoplasmalike organism from *Colladonus montanus* leafhopper vectors. *Appl. Entomol. Zool.* 9:115–26

108. Nienhaus, F. 1976. Rickettsien, Mycoplasmen und Spiroplasmen als Erreger von Pflanzenkrankheiten. *Ber. Dtsch. Bot. Ges.* 89:531–45

109. Nienhaus, F., Brüssel, H., Schinzer, U. 1976. Soil-borne transmission of rickettsia-like organisms found in stunted and witches' broom diseased larch trees (*Larix decidua*). *Z. Pflanzenkr. Pflanzenschutz* 83:309–16
110. Nienhaus, F., Rumbos, I. 1979. Rickettsialike organisms in grapevines with yellows disease in Germany. *Meet. Int. Conf. Study Viruses Virus Dis. Grapevine (ICVG), 6th, Cordoba* 1976: In press
111. Nienhaus, F., Rumbos, I., Green, S. 1978. Rickettsialike organisms isolated from plants cultivated in chick embryo. *Int. Congr. Plant Pathol., 3rd, Munich* 1978:62 (Abstr.)
112. Nienhaus, F., Rumbos, I., Greuel, E. 1978. First results in the cultivation of rickettsia-like organisms of yellows diseased grapevines in chick embryos. *Z. Pflanzenkr. Pflanzenschutz* 85:113–17
113. Nienhaus, F., Schmutterer, H. 1976. Rickettsialike organisms in latent rosette (witches' broom) diseased sugar beets (*Beta vulgaris*) and spinach (*Spinacia oleracea*) plants and in the vector *Piesma quadratum* Fieb. *Z. Pflanzenkrankr. Pflanzenschutz* 83:641–46
114. Nienhaus, F., Steiner, K. G. 1976. Mycoplasmalike organisms associated with Kainkope disease of coconut palms in Togo. *Plant Dis. Reptr.* 60:1000–2
115. Nyland, G., Goheen, A. C., Lowe, S. K., Kirkpatrick, H. C. 1973. The ultrastructure of a rickettsialike organism from a peach tree affected with phony disease. *Phytopathology* 63:1275–78
115a. Nyland, G., Raju, B. C. 1978. Isolation and culture of a spiroplasma from pear trees affected by pear decline. *Phytopathol. News* 12:216 (Abstr.)
116. Ou, S. H. 1972. *Rice Diseases*, p. 22–25, London: Commonwealth Mycol. Inst., Kew. 368 pp.
117. Parthasarathy, M. V. 1974. Mycoplasmalike organisms associated with lethal yellowing disease of palms. *Phytopathology* 64:667–74
118. Pereira, A. L. G., Oliveira, B. S. 1971. Isolation and direct transmission of the causal agent of the corn stunt cultivated in artificial medium. *Arq. Inst. Biol. Sao Paulo* 38:191–200
119. Petzold, H., Marwitz, R. 1973. Mycoplasmen und rickettsienähnliche Bakterien als Erreger von Pflanzenkrankheiten. *Mitt. Biol. Bundesanst. Land Forstwirtsch. Berlin Dahlem* 151:159–78
120. Petzold, H., Marwitz, R. 1976. Versuche zur Infektion von Apfelbäumen mit dem möglichen Erreger der Triebsucht des Apfels. *Phytopathol. Z.* 86:365–69
121. Petzold, H., Marwitz, R. 1978. Light microscopical investigations for a simplified detection of plant infection by mycoplasma-like organisms. *Int. Congr. Plant Pathol., 3rd, Munich* 1978:63 (Abstr.)
122. Petzold, H., Marwitz, R., Özel, M., Goszdziewski, P. 1977. Versuche zum rasterelektronenmikroskopischen Nachweis von mykoplasmaähnlichen Organismen. *Phytopathol. Z.* 89:237–48
123. Plavšić, B., Buturović, D., Krivokapić, K., Erić, Ž. 1978. Some characteristics of mycoplasma-like infection and the effects of kinetin on the MLO-infected plants. *Int. Congr. Plant Pathol., 3rd, Munich* 1978:81 (Abstr.)
124. Ploaie, P. G. 1973. *Micoplasma. Si Bolile Proliferative La Plante.* Bucharest: Intreprinderea Poligrafiča. 178 pp. Ed. 'Ceres'
125. Ploaie, P. G. 1973. Structures resembling rickettsia in plant cells. *Int. Congr. Plant Pathol., 2nd, Minneapolis* 1973:0643 (Abstr.)
126. Purcell, A. H. 1977. Cold therapy of Pierce's disease of grapevines. *Plant Dis. Reptr.* 61:514–18
127. Raine, J., Forbes, A. R., Skelton, F. E. 1976. Mycoplasma-like bodies, rickettsia-like bodies, and salivary bodies in the salivary glands and saliva of the leafhopper *Macrosteles fascifrons* (Homoptera: Cicadellidae). *Can. Entomol.* 108:1009–19
127a. Raju, B. C., Nyland, G. 1978. Effects of different media on the growth and morphology of three newly isolated plant spiroplasmas. *Phytopathol. News* 12:216 (Abstr.)
128. Rana, G. L., Kaloostian, G. H., Oldfield, G. N., Granett, A. L., Calavan, E. C., Pierce, H. D., Lee, I. M., Gumpf, D. J. 1975. Acquisition of *Spiroplasma citri* through membranes by homopterous insects. *Phytopathology* 65:1143–45
129. Raychaudhuri, S. P. 1974. Mycoplasmal diseases of plants. *Acta Bot. Indica* 2:13–16
130. Razin, S. 1978. The mycoplasmas. *Microbiol. Rev.* 42:414–70
131. Rosenberger, D. A., Jones, A. L. 1977. Seasonal variation in infectivity of inoculum from X-diseased peach and chokecherry plants. *Plant Dis. Reptr.* 61:1022–24

132. Rumbos, I. 1978. *Untersuchungen über Rickettsien-ähnliche Organismen in vergilbungskranken Weinreben (Vitis vinifera L.)*. PhD thesis. Univ. Bonn. 146 pp.

133. Rumbos, I., Sikora, R. A., Nienhaus, F. 1977. Rickettsia-like organisms in *Xiphinema index* Thorne & Allen found associated with yellows disease of grapevines. *Z. Pflanzenkr. Pflanzenschutz* 84:240–43

134. Russell, W. C., Newman, C., Williamson, D. H. 1975. A simple cytochemical technique for demonstration of DNA in cells infected with mycoplasmas and viruses. *Nature* 253:461–62

135. Saglio, P., Laflèche, D., Bonissol, C., Bové, J. M. 1971. Isolement et culture in vitro des mycoplasmes associées au 'stubborn' des agrumes et leur observation au microscope électronique. *C. R. Acad. Sci. Ser. D* 272:1387–90

136. Saglio, P., Laflèche, D., Bonissol, C., Bové, J. M. 1971. Isolement, culture et observation au microscope électronique des structures de type mycoplasme associées à la maladie du stubborn des agrumes et leur comparison avec les structures observées dans le cas de la maladie du greening des agrumes. *Physiol. Vég.* 9:569–82

137. Samuel, G., Bald, J. G., Eardley, C. M. 1933. 'Big bud,' a virus disease of the tomato. *Phytopathology* 23:641–53

138. Seemüller, E. 1976. Fluoreszenzoptischer Direktnachweis von mykoplasmaähnlichen Organismen im Phloem pear-decline—und triebsuchtkranker Bäume? *Phytopathol.* 85:368–72

139. Sethi, K. K., Nienhaus, F. 1974. Attempts at reisolation of *Acholeplasma laidlawii* after experimental inoculation in plants. *Phytopathol. Z.* 80:88–90

140. Severin, H. H. P. 1929. Yellows disease of celery, lettuce, and other plants, transmitted by *Cicadula sexnotata* (Fall.). *Hilgardia* 3:543–83

141. Severin, H. H. P., Freitag, J. H. 1945. Additional ornamental flowering plants naturally infected with California aster yellows. *Hilgardia* 16:599–618

142. Seymour, C. P. 1964. Testing for phony peach disease. *Plant Pathol. Circ. No. 25, Fla. Dep. Agric. Div. Plant Ind.*

143. Sinha, R. C. 1974. Purification of mycoplasma-like organisms from China aster plants affected with clover phyllody. *Phytopathology* 64:1156–58

144. Spaar, D., Kleinhempel, H., Müller, H. M., Stanarius, A., Schimmel, D. 1974. Culturing mycoplasmas from plants.

Coll. Inst. Nat. Sante Rech. Med. 33:207–13

145. Stanarius, A., Müller, H. M., Kleinhempel, H., Spaar, D. 1976. Elektronenmikroskopischer Nachweis mykoplasmaähnlicher Organismen in den Speicheldrüsen von *Euscelis plebejus* Fall. nach Injektion von Preßsaft aus proliferationskranken *Catharanthus roseus* (L.) G. Don. *Arch. Phytopathol. Pflanzenschutz* 12:245–52

146. Steiner, K. G. 1976. Remission of symptoms following tetracycline treatment of coconut palms affected with Kainkope disease. *Plant Dis. Reptr.* 60:617–20

147. Steiner, K. G., Nienhaus, F., Marschall, K. J. 1977. Rickettsia-like organisms associated with a decline of coconut palms in Tanzania. *Z. Pflanzenkr. Pflanzenschutz* 84:345–51

148. Strobel, G. A. 1977. Bacterial phytotoxins. *Ann. Rev. Microbiol.* 31:205–24

149. Su, H. J., Lei, J. D., Chen, T. A. 1978. Spiroplasmas isolated from green leaf bug (*Trigonotylus ruficornis*) and rice plant. *Int. Congr. Plant Pathol., 3rd, Munich* 1978:61 (Abstr.)

150. Suto, Y., Ishiie, T. 1970. Inoculation of mulberry dwarf disease agent by the insect-infection method. *J. Sericult. Sci. Jpn.* 39:451–57

151. Teranaka, M., Otsuka, K. 1973. Culture of mycoplasma-like organisms associated with several witches' broom and aster yellow plants on artificial media. *College Agric. Utsunomiya Univ. Bull.* 8:11–25

152. Thomson, S. V., Davis, M. J., Kloepper, J. W. 1978. Alfalfa dwarf: relationship to the bacterium causing Pierce's disease of grapevines and almond leaf scorch disease. *Int. Congr. Plant Pathol., 3rd, Munich* 1978:65 (Abstr.)

153. Townsend, R., Burgess, J. 1978. Scanning electron microscopy of *Spiroplasma citri. Ann. Rep. John Innes Inst., 68th* 1977:100–2

154. Townsend, R., Markham, P. G., Plaskitt, K. A., Daniels, M. J. 1977. Isolation and characterization of a non-helical strain of *Spiroplasma citri. J. Gen. Microbiol.* 100:15–21

155. Townsend, R., Plaskitt, A. 1978. The occurrence of actin-like protein in *Spiroplasma citri. Ann. Rep. John Innes Inst. 68th* 1977:102–4

156. Tully, J. G., Whitcomb, R. F., Bové, J. M., Saglio, P. 1973. Plant mycoplasmas: Serological relation between agents associated with citrus stubborn

and corn stunt diseases. *Science* 182:827–29

157. Tully, J. G., Whitcomb, R. F., Clark, H. F., Williamson, D. L. 1977. Pathogenic mycoplasmas: Cultivation and vertebrate pathogenicity of a new spiroplasma. *Science* 195:892–94

158. Ulrychová, M., Petrů, E. 1975. Elimination of mycoplasma in tobacco callus tissues (*Nicotiana glauca* Grah.) cultured *in vitro* in the presence of 2,4-D in nutrient medium. *Biol. Plant.* 17:103–8

159. Ulrychová, M., Vanek, G., Jokeš, M., Klobáska, Z., Králík, O. 1975. Association of rickettsialike organisms with infectious necrosis of grapevines and remission of symptoms after penicillin treatment. *Phytopathol. Z.* 82:254–65

160. Urbina-Vidal, C. 1974. Non-viral agents associated with sugar beet yellow wilt in Chile. *Phytopathol. Z.* 81:114–23

161. Vago, C., Giannotti, J. 1972. Les mycoplasmes chez les végétaux et chez les vecteurs. *Physiol. Veg.* 10:87–101

162. Valenta, V. 1959. Zwei bisher unbekannte, Kartoffelwelke verursachende Viren aus Mitteleuropa. *Phytopathol. Z.* 35:271–76

163. Valenta, V., Musil, M. 1963. Investigations on European yellows-type viruses. II. The clover dwarf and parastolbur viruses. *Phytopathol. Z.* 47:38–65

164. Valenta, V., Musil, M., Mišiga, S. 1961. Investigations on European yellows-type viruses. I. The stolbur virus. *Phytopathol. Z.* 42:1–38

165. Vanek, G., Ulrychová, M., Blattný, C., Jokeš, M., Novák, M. 1976. Rickettsialike structures in hop plants with crinkle disease and therapeutic effect of antibiotics. *Phytopathol. Z.* 87:224–30

166. Van Vuuren, S. P., Moll, J. N., da Graca, J. V. 1977. Preliminary report on extended treatment of citrus green-ing with tetracycline hydrochloride by trunk injection. *Plant Dis. Reptr.* 61: 358–59

167. Waters, H., Eden-Green, S. J., Dabek, A. J. 1978. Coconut lethal yellowing and diseases of other plants associated with mycoplasma-like organisms in Jamaica. *Int. Congr. Plant Pathol., 3rd, Munich* 1978:79 (Abstr.)

168. Westwood, M. N., Cameron, H. R. 1978. Environment-induced remission of pear decline symptoms. *Plant Dis. Reptr.* 62:176–79

169. Whitcomb, R. F. 1973. Diversity of procaryotic plant pathogens. *Proc. North Cent. Branch Entomol. Soc. Am.* 28: 38–60

170. Whitcomb, R. F., Tully, J. G., Bové, J. M., Saglio, P. 1973. Spiroplasmas and acholeplasmas: multiplication in insects. *Science* 182:1251–53

171. Wilhelm, H., Knösel, D. 1976. Penetration und Translokation von ^{3}H-Tetracyclin-hydrochlorid in pflanzlichem Gewebe. *Z. Pflanzenkr. Pflanzenschutz* 83:241–52

172. Williamson, D. L., Whitcomb, R. F. 1975. Plant mycoplasmas: A cultivable spiroplasma causes corn stunt disease. *Science* 188:1018–20

173. Wolanski, B. S. 1973. Negative staining of plant agents. *Ann. NY Acad. Sci.* 225:223–35

174. Wroblewski, H., Johansson, K. E. 1976. Spiraline, the main protein from the *Spiroplasma citri* membrane: Purification and characterization. *Proc. Soc. Gen. Microbiol.* 3:165 (Abstr.)

175. Zelcer, A., Loebenstein, G., Bar-Joseph, M. 1972. Effects of elevated temperature on the ultrastructure of mycoplasmalike organisms in periwinkle. *Phytopathology* 62:1453–57.

Ann. Rev. Phytopathol. 1979. 17:59–73
Copyright © 1979 by Annual Reviews Inc. All rights reserved

YELLOW RUST
EPIDEMIOLOGY

❖3701

F. Rapilly

Institut National de la Recherche Agronomique, CNRA Station de Pathologie
Végétale, Etoile de Choisy, Route de Saint-Cyr, 78,000 Versailles, France

INTRODUCTION

Yellow rust epidemics, caused by *Puccinia striiformis* West., have attracted
considerable attention because they result in very high decreases in yield.
Losses in wheat have been assessed by Roelfs at 8–75% (48), and by Doling
& Doodson at 8–20% (7). The parasite that causes yellow rust has been
reported all over the world [Hassebrauk (18; see also 19–20)], except in
New-Zealand [Smith, Smith & Marshall (58)]. The most severe epidemics
of this disease occur in temperate-cool and wet climates, affecting wheat,
barley, and *Triticale*.

This review deals with quantitative epidemiology and in particular the
effects of climate, host, host-population, and the parasite itself on the various events comprising the yellow rust infectious cycle and its continuation
to produce epidemics. Because of space limitations I cannot deal here with
qualitative epidemiology, and especially with the acquisition or loss by the
parasite of virulence genes; noted, however, is the importance of such
studies for the breeder and for the definition of control strategies.

OVERWINTERING AND OVERSUMMERING

Puccinia striiformis is a *brachycyclic* rust; the absence of an aecidial host
raises the question whether the parasite can survive in the absence of
cereals. Even if an aecidial host is discovered, however, it is questionable
whether this spore stage would be produced in nature. As a matter of fact,
Wright & Lennard (68) reported that teliospore dormancy is short and
basidiospores are produced very quickly. Thus their ability to survive and

59

0066-4286/79/0901-0059$01.00

to infect an alternate host seems unlikely. Furthermore, on the basis of a cytological study, Goddard (14) hypothesized that this rust would be heterothallic.

Two hypotheses that are proposed to explain the oversummering of the yellow rust primary inoculum are dealt with below.

Local Origin

The hypothesis that tillers and wild weed grass, especially couch grass, have a role in the oversummering of the yellow rust primary inoculum has been dealt with by numerous authors: Chumakov (5), Hassebrauk (18), Viennot-Bourgin (67), and Zadoks (71). Oversummering on weed grasses is common in mountain areas, in particular in western Pakistan [Hassan (17)] and western India [Joshi & Palmer (30; see also 29)]; in the United States the importance of weed grasses and tillers is reported also by Shaner & Powelson (55).

Long-Distance Origin

The second hypothesis to explain the oversummering suggests that the primary inoculum originates at a site far from the site of infection and initial infection results from long-range dispersal of infectious urediospores. This hypothesis is based on the study of the geographical distribution of the physiological races. The study of the yellow rust epidemic on the wheat variety Heine VII in 1955 in the Netherlands and of its spread to other countries in Western Europe suggested to Zadoks (71, 72) that the inoculum could spread as far as 800 km from its point of origin. Hermansen & Stapel (24) have suggested that the 1972 epidemic on the variety Kranich in Denmark was caused by a race spread from Great Britain.

Choosing between either hypothesis is difficult because the genetic uniformity of some varieties, as far as resistance is concerned, enables a rapid spread of epidemics.

However, various reasons suggest that yellow rust is an endemic disease. Shaner & Powelson's observations (55) showed that urediospore survival is very short: one month on dry ground, four to six weeks on dead leaves. Nevertheless, the considerable susceptibility of spores to UV light, which is three times higher than that of black rust spores, according to Maddison & Manners (35), reduces the survival chance of this parasite. A one-day-exposure of yellow rust spores to sunshine reduces the germination of urediospore to 0.1%. Moreover, efficient dispersal conditions correspond to periods of high relative humidity, but spore clumps produced when humidity is high would not spread great distances. (see section on dissemination conditions) [Rapilly, Fournet & Skajennikoff (45)]. Finally, Zadoks (73) observed that artificial yellow rust foci used by plant breeders do not result in generalized epidemics, which suggests that efficient dispersal is limited to short distances.

CATCH AND RETENTION

Few studies have been carried out about the settling of a fungal parasite on aerial parts of plants and about the factors affecting whether plants contract or escape disease. Rapilly & Foucault [43; see also 2 (Azmeh)] point out that the adhesion of yellow rust urediospores on smooth surfaces is significantly greater than on rough surfaces. The adhesive force between the spore and the receptive surface is significantly greater with high relative humidity, thus increasing the efficiency of spore deposition. The presence of a mucilaginous layer on the spore surface that is sensitive to relative humidity variations may explain this phenomenon [Rapilly (40) Stanbridge & Gay (59)].

The distribution of spores on leaves was studied by Russell (49), who observed the most efficient catch on the abaxial face in the middle of the lamina. Russell also indicated that the erect leaves of W1343, unlike the prostrate leaves of the varieties Capelle-Desprez or Little Joss, allow for a threefold decrease in the number of spores retained by W1343 leaves.

GERMINATION

Germination involves highly complex interactions between climate, host, and the parasite itself. The germination of yellow rust urediospores has been observed to be haphazard. The climatic conditions during spore formation and spread play a major role in the potential of a spore to germinate [Gassner & Straib (12), Sharp (56)], which explains the considerable variation observed between spore collections (Table 1).

Georgievskaja (13) reported that spores produced between 5°C and 10°C germinated best whereas spores produced at 30°C and over were unable to germinate; this observation may explain the correlation observed between the decrease in epidemics and high temperatures.

Table 1 The effect of daily temperature average in the field during the four days before harvest on germination of yellow rust urediospores on water agar at 15° C

Average temperature sum[a]	−13.6	−0.5	+7.4	+17.1	+32.3
Germination percentage	13.1	63	22.5	26.4	11.4

[a] Sum of the average daily temperatures for the four days prior to the day of spore collection.

Effects of Climatic Factors

Urediospores germinate poorly in free water, but require a relative humidity near to saturation, which results in the formation of condensation droplets, which the germ tubes grow around [Burrage (4)]. The condensation must last at least 3 hr before germination begins [Tu & Hendrix (64); see also Hermansen (23)]. Prewetting spores increase the germination percentage; germination, however, is definitely stopped if a period of desiccation occurs. Germination is not inhibited by moderate frost [Hassebrauk & Schroeder (21; see also 45)] and is promoted by thermal shocks [MacDonald & Strange (32)].

In the presence of dew, germination and formation of appressoria occur in the temperature range of 2°C to 15°C [Sharp (56)], with an optimum at 7°C. Simulating yellow rust epidemics, Zadoks & Rijsdijk (76) chose daily mean temperatures between 4°C and 25°C. Shrum (57), taking the data given by Schmitt et al as a basis (53), considered a temperature range from −2°.8 to 21°.7 with the optimum 9.7°C; in the presence of free water the germination percentage increased by 1.2% for each degree hour exceeding 6.1°C Artificial infections generally occur at 15°C in a moist chamber with controlled temperature.

The percentage of single urediospores that germinate decreases with an increase in time of exposure to sunshine. [Maddison (34), Maddison & Manners (35)] and decreases to almost nil after one day of exposure. This marked sensitivity of yellow rust to sunshine has major epidemiological consequences for the assessment of the efficiency of long-distance dispersal.

Effects of Parasite and Host

Self-inhibiting or stimulating effects in fungi have often been reported; for yellow rust urediospores, Hassebrauk & Schroeder (21) and Tollenaar & Houston (62, 63) were the first to report this phenomenon. Yellow rust urediospores in high concentrations on water agar germinate poorly, whereas in low concentrations they germinated readily. Rapilly et al (45) observed that germination of samples of spores from the same collection varied from 3.3% to 22% at 10°C. Macko, Trione & Young (33) identified the water-soluble substance responsible for this phenomenon as *cis*-3,4-dimethoxycinnamate, which is present in various *Puccinia* spp.

From an epidemiological point of view, the presence of this substance in urediospores confirms that for a better understanding of yellow rust epidemics, and especially for a pathological assessment of dispersal gradients, it is necessary to consider spreading units of dissemination (UD) [Rapilly et al (45), Rapilly (42)] or infectious entities [Van der Plank (66)] and not the quantity of propagules (here urediospores).

The germination percentage also varies with parasite races. Matveenkoo (36) thus reported differences varying by a factor of one to two.

The host itself influences the germination percentage and number of appressoria formed. Russell (50) reported differences between leaf levels. On the flag leaf, the germination was always lower, especially for the wheat variety Holdfast; whatever the foliar level, germination was always higher on the adaxial face of the lamina than on the abaxial face. On *Triticum spelta* var. *album* the germination of race 60 was low [Stubbs & Plotnikova (61)]. But among the different varieties no correlation with the length of epidemic hairs or their density could be proved. Similarly, the number of appressoria seems to be independent of the stomate density. Furthermore, penetration is achieved through stomata or at the junction of epidermal cells depending on the varieties.

LATENT PERIOD

In an epidemic wherein cycles overlap and susceptibility often corresponds with the duration of the crop, two sequences greatly condition the rate of epidemic progression, namely the latent period and the sporulation. The latent period is the time, generally expressed in days, which separates germination and penetration from the appearance of newly formed sporulating sori; this period depends on the climate, host, and the parasite itself. Both sequences are generally considered as factors of aggressiveness and not of virulence.

Effects of Climatic Factors

The latent period of yellow rust depends mainly on temperature. Zadoks's works (71–74) led me to propose a mathematical relationship between the latent period expressed in days and the temperature expressed in degrees centigrade. This quantified formulation of the latent period follows the second rule defined by Zadoks (75) for studies on quantitative epidemiology. According to this author, this formula only pertains to temperatures between 4°C and 19°C.

The tests we were able to carry out in the field led us to check the predictive value of this formula (Table 2); however, we considered that temperatures between 0°C and −10°C inhibit but do not halt parasite development as long as the temperature does not go below −10°C [Rapilly (41)].

This example concerns a first attempt to quantify a yellow rust epidemic sequence. This formula for the duration of latency, which, as it seems, holds good only for varieties with a low level of horizontal resistance, was employed by Rijsdijk (46) in his yellow rust simulation model.

Table 2 Comparison of the latent period observed in the field on the wheat line 53–33 with a theoretical duration

Date of contamination	Number of days of latent period	
	Observed[a]	Theoretical[b]
27 October 1967	20–25	24
22 December 1967	50–55	51
5 December 1968	52–57	56
15 January 1969	46–51	50
5 February 1969	39–44	43
3 March 1969	29–34	31
10 October 1975	82–86	84

[a] A range of values is given for the observed number of days as assessments were made only every five days. (Climatic data of Versailles).

[b] Theoretical estimation is obtained by application of the formula given by Zadoks (71–74).

Shrum (57) considered in his model that the latent period is completed after a total of degree-hours equal to 5568 at a temperature between –2°C and +28°C. If the minimum temperature during latency becomes lower than –2°C, this should be taken into account; otherwise the degree-hours estimation for completion of the latent period will be too low.

Temperatures over 30°C hinder sporulation, and consequently prolong the latent period; Georgievskaja (13) showed, however, that once established in the plant the fungus could stand peaks to 38°C without being destroyed.

Contrary to other parasites with a marked necrotrophic character like *Septoria nodorum* Berk., the glume blotch of wheat, or *Phoma lingam* (Tode ex Fr) Desm., the black rot of rape seeds, yellow rust in the latent state is not influenced by relative humidity. On the other hand, strong light, by modifying the host reaction type, indirectly affect latency by suppressing sporulation [Pochard, Goujon & Vergara (37); see also Stubbs (60)].

Effects of Host and Parasite

The duration of the latent period does not depend on the number of infection sites, but can be affected by the physiological state of the host. Thus, in the variety Little Club any latency period can take place before heading because contamination is possible only after heading; on wheat line L 7 sporulation is only observed on seedlings (Y. Cauderon, personal communication). As early as 1934 Gassner & Straib (12) noted differences in behavior among varieties. Young & Powelson (70) stated that compared with other wheat varieties, Yamhill showed increased duration in latency from 3 to 23 days according to the temperature rhythm affecting plants. More obvious differences among varieties due to temperature were observed with brown rust and for barley with powdery mildew. However, it is presently difficult to take advantage of these differences to increase the horizontal resistance of varieties to obtain slow rusting epidemics, especially because Fuchs (10) reported that the behavior of a race on a host is not always consistent; sometimes it is slow, sometimes it is fast in latency.

SPORULATION

Sporulation, which corresponds to the contagious period from the epidemiological point of view, must be assessed quantitatively and with regard to the quality of dispersed spore. Statics, as well as dynamics (by including the duration of spore production), can be applied to the study of sporulation.

Determining the number of spores produced raises numerous technical difficulties; counts with Thoma's cell are not accurate enough. Priestley & Doling (38) suggested using the Coulter Counter for measuring sporulation. Johnson & Bowyer (27) compared Thoma's cell, microbalance, and spore suspension turbidity and pointed out the advantages of the microbalance method. As for us, we evaluate sporulation by measuring transmission (wave length $= 400$ nm) through a spore suspension in 1% water-agar. The differences that can be detected are about 2000 spores per cm^2 of sporulating surface (unpublished results).

Static Sporulation

EFFECT OF MOISTURE   For sporulation to occur, the relative humidity after the latent period is completed must exceed 50%. The number of spores formed per unit of time is related exponentially to the increase in relative humidity [Rapilly & Fournet (44)]; however, liquid water stops sporulation.

INTERACTIONS BETWEEN HOST AND PARASITE   Strong light over 30,000 to 40,000 Lux modifies the reaction of the plant: The mycelium density tends to decrease, which results in fewer and fewer sporulating lesions (see 13, 36a, 37, 60). This observation is related to the fact that in the field sporulation tends to decrease with an increase in duration of daylight. In fact it is difficult to separate the effect of light from that of temperature which stops sporulation at 33°C and above. According to Brown & Sharp (3) the effect of temperature on infection type is controlled by minor genes in the wheat variety Rego.

The linear distribution of soris on the infected leaves gave this disease its name of stripe rust; on triticale, however, this linear distribution occurs with the same frequency as distribution in concentric circles around the infection site.

In varieties with lesions of type 4, differences in spore numbers are noted; they occur in proportions from 1–11 [Priestley & Doling (38)]; Pyzhikova (39) reported one- to two-fold differences between the wheat varieties Svokaya and Melianopus 26.

Vernalization induces a threefold increase in sporulation for the varieties Maris Widgeon and Nord Desprez, but has no influence on the variety Cappelle-Desprez [Russell & Hudson (51)]. Except for Nord-Desprez the number of spores produced varies with the leaf level and tends to diminish as the variety ages (52); in the variety Kolibri this phenomenon results in an apparent resistance which increases as the plant ages [Heitefuss, Dehne and Einfeld (22)].

With respect to races [Johnson & Bowyer (27)], the sporulation intensity varies within the same variety. But the greatest sporulation decrease is observed in the case of superimposed inoculations on the same host of an avirulent race, then of a virulent race four days later [Johnson & Allen (26), Johnson & Taylor (28)]. This decrease, which can reach twofold, varies with the variety; for example, Maris Ranger shows few variations. This effect should play a role in the field for the epidemic development on multilines or on composite varieties, although Shrum's simulation trials (57) on a homogenous population indicate that a twofold reduction of the number of spores formed has a negligible effect on the final epidemic severity.

Dynamic Sporulation

The total number of spores produced depends on the duration of their production and on the speed of urediospore formation and sorus enlargement.

In a nearly saturated atmosphere a sorus emptied of spores by the wind again produces viable urediospores after 2 hr. At 27°C a sorus produces urediospores for eight days [Schmitt et al (53)]; after four days, maximal sporulation is reached. Zadoks (71) reported spore production that lasted for three weeks in winter. Pyzhikova (39) noted that at 16°C with a relative humidity of 55% production of spores lasted from 8 to 10 days, with maximum production occurring between the fourth and the fifth day, which would correspond to 1.8×10^3 spores per day and per sorus. For a leaf (second leaf) the sporulation time would range from 39 to 40 days with a mean production per day of $80–140 \times 10^3/cm^2$ spores on the variety Melianopus. Young (69) reported a daily production of 3,000–35,000 urediospores per cm^2.

At 21.4°C, the mycelium can spread within the leaf at the rate of 8.9 mm per day [Shaner (54)]. Lesions only enlarge in length as veins prevent their lateral extension. The speed of enlargement is linear and decreases with maturation of the host [Emge, Kingsolver & Johnson (8)]. In the case of plants at 20°C ± 2, the sorus area increases between 8.8 and 18.8 mm² per day. Zadoks (71) indicated a linear enlargement of 1 mm per day on the variety Leda and of 5 mm per day on the variety Michigan Amber. Young (69) noted no difference between wheat varieties and indicated a mean enlargement of 1.5 to 2 mm per day.

Both the growth of the lesion and the decrease in the amount of sporulation after four or five days imply that the number of spores formed per day is variable. This is why the daily multiplication factor, DMFR, defined by Zadoks (74), is a variable, and the spore removals discussed by Van der Plank (65) probably have little influence on the rate of development of yellow rust epidemics. This could explain why it is hardly possible to stop a yellow rust epidemic with a fungicide application that is too late.

As stated previously, numerous interactions between host, pathogen, and environment condition sporulation; consequently if the number of spores produced is considered a criterion of pathogen aggressiveness it seems very difficult to distinguish between horizontal pathotypes.

DISSEMINATION CONDITIONS

Knowledge of factors affecting the dispersal of yellow rust spores is required for understanding epidemic development, and thus for defining control strategies. The release of spores and the efficiency of the dissemination of the inoculum are dealt with in this section. By analogy with brown and black rusts, yellow rust urediospores have long been thought to be released only by wind; however, rain can play a significant role in the dispersal of spores not only of yellow rust but of the other rusts as well.

Dispersal by Rain

Raindrops release urediospores either by direct impact or by splashing [Rapilly, Fournet & Skajennikoff (45)]. With rains of 5 to 10 mm per hour, sori are emptied within about 1 hr; it takes 3–5 hr before a significant spread of new spores can be observed. This delay depends on how quickly the sori are emptied and consequently on the intensity of the rain. Dispersal by rain, though limited in distance, can be very efficient because spore clusters, the units of dissemination, by rain have a very high germination potential.

Dispersal by Wind and Dispersal Gradient

Spread of rust by wind has been the topic of numerous studies. Yellow rust, however, has certain unusual characteristics. Earlier in this chapter, we prove that release is possible only with a wind speed equal or more than 1m/sec. The number of urediospores released increases exponentially with increase in wind speed, and wind blasts thus play a major role in spread of rust. However, results obtained by use of models [Itier & Pauvert (25)] indicate, that with regard to the canopy roughness, spore transport is nil for speeds under 0.25 m/sec, whereas spores must be considered as a gas for wind speeds over 2.5 m/sec.

Equations in the various simulation models, EPIDEMIC [Shrum (57)], EPISIM 72 [Zadoks & Rijsdijk (76), Rijsdijk (46)], and EPIMUL [Kampmeijr & Zadoks (31)] take wind dissemination into account but not the essential role of relative humidity in release of spores by the wind. In fact, Rapilly & Fournet (44; see also 45) showed that depending on the relative humidity of the atmosphere, urediospores were spread individually or in clusters. Clusters were noted to increase in size as the humidity increased. These observations led us to consider each single spore or spore cluster as the true spreading units, i.e. units of dissemination (UD); this notion may be applied to other fungal parasites and allows us to consider the problem of dispersal from an epidemiological point of view, because it takes into account differences in infection potentials of the units [Rapilly (42)].

This viewpoint has been established by using passive spore traps such as rotorod in the field; in fact, there is a linear relationship between the size of the UD and the relative humidity. For example the mean size of the UD is 1.6 urediospores at a 70% relative humidity, but is 3 urediospores at a 95% relative humidity. This phenomenon does not occur with brown and black rusts, but does occur with yellow rusts because of the characteristic morphology of yellow rust urediospores (see above). The germination potential of these UD includes the self-stimulating effects among spores, already mentioned in the section on germination; consequently the infection potential of the UD that are spread at a high relative humidity is doubled.

The introduction of the notion of UD allows a better interpretation of the gradient of infectious dispersal. We noted above that duration of spore viability is increased when sunshine is reduced, and thus when the ambient humidity is high, which corresponds to UD in the form of spore clusters. Because of cluster weight, the transport occurs only over short distances, which might explain the weak focus progressions from 1 m to 4.30 m that Corbaz (6) observed after 6 days of dispersal. Earlier in this chapter we noted that high humidity leads to a stronger adhesion of UD on leaves. All of these factors indicate that a quick development of a yellow rust epidemic requires a random distribution of inoculum foci. This characteristic has never been observed with black and brown rusts, for which Roelfs (47) showed that dispersal gradients are flat, as the UD are single urediospores.

EPIDEMIC DEVELOPMENT

The rate of spread of yellow rust epidemics depends on numerous factors. One of them is the variety, which affects the rate of spread according to its resistance level. In the variety Nugaine, Young (69) reported a rate $r = 0.002$ whereas in the variety Yamhill the rate r was $= 0.10–0.20$. Climatic factors are important at each stage of the epidemic cycle, but the rate of spread of primary inoculum is an essential parameter. The interpretation of this rate depends on whether the field or area is sown with only one or more varieties with the same resistance genes or whether the field or area is sown with plants with different resistance genes (multiline, composite varieties, etc).

The occurrence of primary foci and their extension are required for any release of an epidemic. Under equal climatic conditions, the development of a yellow rust epidemic is very much dependent on this aspect.

In the case of plots sown with only one variety, the rate of spread from only one focus has been observed by several authors. Zadoks (71) reported that in winter after four successive generations a focus corresponding to one contaminated leaf reaches 1 m². Joshi & Palmer (30) noted a progression of 100 m in India in January and February (within 55 to 75 days depending on the year). Emge & Shrum (9) reported that windward or leeward from the focus the speeds are practically identical: $r = 0.279$.

The genetic conditions for the appearance of a new race compatible with a variety hitherto resistant led us to think that the foci distribution is random, but becomes denser as the selection pressure increases. A massive supply of an exodemic inoculum due to a long-distance transport seems to us unlikely in the case of yellow rust. We think that the origins of epidemics are endemic multiple foci corresponding to a few infected plants or even only to a few sori.

The multiple foci allow the epidemic to progress 2.5 times faster than from only one focus with the same infectious potential. Furthermore this random distribution requires four times less inoculum to reach the same final contamination rate than from only one focus [Rapilly, Fournet & Skajennikoff (45)]. In his simulation trials, Shrum (57) noted that a reduction by two of the primary inoculum has no special influence, except if climatic conditions are detrimental, but this reduction is quantitative and does not concern the inoculum distribution. Kampmeijr & Zadoks (31) showed in their simulation that the same inoculum quantity randomly distributed among 400 or 16 foci produced the same rate of epidemic development, but that the rate was much slower than when an epidemic originated from only one focus. The same authors showed that the main effect of multilines is the damping of the focus extension.

On wheat, foci can appear only during autumn or winter periods, when the relative humidity is often very high, which allows only a short range for efficient spread of urediospores. A generalized epidemic, which usually occurs only in spring, can take place, if numerous foci, which are difficult to discern, are present or if the parasite can profit by a sufficient delay, of approximately one year, during which secondary foci are formed that allow its geographical repartition over a large area. This is why the study of winter climatic conditions to evaluate the possibilities of parasite multiplication partially enables us to forecast epidemic changes in spring [Rapilly (41)].

CONCLUSION

Since 1961 considerable progress has been made in the description of yellow rust epidemiology. During the last few years a greater understanding of the disease has been achieved by quantifying each particular event of the infection cycle and by taking into account the numerous interactions between the host, pathogen, and the environment. This quantification has made possible the development of various models [Zadoks & Rijsdijk (76), Rijsdijk (46), Shrum (57), Kampmeijr & Zadoks (31)] which enable researchers to perform prospective studies that help breeders and farmers.

For example, forecasting is now possible so as to give advice to optimize fungicide treatments [Garadagi (11), Rijsdijk (41), Rapilly (46)], but the major utility of forecasting may suggest new methods to select for disease resistance and how better to use control systems that are already recommended in some countries [Harvey (16; see also 1)]. Such advances also will make possible the study of the genetic control of slow rusting epidemics now that the infectious cycle sequences in relation to important epidemiological consequences can be determined.

It is likely that the next step will connect quantitative and qualitative epidemiology, including all the competition and preimmunization aspects among the parasite races. These studies foster new interest in multilines and composite varieties and take into account the interdependencies between horizontal and vertical resistance.

ACKNOWLEDGMENT

Many thanks are due to M. F. Commeau for linguistic help.

Literature Cited

1. Anonymous 1978. Varietal diversification to reduce risk of yellow rust in winter wheat. *Farmers Leaflet,* No. 8
2. Azmeh, F. 1976. *Contribution à l'étude de la dissémination de spores fongiques anémophiles: Adhésion des spores.* Thesis Bordeaux I No. 12762. 90 pp.
3. Brown, J. F., Sharp, E. L. 1969. Interactions of minor host genes for resistance to *Puccinia striiformis* with changing temperatures regimes. *Phytopathology* 59:999–1001
4. Burrage, S. W. 1969. Dew and the growth of the uredospore germ tube of *Puccinia graminis* on the wheat leaf. *Ann. Appl. Biol.* 64:495–501
5. Chumakov, A. Y. 1968. The problem of grain rusts. *Zashch. Rast. Moscow* 4:19–21 (In Russian)
6. Corbaz, R. 1966. Notes sur la Rouille Jaune du Froment en Suisse Romande. *Phytopathol Z.* 36:40–53
7. Doling, D. A., Doodson, J. K. 1968. The effect of yellow Rust on the yield of spring and winter wheat. *Trans. Br. Mycol. Soc.* 51:427–34
8. Emge, R. G., Kingsolver, C. H., Johnson, D. R., 1975. Growth of the sporulating zone of *Puccinia striiformis* and its relationship to Stripe Rust epiphytology. *Phytopathology* 65:679–81
9. Emge, R. G., Shrum, R. D. 1976. Epiphytology of *Puccinia striiformis* at five selected locations in Oregon during 1968 and 1969. *Phytopathology* 66: 1406–12
10. Fuchs, E. 1972. Some observations with different races of *Puccinia striiformis. Proc. Eur. Mediterr. Cereal Rusts Conf. Praha* 2:135–38
11. Garadagi, S. 1975. Short-term forecast and chemical treatments in control of yellow rust of wheat. *Ref. Zh. Biol.* 4: 4–464 (In Russian)
12. Gassner, G., Straib, W. 1934. Recherches expérimentales sur l'Epidémiologie de la rouille Jaune (*Puccinia glumarum* Schm. (Erikss. et Henn.). *Phytopathol. Z.* 7:285–302
13. Georgievskaja, N. A. 1966. Quelques lois sur le développement de la rouille Jaune du blé. *Tr. Vses. Nauchno Issled. Inst. Zabtsh. Rôst. Leningrad* 26:55–63 (In Russian)
14. Goddard, M. V. 1976. Cytological studies of *Puccinia striiformis* (yellow rust of wheat). *Trans. Br. Mycol. Soc.* 66:433–37
15. Deleted in proof
16. Harvey, G. 1978. The Cambridge strategy. *New Sci.* 77:428–30

17. Hassan, F. 1968. Cereal rusts situation in Pakistan. *Proc. Eur. Mediterr. Cereal Rusts Conf. Oeïras,* Portugal, pp. 124–25
18. Hassebrauk, K. 1965. Nomenclatur, Geographische Verbreitung und Wirtsbereich des Gelbrostes, *Puccinia striiformis* West. *Mitt. Biol. Bundesanst. Land Forstwirtsch. Berlin Dahlem* 116: 75 pp.
19. Hassebrauk, K., Robbelen, C. 1974. Der Gelbrost *Puccinia striiformis* West. III. Die Spezialisierumg. *Mitt. Biol. Bundesanst. Land Forstwirtsch Berlin Dahlem* 156:62–69
20. Hassebrauk, K., Robbelen, C. 1975. Der Gelbrost *Puccinia striiformis* West. IV Epidemiologie Bekampfungsmassnahmen. *Mitt. Biol. Bundesanst. Land Forstwirtsch. Berlin Dahlem* 164:1–53
21. Hassebrauk, K., Schroeder, J. 1964. Studies on germination of yellow rust uredospores *Puccinia striiformis. Proc. Cereal Rust Conf. Cambridge,* pp. 12–18
22. Heitefuss, R., Dehne, D., Einfeld, E. 1977. Evaluation of resistance and tolerance in wheat against *Puccinia striiformis* and in barley against *Erysiphe graminis. Proc. Vienne Conf. Induced Mutations Against Plant Diseases* IAEA, Vienna. 65 pp. 89–95
23. Hermansen, J. E. 1968 Studies on the spread and survival cereal rust and mildew diseases in Denmark. Thesis Copenhague. *Nord. Mykol T.* 8:206 pp.
24. Hermansen, J. E., Stapel, C. 1973. Notes on the yellow rust Epiphytotic in Denmark in 1972. *Cereal Rusts Bull.* 1:5–8
25. Itier, B., Pauvert, P. 1976. Modélisation de transports horizontaux. *Conf. OEPP-/OILB. L'établissement de Modèles dans le Cadre de la Protection des Plantes,* Paris (Abstr.)
26. Johnson, R., Allen, D. J. 1975. Induced resistance to rust diseases and its possible role in the resistance of multiline varieties. *Ann. Appl. Biol.* 80:359–63
27. Johnson, R., Bowyer, D. E. 1974. A rapid method for measuring production of yellow rust spores on single seedlings to assess differential inter-actions of wheat cultivars with *Puccinia striiformis. Ann. Appl. Biol.* 77:251–58
28. Johnson, R., Taylor, A. J. 1976. Effects of resistance induced by non-virulent Races of *Puccinia striiformis.* See Ref. 36a, pp. 49–51
29. Joshi, L. M., Goel, L. B., Sinha, V. C. 1976. Role of the Western Himalayas in

the annual occurrence of yellow rust in Northern India. *Cereal Rusts Bull.* 4:27–30

30. Joshi, L. M., Palmer, L. T. 1973. Epidemiology of stemleaf and stripe rusts of wheat in Northern India. *Plant. Dis. Reptr.* 57:8–12

31. Kampmeijr, P., Zadoks, J. C. 1977. *EPIMUL,* a simulator of foci and epidemics in mixtures of resistant and susceptible plant mosaics and multilines. *Simulation Monographs,* Wageningen: Pudoc. 50 pp.

32. MacDonald, E. A., Strange, R. N. 1976. Effects of temperature shocks, hydration and leaching on the subsequent germination of uredospores of *Puccinia striiformis. Trans. Br. Mycol. Soc.* 66:555–57

33. Macko, V., Trione, E. J., Young, S. A. 1977. Identification of the germination self-inhibitor from uredospores of *Puccinia striiformis. Phytopathology* 67: 1473–74

34. Maddison, A. C. 1970. *Some aspects of the epidemiology of Puccinia striiformis West. with special reference of the effect of U.V. radiation on long distance transport.* PhD thesis. Southampton Univ.

35. Maddison, A. C., Manners, J. G. 1972. Sunlight and viability of cereal rust uredospores. *Trans. Br. Mycol. Soc.* 59:429–43

36. Matveenko, A. N. 1974. The rate of germination of uredospores of *Puccinia striiformis* West. *Mikol. Fitopatol.* 8:146–47 (In Russian)

36a. McGregor, Manners, J. C. 1976. Some effects of lights on the growth of yellow rust on wheat. *Proc. Eur. Mediterr. Cereal Rusts Conf. Interlaken,* pp. 56–66

37. Pochard, E., Goujon, C., Vergara, S. 1962. Influence de la température, de l'éclairement et du stade de la plante sur l'expression de la sensibilité à la rouille jaune de quelques variétés Françaises de blé. *Ann. Amelior. Plant.* 12:45–58

38. Priestley, R. H., Doling, D. A. 1972. A technique for measuring the spore production of yellow rust on wheat varieties. See. Ref. 10, pp. 219–23

39. Pyzhikova, G. V. 1975. Spore forming activity of *Puccinia striiformis* West f. sp. *tritici* a causal agent of yellow rust. *Mikol. Fitopatol.* 9:228–30 (In Russian)

40. Rapilly, F. 1968. Quelques remarques sur la morphologie des urédospores de *Puccinia striiformis* f. sp. *tritici. Bull. Trimest. Soc. Mycol. Fr.* 84:493–96

41. Rapilly, F. 1976. Essai d'explication de l'épidémie de rouille jaune sur blé en 1975. *Sél. Fr.* 22:47–52

42. Rapilly, F. 1977. Réflexions sur les notions de propagule et d'unité de dissémination en Epidémiologie Végétale. *Ann. Phytopathol.* 9:161–76

43. Rapilly, F., Foucault, B. 1976. Premières études sur la rétention de spores fongiques par des épidermes foliaires. *Ann. Phytopathol.* 8:31–40

44. Rapilly, F., Fournet, J. 1968. Observations sur la dissémination de *Puccinia striiformis* en fonction de l'humidité relative, relation avec la structure morphologique des urédospores. See Ref. 17, pp. 26–29

45. Rapilly, F., Fournet, J., Skajennikoff, M. 1970. Etudes sur l'Epidémiologie et la Biologie de la rouille jaune du blé: *Puccinia striiformis* West. *Ann. Phytopathol.* 2:5–31

46. Rijsdijk, F. H. 1975. A simulator of yellow rust of wheat. *Semaine hygiène des plantes. Bull. Rech. Agro. Gembloux,* Sept. 12, pp. 411–18

47. Roelfs, A. P. 1970. *Gradients in horizontal dispersal of cereal rusts uredospores.* PhD thesis. Univ. Minn., Minneapolis. 77 pp.

48. Roelfs, A. P. 1978. Estimated losses causes by rust in small grain in cereals in United States 1918–1976. *US Dep. Agric. Misc. Publ.* 1363. 85 pp.

49. Russell, G. E. 1975. Deposition of *Puccinia striiformis* uredospores on adult wheat plants in laboratory experiments. *Cereal Rusts Bull.* 3:40–43

50. Russell, G. E. 1976. Characterization of adult plant resistance to yellow rust in wheat. See Ref. 36a, pp. 21–23

51. Russell, G. E., Hudson, L. R. L. 1973. Effects of vernalisation on yellow rust in certain winter wheat varieties. *Cereal Rusts Bull.* 1:13–15

52. Russell, G. E., Hudson, L. R. L. 1974. Sporulation of *Puccinia striiformis* on nine winter wheat varieties at different growth stages. *Cereal Rusts Bull.* 2:39–43

53. Schmitt, C. G., Hendrix, J. W., Emge, R. G., Jones, N. W. 1964. Stripe rust, *Puccinia striiformis* West. *Tech. Rept. 43* Plant Sci. Lab., Fort Detrick, Md. 111 pp.

54. Shaner, G. E. 1969. Epidemiology of stripe rust (*Puccinia striiformis* West.) of wheat in Oregon. *Diss. Abstr.* 29:31158 B (Abstr.)

55. Shaner, G., Powelson, R. L. 1973. The oversummering and dispersal of inoculum of *Puccinia striiformis* in Oregon. *Phytopathology* 63:13–17

56. Sharp, E. L. 1965. Prepenetration and postpenetration environment and devel-

opment of *Puccinia striiformis* on wheat. *Phytopathology* 55:198–203

57. Shrum, R. 1975. Simulation of wheat stripe rust *Puccinia striiformis* West. using Epidemic, a flexible plant disease simulator. *Pa. State Univ. Agric. Exp. Stn. Prog. Rep.* 347. 41 pp.

58. Smith, H. C., Smith, N., Marshall, M. I. 1973. Powdery mildew is not a pathogen of oats in New Zealand. *NZ. J. Exp. Agric.* 4:387–89

59. Stanbridge, B., Gay, J. L. 1969. An electron microscope examination of the surfaces of the uredospores of four races of *Puccinia striiformis. Trans. Br. Mycol. Soc.* 53:149–53

60. Stubbs, R. W. 1967. Influence of light intensity on the reactions of wheat and barley seedlings to *Puccinia striiformis. Phytopathology* 57:615–17

61. Stubbs, R. W., Plotnikova, J. M. 1972. Uredospore germination and germ tube penetration of *Puccinia striiformis* in seedling leaves of resistant and susceptible wheat varieties. *Neth. J. Pathol.* 78:258–64

62. Tollenaar, H., Houston, B. R. 1965. Effect of spore density on *in vitro* germination of uredospores of *Puccinia striiformis. Phytopathology* 55:1080 (Abstr.)

63. Tollenaar, H., Houston, B. R. 1966. *In vitro* germination of *Puccinia graminis* and *P. striiformis* at low spore densities. *Phytopathology* 56:1036–39

64. Tu Jui-Chang, Hendrix, J. W. 1967. The summer biology of *Puccinia striiformis* on Southwestern Washington. 1. Induction of infection during the summer. *Plant. Dis. Reptr.* 51:911–14

65. Van der Plank, J. E. 1963. *Plant Diseases: Epidemics and Control.* New York & London: Academic. 349 pp.

66. Van der Plank, J. E. 1975. *Principles of Plant Infection.* New York & London: Academic. 216 pp.

67. Viennot-Bourgin, G. 1940–1941. La rouille jaune des graminées. *Ann. Ec. Natl. Grignon* 3:129–217

68. Wright, R. G., Lennard, J. H. 1978. Mitosis in *Puccinia striiformis.* light microscopy. *Trans. Br. Mycol. Soc.* 70:91–98

69. Young, S. A. 1977. Quantitative epidemiology of stripe rust *Puccinia striiformis* West. in wheat cultivars *triticum aestivum* Vill. with general resistance. *Diss. Abstr.* 2455-B (Abstr.)

70. Young, S. A., Powelson, R. L. 1976. Quantitative epidemiology of non specific resistance in wheat to yellow rust. See Ref. 36a, p. 68

71. Zadoks, J. C. 1961. Yellow rust on wheat: Studies in epidemiology and physiologic specialisation. *Tijdschr. Planteziekten* 67:69–256

72. Zadoks, J. C. 1965. Epidemiology of wheat rusts in Europe. *FAO Plant. Prot. Bull.* 13:1–12

73. Zadoks, J. C. 1966. On the dangers of artificial infection with yellow rust to the barley crop of the Netherlands; A quantitative approach. *Neth. J. Plant Pathol.* 72:12–19

74. Zadoks, J. C. 1971. Systems analysis and the dynamics of epidemics. *Phytopathology* 61:600–10

75. Zadoks, J. C. 1972. Methodology of epidemiological research. *Ann. Rev. Phytopathol.* 10:253–76

76. Zadoks, J. C., Rijsdijk, F. H. 1972. Epidemiology and forecasting of cereal rust studied by means of a computor simulator named EPISIM. See Ref. 10, pp. 293–96

Ann. Rev. Phytopathol. 1979. 17:75–96
Copyright © 1979 by Annual Reviews Inc. All rights reserved

MODIFICATION OF HOST-PARASITE INTERACTIONS THROUGH ARTIFICIAL MUTAGENESIS[1]

❖3702

M. D. Simons

Agricultural Research, SEA, US Department of Agriculture, Department of Botany and Plant Pathology, Iowa State University, Ames, Iowa 50011

INTRODUCTION

The use of mutagenic agents to artificially modify the reactions of crop plants to disease organisms was started in the 1930s, and the first account of success was published in 1942 (22). Since that time, many investigators around the world have applied mutagenic agents to both host plants and parasitic organisms. They have had different objectives, but most were interested in creating useful disease resistance in crop plants. The practical success of these efforts is attested by the release of a number of cultivars of different crop species having improved disease resistance attributable to the application of mutagenic agents (18, 65, 66).

Another aspect of the use of mutagenic agents has been emphasized recently by Nilan, Kleinhofs & Konzak (51), who suggested that induced mutations will become increasingly important sources of genetic variability for crop improvement because the sources of natural genetic variability, for some species in particular, are at various stages of depletion. For example, natural populations of the wild relatives of the important cereal species are being rapidly reduced because of the extension and intensification of agriculture in their areas of adaptation. The use of artificial mutagenic agents could help to substitute for the loss of this potentially valuable variability.

[1]Cooperative Research between AR, SEA, US Department of Agriculture and the Iowa Agriculture and Home Economics Experiment Station, Project No. 1752, Journal Paper No. 9323.

75

0066-4286/79/0901-0075$01.00

This review was undertaken to assess the current status of the use of artificial mutagens on host plants and on their pathogens. Therefore, the review consists mainly of a critical summary of the relatively recent literature. Older literature has been included only when needed to provide perspective necessary to understanding of the topic under consideration.

MUTAGENESIS OF HOST PLANTS

Specific Resistance

The first report of successful induction of a mutation for disease resistance was made by Freisleben & Lein (22) in Germany. They treated Haisa barley with X rays and screened 12,000 progenies for resistance to powdery mildew (*Erysiphe graminis*). A mutant having resistance to three races of the pathogen was isolated. Hänsel (28) also used X rays to produce a mutation for mildew resistance in barley. Genetic studies showed that the resistance of this mutant was monogenic and recessive and that it was identical with the mutant reported earlier by Freisleben & Lein.

Jørgensen (33, 34) carried out a detailed study of 10 mutant genes for resistance to powdery mildew of barley, and, even though they originated in 6 different barley cultivars, they were all found to be recessive and at the same locus (*ml-o*). All 10 had a similar, unique, and seemingly universal range of resistance to all pathogenic forms of the fungus, and all were associated with leaf necrosis and reduced grain yields. Intercrossing between certain plants having these mutant genes produced a few susceptible plants, suggesting that at least some of the alleles at this locus differed at sites within the locus.

The *ml-o* locus for powdery mildew reaction in barley seems highly mutable. Hentrich (31) obtained 95 resistant mutants with various mutagenic chemicals, and all were found to be at this locus. He too, however, found quantitative differences in the type of infection and the amount of necrotic leaf spotting, indicating that a number of distinct alleles were involved. One of these alleles was of special interest because it conditioned moderate resistance and the pleiotropic chlorotic-necrotic effects were lacking or nearly so. Not all mutations for barley mildew resistance involve changes at the *ml-o* locus. Moës (46) applied X rays to seed of the cultivar Piroline, which was resistant to two of the four races of mildew he used. A mutant was recovered that was resistant to all four races. The resistance of the mutant was shown to be monogenic and dominant, in contrast to the resistance of the parent, which was recessive.

With wheat, chemical mutagen treatment of the susceptible cultivar Omar resulted in mutants with increased resistance to stripe rust (*Puccinia striiformis*) (41). Both race-specific and race-nonspecific resistance was induced. Several mutant lines were moderately susceptible at low temperatures in all stages of growth, but were moderately resistant at high temperatures in later stages of growth. Successful induction of specific resistance to the rice blast fungus (*Pyricularia oryzae*) (70) and to bacterial blight (*Xanthomonas malvacearum*) of cotton (9) has also been achieved.

General Resistance and Tolerance

Even though it has been a quarter of a century since Frey (23) showed that X rays could increase the field tolerance of oats to crown rust (*Puccinia coronata*), opinion is still divided as to the practical feasibility of using artificial mutagenesis to create or modify general resistance and tolerance to disease. The subject can be argued both ways on theoretical grounds. Knott (37), for example, was not optimistic about the possibilities of using mutagens to induce general resistance. His genetic studies of different sources of field resistance to stem rust (*Puccinia graminis*) of wheat indicated that this character was usually inherited in a quantitative manner, with a number of generally recessive genes, each with a small additive effect, conditioning reaction. He interpreted this to mean that in an induced mutation program, no individual plant would be likely to have enough mutant genes to appear resistant.

The opposite view was taken by Micke (45), who concluded that attempting to induce general resistance was probably more promising than attempting to induce specific resistance. He discounted the potential difficulty of inducing changes in a polygenic system. Frey (24) noted that qualitatively inherited mutations tend to be predominately deleterious, whereas mutations that affect quantitative traits form a rather normal distribution of plus and minus effects around a mean. He believed therefore that mutation breeding should be more successful for quantitatively than for qualitatively inherited traits. Similarly, Gaul (25) discussed the concept of macro- and micromutations. He thought that micromutations might be useful in plant breeding programs for two reasons: (*a*) They might occur much more frequently than macromutations, and (*b*) they might not so often result in reduced vitality as do macromutations. Further, the micromutations have the advantage of possibly being used directly, while this is usually not the case with macromutations. On the other hand, he noted that micromutations are difficult to detect and manipulate, usually requiring replicated measurements.

Regardless of theoretical considerations, numerous investigators have recently reported varying degrees of success in inducing mutations for general resistance. In barley, which has long been a favorite subject for studying qualitative mutations concerned with both disease reaction and other traits, 46 mutants for reaction to powdery mildew (*E. graminis*) were obtained from material treated with ethyl methanesulfonate (EMS) (58). Detailed studies in both the field and the growth chamber with eight of these mutant isolates failed to show clear-cut specific resistance relationships between the mutants and single races of the fungus. Instead, the level of resistance found in the parents was shifted either higher or lower in degree.

Working with powdery mildew (*E. graminis*) of wheat, Király & Barabás (36) were unable to find any resistance in 58,000 seedlings tested in the greenhouse in the M_1 generation. About 120,000 adult plants in the M_2 were then evaluated in the field. A small percentage of these plants were rated as field resistant. Borojević (7), working with wheat derived from material treated with gamma rays, found resistance to leaf rust (*Puccinia recondita*) in advanced generations. The resistance was expressed as a lower severity of infection, as a more resistant pustule type, and as tolerance, the latter expressed as variation in the ratio of kernel weight in rusted and nonrusted plots. These characters are ordinarily regarded as expressions of general, as opposed to specific, resistance. Artificial mutagenesis was also effective in obtaining lines of oats with increased tolerance to *P. coronata* (67) and lines of wheat with increased tolerance to *Septoria nodorum* (10, 42).

Attempts of Simons & Frey (68) to use EMS to induce a higher degree of field resistance to crown rust of oats in a cultivar rated as moderately resistant to certain races were unsuccessful. The same material also was tested for resistance to a race of crown rust to which the parent cultivar was highly susceptible. Certain lines were significantly more damaged and others significantly less damaged by this race than was the parent. In the absence of rust, only 5 of the 130 lines tested were not significantly below the parent for yield, but one of these 5 had significantly improved tolerance. EMS treatment of a highly susceptible tetraploid line of oats also resulted in both significantly increased and significantly reduced tolerance to crown rust.

Nakai & Goto (49) treated four cultivars of rice with gamma rays, thermal neutrons, and ethyleneimine. There was greater variability in reaction to bacterial leaf blight (*Xanthomonas oryzae*) in the treated population than in the controls. This increased variability, which was assumed to be due to polygenic mutations, was expressed as both greater resistance and greater susceptibility than found in the parental cultivars.

Less work has been done with legumes than with cereals, but there are definite indications that general resistance to diseases of legumes can be modified by mutagenic treatment. Thus dry bean (*Phaseolus vulgaris*) seed treated with EMS gave rise to one M_2 progeny that showed milder golden mosaic symptoms than the parental check (74). The field bean (*Vicia faba*) is a major crop and an important source of protein for human consumption in Egypt. Over 20,000 M_2 plants from seed of field beans treated with gamma rays were screened for reaction to chocolate spot (*Botrytis fabae*) and rust (*Uromyces fabae*) (1). A total of 27 M_2 plants showed significantly reduced amounts of both chocolate spot and rust as compared with the parent, but no plants were found that could be rated as highly resistant. Lesser degrees of resistance such as this are a common manifestation of general resistance.

Grapes (*Vitis vinifera*) with satisfactory resistance to mildew (*Plasmopara viticola*) would be of great value in Europe. Resistance, however, is quantitative in nature and is controlled polygenically. The problem is further complicated by a lack of resistance in the cultivated grape and by certain genetic correlations between resistance and poor quality of the fruit. Seedlings from grape seed treated with X rays and neutrons were screened for resistance to mildew under conditions of both natural and artificial inoculation (14). Certain plants expressed resistance in the form of smaller lesions and reduced sporulation. By using these criteria, plants representing approximately 0.001 to 0.0025% of the total number of plants screened were isolated for further study.

Mutation Rates

A knowledge of mutation rates that might be expected when attempting to induce mutations for disease resistance would be useful to anyone contemplating research or practical work in this area. In view of the relatively large number of papers published on the subject, it would seem that reliable information should be readily available. This, however, is not the case. One problem is that investigators usually publish their work only when it has resulted in the production of disease-resistant plants. Also, there is no general agreement on what constitutes a mutation. An investigator who makes accurate measurements will obviously report a higher mutation rate than one who simply "eyeballs" his material. Much of the data that are reported are given in terms of the number of plants tested, and these are often plants in the M_2, M_3, or later generations. A mutation rate in terms of percentage of mutants in the treated material is often difficult or impossible to deduce from such data. Still another factor is that mutation rates usually are related to mutagen dosage and the dosage used is often inadequately described.

Rates of mutation reported in the literature vary tremendously. They range from zero to "extremely high." The sizes of populations tested also differ greatly. In one recently reported study (29) about 7 million oat plants derived from either irradiated or chemically treated material were tested for reaction to the stem rust fungus (*P. graminis*). Only one resistant mutant was obtained, and its resistance was associated with poor plant type. The same material was subjected to natural crown rust (*P. coronata*) infection in the field, and no resistance to this rust was found either (44). In another ambitious study that extended over a period of years (31), 2.5 million M_2 seedlings of barley derived from seed treated with mutagenic chemicals were screened for reaction to a mixture of races of powdery mildew (*E. graminis*) under greenhouse conditions. A total of 95 resistant plants were found, but all resulted from mutations at the *ml-o* locus. It seems reasonable to assume that other loci for mildew reaction are far more difficult to alter than is the *ml-o* locus. Thus when estimating mutation rates, even when the same crop cultivar and mutagen treatment are involved, the investigator must recognize that rates at different loci may vary greatly.

At the other extreme for rates of mutation, Simons & Frey (68) reported an experiment in which only 110 pure lines of oats, each derived from a different EMS-treated seed, were tested for tolerance to the crown rust fungus (*P. coronata*). No visible differences between the treated lines and the check were evident. However, about 25% of the 110 lines showed a significantly greater tolerance to crown rust infection, in terms of reduction in seed weight caused by the disease, than did the parental variety. Similarly, Offutt & Riggs (53) found that 17% of 77 lines of mutagen-derived lespedeza (*Lespedeza stipulacea*) were resistant to one strain of the root-knot nematode (*Meloidogyne incognita*) and 12% were resistant to another strain. About half of the 77 were equal or superior in resistance to a standard cultivar that was rated as moderately resistant.

More typical information on mutation rates has been given by Abdel-Hak & Kamel (1), who grew 200,000 M_2 broad bean (*V. faba*) plants that had been derived from 9,000 M_1 plants treated with gamma rays. They were screened for reaction to chocolate spot (*B. fabae*) and rust (*U. fabae*). A total of 27 M_2 plants showed significantly less disease than the parent. They also treated 9000 seeds of four wheat cultivars with gamma rays and increased and tested more than 275,000 M_2 plants for resistance to leaf rust (*P. recondita*). Thirty-eight of the M_2 plants were classified as completely resistant. Király & Barabás (36) screened about 58,000 wheat seedlings from seed treated with gamma rays for resistance to powdery mildew (*E. graminis*) in the greenhouse in the M_1 generation. No resistance was found in this test. About 120,000 adult plants were then screened under field conditions for resistance in the M_2 generation, and these showed mutation rates estimated at 0.6 to 0.9%. Saccardo & Ramulu (61) treated seeds of pepper (*Capsicum annuum*) with fast neutrons and EMS and screened M_2 plants for reaction to cucumber mosaic virus. Mutation rates of 0.54% from the neutron-treated material and of 0.05% from the EMS-treated material were measured.

Intuitively, it would seem that mutation rates for disease resistance should not differ greatly among cultivars of the same crop species. Nevertheless, reports of marked differences are common. Thus, significant differences were found in the rates of mutation of four rice cultivars that had been treated with gamma rays, thermal neutrons, or ethylene amine and screened for bacterial blight (*X. oryzae*) resistance (49). Simons & Frey (68) found marked differences in mutation rates between the oat cultivars Richland and CI 6665 for tolerance to crown rust (*P. coronata*).

One of eight lines of mung bean (*Vigna radiata*) treated with gamma rays yielded 1.99% resistant mutants and 4.15% mutants with moderate resistance to yellow mosaic (64). The frequency of resistant mutants in the other lines ranged only from 0.16% to 0.26%. Similarly resistance to stem (*P. graminis*) and stripe (*P. striiformis*) rust was obtained in only one out of four cultivars of wheat treated with various mutagens (69). Murty (48) detected substantial differences in the response of various mutagen-treated inbred lines of millet (*Pennisetum glaucum*) in terms of frequency of mutations for resistance to downy mildew (*Sclerospora graminicola*). Such results suggest that success of a mutation breeding project may well depend on choice of the proper cultivar. Making such a choice, however, entails expensive and time-consuming preliminary testing.

Efforts to alter the disease reactions of crop plants to many different pathogens by mutagenic agents have been reported. A casual examination of that literature suggests that resistance to some diseases can be induced much more readily than resistance to others. Few experiments have been done in such a way, however, that direct comparisons can be made. Those that have generally bear out the idea that resistance to some pathogens is more easily induced than is resistance to others. Skorda (69) screened M_1 plants of a wheat cultivar that had been treated with thermal neutrons of gamma rays for resistance to stripe rust, stem rust, and leaf rust (*P. recondita*). Of 986 plants, 71 were resistant to stripe rust, 54 to stem rust, and none was resistant to leaf rust. In a parallel study (26) the frequencies of leaf and stem rust resistance mutants were about the same.

Detection of Outcrosses and Accidental Seed Mixtures

When the artificial induction of mutations for disease resistance and other traits of crop plants first became widely popular in the 1950s, one of the major concerns of the individual investigator was to prove that he really had created a new gene. Much of this work was done with the small grains that are naturally self-fertilized. What was not so well known was that artificial mutagens commonly cause severe male sterility in the M_1 generation. The glumes on such male-sterile plants often open in time for stray pollen from nearby plants to effect fertilization. Skeptics were quick to note that, when such pollen came from a plant with genes for disease resistance, the investigator could not distinguish between such outcrosses and genuine mutations.

It is now common practice to grow M_1 material in some sort of isolation. Even so, few investigators can say with absolute certainty that no outcrossing occurred in their material. Occurrence of accidental seed mixtures in genetic stocks is a related problem.

In addition to simply taking common sense precautions against outcrossing and mixture, the investigator who is concerned with the authenticity of his mutants may also attempt to show that his mutants differ from any known possible sources of contamination. Thus Jørgensen (33) compared two artificially induced genes for resistance to powdery mildew of barley with previously known genes for resistance with regard to the location of the genes on the chromosomes, their patterns of inheritance, and the range of resistance conferred. With these criteria, no spontaneous genes identical with the mutant genes were found. Haniš et al (27) compared electrophoresis bands of grain gliadins from parent and mutant lines of wheat. The bands of several disease-resistant mutants and their respective parents were similar and this was regarded as evidence that the resistance had arisen by mutation and had not resulted from some sort of contamination. Refinement of this approach, including use of bands from isoenzymes, holds promise for the future, in which critical experimental work might require maximum insurance against confounding by contaminants.

Selection of Mutations for Disease Resistance in Relation to Other Traits

Much work has been carried out in which crop plants have been treated with some sort of mutagenic agent and then have been screened for mutant traits other than disease resistance, including height, yield, seed weight, and maturity. If we view any kind of a desirable mutation as a relatively rare event, it would seem that there would be little chance of finding two desirable mutations in the same plant as a result of a single mutagenic treatment. Nevertheless, many cases have been reported in which plants selected for some trait such as height were subsequently found also to carry new genes for disease resistance.

Bozzini (8) tested eight mutant lines of durum wheat that had been selected originally for short straw or earliness for resistance to bunt (*Tilletia triticoides*). Two of these lines showed a higher level of resistance to bunt than the controls, while one was more susceptible. Using 130 mutant lines of wheat from leaf rust (*P. recondita*) susceptible cultivars that had been selected for nonpathogenic quantitative characters, Borojević (6) found four lines that were more resistant than the original parents in terms of infection type. Lines in which leaf rust resistance was expressed as a lower intensity of infection rather than a resistant infection type were more frequent. In a study designed mainly to improve certain agronomic traits of a cultivar of rice, Bernaux & Marie (4) treated seed with gamma rays and eventually selected 28 lines superior to the control for these traits. The lines were then tested for reaction to *Sclerotium oryzae* and *S. hydrophilum*. Eight of the 28 were resistant to *S. oryzae* and 21 to *S. hydrophilum;* seven were resistant to both pathogens.

Brönnimann & Fossati (11) started with a dwarf wheat that was suscepti-ble to *S. nodorum* and treated it with chemical or physical mutagens. Mass selection was used to screen for tolerance to the fungus in the M_2 and M_3 generations. There was a general tendency for tolerance (as indicated by thousand kernel weight) to increase but, unfortunately, at the same time there was a tendency for the plants to get taller, an undesirable response. Plants having both tolerance and short stature did not yield as well as the parent cultivar, but they were used in crossing programs. A suggested alternative was to apply mutagens to a tall cultivar that had good tolerance and select for short-statured lines that retained the tolerance of the tall parent.

Bálint, Bedö & Kiss (2) tested 29 mutagen-derived lines of corn, origi-nally selected for stalk strength, for resistance to fusarium rot. Thirteen of the 29 proved to be more resistant to fusarium rot than the parent line. Four of these lines were especially interesting because they showed a particularly high resistance to fusarium in combination with both improved stalk strength and high protein percentage.

Barabás (3) pointed out that many of the older wheat cultivars no longer being grown in Hungary had some excellent traits such as high protein percentage and good baking quality. They also, however, had weaknesses such as disease susceptibility and weak straw. He suggested that such cultivars could be "saved" by application of mutagens to increase disease resistance or to strengthen straw, while retaining all the good traits of the old cultivar.

Linkage of Mutations for Disease Resistance to Other Traits

The application of mutagenic agents to crop plants commonly has a delete-rious effect on traits such as yield or vigor. Although not usually reported, there is good reason to think that even mutations for disease reaction are more often than not in the direction of greater susceptibility rather than greater resistance (68, 69). Deleterious mutations for either pathogenic or nonpathogenic characters, of course, complicate the practical utilization of genes for disease resistance in breeding programs.

Skorda (69) was successful in using thermal neutrons and gamma rays to induce mutations for resistance to stem rust (*Puccinia graminis*) and stripe rust (*P. striiformis*) in wheat. Most of these mutants, however, could not be used directly as new cultivars because of concomitant deleterious effects, mainly in quantitative traits. Almost all of 140 rust resistant lines of wheat and 37 mildew resistant lines of barley isolated by other investigators (26, 27) also showed undesirable changes in other characters. Further, the researchers were unable to eliminate the undesirable associations through crossing of several of the mutant lines with diverse genotypes.

Using M_5 derived lines that had originally been treated with EMS, Simons & Frey (68) reported that 125 of a total of 130 mutagen-derived lines of oats were significantly lower in yield than the parent, regardless of the disease reaction. A mutagen-derived line of barley that was increased because of its mildew resistance yielded 4.4% less than its susceptible parent and also had an undesirable necrotic leaf spotting as such mildew-resistant lines often have (28). Working over a period of years, Hentrich (31) obtained 95 mildew-resistant lines of barley derived from mutagen-treated material. About half of them differed from the parents in plant height and size. A mutation for resistance to stem rust in wheat always occurred in association with light green plant color (15).

Because it is now common knowledge that mutagens often have a deleterious effect on yield and other quantitative traits, it has become a common practice for investigators to include data on such characters when they report success in inducing disease resistance. Skorda (69) found that a few of the many disease-resistant mutants he located in wheat yielded as well as the parent cultivar and one of them yielded better. It also yielded even better than the newer higher-yielding cultivars that were being developed for the region. The yielding ability of the mutant was stable over a wide range of environments and in different years. Parodi & Nebreda (56) also reported that some of their rust resistant wheat mutants, on the basis of repeated trials, had grain and protein yields superior to those of the parent cultivar.

Mutant strains of rice resistant to bacterial leaf blight (*Xanthomonas oryzae*) or to blast (*Pyricularia oryzae*) reportedly did not differ significantly in yield or other characters from their parent cultivars (35, 40, 54).

Shakoor et al (64) carried out intensive studies on 6 of 128 disease-resistant, mutagen-derived lines of mung bean (*Vigna radiata*), and found no deleterious pleiotropic effects on various yield components, except for reduction in seed size. Some lines were slightly higher in yield than the parents, and five of the six were shorter than the parents. Tulmann Neto, Ando & Costa (74) isolated a mutant line of bean (*Phaseolus vulgaris*) with resistance to bean golden mosaic that differed from the parent cultivar in leaf shape and seed size but was similar in time to flowering and ripening.

Vegetatively Propagated Crop Plants

Disease control in temperate fruit crops is a major concern and the chemicals commonly used for this purpose are coming under increasing criticism. Therefore, disease resistance has become an important objective in conventional fruit breeding programs. Progress, however, is slow because of the long generation time and the genetic complexity of most tree fruit species. Hence, the prospect of changing a few traits in an otherwise unchanged genetic background, as might be achieved by somatic mutation, is very attractive. The occurrence of natural mutations in fruit cultivars suggests that artifical mutagenesis might be successful in inducing disease resistance (12). Experimental work spans a wide range of crop species and includes, of course, the now classical work on peppermint by Murray (47), which is discussed in more detail later.

An example of success with temperate fruit crops has been furnished by Sigurbjörnsson & Micke (66) who treated shoots of McIntosh apple with gamma rays to produce a new cultivar that was named McIntosh 8F-2-32. This cultivar was resistant to both *Podosphaera leucotricha* and *Venturia inaequalis*. Another tree, the mulberry, has also been successfully treated with mutagens to induce disease resistance (50). The dogare disease reportedly caused by *Diaporthe nomurai* seriously reduces the spring leaf yield of the mulberry in parts of Japan. Branches of mulberry, which had been under chronic irradiation for 3 years in a gamma field, were propagated by grafting. The grafted plants were set out in a severely infested test field. Eight highly resistant mutants were found. Two of these were from a susceptible cultivar, and six were from a moderately resistant cultivar.

Tarasenko (71) treated the Lorch potato with X rays to induce resistance to *Synchytrium endobioticum*. A procedure of repeated vegetative propagation was used to stabilize chimeras. Of 498 clones eventually tested, 17 were resistant. Resistance was maintained over the several years that testing was carried out.

The outstanding success achieved in inducing resistance to *Verticillium albo-atrum* in peppermint (*Mentha piperita*) (47, 73) proved conclusively that the use of mutagens to induce disease resistance in vegetatively propagated plants was practical. Stolons of the susceptible Mitcham peppermint cultivar was radiated with neutrons or X rays. Over 100,000 plants were set in severely wilt-infested soil, resulting in over 6 million plants the following year. Four years of natural selection reduced the stand by about 99%. The 58,000 surviving plants were subjected to intensive additional testing, leading eventually to isolation of seven highly wilt-resistant strains with smaller leaves, earlier maturity, and a more erect habit than the parent variety; five moderately resistant, vigorous strains with coarse leaves and stems; and 50 slightly resistant strains with near normal appearance. All five of the moderately resistant vigorous strains had radiation-induced heterosis for herbage weight and at least three had heterosis for oil yield. The Scotch cultivar of spearmint (*Mentha cardiaca*) also has been successfully treated with mutagens to induce disease resistance (32).

The possibility of using mutagens to treat cell suspension cultures has been opened up by the work of Heinz (30). He used colchicine to treat cell suspension cultures of sugarcane (*Saccharum officinarum*). Plants derived from such cultures were tested for reaction to eye spot disease (*Bipolaris sacchari*). The parental clone was moderately susceptible to the disease. There was a relatively high rate of "natural mutation" toward greater resistance in the untreated controls, but the mutation rate was even higher (21% versus 14%) in plants from the colchicine-treated cell suspensions.

Mutagenesis and Cytoplasmic Sterility

The epidemic of southern corn leaf blight (*Bipolaris maydis*) in the United States in 1972 focused attention on the importance of cytoplasmic male sterility in the production of commercial seed and on possible relationships of this trait to disease susceptibility of important field crops. Cornu et al (13) treated seed of an inbred line of corn having Texas male sterile cytoplasm with EMS. Resulting M_3 seedlings were screened for resistance to *B. maydis*. A number of families exhibiting resistance were isolated, thus indicating a degree of success in the use of artificial mutagenesis where disease susceptibility is associated with a specific type of male sterility.

Downy mildew (*Sclerospora graminicola*) is a serious disease of millet in India, and the problem of inducing resistance is complicated by the fact that commercial seed consists of hybrids produced by using male-sterile female parents (48). Gamma rays were applied to seed of the fertile counterpart of an important male-sterile line, and three mutant lines with very high resistance and complete maintenance of male sterility were isolated and carried to the M_5 generation. The hybrids reconstituted by using the resistant mutant male steriles and corresponding male counterparts yielded 35 to 80% more than susceptible controls under conditions of severe disease.

Choice of Mutagenic Agents

An investigator interested in the induction of mutations for disease resistance has a wide choice of proven mutagenic agents at his disposal. These include X rays, neutrons, gamma rays, and a variety of chemicals. At this time there seems to be general agreement that X rays, neutrons, and gamma rays induce a predominance of chromosomal changes, whereas the variability induced by chemicals is largely genic (24). There is, however, no consensus as to what agent is "best." Thus, it is not surprising that many investigators use two or more mutagenic agents on a project. Some direct comparisons of different agents have been made recently.

Neutrons and EMS were compared for efficiency by treating seeds of pepper (*Capsicum annuum*) and screening M_2 plants for reaction to cucumber mosaic virus (61). The mutation rate, expressed as percentage of symptomless plants was 0.54% for neutrons and for EMS treatment, 0.05%. In comparing mutagenic chemicals for inducing resistance to powdery mildew in barley, Röbbelen, Abdel-Hafez & Reinhold (58) were successful with EMS and with sodium azide (NaN₃), but did not obtain any mutants by using N-methyl-N'-nitro-N-nitrosoguanidine.

Konzak et al (38) considered the problem of choosing a chemical mutagen. EMS has long been known to be an effective mutagen on a wide range of crop species. They believed, however, that NaN₃ was the potentially most interesting of all the available chemicals, but noted that it was one of those least tested. It is known to induce exceptionally high frequencies of mutations in barley, diploid wheat, peas, and bacteria. It does not seem to induce gross chromosome aberrations, nor does it seem to be carcinogenic. Methyl and ethyl nitrosourea are the most effective mutants they used on wheat. Of these, methyl nitrosourea was slightly more effective. Both, however, are potent carcinogens and are potentially explosive. Post-treatment of methyl nitrosourea-treated seeds with 0.001 M NaN₃ had been a very effective treatment in their experience.

Nakai & Goto (49) noted that the frequency of chlorophyll mutations was not a reliable guide to choosing treatments to modify disease reaction. Variation in disease severity induced by thermal neutron radiation was significantly greater than that induced by gamma ray radiation, whereas the reverse was true for rates of chlorophyll mutations.

Recurrent Mutagenic Treatment

The application of successive cycles of mutagen treatment to host plants along with concurrent selection for disease resistance is an intriguing idea. Different types of mutagenic agents might differentially affect gene systems. There is also the possibility that plants subjected to one mutagenic treatment might be more sensitive to subsequent change by another treatment with the same or a different mutagenic agent (38). An improvement, particularly in a quantitative character, might thus be achieved that would otherwise be impossible. There is little information available on the subject of recurrent cycles of mutagen treatment for disease resistance, but an example has been furnished by Simons & Frey (68) who originally treated an oat cultivar with EMS and selected lines that were more tolerant to the crown rust fungus than the parent cultivar. These mutant lines were treated with EMS, and one line was selected that showed a significant increase in degree of tolerance over the "first generation" mutant line.

Artificial Mutagenesis and Basic Research on Host-Parasite Relationships

The creation of new forms of host-plant resistance has great potential as a source of material that can be used in fundamental studies of various aspects of host-parasite relationships. El-Sayed (17), for example, was interested in the relationship of the resistance of tomato (*Lycopersicon esculentum*) to *Phytophthora infestans*, as expressed in the formation of rishitin, a phytoalexin formed in tomatoes infected with *P. infestans*. Tomato seeds were treated with gamma rays, and young M_1 plants were screened for resistance. Five mutants showing different degrees of resistance were obtained. Rishitin formed in infected leaf tissue in direct proportion to the degree of resistance. The correlation was statistically significant. This work corroborated other basic studies of rishitin formation and its relationship to disease resistance.

MUTAGENESIS OF PATHOGENS

Introduction

The strategy for using resistance for the control of plant diseases is often based on hypotheses about the development of the genetic potential of the pathogen when it encounters the resistance in question (52). In the context of a program for mutation breeding for disease resistance, mutations to virulence in the pathogen are as important as mutations to resistance in the host (19, 21). If new races of pathogens could be produced experimentally before they occur in nature, they could, in theory, be used to screen mutagen-treated host populations in advance of the natural occurrence of such races (39).

From another approach, a determination of the relative mutation rates of known genes for pathogenicity could be very useful in choosing those resistance genes against which mutations to virulence in the pathogen occur in lowest frequency. If it is assumed that the rates of mutation for given genes obtained by using artificial mutagens are correlated with the natural mutation rates of the genes, the application of mutagens to pathogens would provide such information. Teo & Baker (72) expanded this line of reasoning and believed that inducing variation in pathogenic fungi by mutagenic agents has many applications. Among these, they listed (*a*) elucidation of the extent, causes, and nature of spontaneous variability, (*b*) creation of cultures for assessing the role of somatic hybridization in creating new pathogenic forms, (*c*) ability to test simultaneously the reactions of segregating host populations with two or more pathogenic cultures differing in virulence and spore color, and (*d*) creation in the case of the rusts, of cultures that produce telia readily for use in genetic studies.

Mutations to Wider Virulence

The effects of mutagenic agents on plant pathogens were being seriously studied as early as the 1950s. Flor (20) applied ultraviolet radiation to urediospores of the flax (*Linum usitatissimum*) rust fungus (*Melampsora lini*) until only about 10% survived. He obtained seven mutants representing four different races. His mutation rates were very low, as estimated from the number of uredia developing on susceptible control cultivars. At about the same time, Schwinghamer (62) obtained over 4300 pathogenic mutants of the same organism on the two resistant flax cultivars Dakota and Koto. He used X rays, ultraviolet, and thermal neutron radiation. Flor (21) then switched from ultraviolet to X ray treatment, which he applied to a culture of the flax rust fungus known to be heterozygous for virulence toward 15 cultivars of flax. Because avirulence is dominant, the 15 cultivars were resistant. About 200,000 treated spores were screened on each of these cultivars. No virulent mutants occurred on seven of them, but 154 mutant uredia developed on the other eight. Only 94 of these, however, were sufficiently vigorous to permit further testing. All but two of the 94 fell into one of six virulence patterns, each differing from the original culture by a single virulence gene. The remaining two cultures each differed by two genes. The number of mutant uredia occurring on the eight cultivars ranged from 2 on Koto to 49 on Wilden, suggesting that loci in the pathogen differed greatly in mutability under natural conditions.

Rowell, Loegering & Powers (59) also used X rays in their studies of pathogenicity of the wheat stem rust fungus (*P. graminis*). They used a culture known to be heterozygous (and therefore avirulent) at four loci for pathogenicity. Treated spores were screened on Marquis, which had resistance genes corresponding to two of the four heterozygous loci in the fungus. One of these resistance genes conditioned a high type resistance that was epistatic to the moderate resistance conditioned by the other. A total of 136 cultures was obtained that induced moderate resistance on Marquis, indicating a mutation to virulence at the locus conditioning high resistance and thus allowing the gene for moderate resistance in the host to express itself. Further work with other host cultivars indicated that additional mutations had occurred in some of the mutant cultures.

More recently, Luig (43) used EMS to treat a culture of stem rust in the course of a study of a possible association of virulence of two wheat cultivars. He obtained two double mutants, each with increased virulence on the two wheat cultivars. He theorized that this could mean that the loci conditioning virulence were very closely linked, with the result that a change in one somehow affected the other. The same experiment yielded three mutants for virulence on Einkorn, one for virulence on Vernal Emmer, and four color mutants. Initial treatment of a culture of the closely related oat stem rust (*P. graminis*) with EMS failed to produce any mutations for pathogenicity (72). Mutants virulent on the oat cultivar Saia were found, however, in the second and third cycles of recurrent EMS treatment with four mutants appearing in the second cycle and two in the third cycle. None of the six mutants produced changes in infection type on other differential cultivars. On universally susceptible cultivars, however, they produced moderately susceptible reactions rather than the susceptible reactions induced by the parent culture.

Success in inducing pathogenic mutants in a culture of the rice blast fungus (*Pyricularia oryzae*) by treatment with gamma rays or nitrosoguanidine has been reported by Notteghem (52). Some variants were isolated that were similar to cultures found in nature, and a multivirulent strain also was isolated. Notteghem concluded that artificially inducing mutants for pathogenicity in the rice blast fungus was an effective method for studying the variability of the pathogen and the behavior of resistant cultivars in the face of this variability. Kwon & Oh (40) applied X rays to a race of the rice blast fungus incapable of attacking the new resistant cultivar Tongil. Mutants were screened on Tongil and monosporidial isolates were taken from this cultivar. There were no striking differences in cultural characters of the colony such as hyphal growth, color, or conidia formation. Some isolates showed decreased virulence; others increased virulence; and some were avirulent. Mutation frequency was proportional to X ray dosage up to a high level. Seven mutant races were identified. Six of these corresponded to known races. The seventh differed greatly from the original culture and had not been seen before in the area where the work was done. It was highly virulent on Tongil.

Mutagenic Agents and Mutation Rates

Schwinghamer (63) compared the effectiveness of fast neutrons, X rays, and ultraviolet radiation as mutagenic agents to induce mutations for virulence in the flax rust fungus. The mutation frequency was proportional to the dosage with ultraviolet and neutron treatments and varied with the square of the dosage with X ray treatments. The mutations induced by the ionizing radiation treatments were believed to be predominately deletions involving loss of a chromosome segment. Ultraviolet radiation, on the other hand, induced primarily point mutations. The average maximum frequency of mutations to virulence, based on almost 3 million uredia developing on susceptible controls, induced by fast neutrons, X rays, and ultraviolet, was 2.0, 1.5, and 0.3% respectively. Flor (20), working with the same fungus, also found that mutation rates induced by ultraviolet radiation were low. In studying recurrent cycles of EMS treatment of the oat stem rust fungus, Teo & Baker (72) calculated a mutation rate in the second cycle of treatment of about 0.016% and in the third cycle, of about 0.008%.

Rates of mutation have been found to vary significantly among different races of the same pathogen (*P. oryzae*) by Notteghem (52) in studies using nitrosoguanidine and gamma rays. Flor (21) similarly found that different races of the flax rust fungus varied in the ease with which they could be modified for pathogenicity. These differences among races may be analogous to differences in mutability among cultivars of the same host species.

Mutations for Pathogenicity in Relation to Other Traits

As just discussed, mutations for disease resistance in crop plants are often associated with changes in other traits, particularly quantitatively inherited traits. It seems reasonable to assume that the same situation would hold for pathogenicity mutants of pathogens. Such traits in pathogens would not ordinarily be observed, or at least would not be as readily seen as in higher plants. A few examples, however, have been recorded. Teo & Baker (72) noted that mutants toward wider virulence in the oat stem rust fungus had 2 to 3 day longer incubation periods than the original culture and also sporulated less vigorously. They also described mutants that differed from the parental cultures in urediospore color and in the interval between the uredial and telial stages. All their mutant cultures remained stable over several generations of subculturing. Rastegar (57), on the other hand, treated barley leaf rust (*Puccinia hordei*) with EMS and obtained three mutants that differed in color from the parent but that were otherwise identical.

Risk of Producing Potentially Dangerous New Races

There is some uneasiness among investigators who are using mutagens to produce new pathogenic forms of parasitic fungi, and they sometimes describe precautions they have taken to insure that new and potentially dangerous races do not escape to the field (52). In a recent discussion of this topic (40), it was believed by some that the risk that a new mutant race may escape to the field and endanger crops is minimal because in most cases the mutant isolates will carry other mutated genes that will make them less fit than spontaneous mutants. Teo & Baker (72) noted that their mutant races of the oat stem rust fungus were less vigorous than the parent races. Their general conclusion was that the mutants they had created were similar to those reported to arise from spontaneous mutations or genetic recombination. Thus, they should pose no special threat even if they did escape to the wild. The experience of others, however, has shown that, while some mutant pathogens were less aggressive than the original cultures, there were others that possessed aggressiveness identical with the parent strains. It also was noted that most of the mutagenic agents in use are not natural and, therefore, might cause mutations that would never appear in nature (40).

Induction of Mutant Forms for Use in Basic Research

The use of artificial mutagenic agents to create forms of plant pathogens that will be of value in basic research, as opposed to practical development of disease-resistant cultivars, is an interesting facet of mutagenesis. There are many possible applications, and several successful examples can be cited.

In assembling material for their quadratic check (for studying basic physiology of host-parasite interactions), Rowell, Loegering & Powers (60) needed two clones of wheat stem rust (*P. graminis*) that were isogenic except for a single gene for virulence toward a specific host-resistance gene. They could not simply select two clones, one of which attacked plants with the resistance gene and one of which did not, because two such clones would also differ in virulence genes at other loci as well as in genes for nonpathogenic traits. To obtain the clones they wanted, they applied X rays to a clone known to be heterozygous for the virulence locus corresponding to the host-resistance gene. They were able to select a clone from the radiated material that was homozygous for virulence toward the host resistance gene. Testing on host cultivars having other resistance genes indicated that the mutant and original clones were isogenic, except that the original was heterozygous and therefore avirulent, while the other was homozygous for the recessive allele for virulence.

Padwal-Desai, Ghanekar & Sreenivasan (55) were interested in basic aspects of the physiology of *Aspergillus flavus*. They applied gamma rays to two cultures of the fungus, one of which produced aflatoxin and one of which did not. The two strains were shown to differ in tolerance to the radiation. *Whetzelinia sclerotiorum*, cause of white mold in legumes, produces sclerotia that overwinter the fungus in the soil. Blanchette & Tourneau (5) noted that study of sclerotia formation, in relation to possible control practices, would be facilitated if strains of the fungus that lacked or varied in their ability to produce sclerotia were available. They treated sclerotia with gamma rays and obtained a wide variety of abnormal forms among mycelial colonies arising from the treated sclerotia. These included pigmentation in the form of yellow and gray patches, extensive aerial mycelia, subsurface mycelia, and mycelia with limited growth. Abnormal development of sclerotia, in terms of numbers of sclerotia produced and pattern of sclerotia formation, also occurred. Eleven nonsclerotia-forming isolates were found.

In another application, Ellingboe & Gabriel (16) investigated the use of the mutagenic chemical nitrosoguanidine to induce temperature-sensitive mutants in plant pathogens that would be useful for basic studies of host-pathogen interactions. They chose *Colletotrichum lindemuthianum* and *Phyllosticta maydis*. All the mutants they classified grew at the normal temperature of 22°C, but fell into different patterns for ability to grow or not grow at 28°C on media and for their ability to parasitize their hosts at 22° or 28°C.

CONCLUSIONS

The application of artificial mutagenic agents to crop plants to improve disease resistance has been widely practiced since the 1950s. New cultivars (representing species propagated vegetatively and by seed) with improved resistance have been produced and released, and the feasibility of the procedure is now generally recognized. There is no question that artificial mutagens can be used to induce mutations for specific resistance, but there is a disagreement as to their effectiveness in inducing general resistance. The evidence suggests to me that mutagens can modify general resistance, and, indeed, the greatest promise for the future use of mutagens may lie in this direction.

As this is written, there are a number of problems and questions that need to be resolved before more rapid progress can be made in utilizing artificial mutagenesis. A reasonable idea of what rates of mutation might be expected, for example, would be very useful. Information presently available on the subject is so variable that it is almost worthless. The problem of the association of mutations for disease resistance with other traits, particularly deleterious traits, such as low yielding ability, which were also induced by the mutagenic treatment, has plagued many investigators. There is reason to think that this can be overcome, however, if for no other reason than that other investigators have had no trouble with such linkages.

Many mutagenic agents, all proven more or less effective, are now available. The problem here is choice of the one most apt to yield the desired results. Currently, there are indications that some of the less well known chemical mutagens may have advantages over the ones commonly used. Further study may reveal new materials that are even better. The search for such materials and the study of combination or recurrent mutagen treatments should be fruitful areas of research that will improve the effectiveness of artificial mutagenesis.

The use of artificial mutagens to alter the pathogenicity of organisms that cause disease in plants has been clearly shown to be feasible. When such pathogens have wider virulence than that found in nature, they can be used to screen mutagen-treated host populations in advance of the natural occurrence of such virulence. The techniques can also be used to produce pathogens with specific characteristics needed for various lines of fundamental research. The deliberate creation of new forms of pathogens, however, is not without some risk. Research is needed on procedures to produce more efficiently the types of pathogen wanted and to assess the risk of new forms escaping to the wild where they might threaten crops.

Literature Cited

1. Abdel-Hak, T. M., Kamel, A. H. 1977. Mutation breeding for disease resistance in wheat and field beans. *Induced Mutations Against Plant Diseases,* pp. 305–14. Vienna: Int. At. Energy Agency. 581 pp.
2. Bálint, A., Bedö, Z., Kiss, E. 1977. Examination of wf 9 mutant sublines for lodging and *Fusarium* resistance. See Ref. 1, pp. 489–94
3. Barabás, Z. 1977. Prevention of gene erosion of old wheat varieties by backcrossing and X-ray irradiation. See Ref. 1, pp. 29–32
4. Bernaux, P., Marie, R. 1977. Mutants induits chez le riz (*Oryza sativa* L.) pour la reponse A *Sclerotium oryzae* Catt. et *Sclerotium hydrophilum* Sacc. See Ref. 1, pp. 157–70
5. Blanchette, B., Tourneau, D. L. 1977. The effects of gamma radiation on sclerotia of *Whetzelinia sclerotiorum. Environ. Exp. Bot.* 17:49–54
6. Borojević, K. 1974. Screening induced mutations for resistance to leaf rust in wheat. *Induced Mutations for Disease Resistance in Crop Plants,* pp. 133–46. Vienna: Int. At. Energy Agency. 193 pp.
7. Borojević, K. 1977. Studies on resistance to *Puccinia recondita tritici* in wheat population after mutagenic treatments. See Ref. 1, pp. 393–401
8. Bozzini, A. 1971. First results of bunt resistance analysis in mutants of durum wheat. *Mutation Breeding for Disease Resistance,* pp. 131–38. Vienna: Int. At. Energy Agency. 249 pp.
9. Brinkerhoff, L. A., Verhalen, L. M., Mamaghani, R., Johnson, W. M. 1978. Inheritance of an induced mutation for bacterial blight resistance in cotton. *Crop Sci.* 18:901–3
10. Brönnimann, A., Fossati, A. 1977. Tolerance A *Septoria nodorum* Berk. Chez le ble: Methodes d'infection et selection par mutagenese. See Ref. 6, pp. 117–23
11. Brönnimann, A., Fossati, A. 1977. Tolerance to *Septoria nodorum* in wheat: Evaluation of the infection and selection method and the mutagenesis programme. See Ref. 1, pp. 403–8
12. Campbell, A. I., Wilson, D. 1977. Prospects for the development of disease-resistant temperate fruit plants by mutation induction. See Ref. 1, pp. 215–26
13. Cornu, A., Cassini, R., Berville, A., Vuillaume, E. 1977. Recherche par mutagenese d'une resistance A *Helminthosporium maydis,* race T, chez les mais a cytoplasme male-sterile Texas. See Ref. 1, pp. 479–88
14. Coutinho, M. P. 1977. Utilisation des rayonnements pour l'amelioration de la vigne au point de vue de la resistance au mildiou. See Ref. 1, pp. 233–40
15. Edwards, L. H., Williams, N. D., Gough, F. J., Lebsock, K. L. 1969. A chemically induced mutation for stem rust resistance in 'Little Club' wheat. *Crop Sci.* 9:838–39
16. Ellingboe, A. H., Gabriel, D. W. 1977. Induced conditional mutatants for studying host-pathogen interactions. See Ref. 1, pp. 35–46
17. El-Sayed, S. A. 1977. Phytoalexin-generating capacity in relation to late blight resistance in certain tomato mutants induced by gamma irradiation of seeds. See Ref. 1, pp. 265–74
18. Favret, E. A. 1976. Breeding or disease resistance using induced mutations. *Induced Mutation in Cross-Breeding,* pp. 95–117. Vienna Int. At. Energy Agency. 255 pp.
19. Flor, H. H. 1956. The complementary genic systems in flax and flax rust. *Adv. Genet.* 8:29–54
20. Flor, H. H. 1956. Mutations in flax rust induced by ultraviolet radiation. *Science* 124:888–89
21. Flor, H. H. 1958. Mutation to wider virulence in *Melampsora lini. Phytopathology* 48:297–301
22. Freisleben, R., Lein, A. 1942. Über die Auffindung einer mehltauresistenten Mutants nach Röntgenbestrahlung einer anfälligen runen Linie von Sommergerste. *Naturwissenschaften* 30:608
23. Frey, K. J. 1954. Artifically induced mutations in oats. *Agron. J.* 46:49
24. Frey, K. J. 1965. Mutation breeding for quantitative attributes. *The Use of Induced Mutations in Plant Breeding,* pp 465–75. London: Pergamon. 832 pp.
25. Gaul, H. 1965. The concept of macro- and micro-mutations and results on induced micro-mutations in barley. See Ref. 24, pp 407–28
26. Haniš, M. 1974. Induced mutations for disease resistance in wheat and barley. See Ref. 6, pp. 49–56
27. Haniš, M., Hanišová, A., Knytl, V., Černý, J., Benc, S. 1977. Induced mutations for disease resistance in wheat and barley. See Ref. 1, pp. 347–57
28. Hänsel, H. 1971. Experience with a mildew-resistant mutant (Mut. 3502) of 'Vollkorn' barley induced in 1952. See Ref. 8, pp. 125–29

29. Harder, D. E., McKenzie, R. I. H., Martens, J. W., Brown, P. D. 1977. Strategies for improving rust resistance in oats. See Ref. 1, pp. 495–98
30. Heinz, D. J. 1973. Sugar-cane improvement through induced mutations using vegetative propagules and cell culture techniques. *Induced Mutations in Vegetatively Propagated Plants,* pp. 53–59. Vienna: Int. At. Energy Agency. 222 pp.
31. Hentrich, W. 1977. Tests for the selection of mildew-resistant mutants in spring barley. See Ref. 1, pp. 333–41
32. Horner, C. F., Melouk, H. A. 1977. Screening, selection and evaluation of irradiation-induced mutants of spearmint for resistance to verticillium wilt. See Ref. 1, pp. 253–62
33. Jørgensen, J. H. 1971. Comparison of induced mutant genes with spontaneous genes in barley conditioning resistance to powdery mildew. See Ref. 8, pp. 117–24
34. Jørgensen, J. H. 1974. Induced mutations for powdery mildew resistance in barley. A progress report. See Ref. 6, p. 67
35. Kaur, S., Padmanabhan, S. Y., Kaur, P. 1977. Induction of resistance to blast disease (*Pyricularia oryzae*) in the high yielding variety, *Ratna* (IR 8 X TKM 6). See Ref. 1, pp. 147–56
36. Király, Z., Darabás, Z. 1974. Report on mutation breeding for mildew and rust resistance of wheat in Hungary. See Ref. 6, pp. 85–88
37. Knott, D. R. 1977. Studies on general resistance to stem rust in wheat. See Ref. 1, pp. 81–88
38. Konzak, C. F., Line, R. F., Allan, R. E., Schafer, J. F., 1977. Guidelines for the production, evaluation, and use of induced resistance to stripe rust in wheat. See Ref. 1, pp. 437–59
39. Kwon, S. H. 1974. Selection for blast-resistant mutants in irradiated rice populations. See Ref. 6, pp. 103–15
40. Kwon, S. H., Oh, J. H. 1977. Selection for blast-resistant mutants in irradiated rice populations. See Ref. 1, pp. 131–45
41. Line, R. F., Konzak, C. F., Allan, R. E. 1974. Evaluating resistance to *Puccinia striiformis* in wheat. See Ref. 6, pp. 125–32
42. Little, R. 1971. An attempt to induce resistance to *Septoria nodorum* and *Puccinia graminis* in wheat using gamma rays, neutrons and EMS as mutagenic agents. See Ref. 8, pp. 139–49
43. Luig, N. H. 1978. Close association of two factors for avirulence in *Puccinia graminis tritici. Phytopathology* 68:936–37
44. McKenzie, R. I. H., Martens, J. W. 1974. Breeding for stem rust resistance in oats. See Ref. 6, pp. 45–48
45. Micke, A. 1974. Scope and aims of the co-ordinated research programme on induced mutations for disease resistance in crop plants. See Ref. 6, pp. 3–7
46. Moës, A. J. 1977. Le mutant 5455, orge resistant a l'oidium. See Ref. 1, pp. 343–46
47. Murray, M. J. 1969. Successful use of irradiation breeding to obtain *Verticillium*-resistant strains of peppermint, *Mentha piperita* L. *Induced Mutations in Plants,* pp. 345–71. Vienna: Int. At. Energy Agency. 748 pp.
48. Murty, B. R. 1974. Mutation breeding for resistance to downy mildew and ergot in *Pennisetum* and to *Ascochyta* in Chickpea. See Ref. 6, pp. 89–100
49. Nakai, H., Goto, M. 1977. Mutation breeding of rice for bacterial leaf-blight resistance. See Ref. 1, pp. 171–86
50. Nakajima, K. 1973. Induction of useful mutations of mulberry and roses by gamma rays. See Ref. 30, pp. 105–16
51. Nilan, R. A., Kleinhofs, A., Konzak, C. F. 1977. The role of induced mutation in supplementing natural genetic variability. *Ann. NY Acad. Sci.* 287:367–84
52. Notteghem, J. L. 1977. Etude de la resistance d'*Oryza sativa* A *Pyricularia oryzae* par application au parasite des techniques de mutageneses. See Ref. 1, pp. 121–30
53. Offutt, M. S., Riggs, R. D. 1970. Radiation-induced resistance to root-knot nematodes in Korean lespedeza. *Crop Sci.* 10:49–50
54. Padmanabhan, S. Y., Kaur, S., Rao, M. 1977. Induction of resistance to bacterial leaf-blight (*Xanthomonas oryzae*) disease in the high-yielding variety, *Vijaya* (IR 8 X T90). See Ref. 1, pp. 187–98
55. Padwal-Desai, S. R., Ghanekar, A. S., Sreenivasan, A. 1976. Studies on *Aspergillus flavus.* I. Factors influencing radiation resistance of non-germinating conidia. *Environ. Exp. Bot.* 16:45–51
56. Parodi, P. C., Nebreda, M. 1977. Reaccion a enfer-medades de seis genotipos de trigo (*Triticum* spp.) tratados con rayos gamma. See Ref. 1, pp. 375–83
57. Rastegar, M. F. 1976. Competitive ability of an induced mutant race of *Puccinia hordei* Otth. in mixtures. *Proc. 4th Eur. Mediterr. Cereal Rusts Conf.,* pp.

58–59. Zurich: Eur. Mediterr. Cereal Rusts Found. 175 pp.

58. Röbbelen, G. P., Abdel-Hafez, A. G., Reinhold, M. 1977. Use of mutants to study host/pathogen relations. See Ref. 1, pp. 359–74

59. Rowell, J. B., Loegering, W. Q., Powers, H. R. Jr. 1960. Mutation for pathogenicity in *Puccinia graminis* var. *tritici. Phytopathology* 50:653 (Abstr.)

60. Rowell, J. B., Loegering, W. Q., Powers, H. R. Jr. 1963. Genetic model for physiologic studies of mechanisms governing development of infection type in wheat stem rust. *Phytopathology* 53:932–37

61. Saccardo, F., Ramulu, S. K. 1977. Mutagenesis and breeding for disease resistance in *Capsicum.* See Ref. 1, pp. 275–80

62. Schwinghamer, E. A. 1957. Radiation-induced mutations in race 1 of the flax rust fungus. *Phytopathology* 47:31 (Abstr.)

63. Schwinghamer, E. A. 1959. The relation between radiation dose and the frequency of mutations for pathogenicity in *Melampsora lini. Phytopathology* 49:260–69

64. Shakoor, A., Ahsanul-Haq, M., Sadiq, M., Sarwar, G. 1977. Induction of resistance to yellow mosaic virus in mungbean through induced mutations. See Ref. 1, pp. 293–302

65. Sigurbjörnsson, B., Micke, A. 1969. Progress in mutation breeding. See Ref. 47, pp. 673–98

66. Sigurbjörnsson, B., Micke, A. 1973. List of varieties of vegetatively propagated plants developed by utilizing induced mutations. See Ref. 30, pp. 195–202

67. Simons, M. D. 1971. Modification of tolerance of oats to crown rust by mutation induced with ethyl methanesulfonate. *Phytopathology* 61:1064–67

68. Simons, M. D., Frey, K. J. 1977. Induced mutations for tolerance of oats to crown rust. See Ref. 1, pp. 499–512

69. Skorda, E. A. 1977. Stem and stripe rust resistance in wheat induced by gamma rays and thermal neutrons. See Ref. 1, pp. 385–92

70. Tanaka, S. 1969. Some useful mutations induced by gamma irradiation in rice. See Ref. 47, pp. 517–27

71. Tarasenko, N. D. 1977. Obtaining potato mutants of the variety Lorch resistant to *Synchytrium endobioticum.* (In Russian, with English summary) See Ref. 1, pp. 247–51

72. Teo, C., Baker, E. P. 1975. Mutagenic effects of ethyl methanesulphonate on the oat stem rust pathogen (*Puccinia graminis* f. sp. *avenae*). *Proc. Linn. Soc. NSW* 99:166–73

73. Todd, W. A., Green, R. J. Jr., Horner, C. E. 1977. Registration of Murray Mitchum peppermint. *Crop Sci.* 17:188

74. Tulmann Neto, A., Ando, A., Costa, A. S. 1977. Attempts to induce mutants resistant or tolerant to golden mosaic virus in dry beans (*Phaseolus vulgaris*). See Ref. 1, pp. 281–90

Ann. Rev. Phytopathol. 1979. 17:97–122
Copyright © 1979 by Annual Reviews Inc. All rights reserved

CALCIUM-RELATED ♦3703
PHYSIOLOGICAL DISORDERS
OF PLANTS

F. Bangerth

Universität Hohenheim (05700), 7000 Stuttgart 70, West Germany

INTRODUCTION

Many of the physiological disorders afflicting both storage organs, such as fruits, certain vegetables, roots, and young, enclosed leafy structures are related to the Ca content of the respective tissues. Improving their Ca content normally diminishes the occurrence of the respective disease. Shear (157) has listed more than 30 Ca-deficiency disorders and this list probably will be extended as research in this field proceeds. These disorders are believed to be due to an inefficient distribution of Ca rather than poor Ca uptake. This problem is illustrated by the observation that leaves contain considerably more Ca than storage organs or young enclosed tissues from the same plant (19, 23, 126). Poor Ca distribution within the plant can also explain both the frequent appearance of Ca-deficiency disorders even on Ca-rich soils and the failure of Ca fertilization, although often not done (123), to solve this problem.

Ca has received considerable attention in recent years not only because of its relationship to physiological disorders but also because of other desirable effects, particularly in fruits where it can reduce respiration (13, 26), delay ripening (155), extend storage life (155), increase firmness (13, 40) and vitamin C content (12), and reduce storage rot (155). The continued increase in interest in the effects of Ca on different physiological processes in plants as well as in its uptake and distribution within the plant is indicated by some international symposia on this subject. The reports of these symposia, together with some comprehensive reviews, cover a great deal of the relevant literature (28, 31, 33, 49, 52, 57, 70, 80, 108, 111, 157, 203).

0066-4286/79/0901-0097$01.00

Therefore, no attempt is made here to present a complete survey of the existing literature. Rather the aim is to concentrate on recent developments and on basic principles involved in uptake, distribution, and functions of Ca in relation to disease. It is hoped that a better understanding of these phenomena will help overcome present difficulties in reducing or preventing Ca-deficiency disorders.

CA UPTAKE

According to Fried & Shapiro (55), the calcium concentration in most soil solutions varies between 3.4 and 14 mM whereas at the root surface 0.1 to 1 mM appears to be adequate provided that the concentrations of other ions are not excessive (199). It would seem, therefore, that the majority of soils at least in the temperate zones provides enough calcium to meet the plant's demand. The relatively high Ca content of soils, when compared to the demand by most plant species, implies that the mass flow of calcium with the soil solution to the root surface is the predominant transport process in soil (17). In consequence, Ca movement to the roots appears to depend more on the transpiration rate of the plant than on root elongation and interception.

After reaching the root surface, Ca moves across the root cortex either by diffusion or more likely by displacement exchange in the free space. Before entering the stele and the xylem vessels, ions and water have to pass through the endodermis. The suberized Casparian strip of endodermal cells, however, effectively blocks transport through the apoplast and thereby forces water and ion transport to proceed in the symplast. Because symplasmic calcium transport is limited (see below), its transport into the stele and xylem occurs preferably at the root tip and temporarily at the site of branch root formation, where the development of the Casparian strip appears to lag behind endodermal cell division (44, 53, 90). Thus, it would appear that root growth and root branching are important in providing sites for Ca uptake where the Casparian strip is not fully developed. Consistent with this view is the finding that the Ca uptake rate is proportional more to the growth rate of roots than of shoots (145).

The question whether Ca uptake is an active energy-dependent process or not remains unanswered. Respiration inhibitors and low temperatures have been found to depress Ca uptake (75, 90), especially at low Ca concentrations in nutrient solution. However, not all species respond in this way (159, 176), and compared to most other nutrient ions Ca uptake is still considered to be mainly passive (68, 81).

Interactions with other ions can strongly influence Ca uptake. Non-specific cation competition (cation antagonism) notably by K, Mg, and

NH_4 can substantially depress Ca uptake (81, 201), but their depressive effect depends on their concentration in the soil solution. Stimulated Ca uptake by other ions such as NO_3 or PO_4 (synergism) has been shown; this is also a nonspecific process (77, 82). A great number of other ionic interactions may affect Ca uptake.

Phytohormones, especially abscisic acid (ABA) and cytokinins, can affect ion uptake and translocation to the shoot (79, 89). ABA decreases and cytokinins increase the resistance of the root to water uptake and they therefore influence ion uptake also. It seems, however, that these hormones act on the cytoplasmic transport pathway. In accord with this view is the fact that Ca fluxes are only marginally affected, particularly when compared with, say, K fluxes (39).

In the root, as in the shoot, Ca moves only in an upward direction and not to the root tip (115). The root tip, therefore, must be supplied continuously with Ca; otherwise its meristematic activity is disrupted and it dies (78). Therefore, to maintain adequate root growth a higher and more evenly distributed Ca concentration in the soil is necessary, more so perhaps than for the uptake and translocation processes themselves.

CA TRANSLOCATION AND DISTRIBUTION

As previously mentioned, the low Ca content of particular plant organs is in general not the result of insufficient Ca uptake, but rather a problem of calcium distribution. The mode of calcium translocation and distribution is therefore of paramount importance.

DOES LONG-DISTANCE TRANSPORT OCCUR VIA XYLEM OR PHLOEM?

Ca is considered to move preferentially in the xylem. In a few experiments, mainly with young apple trees, phloem transport was found to predominate (158, 165); however, because this contradicts almost all other such experiments, further verification is necessary before making any conclusions.

Evidence for a low phloem mobility of Ca is its rather low concentration in the phloem sap (84, 208), its lack of accumulation, above a girdle (118), and its lack of movement when applied to leaves (8, 30), and during leaf senescence (209). This, as well as other evidence, leads to the conclusion that Ca generally does not move via the phloem.

Nevertheless, phloem transport should not be ignored entirely; several authors (65, 137, 138, 178, 200) have interpreted their results to indicate that at least a fraction of the calcium moves in the phloem. However, none of the methods currently available for distinguishing between xylem and phloem transport seems free of criticism and some authors doubt that phloem-transported Ca contributes more than a small amount in comparison with the total Ca that is transported (85, 126, 193). The results that currently are available conflict and, therefore, do not permit any firm conclusions. This is not surprising if one considers the widely different species, methods, and environments used in the various experiments. Indeed, it has been shown that the Ca concentration in the phloem sap of different plant species can differ enormously (43, 137, 184) and that it can be altered by changing the nutrients in the solution in which the plants are grown (120, 200). Because minor improvements in the phloem mobility of Ca would be of considerable importance for the Ca nutrition of storage organs, further critical investigations along these lines should be encouraged.

Because xylem transport is considered to be the main route of Ca translocation, many experiments have been undertaken to investigate the factors that could affect this process. Unlike the monovalent cations, Ca does not move with the bulk solution (mass flow) but rather by a process of exchange with negatively charged molecular groups (pectins and lignins) in the xylem (20, 75, 158, 171). Ca translocation, therefore, lags behind the water or K movement. According to measurements and calculations, these exchange sites are located in an extended volume rather than only at the inner surface of the vessel walls (183) and their density changes during the growing season and development of the plant (54). In order to desorb Ca from its adsorption sites the transpiration stream must contain a sufficient amount of certain cations. Only Ca itself and Mg would be sufficiently effective in this respect (20, 75) judging from their concentration in the xylem sap and their exchange energy (position in the lyotropic series). Chelating agents that form uncharged or even negatively charged complexes with Ca can prevent this adsorption process and expedite Ca translocation (54, 76). Malic and citric acids would be the likely candidates for Ca sequestration in most plants because of their concentrations in xylem sap and the stability constants of their Ca complexes. Whereas Ferguson & Bollard (54) do not expect that more than a small amount of the Ca in the xylem of a young apple tree would be in a chelated form, Bradfield (24) concluded from measurements with an ion selective electrode that roughly 50% of the Ca was probably in a sequestered form.

During its passage through the xylem, part of the Ca is readily transported laterally and often permanently lost probably as Ca oxalate precipitate (20, 193, 195). In the bark of trees and the pedicels of apple fruits, particularly high amounts of immobilized Ca have been found (100, 166, 170, 172, 195). Different species and even cultivars do show considerable variation in their capacity to immobilize Ca, and this is claimed to affect the susceptibility of plants or plant organs to Ca deficiency (27, 59, 100,

104). Whether this immobilization is as permanent as is often considered seems doubtful, since dissolution of Ca-oxalate crystals has been observed in several tissues (59, 101). Quantification of this Ca remobilization, particularly in vegetative plant parts would be desirable, especially in view of recent experiments (170) which have shown, by X-ray diffraction analysis, that the immobilized Ca crystals in apple trees are not Ca oxalate. The true nature of the crystals has not yet been established.

A more temporary fixation of Ca, particularly in woody perennials, has been observed usually later in the growing season and in the basal part of the tree (170, 172, 195). Almost 20% of the Ca in an apple fruit can originate from these temporary Ca reservoirs laid down in the previous season (56, 170, 197). It seems reasonable to assume that this temporary fixation occurs in the xylem (170) and is due to an increase in exchange sites (54) and/or a decrease in the concentration of desorbing ions or chelating agents in the xylem.

SHORT-DISTANCE CA TRANSPORT

Apart from Ca uptake in the roots, nonvascular Ca transport is certainly significant in the Ca nutrition of certain storage organs such as potatoes or groundnuts that grow in the soil (21, 84). It is also important in the redistribution of Ca from pods to seeds in legumes (127, 138) and in the Ca distribution in leaf mesophyll (53, 192) and fruit parenchyma cells (53) where specialized vascular elements are not always present. Ca transport in the most apical part of the terminal bud (114) and lateral transport in stems (20) also belong to this transport category. Some of these short-distance transport processes are as important as long-distance Ca translocation. Examples of inadequate short-distant transport are the localized appearance of many Ca-deficiency disorders in storage organs and the uneven distribution of Ca in these tissues. Nevertheless, very little is known about this transport pathway. The most likely route for such Ca translocation seems to be the apoplast, and displacement exchange in the cell walls is probably as important as in xylem transport. Simple diffusion seems less likely because a more uniform distribution and a higher transport rate, e.g. from one side of an apple fruit to the other, would be expected (8). Ca transport in the symplast in large quantities is not very probable because for the same reason as phloem transport, the solubility of Ca in the cytoplasm is very low (146). Since, however, polar auxin transport is in some unknown way linked to Ca translocation (see below) and this kind of auxin transport occurs alternately in the cytoplasm and the cell wall (62), a cytoplasmic (not symplasmic) transport should not be excluded completely at least for certain tissues.

GENETIC VARIATION AND POSSIBLE REGULATORY MECHANISMS FOR CA UPTAKE AND TRANSLOCATION

The difference in Ca content between calcicole and calcifuge plants growing on high and low calcium soils respectively is well known (91) and is under genetic control (73). A characteristic of calcicole plants is their high soluble Ca concentration, whereas calcifuge plants have low soluble Ca contents due mainly to high concentrations of oxalic acid. Unfortunately, most of the pertinent experiments have not been done on cultivated plant species. Nevertheless cultivar differences in Ca uptake and distribution as well as in susceptibility to Ca-deficiency disorders have been reported for tomatoes (64), lettuce (37, 38), peanuts (21, 190), brussels sprouts (124), lupins (71), cabbage (131), and rootstocks and scions of several fruit species (45, 47, 83, 153). Utilization of this genetic variation for the breeding of cultivars with a higher tolerance to Ca-deficiency disorders seems promising, at least for some annual or biennial crops, whereas for woody perennials the opportunity to achieve this objective is less likely because of the long-term breeding programs involved. Mutation breeding, however, might overcome this difficulty (94).

Only a few experiments have been undertaken to study the biochemical and physiological mechanisms underlying genetic variation, although a better understanding would be valuable in several respects. Screening methods (37) as well as cultural procedures to improve the Ca status of susceptible plants and organs could probably be based on such an understanding.

Does the observed genetic variation imply that plants have developed mechanisms to regulate their Ca uptake and translocation? Based on their displacement exchange hypothesis, Bell & Biddulph (20) concluded that the different plant organs obtain their Ca according to their physiological demand rather than in proportion to their transpiration rate. Recent experiments have shown, however, that the demand of above-ground plant parts does not regulate either Ca uptake or translocation (129, 145, 176). Other observations, e.g. the absence of a selective and active uptake mechanism, the inability of Ca to move into the phloem, which can be controlled more easily, by the plant (167), the frequent lack of any relationship between CEC and Ca content of particular organs (126), and above all the occurrence of Ca deficiency in certain tissues on a plant otherwise well supplied with Ca, all indicate that a regulatory mechanism seems unlikely (cf 18).

On the other hand, it may be possible to exert some kind of control on Ca uptake perhaps via root growth (see above) by regulating the supply of respiratory substrates to the roots. Apart from this possibility of regulated uptake, the effect of a hormone (auxin)-directed Ca distribution (discussed later) could be to control Ca translocation at least in certain tissues.

An interesting hypothesis about the regulation of Ca distribution was introduced by Marschner and co-workers (126). These authors suggested that if there is any regulation of Ca uptake and translocation its function may be to prevent a possible surplus of Ca especially in fast growing storage organs, where high Ca concentrations might interfere with influx of assimilates.

EFFECTS OF GROWTH RATE ON CA IMPORT AND TISSUE SENSITIVITY

It is recognized that a high growth rate of susceptible tissue can contribute to the development of Ca deficiency disorders in fruits (9, 157, 198) as well as in leafy tissues (4, 132, 173). It is not clear, however, whether this happens because the Ca translocation into rapidly expanding storage tissues is reduced, and/or because an accelerated rate of cell division, cell expansion, and metabolism requires more Ca and therefore renders the tissue more susceptible to shortages of Ca.

According to Wiersum and Ansiaux's hypothesis (6, 198), a fast growing, low transpiring tissue gets more water via the phloem and less via the xylem and hence less Ca compared to a slow growing organ. There are few detailed experiments investigating this possibility, but they do show that with an accelerated growth rate a decrease in the Ca concentration can indeed be observed (125, 198). The same conclusion could be drawn from results showing that large fruits have less Ca (16, 139), because their larger size would be the outcome of an increased growth rate at some time during their development. The importance of growth rates on the resulting Ca content is further strengthened by the characteristic Ca uptake curves found e.g. for fruits (202). A rapid increase in Ca was generally noted during the earliest stage of fruit development, when growth rate is slow. At later stages, however, a remarkable reduction in the rate of Ca accumulation may be observed, when the fruit grows at a much faster rate. Some exceptions to this general rule are, for example, a continuous linear increase (177, 178), a rapid increase when the fruit approaches maturity (133), or even a decrease in Ca at that time (126, 202). Such results show that Ca uptake is not determined by growth rate alone. This could in some way be expected, because conditions that affect the growth rate of storage organs might also affect vegetative plant parts. This probably creates competition between vegetative and storage organs which could well have influenced Ca distribution. In order to study such phenomena, it is necessary to control selectively the growth of vegetative tissue and storage organs on intact plants, and this is difficult to accomplish experimentally. Thinning, for example, not only accelerates fruit growth rate, but also stimulates the growth of vegetative tissue. In the past these and other side effects have not always been taken into consideration.

It was established earlier that a high growth rate of a storage organ usually diminishes its Ca concentration. This, however, does not exclude additional factors which may increase the susceptibility of tissue to low levels of Ca but which should not be confused with differences in susceptibility due to genetic variation. To prove any enhanced susceptibility would be extremely difficult. It would be necessary to have both fast and slow growing tissues under the same environment and with approximately the same "active" Ca concentration during the developmental period under consideration. As far as I know, this demand has not been met in most of the reports where a higher tissue suceptibility to low Ca concentrations has been claimed.

WATER AND CA DISTRIBUTION

There is no proportional relationship between Ca and water influx into various organs (125). Nevertheless water is one of the decisive factors for Ca distribution in the above-ground plant parts (9, 121, 198).

When adequate Ca and Mg concentrations are present in the xylem, however, the relationship between Ca distribution and transpiration becomes closer (75). Several experimental results point to the crucial role water and transpiration play in supplying the various plant tissues with Ca. Ca deficiency is normally restricted to organs and tissues with a low transpiration rate but a high demand for assimilates. Shortage of water or an irregular water supply results in a reduced Ca translocation into these organs and a severe increase in Ca deficiency disorders such as bitter pit in apples (16), blossom-end rot in tomatoes (9, 61, 186), and tipburn in cabbage (92). In contrast, withholding water has little effect on Mg and almost none on K influx into these tissues with the result that the $Mg+K/Ca$ ratio is sometimes increased considerably (16, 86, 125). The most likely explanation for this would be that, because of competition for transpirational water between leaves and storage organs, xylem transport with its high Ca concentration to the storage organs is restricted. From this it seems logical that a selective increase in the transpiration rate of storage organs should increase their Ca content and reduce the $K+Mg/Ca$ ratio. This was indeed shown for tomatoes (198), potato tubers (85, 198), and pepper and bean fruits (125). A selective decrease in the transpiration rate, on the other hand, reduced the Ca content of several fruits (138, 198).

Diurnal changes in the water potential of the plant, created by opening and closure of stomata and by changes in the evaporative capacity of the atmosphere, were found to play a significant role in Ca transport insofar as during the night the growth rate of many storage organs is much higher than during daytime. The result was a considerable increase in xylem water and therefore Ca transport into these organs (191, 192). A somewhat different hypothesis for this daily fluctuation in Ca influx into certain storage organs suggests that it is root-pressure flow that pushes water and Ca through the xylem into these tissues during the night (25, 136, 173). Irrespective of the hypothesis, the conclusion would probably be the same: An increase in the diurnal transpiration amplitude created by dry days, nights with low evaporative capacity, and good water supply to the roots should increase the Ca supply to storage organs and weakly transpiring leaves. It also would be of interest to ascertain whether fruits show this diurnal change in Ca influx particularly since it is known that they also have higher growth rates during the night (22).

A nonselective increase in the transpiration of the whole plant improved the Ca content of storage organs (125), had no effect (178), or even reduced it (61, 198). Different experimental conditions may be responsible for these variable results. For example, in water culture, water stress due to relatively high evaporative capacities of the atmosphere seems less likely than in field soils, where a resulting competition for water between leaves and storage organs can lead to Ca shortages. With high transpiration rates and increase in water deficit of the leaves, an efflux of water and ^{45}Ca from fruits was observed (125, 151, 207). This may explain the decrease sometimes observed in total Ca in apple (202) and quince fruits (126).

This short review of experimental results demonstrates that water and the various water potentials of the different plant parts are some of the most decisive factors in Ca distribution in the plant. Often they are the real causes of the effects of other factors on Ca distribution such as light and temperature.

PHYTOHORMONES AND SYNTHETIC GROWTH REGULATORS: THEIR EFFECT ON CA TRANSLOCATION

Phytohormones and growth regulators have marked regulatory effects on almost all developmental processes of plants, and it seems only logical that they also participate in controlling mineral uptake and translocation (188). The relevant literature, however, is rather limited especially if one restricts the topic to the effect of these substances on the uptake and translocation of Ca.

Variable results on the Ca content of aboveground plant parts have been obtained after the application of several commercially available growth regulators (147, 153). With the exception of auxin transport inhibitors (see below), however, no generalization for Ca translocation in response to a particular treatment seems possible.

As far as endogenous phytohormones are concerned, three of them (gibberellins, cytokinins, and auxins) have distinctive effects on the Ca content of aerial plant organs. A number of investigations have shown that for a variety of plant species root or shoot application of GA_3 reduced the Ca content of aboveground plant parts by interference with Ca translocation (9, 92, 194, 206). Ca-deficiency disorders were observed on gibberellin-treated vegetative plant parts as well as on fruits. The observed difference in Ca translocation was not correlated with the GA_3-induced growth response (194). It would be interesting to study whether a reduction in endogenous gibberellin content, e.g. by the application of gibberellin synthesis inhibitors such as amo 1618 or phosphon D, would have a stimulating effect on Ca translocation.

Spraying young apple trees with two synthetic cytokinins (benzyladenine and kinetin) increased the movement of ^{45}Ca into mature leaves (158). Concerning the mechanism of this cytokinin effect, one can only speculate that senescence of these leaves was retarded (young leaves are stronger sinks for Ca) or that their transpiration rates were increased (74) which would have increased Ca influx. A very interesting effect was observed by LéJohn et al (93) in the fungus, *Achlya,* where cytokinins released Ca from a certain binding site in cell walls and subsequently stimulated Ca intake. Experiments with mung bean hypocotyls also indicate a higher Ca uptake in the presence of cytokinins (88). The higher Ca intake could probably be explained by a modified phosphorylation of membrane proteins caused by these hormones (144). It would be interesting to investigate the possibility that these effects of cytokinins can explain the increase of Ca movement into mature leaves following hormone application.

Spray applications of TIBA (2,3,5-triiodobenzoic acid), a well-known inhibitor of basipetal auxin transport, to apple trees or tomato plants increased bitter pit and blossom-end rot and reduced the Ca content of the fruits (9, 14, 133, 163, 164). Ca translocation to the shoots of pea and bean plants was likewise inhibited (194). More detailed experiments (114) revealed that TIBA interfered only with the Ca translocation into the tip of the apical bud and into the youngest leaves. From their results, especially from the almost parallel reduction in cation exchange capacity (CEC), these authors concluded that a change in growth rate caused by TIBA or IAA (indoleacetic acid) is responsible most probably for the observed change in Ca translocation. An alternative hypothesis (11) suggests that it is the basipetal auxin transport which forces Ca to be translocated acropetally. No obvious correlation could be found in these experiments between growth phenomena induced by these chemicals and the observed effect on Ca translocation. Lateral redistribution of Ca in IAA-, photo-, and geotropically stimulated plant parts are in agreement with this (7, 63). It also explains observations that only auxins with high basipetal transport rates (indole auxins and NAA (naphthalene acetic acid), but not phenoxy type auxins) seem to stimulate Ca translocation into parthenocarpic fruits (S. Bünger and F. Bangerth, in preparation). The physiological mechanism of this proposed reciprocal transport of auxins and Ca is as yet not known; neither is there any indication whether this transport occurs in the xylem, the phloem, the apoplast, or the symplast (however, see above).

Competition for Ca between leaves and fruits could probably be reduced by the application of growth retardants (130). Due to variable results and many side effects, however, this possibility seems to be limited (153; F. Bangerth, unpublished). From the results mentioned above, applications of IAA or NAA would be expected to increase Ca concentration in fruits, and a few reports confirm this hypothesis (153). Martin et al (117), however, were unable to reproduce these results. This is not surprising, since an unselective treatment of fruits and vegetative tissue would probably affect Ca transport into both. Selectively disturbing the shoot tips may be an alternative to auxin applications, because this could reduce or eliminate the strong competition of these organs for Ca (143), which is probably correlated with their high auxin production.

FUNCTIONS OF CA IN PLANTS

Several reviews concerning the functions of Ca in plants have been written in recent years (31, 32, 35, 78) and, therefore this survey is restricted to those more recent findings that relate the functions of Ca to calcium deficiency disorders. Essentially, there are four biological functions of Ca that may be associated with the development of the disorders: (a) effects on membranes, (b) effects on enzymes, (c) effects on cell walls, and (e) Ca-phytohormone interactions.

According to Williams (204), who has considered the biological functions of Ca in relation to its chemistry, the most important attribute of this cation seems to be its coordination ability which can provide reversible cross-links that can respond rapidly to changes in conditions. Williams defined the action of Ca as a function of its concentration, binding strength, structure (of Ca and ligand), and rate constants.

Determining the Ca concentration at its different sites of action is extremely difficult. Total calcium (which is determined most frequently) certainly exceeds considerably the free ion concentration; Ca in the cell wall and vacuole exceeds that in the cytosol by several orders of magnitude. Estimations of Ca in the cytosol vary between 10^{-3} and 10^{-7}M (106, 146, 204) depending on the tissue in which the Ca is measured. Suggestions for the possible functions of Ca that are based only on binding constants and that do not take into account this low concentration are of little value. Consequently, current ideas about the basic mechanisms of Ca functions in plant cells are based more on speculation than on fact.

CA AND MEMBRANE FUNCTIONS

Many Ca-deficiency disorders, particularly in fruits, are probably due to impaired membrane functions and a subsequent disruption in metabolic compartmentation (10). There are three pieces of evidence that show the importance of Ca for the stabilization of membranes: 1. Selective ion uptake is mediated by Ca and this has been found to be localized at the plas-malemma (46, 78). 2. Leakiness of cells is affected greatly by the concentration of Ca surrounding these cells (78). 3. Studies using electron microscopy have revealed the unique importance of Ca for the stabilization of membranes (67, 110). Even membranes that have become highly disorganized can be restored by the addition of Ca (112).

Leakiness of Ca-deficient tissue is a phenomenon that has been studied widely with quite different plant species and organs (66, 156, 160, 187). It was van Goor (185) who first showed that low-Ca tomato fruits had a higher tissue permeability than high-Ca tomato fruits. This was confirmed for apple fruits by other investigators using different methods (40, 149).

Further confirmation of these observations comes from electron microscope studies. Considerable differences in membrane integrity have been observed in apples and tomatoes either treated with Ca or deficient in Ca (9, 58, 107), with the endoplasmic reticulum, tonoplast, and plasmalemma deteriorating first.

Mitochondria, which can accumulate more Ca than the ground cytoplasm (78), have a far higher resistance to breakdown. Even though fruit tissue is by no means an ideal material for studies by electron microscopy the extent of membrane disintegration observed certainly reflects alterations in Ca-deficient membranes.

Several physiological events in Ca-deficient fruit cells might be related to impaired membrane function. For example the less efficient uptake of sorbitol into fruit cells, probably one of the reasons for watercore and internal breakdown development in apple (5, 52), can be improved by the application of Ca which obviously reduces both sorbitol leakage and the development of these disorders (13).

The well-established correlation between the Ca content of fruits and their respiration rate after harvest (13, 26, 40, 51, 174) can also be explained in part by an altered membrane permeability. In this case Ca may reduce endogenous substrate catabolism by limiting the diffusion of substrate from the vacuole to the respiratory enzymes in the cytoplasm (13). Poor storage ability (155) as well as the development of certain toxic volatiles (9) may be the result of such an accelerated respiration rate induced by the shortage of Ca. Ca deficiency can also be the cause of a higher respiration rate in leafy tissue (135).

A brown discoloration of the tissue occurs finally in most Ca-deficiency disorders and this could be brought about by increased leakage of phenolic precursors from the vacuole into the cytoplasm with subsequent oxidation by polyphenoloxidases (49). Polyphenols, however, also can damage enzymes, mitochondria, etc. Therefore they not only are a result of these disorders but also may be involved in the development of them.

The action of Ca on membranes can be seen as a continuous interaction with other ions. Some cations, depending on their concentrations, can replace Ca from its binding sites in the membranes. Only Mn and Sr, however, can displace Ca without causing a great increase in leakiness and loss of compartmentation (60, 160, 187). Consequently Sr sprays have been found to reduce blossom-end rot in tomatoes (9), internal breakdown in apples (205), and blackheart in celery (169). Lidster et al (99), however, were unable to demonstrate an effect of Sr on internal breakdown. In most plant tissues, however, only the concentrations of Mg, K, and H are such that they are potentially antagonistic to the effect of Ca (10). Their ability to stabilize membranes, however, is limited and after replacing Ca they can greatly increase permeability. This is particularly true for H (113) and because the concentration of this cation is increased at higher respiration rates, Ca, by reducing respiration, may prevent its own displacement. The relation Mg+K(+H)/Ca mentioned earlier is therefore not only significant in Ca uptake but also in regulating membrane permeability. Because Mg and K have little harmful effect on the development of the disorders if Ca concentrations in the tissue are high (40, 116, 139), Ca displacement in membranes seems significant only when the Ca concentration falls below a "critical" level.

Several investigators have concluded that membrane changes are also involved in Ca-dependent senescence processes (140, 155, 189). Some disorders in stored fruits like senescent breakdown of apples, which respond clearly to Ca applications, may belong to this category together with accelerated ripening and fruit senescence. Unlike other workers (155; F. Bangerth, unpublished) Bramlage et al (26) were unable to detect an advanced ripening of low-Ca fruits; there are no reports of investigations to find whether senescence has a role in other Ca-deficiency disorders, particularly in vegetative tissue.

Even though Ca is generally recognized as important in maintaining cell permeability and compartmentation, little is known about the biochemical reactions underlying these Ca-membrane interactions. By binding to certain membrane components, such as phospholipids (134), cholesterol (109), or specially arranged carboxyl groups (204), Ca could possibly modify the pore radius or trigger conformational changes within the membrane (35, 60, 204).

There are a number of possible ways to determine whether the effect of calcium on membrane permeability is or is not one of the most crucial factors in the development of calcium-deficiency disorders. One possibility is the more extensive use of Sr which can compensate for Ca in maintaining membrane structure almost completely (see above) though not in many other physiological processes (122). Other possibilities include the application of substances that interact with Ca-affected membrane permeability such as victorin (150, 162), poly-L-lysine (161), α-tomatine (148), and probably malformin (42).

CA AND CELL WALLS

Observations by both light and electron microscopy on tissues containing low concentrations of calcium have revealed considerable disorganization of the cell walls (32, 58, 110). It would appear that the maintenance of cell wall structure depends upon Ca cross-linkage particularly with the pectin components of the middle lamella. However, other electron microscopy investigations (9, 67, 112) have shown considerable disruption within calcium-deficient tissues whereas the integrity of the cell walls and middle lamella was maintained. It seems doubtful therefore that cell wall breakdown is a causal event in the development of Ca-related disorders.

Weakening of cell walls and middle lamellae in the partial absence of calcium, however, may occur and may have other undesirable effects. Fuller (58) related cell wall breakdown in apples to the fruit calcium content and to the appearance of mealiness; Simon (162) claimed that this weakness in cell walls may facilitate cell bursting which could, for example, lead to watercore and internal breakdown. Hypobaric storage reduces or prevents internal breakdown even when severe watercore had appeared (F. Bangerth, unpublished). This storage method reduces respiration and accumulation of toxic volatiles but is not likely to affect cell bursting. Cell bursting induced by Ca deficiency therefore seems less likely to be involved in the development of this disorder.

On the other hand, a decrease in cell wall rigidity due to a shortage of calcium may well be an important factor in tissue resistance to fungal attack, e.g. gloeosporium rot (155), as well as to fruit cracking (9, 162). These effects of Ca are probably much less specific than the interactions with membranes mentioned above. In fruit cracking for example, some divalent cations can substitute for Ca and trivalent cations are even more effective (9, 34).

EFFECTS OF CA ON ENZYMES

Ca binds specifically to some 70 different proteins (86). From the data available, however, it is difficult to decide when this binding is of physiological significance. Because the concentration of Ca in the ground cytoplasm is low, the affinity of most cytosol enzymes is certainly too low to be effective. Membrane-bound enzymes are therefore probably more important particularly since some membranes have a "Ca-rich" environment on at least one of their sides (e.g. plasmalemma, tonoplast, mitochondria, chloroplast).

Membrane-bound ATPases belong to this category. They are probably involved in the transport of ions and other solutes across membranes (72), and the activity of at least one or two of them is influenced by Ca (87). By affecting membrane transport they must certainly influence a number of metabolic processes; therefore a role in the development of Ca-deficiency disorders is possible. The same is true for a very interesting group of membrane-bound enzymes, the protein kinases. It was shown recently by Ralph et al (144) that Ca could inhibit protein kinase–mediated phosphorylation of membrane proteins which could have far-reaching effects on membrane transport and cell metabolism. The effect of Ca on these two groups of enzymes could account for many of the observed metabolic events in Ca-deficient tissue; this hypothesis should be tested. In addition, Ca has pronounced effects on the integrity of membranes (see above) and on the colloidal structure of the cytoplasm so that nonspecific effects on many membrane-bound and cytosol enzymes can be expected. The observed decrease in respiration after Ca application (see above) could also be explained by the interaction of Ca with ATPase and/or protein kinase–mediated changes in membrane permeability. However, it was shown for fruits and vegetative tissue that this effect of Ca on respiration could not be explained completely by membrane-regulated processes, and a direct effect on respiratory enzymes was suggested (51, 135). Pyruvate kinase is inhibited by Ca (175; unpublished results from our laboratory for apple fruit pyruvate kinase), and because it is a regulatory enzyme for a number of metabolic processes (179) it could well be involved in some "Ca effects," such as reduction in respiration or the formation of toxic volatiles. It is, however, questionable whether the low concentration of Ca in the cytosol can affect this enzyme in vivo.

α-Amylase, an enzyme with a strong requirement for Ca, is activated at a very low Ca concentration (10^{-7}M), but whether it is causally involved in Ca-deficiency disorders is not yet known.

CA-PHYTOHORMONE INTERACTIONS

The phytohormone ethylene (C_2H_4) like Ca deficiency, induces an enhancement of membrane permeability, respiration, ripening, and senescence (1, 168). In addition, other interactions seem to exist: C_2H_4 production is stimulated in Ca-deficient tissue (50) while the enzyme system for ethylene synthesis is obviously located in a cell wall–cell membrane complex (119) where the very first Ca-deficiency symptoms can be demonstrated by electron microscopy (67). This allows one to suppose that C_2H_4 might be intimately involved in the development of Ca-deficiency disorders. A direct causal involvement is unlikely, however, at least in some of these disorders, since hypobaric ventilation reduces but does not prevent bitter pit and blossom-end rot (F. Bangerth, unpublished). Yet the participation of C_2H_4 in disease development, particularly in the development of necrosis in the final stage of almost all Ca-deficiency disorders, cannot be totally excluded. This interaction between Ca and C_2H_4 is widely unexplored but nevertheless is a promising field for further research not only with respect to Ca-deficiency disorders but also with respect to ripening and senescence which can be retarded by Ca applications or C_2H_4 removal (1, 140, 155). Interactions between Ca and all the other phytohormones have also been reported in the literature (96). In the past, Ca-auxin relations in cell extension growth have been extensively studied (31) and it was reported recently (41) that these relations may be significant in the development of Ca-deficiency disorders. These authors suggested that one of the causal agents in the development of lettuce tipburn might be the presence of supraoptimal levels of the auxin IAA. If, during the auxin-mediated tissue extension phase the demand for Ca, especially for the maintenance of plant cell walls, exceeds supply, cells collapse and necrosis results. High levels of auxin can arise because the enzyme IAA oxidase can be inactivated, for example, by chlorogenic acid and it was shown by Collier et al (37) that of the lettuce cultivars tested, those most susceptible to tipburn had indeed the highest concentration of this polyphenol.

Other effects of Ca on auxin mediated processes, e.g. its influence on auxin binding (141) or on acidification of auxin-treated tissue (36), are less obviously related to Ca-deficiency disorders. Whether the Ca auxin-induced ultrastructural changes of plasma membranes (128) are significant in this respect remains to be seen, but in view of the unphysiologically high concentration of Ca needed for this effect (0.5 M) it appears doubtful (see also 188).

PREVENTING CA-DEFICIENCY DISORDERS

Due to the limited space available it is not possible to deal in detail with the many consequences and possibilities that could arise from the above-mentioned results on Ca uptake and distribution studies. This might not even be necessary, since a number of extensive reviews on the possibilities of preventing Ca-deficiency disorders have been written recently (33, 70, 81, 111, 139, 157, 182, 200, 203). Only a few examples will be given which should demonstrate that experimental results—which are often derived from solution or sand culture studies under controlled environments and which provide invaluable information about the fundamental processes— cannot always be extrapolated to the greenhouse or field situation.

It has been shown above and in experiments with photosynthetically active herbicides (48) that root growth and root activity are a necessary prerequisite for adequate Ca uptake and translocation and that on the other hand a high Ca concentration is needed to ensure good root growth (115). From this it would be expected that Ca fertilization and conditions that favor root activity may help to reduce Ca-deficiency disorders in the field. This, however, has not always been the case. Ca manuring, although beneficial in some instances (123, 181), has not shown consistent positive effects in many other field experiments. Sometimes liming even increased these Ca-deficiency disorders temporarily (15, 180) probably by increasing K in the soil solution. Heavier liming can therefore not always be recommended and Ca applications in more frequent and smaller doses by using Ca containing N and P fertilizers may be more advisable. Not only liming, but also active root growth and root activity might have detrimental effects in certain circumstances. For example, Jakobsen (77) has shown that a high root respiration rate might diminish Ca uptake and that vigorous root growth can enhance shoot growth, which accelerates the competition between leaves and storage organs for Ca. Consequently, cutting part of the roots of strong growing rootstocks is a measure used to reduce bitter pit in some German orchards. Moreover, summer pruning, which is known to reduce bitter pit, also reduces vegetative growth probably via diminished root growth (15, 142, 153).

Another example where results obtained in solution or sand culture differ widely from the results of field experiments is with K nutrition. It has often been shown that this cation reduces Ca uptake and Ca concentrations in aerial plant parts (2, 3, 105) whereas the K concentration e.g. in fruits can be increased considerably (98, 105). K also changes Mg distribution such that a larger part of the Mg is diverted into fruits or other storage organs (3, 105). Furthermore, K can accelerate phloem transport (120), which could reduce the Ca content in storage organs by bringing about (*a*) a larger supply of roots with respiratory substrates, which could reduce Ca uptake (77) and (*b*) a stimulated phloem transport and hence reduced xylem and Ca transport into these organs. On the whole, this should drastically increase the K+Mg/Ca ratio with its already mentioned unfavorable effects. Indeed Ca-deficiency disorders have been induced by increased K-applications in such experiments (29, 98). Nevertheless, reports about an increase in physiological disorders following the application of higher rates of K fertilizer in field or orchard experiments have been rather scarce and in many instances even absent when fruit size was not affected (15, 116, 139).

Other decisive factors that have been shown experimentally to affect these disorders have produced contradictory results in the field. Increased transpiration of the whole plant for example, was said to decrease the Ca content of storage organs, and consequently shelterbelts have been suggested as possible means of reducing wind speed and transpiration (200). It was demonstrated for apples (97), however, that increased wind speed reduced bitter pit and increased Ca content of fruit, probably by reducing vegetative growth and the competition for Ca. Although drought periods certainly aggravate Ca-deficiency disorders in controlled environments as well as in the field, irrigation has not always given beneficial results (15, 69). It was shown for apple fruits that the beneficial effect of irrigation or rain may be offset by excessive fruit growth which could even reduce the Ca concentration especially when following a drought period (139). Overhead sprinkling during the night to increase the diurnal transpiration amplitude (see above) should be tested as an alternative.

Even an accelerated growth rate is not unequivocally linked to a reduced Ca content and a higher incidence of Ca-deficiency disorders. Examples where special treatments lead to bigger fruits that contain more Ca and have less bitter pit have been mentioned in the literature (16, 139), and Link (102, 103; see also 95) found weak positive correlations between fruit size and Ca deficiency in the orchard when he compared fruits within the same tree. He claimed that Ca deficiency in apples is correlated much more closely with the cropping level of the tree than with fruit size.

This list of contradictory results could be extended. These few examples demonstrate that results obtained from controlled environment, sand, or solution culture experiments cannot be extrapolated to the orchard or field situation. Only when these results have been verified over a number of seasons, under different environments, and with different cultivars can they become the basis for more general recommendations.

The number of possible interactions that can affect Ca uptake and distribution is so great that in the near future we are unlikely to see the development of cultural practices that will completely eliminate Ca deficiency, without a direct application of Ca to the susceptible organ.

Methods should, therefore, be developed or improved for direct Ca treatments as has been done successfully for apple fruits (154, 180). Such studies should parallel the continuing investigations into the basic principles of plant Ca nutrition.

CONCLUSION

Considering the large number of factors that may affect Ca uptake, distribution, and function and the bewildering complexity of possible interactions in the field, the task of raising the Ca concentration in certain organs seems quite difficult. Yet many of the factors known to affect Ca uptake and distribution are possibly of academic interest rather than of special importance in affecting the final Ca content of a plant organ grown in the field. It therefore seems necessary to set up systematic field experiments to define which factors are decisive for Ca uptake and distribution. Such experiments are certainly far more laborious and expensive than investigations in controlled environments but they are necessary for a more efficient utilization of the already accumulated information of the basic mechanisms of Ca uptake, distribution, and function.

In addition, basic studies on new, still controversial developments such as hormone–Ca interactions should be encouraged to see whether they might provide additional or alternative ways of diminishing Ca-deficiency disorders.

ACKNOWLEDGMENT

The author expresses his appreciation to Dr. G. F. Collier, NVRS, Wellesbourne, Great Britain, for reading the English manuscript and for helpful suggestions. I am also grateful to many authors for access to manuscripts prior to publication.

Literature Cited

1. Abeles, F. B. 1973. *Ethylene in Plant Biology.* New York: Academic. 302 pp.
2. Addiscott, T. M. 1974. Potassium and the absorption of calcium and magnesium by potato plants from soil. *J. Sci. Food Agric.* 25:1165–72
3. Addiscott, T. M. 1974. Potassium and the distribution of calcium and magnesium in potato plants. *J. Sci. Food Agric.* 25:1173–83
4. Algera, L. 1968. Topple disease of tulips. *Phytopathol Z.* 62:251–61
5. Amezquita-Garcia, R. 1972. *Control of internal breakdown in Jonathan apple fruits by preharvest regulation of sorbitol content.* PhD thesis. Mich. State Univ., East Lansing
6. Ansiaux, J. R. 1959. La composition minérale des fruits et la voie de transport des ions alimentaires vers ceux-ci. *Ann. Physiol. Veget. Univ. Bruxelles* 4:53–88
7. Arslan-Çerim, N. 1966. The redistribution of radioactivity in geotropically stimulated hypocotyls of *Helianthus annuus* pretreated with radioactive calcium. *J. Exp. Bot.* 17:236–40
8. Bangerth, F. 1969. Untersuchungen zur Ursache der Entstehung der Stippigkeit bei Apfelfrüchten und Möglichkeiten

zur Verhinderung. *Angew. Bot.* 42:240–62

9. Bangerth, F. 1973. Investigations upon Ca related physiological disorders. *Phytopathol. Z.* 77:20–37

10. Bangerth, F. 1974. Second discussion meeting on bitter pit in apples. *Acta Hortic.* 45:43–52

11. Bangerth, F. 1976. A role for auxin and auxin transport inhibitors on the Ca content of artificially induced parthenocarpic fruits. *Physiol. Plant.* 37:191–94

12. Bangerth, F. 1976. Beziehungen zwischen dem Ca-Gehalt bzw. der Ca-Versorgung von Apfel-, Birnen- und Tomaten-früchten und ihrem Ascorbinsäuregehalt *Qual. Plant.* 26:341–48

13. Bangerth, F., Dilley, D. R., Dewey, D. H. 1972. Effect of postharvest calcium treatments on internal breakdown and respiration of apple fruits. *J. Am. Soc. Hortic. Sci.* 97:679–82

14. Bangerth, F., Firuzeh, P. 1971. Der Einfluß von 2,3,5-Trijodbenzoesäure (Tiba) auf den Mineralstoffgehalt und die Stippigkeit von "Boskoop" Früchten. *Z. Pflanzenkr.* 78:93–97

15. Bangerth, F., Link, H. 1972. Möglichkeiten der Entstehung und Bekämpfung von Stippigkeit und Lentizellenflecken I und II. *Der Erwerbsobstbau* 14:113–16, 138–40

16. Bangerth, F., Mostafawi, M. 1969. Einfluß der Wasserversorgung und des Fruchtgewichtes auf den Mineralstoffgehalt und die Stippigkeit von Apfelfrüchten. *Der Erwerbsobstbau* 11:101–4

17. Barber, S. A., Ozanne, P. G. 1970. Autoradiographic evidence for the differential effect of four plant species in altering the calcium content of the rhizosphere soil. *Soil Sci. Soc. Am. Proc.* 34:635–37

18. Barta, A. L. 1977. Uptake and transport of calcium and phosphorus in *Lolium perenne* in response to N supplied to halves of a divided root system. *Physiol. Plant.* 39:211–14

19. Baumeister, W. 1958. In *Encyclopedia of Plant Physiology,* ed. W. Ruhland, IV:5–36. Berlin: Springer. 1210 pp.

20. Bell, C. W., Biddulph, O. 1963. Translocation of calcium. Exchange versus mass flow. *Plant Physiol.* 38:610–14

21. Beringer, H., Taha, M. A. 1976. ^{45}Ca absorption by two peanut cultivars of groundnut. *Exp. Agric.* 12:1–7

22. Bollard, E. G. 1970. In *The Biochemistry of Fruits and Their Products,* ed. A. C. Hulme, 1:387–425. London & New York: Academic. 620 pp.

23. Bonhier, D., L'Ecluse, R. 1962. Observations sur le Bitter-Pit des pommes. *Acad. Agric. Fr. CR Seances Acad. Agric. Fr.* 48:817–20

24. Bradfield, E. G. 1975. Calcium complexes in the xylem sap of apple shoots. *Plant Soil* 44:495–99

25. Bradfield, E. G., Guttridge, G. G. 1977. Uptake and movement of calcium in plants. *Rept. Long Ashton Res. Stn. 1977,* pp. 48–50

26. Bramlage, W. J., Drake, M., Baker, J. H. 1974. Relationships of calcium content to respiration and postharvest condition of apples. *J. Am. Soc. Hortic. Sci.* 99:376–78

27. Brumagen, D. M., Hiatt, A. J. 1966. The relationship of oxalic acid to the translocation and utilization of calcium in *Nicotiana tabacum. Plant Soil* 24:239–49

28. Bünemann, G. 1972. *Annotated Bibliography on Bitter Pit of Apples,* Vol. 2. Berlin: Bibliogr. Reihe Univ. 170 pp.

29. Bünemann, G., Lüdders, P. 1975. Die Wirkung jahreszeitlich unterschiedlicher Kaliumverfügbarkeit auf Apfelbäume. VI. Einfluß auf Fruchtkrankheiten. *Gartenbauwissenschaft* 40: 208–14

30. Bukovac, M. J., Wittwer, S. H. 1957. Absorption and mobility of foliar applied nutrients. *Plant Physiol.* 32: 428–35

31. Burström, H. 1968. Calcium and plant growth. *Biol. Rev.* 43:287–316

32. Bussler, W. 1963. Die Entwicklung von Calcium-Mangelsymptomen. *Z. Pflanzenernaehr. Bodenkd.* 100:53–58

33. Carolus, R. 1975. Calcium relationships in vegetable nutrition and quality. *Commun. Soil Sci. Plant Anal.* 6:285–98

34. Christensen, J. V. 1972. *Revner i kirsebaer. Nogle Salte og Kemikaliers Virkning Pa Revnetilbojeligheden,* pp. 1–12. Oslo: Sartrykk av Frukt av Baer, Landbruksforlage 1

35. Christiansen, M. N., Foy, C. D. 1979. Fate and function of calcium in tissue. *Commun. Soil Sci. Plant Anal.* In press

36. Cohen, J. D., Nadler, K. D. 1976. Calcium requirement for indoleacetic acid-induced acidification by *Avena* coleoptiles. *Plant Physiol.* 57:347–50

37. Collier, G. F., Huntington, V. C., Cox, E. F. 1979. A possible role for chlorogenic acid in calcium related disorders of vegetable crops. *Commun. Soil Sci. Plant Anal.* In press

38. Collier, G. F., Scaife, M. A., Huntington, V. C. 1976. Nutritional aspects of

physiological disorders. *Wellesbourne, Natl. Veg. Res. Rept.,* pp. 43–44

39. Collins, J. C., Kerrigan, A. P. 1973. Hormonal control of ion movements in the plant root? In *Ion Transport in Plants,* ed. W. P. Anderson, pp. 589–93. London: Academic. 663 pp.

40. Cooper, T., Bangerth, F. 1976. The effect of Ca and Mg treatments on the physiology, chemical composition and bitter pit development of "Cox's Orange" apples. *Sci. Hortic.* 5:49–57

41. Crisp, P., Collier, G. F., Thomas, T. H. 1976. The effect of boron on tipburn and auxin activity in lettuce. *Sci. Hortic.* 5:215–26

42. Curtis, R. W., John, W. W. 1975. Effect of divalent cations on malformin-induced efflux and biological activity. *Plant Cell Physiol.* 16:835–44

43. Dixon, A. F. G. 1975. Aphids and translocation. In *Encyclopedia of Plant Physiology,* 1:154–70. Berlin: Springer. 535 pp.

44. Dumbroff, E. B., Pierson, D. R. 1971. Probable sites for passive movement of ions across the endodermis. *Can. J. Bot.* 49:35–38

45. Eaton, G. W., Robinson, M. A. 1977. Interstock effects upon apple leaf and fruit mineral content. *Can. J. Plant Sci.* 57:227–34

46. Epstein, E. 1961. The essential role of calcium in selective cation transport by plant cells. *Plant Physiol.* 36:437–44

47. Epstein, E. 1972. Mineral nutrition of plants: principles and perspectives. New York: Academic. 412 pp.

48. Faust, M., Korcak, R. 1978. Effect of herbicides on calcium uptake by apple seedlings. *Proc. Int. Coll. Plant Anal. Fert. Probl.,* 8th, Auckland, NZ

49. Faust, M., Shear, C. B. 1968. Corking disorders of apples: A physiological and biochemical review. *Bot. Rev.* 34:441–69

50. Faust, M., Shear, C. B. 1969. Biochemical changes during the development of cork spot of apples. *Qual. Plant. Mater. Veg.* 19:255–65

51. Faust, M., Shear, C. B. 1972. The effect of calcium on respiration of apples. *J. Am. Soc. Hortic. Sci.* 97:437–39

52. Faust, M., Shear, C. B., Williams, M. W. 1969. Disorders of carbohydrate metabolism of apples. (Watercore, internal breakdown, low temperature and carbon dioxide injuries.) *Bot. Rev.* 35:168–94

53. Ferguson, I. B. 1979. The movement of calcium in non-vascular tissue of plants. *Commun. Soil. Sci. Plant Anal.* In press

54. Ferguson, I. B., Bollard, E. G. 1976. The movement of calcium in woody stems. *Ann. Bot.* 40:1057–65

55. Fried, M., Shapiro, R. E. 1961. Soil-plant relationships in ion uptake. *Ann. Rev. Plant Physiol.* 12:91–112

56. Führ, F., Wieneke, J. 1974. Secondary translocation of ^{45}Ca to the fruit of apple trees the year after dormancy. In *Mechanisms of Regulation of Plant Growth,* ed. R. L. Bieleski, A. R. Ferguson, M. M. Cresswell, Bull. 12, pp. 171–75. Wellington: R. Soc. NZ

57. Fukuda, H. 1976. Physiological mechanism and classification of spotting disorders of apple fruits. *Agric. Hortic.* 51:1221–24 (In Japanese)

58. Fuller, M. M. 1976. The ultrastructure of the outer tissues of cold stored apple fruits of high and low calcium content in relation to cell breakdown. *Ann. Appl. Biol.* 83:299–304

59. Gallaher, R. N. 1975. The occurrence of calcium in plant tissue as crystals of calcium oxalate. *Commun. Soil. Sci. Plant Anal.* 6:315–30

60. Garrard, L. A., Humphreys, T. E. 1967. The effect of divalent cations on the leakage of sucrose from corn scutellum slices. *Phytochemistry* 6:1085–95

61. Gerard, C. J., Hipp, B. W. 1968. Blossom-end rot of "Chico" and "Chico grande" tomatoes. *Proc. Am. Soc. Hortic. Sci.* 93:521–31

62. Goldsmith, M. H. M. 1977. The polar transport of auxin. *Ann. Rev. Plant Physiol.* 28:439–78

63. Goswami, K. K. A., Audus, L. J. 1976. Distribution of calcium, potassium and phosphorus in *Helianthus annuus* hypocotyls and *Zea mays* coleoptiles in relation to tropic stimuli and curvatures. *Ann. Bot.* 40:49–64

64. Greenleaf, W. H., Adams, F. 1969. Genetic control of blossom-end rot disease in tomatoes through calcium metabolism. *J. Am. Soc. Hortic. Sci.* 94:248–50

65. Guardiola, J. L., Sutcliffe, J. F. 1972. Transport of materials from the cotyledons during germination of seeds of the garden pea (*Pisum sativum*). *J. Exp. Bot.* 23:322–37

66. Hawker, J. S., Marschner, H., Downton, W. J. S. 1974. Effects of sodium and potassium on starch synthesis in leaves. *Aust. J. Plant Physiol.* 1:491–501

67. Hecht-Buchholz, C. 1979. Calcium deficiency and plant ultrastructure. *Commun. Soil. Sci. Plant Anal.* In press

68. Higinbotham, N. 1973. The mineral absorption process in plants. *Bot. Rev.* 39:15–69

69. Hilkenbäumer, F. 1971. Einfluß zusätzlicher Regengaben auf die vegetative und generative Leistung verschiedener Apfelsorten im Vollertragsstadium. *Erwerbsobstbau* 13:57–59

70. Hilkenbäumer, F., Naumann, G., eds. 1976. Second discussion meeting on bitter pit in apples. *Acta Hortic.* 45:9–75

71. Hocking, P. J., Pate, J. S. 1977. Mobilization of minerals to developing seeds of legumes. *Ann. Bot.* 41:1259–78

72. Hodges, T. K. 1976. ATPases associated with membranes of plant cells, See Ref. 43, II (Part A):260–83

73. Horak, O., Kienzel, H. 1971. Typen des Mineralstoffwechsels bei den höheren Pflanzen *Oesterr. Bot. Z.* 119:475–95

74. Hsiao, T. C. 1973. Plant responses to water stress. *Ann. Rev. Plant Physiol.* 24:519–70

75. Isermann, K. 1970. Der Einfluß von Adsorptionsvorgängen im Xylem auf die Calcium-Verteilung in der höheren Pflanze. *Z. Pflanzenernaehr. Bodenkd.* 126:191–203

76. Isermann, K. 1978. Einfluß von Chelatoren auf die Calcium-Verlagerung im Sproß höherer Pflanzen. *Z. Pflanzenernaehr. Bodenkd.* 141:285–98

77. Jakobsen, S. T. 1979. Interaction between phosphate and calcium in nutrient uptake by plant roots. *Commun. Soil Sci. Plant Anal.* In press

78. Jones, R. G. W., Lunt, O. R. 1967. The function of calcium in plants. *Bot. Rev.* 33:407–26

79. Karmoker, J. L., van Steveninck, R. F. M. 1978. Stimulation of volume flow and ion flux by abscisic acid in excised root systems of *Phaseolus vulgaris* L. cv. redland pioneer. *Planta* 141:37–43

80. Kawasaki, T., Moritsugu, M. 1979. A characteristic symptom of calcium deficiency in maize and sorghum. *Commun. Soil Sci. Plant Anal.* In press

81. Kirkby, E. A. 1979. Maximizing calcium uptake by plants. *Commun. Soil Sci. Plant Anal.* In press

82. Kirkby, E. A., Knight, A. H. 1977. Influence of the level of nitrate nutrition on ion uptake and assimilation, organic acid accumulation, and cation-anion balance in whole tomato plants. *Plant Physiol.* 60:349–53

83. Köksal, A. J. 1973. Wechselwirkungen zwischen Sorten, Unterlagen und Zwischenveredlungen beim Apfel. *Gartenbauwissenschaft* 38:287–310

84. Kraus, A., Marschner, H. 1973. Langstreckentransport von Calcium in Stolonen von Kartoffelpflanzen. *Z. Pflanzenernaehr. Bodenkd.* 136:229–40

85. Kraus, A., Marschner, H. 1974. Einfluß der Tag/Nacht-Periodik auf Knollengewicht und Ca-Verlagerung in Stolonen von Kartoffelpflanzen. *Z. Pflanzenernaehr. Bodenkd.* 137:116–23

86. Kretsinger, R. H. 1976. Calcium-binding proteins. *Ann. Rev. Biochem.* 45:239–66

87. Kylin, A., Kähr, M. 1973. The effect of magnesium and calcium ions on adenosine triphosphatases from wheat and oat roots at different pH. *Physiol. Plant.* 28:452–57

88. Lau, O.-L., Yang, S. F. 1975. Interaction of kinetin and calcium in relation to their effect on stimulation of ethylene production. *Plant Physiol.* 55:738–40

89. Läuchli, A. 1972. Translocation of inorganic solutes. *Ann. Rev. Plant Physiol.* 23:197–218

90. Läuchli, A. 1976. Apoplasmic transport in tissues. See Ref. 43, II (Part B):3–34

91. Läuchli, A. 1976. Genotypic variation in transport. See Ref. 43, II(Part B):372–93

92. Leh, O. H. 1963. Untersuchungen über die Wirkung von Gibberellin auf Entwicklung, Calcium-Aufnahme und Calcium-Transport einiger Pflanzen. *Phytopathol. Z.* 49:71–83

93. LéJohn, H. B., Cameron, L. E., Stevenson, R. M., Meuser, R. U. 1974. Influence of cytokinins and sulfhydryl group-reacting agents on calcium transport in fungi. *J. Biol. Chem.* 249: 4016–20

94. Lennard, J., Lacey, C. N. D. 1977. Clonal differences within bramley fruit. *Long Ashton Ann. Rept.,* p. 20

95. Lenz, F. 1979. Effect of fruit load on calcium distribution in citrus. *Commun. Soil Sci. Plant Anal.* In press

96. Leopold, A. C., Poovaiah, B. W., dela Fuente, R. K. 1974. Regulation on growth with inorganic solutes. In *Plant Growth Substances 1973,* pp. 780–88. Tokyo: Hirokawa. 1242 pp.

97. Lewis, T. L., Martin, D., Cerny, J., Ratkowsky, D. A. 1977. The effects of a sheltered environment on the mineral element composition of Merton Worcester apple fruits and leaves and on the incidence of bitter pit at harvest. *J. Hortic. Sci.* 52:401–7

98. Lewis, T. L., Martin, D., Cerny, J., Ratkowsky, D. A. 1977. The effects of increasing the supply of nitrogen, phosphorus, calcium and potassium to the roots of Merton Worcester apple trees on leaf and fruit composition and on the incidence of bitter pit at harvest. *J. Hortic. Sci.* 52:409–19

99. Lidster, P. D., Porritt, S. W., Eaton, G. W. 1978. Effects of spray applications of boron, strontium and calcium on breakdown development in Spartan apples. *Can. J. Plant Sci.* 58:283–85

100. Liegel, W. 1970. Calciumoxalat-Abscheidungen in Fruchtstielen einiger Apfelvarietäten. *Angew. Bot.* 44:223–32

101. Liegel, W., Buchloh, G. 1976. Ein Beitrag zur Remobilisierung von Ca im Fruchtparenchym zweier Apfelvarietäten. *Gartenbauwissenschaft* 41:15–18

102. Link, H. 1973. Symp. on growth regulators in fruit production. *Acta Hortic.* 34:115–18

103. Link, H. 1978. Gegenwärtiger Stand der Fruchtausdünnung. *Erwerbsobstbau* 20:88–91

104. Lötsch, B., Kienzel, H. 1971. Zum Calciumbedarf von Oxalatpflanzen. *Biochem. Physiol. Pflanzen* 162:209–19

105. Lüdders, P., Bünemann, G., Ortmann, U. 1975. Die Wirkung jahreszeitlich unterschiedlicher Kaliumverfügbarkeit auf Apfelbäume V. Einfluß auf den Mineralstoffgehalt der Früchte. *Gartenbauwissenschaft* 40:151–59

106. Macklon, A. E. S. 1975. Cortical cell fluxes and transport to the stele in excised root segments of *Allium cepa* L. II. Calcium. *Planta* 122:131–41

107. Mahanty, H. K., Fineran, B. A. 1975. The effects of calcium on the ultrastructure of Cox's Orange apples with reference to bitter pit disorder. *Aust. J. Bot.* 23:55–65

108. Malavolta, E., Camargo, P. R., da Cruz, C., da Cruz, V. F. 1975. Calcium and its relation to blossom-end rot in tomato. *Commun. Soil Sci. Plant Anal.* 6:273–84

109. Marfey, P., Chessin, H. 1974. Cholesterol: Complexes with metallic salts. *Biochim. Biophys. Acta* 337:136–44

110. Marinos, N. G. 1962. Studies on submicroscopic aspects of mineral deficiencies. I. Calcium deficiency in the shoot apex of Barley. *Am. J. Bot.* 49:834–41

111. Marschner, H. 1974. Calcium nutrition of higher plants *Neth. J. Agric. Sci.* 22:275–82

112. Marschner, H., Günther, J. 1964. Ionenaufnahme und Zellstruktur bei Gerstenwurzeln in Abhängigkeit von der Calciumversorgung. *Z. Pflanzenernaehr. Bodenkd.* 107:118–36

113. Marschner, H., Handley, R., Overstreet, R. 1966. Potassium loss and changes in fine structure of corn root tips induced by H$^+$ ion. *Plant Physiol.* 41:1725–35

114. Marschner, H., Ossenberg-Neuhaus, H. 1977. Wirkung von 2, 3, 5-Trijodbenzoesäure (TIBA) auf den Calcium-transport und die Kationenaustauschkapazität in Sonnenblumen. *Z. Pflanzenphysiol.* 85:29–44

115. Marschner, H., Richter, C. 1973. Akkumulation und Translokation von K$^+$, N$^+$ und Ca^{2+} bei Angebot zu einzelnen Wurzelzonen von Maiskeimpflanzen *Z. Pflanzenernaehr. Bodenkd.* 135:1–15

116. Martin, D., Lewis, T. L., Cerny, J., Ratkowsky, D. A. 1975. The predominant role of calcium as an indicator in storage disorders in Cleopatra apples. *J. Hortic. Sci.* 50:117 55

117. Martin, D., Lewis, T. L., Cerny, J., Ratkowsky, D. A. 1976. The effect of tree sprays of calcium, boron, zinc, and naphthaleneacetic acid. *Aust. J. Agric. Res.* 27:391–98

118. Mason, T. G., Maskell, E. J. 1931. Further studies on transport in the cotton plant. *Ann. Bot.* 45:125–73

119. Mattoo, A. K., Lieberman, M. 1977. Localization of the ethylene-synthesizing system in apple tissue. *Plant Physiol.* 60:794–99

120. Mengel, K., Haeder, H.-E. 1977. Effect of potassium supply on the rate of phloem sap exudation and the composition of phloem sap of *Ricinus communis. Plant Physiol.* 59:282–84

121. Michael, G., Marschner, H. 1962. Einfluß unterschiedlicher Luftfeuchtigkeit und Transpiration auf Mineralstoffaufnahme und -Verteilung. *Z. Pflanzenernaehr. Bodenkd.* 96:200–12

122. Michael, G., Shilling, G. 1960. Strontium in der höheren Pflanze. *Z. Pflanzenernaehr. Bodenkd.* 91:147–58

123. Millaway, R. M., Wiersholm, L. 1979. Calcium and metabolic disorders. *Commun. Soil Sci. Plant Anal.* In press

124. Millikan, C. R., Hanger, B. C. 1966. Calcium nutrition in relation to the occurrence of internal browning in brussels sprouts. *Aust. J. Agric. Res.* 17:863–74

125. Mix, G. P., Marschner, H. 1976. Calciumgehalte in Früchten von Paprika, Bohne, Quitte und Hagebutte im Verlauf des Fruchtwachstums. *Z. Pflanzenernaehr. Bodenkd.* 139:537–49

126. Mix, G. P., Marschner, H. 1976. Einfluß exogener und endogener Faktoren auf den Calciumgehalt von Paprika- und Bohnenfrüchten. *Z. Pflanzenernaehr. Bodenkd.* 139:551–63

127. Mix, G. P., Marschner, H. 1976. Calcium-Umlagerung in Bohnenfrüchten

während des Samenwachstums. *Z. Pflanzenphysiol.* 80:354–66

128. Morré, D. J., Bracker, C. E. 1976. Ultrastructural alteration of plant plasma membranes induced by auxin and calcium ions. *Plant Physiol.* 58:544–47

129. Mostafa, M. A. E., Ulrich, A. 1973. Calcium uptake by sugar beets relative to concentration and activity of calcium. *Soil Sci.* 116:432–36

130. Naumann, W. D. 1971. Calciumverteilung in Apfelbäumen and Entwicklung von Stippigkeit unter dem Einfluβ von Wachstumsregulatoren. *Gartenbauwissenschaft* 36:63–69

131. Nieuwhof, M. 1960. Internal tipburn in white cabbage. I. Variety trials. *Euphytica* 9:203–8

132. Nieuwhof, M., Garresen, F., Wiering, D. 1960. Internal tipburne in white cabbage. II. The effect of some environmental factors. *Euphytica* 9:275–80

133. Oberly, G. H. 1973. Effect of 2, 3, 5-triiodobenzoic acid on bitter pit and calcium accumulation in "Northern spy" apples. *J. Am. Soc. Hortic. Sci.* 98:269–71

134. Oursel, A., Lamant, A., Salsac, L., Mazliak, P. 1973. Etude comparee des lipides et de la fixation passive du calcium dans les racines et les fractions subcellulaires du *Lupinus luteus* et de la *Vicia faba. Phytochemistry* 12:1865–74

135. Pal, R. N., Rai, V. K., Laloraya, M. M. 1973. Calcium in relation to respiratory activity of peanut and linseed plants. *Biochem. Physiol. Pflan.* 164:258–65

136. Palzkill, D. A., Tibbitts, T. W. 1977. Evidence that root pressure flow is required for calcium transport to head leaves of cabbage. *Plant Physiol.* 60:854–56

137. Pate, J. S. 1975. Exchange of solutes between phloem and xylem and circulation in the whole plant. *Sci. Hortic.* 1:451–73

138. Pate, J. S., Hocking, P. J. 1978. Phloem and xylem transport in the supply of minerals to a developing legume (*Lupinus albus* L.) fruit. *Ann. Bot.* 42:911–21

139. Perring, M. A. 1979. The effects of environment and cultural practices on calcium concentration in the apple fruit. *Commun. Soil Sci. Plant Anal.* In press

140. Poovaiah, B. W., Leopold, A. C. 1973. Deferral of leaf senescence with calcium. *Plant Physiol.* 52:236–39

141. Poovaiah, B. W., Leopold, A. C. 1976. Effects of inorganic solutes on the binding of auxin. *Plant Physiol.* 58:783–85

142. Preston, A. P., Perring, M. A. 1974. The effect of summer pruning and nitrogen on growth, cropping and storage quality of Cox's Orange Pippin apple. *J. Hortic. Sci.* 49:77–83

143. Quinlan, J. D., Preston, A. P. 1973. Effect of shoot tipping on fruiting and apple tree growth. *East Malling Res. Stn. Rept. 1972*, p. 91

144. Ralph, R. K., Bullivant, S., Wojcik, S. J. 1976. Effects of kinetin on phosphorylation of leaf membrane proteins. *Biochem. Biophys. Acta* 412:319–27

145. Rapper, C. D., Patterson, D. T., Parsons, L. R., Kramer, P. J. 1977. Relative growth and nutrient accumulation rates for tobacco. *Plant Soil* 46:473–86

146. Raven, J. A. 1977. H^+ and Ca^{2+} in phloem and symplast: Relative immobility of the ions to the cytoplasmic nature of the transport path. *New Phytol.* 79:465–80

147. Robinson, J. B. D. 1975. The influence of some growth-regulating compounds on the uptake, translocation and concentration of mineral nutrients in plants, *Hortic. Abstr.* 45:611–17

148. Roddick, J. G. 1975. Effect of α-tomatine on the permeability of plant storage tissues. *J. Exp. Bot.* 26:221–27

149. Rousseau, G. G., Haasbrock, F. J., Visser, C. K. 1972. Bitter pit in apples. The effect of calcium on permeability changes in apple fruit tissue. *Agroplantae* 4:73–80

150. Saftner, R. A., Evans, M. L., Hollander, P. B. 1976. Specific binding of victorin and calcium: Evidence for calcium binding as a mediator of victorin activity. *Physiol. Plant Pathol.* 8:21–34

151. Schimansky, C. 1977. Orientierende Untersuchungen über den Transport von Magnesium, Calcium und Kalium in die Fruchtstände von Rebpflanzen nach Kurzzeitapplikation von ^{28}Mg, ^{45}Ca und ^{86}Rb. *Mitt. Klosterneuburg* 27:181–84

152. Schumacher, R., Fankhauser, F. 1972. Beeinflussung der Stippigkeit durch Sommerschnitt und Wachstums-regulatoren. *Schweiz. Z. Obst Weinbau* 108:243–51

153. Sharples, R. O. 1974. *Acta Hortic.* 45:21–24

154. Sharples, R. O. 1976. Post-harvest chemical treatments for the control of storage disorders of apples *Ann. Appl. Biol.* 83:157–67

155. Sharples, R. O., Johnson, D. S. 1977. The influence of calcium on senescence changes in apple *Ann. Appl. Biol.* 85:450–53

156. Shay, F. J., Hale, M. G. 1973. Effects of low levels of calcium on exudation of

sugars and sugar derivatives from intact peanut roots under axenic conditions *Plant Physiol.* 51:1061–63

157. Shear, C. B. 1975. Calcium related disorders of fruits and vegetables. *HortScience* 10:361–65

158. Shear, C. B., Faust, M. 1970. Calcium transport in apple trees. *Plant Physiol.* 45:670–74

159. Shone, M. G. T., Clarkson, D. T., Sanderson, J., Wood, A. V. 1973. A comparison of the uptake and translocation of some organic molecules and ions in higher plants. See Ref. 39, pp. 571–82

160. Siegel, S. M. 1970. Further studies on regulation of betacyanin efflux from beetroot tissue: Ca-ion-reversible effects of hydrochloric acid and ammonia water. *Physiol. Plant.* 23:251–57

161. Siegel, S. M., Daly, O. 1966. Regulation of betacyanin efflux from beet root by poly-L-lysine, Ca-ion and other substances. *Plant Physiol.* 41:1429–34

162. Simon, E. W. 1978. The symptoms of calcium deficiency in plants *New Phytol.* 80:1–15

163. Stahly, E. A., Benson, N. R. 1970. Calcium levels of "Golden Delicious" apples sprayed with 2, 3, 5-triiodobenzoic acid. *J. Am. Soc. Hortic. Sci.* 95:726–27

164. Stahly, E. A., Benson, N. R. 1976. Calcium levels of "Golden Delicious" apples as influenced by calcium sprays, 2, 3, 5-triiodobenzoic acid and other plant growth regulator sprays. *J. Am. Soc. Hortic. Sci.* 101:120–22

165. Stebbins, R. L., Dewey, D. H. 1972. Role of transpiration and phloem transport in accumulation of ^{45}calcium in leaves of young apple trees. *J. Am. Soc. Hortic. Sci.* 97:471–74

166. Stebbins, R. L., Dewey, D. H., Shull, V. E. 1972. Calcium crystals in apple stem, petiole and fruit tissue. *HortScience* 7:492–93

167. Sutcliffe, J. F. 1976. Regulation in the whole plant. See Ref. 43, II (Part B): 394–417

168. Suttle, J. C., Kende, H. 1978. Ethylene and senescence in petals of Tradescantia. *Plant Physiol.* 62:267–71

169. Takatori, F. H., Lorenz, O. A., Cannell, G. H. 1961. Strontium and calcium for the control of blackheart of celery. *Proc. Am. Soc. Hortic. Sci.* 77:406–14

170. Terblanche, J. H., Wooldridge, L. G., Hesebeck, I., Joubert, M. 1979. The redistribution and immobilisation of calcium in apple trees with special reference to bitter pit. *Commun. Soil. Sci. Plant Anal.* In press

171. Thomas, W. A. 1967. Dye and calcium ascent in dogwood trees. *Plant Physiol.* 42:1800–2

172. Thomas, W. A. 1970. Retention of calcium-45 by dogwood trees. *Plant Physiol.* 45:510–11

173. Tibbitts, T. W., Palzkill, D. A. 1979. Requirement for root pressure flow to provide adequate calcium to low-transpiring tissue. *Commun. Soil Sci. Plant Anal.* In press

174. Tingwa, P. O., Young, R. E. 1974. The effect of calcium on the ripening of avocado (*Persea americana*) fruits. *J. Am. Soc. Hortic. Sci.* 99:540–42

175. Tomlinson, J. D., Turner, J. F. 1973. Pyruvate kinase of higher plants. *Biochem. Biophys. Acta* 329:128–39

176. Tromp, J. 1978. The effect of root temperature on the absorption and distribution of K, Ca and Mg in three rootstock clones of apple budded with Cox's Orange Pippin. *Gartenbauwissenschaft* 43:49–54

177. Tromp, J. 1979. The intake curve for calcium into apple fruits under various environmental conditions. *Commun. Soil Sci. Plant Anal.* In press

178. Tromp, J., Oele, J. 1972. Shoot growth and mineral composition of leaves and fruits of apple as affected by relative air humidity. *Physiol. Plant.* 27:253–58

179. Turner, J. F., Turner, D. H. 1975. The regulation of carbohydrate metabolism. *Ann. Rev. Plant Physiol.* 26:159–86

180. van der Boon, J. 1974. *Acta Hortic.* 45:9–14

181. van der Boon, J., Das, A., van Schreven, A. C. 1966. A five-year fertilizer trial with apples on a sandy soil; the effect of magnesium deficiency, foliage and fruit composition and keeping quality *Neth. J. Agric. Sci.* 14:1–31

182. van der Boon, J., Van Goor, B. J., Wiersum, L. K. eds. 1969. Discussion meeting on bitter pit in apples. *Acta Hortic.* 16:1–30

183. van der Geijn, S. C., Petit, C. M., Roelofsen, H. 1979. Measurement of the cation exchange capacity of the transport system in intact plant stems. *Commun. Soil Sci. Plant Anal.* In press

184. van Die, J., Tammes, P. M. L. 1975. Phloem exudation from monocotyledonous axes. See. Ref. 43, I:196–222

185. van Goor, B. J. 1968. The role of calcium and cell permeability in the disease blossom-end rot of tomatoes. *Physiol. Plant.* 21:1110–21

186. van Goor, B. J. 1974. Influence of restricted water supply on blossom-end rot and ionic composition of tomatoes

grown in nutrient solution *Commun.
Soil. Sci. Plant Anal.* 5:13–24

187. van Steveninck, R. F. M. 1965. The significance of calcium on the apparent permeability of cell membranes and the effects of substitution with other divalent ions. *Physiol. Plant.* 18:54–69

188. van Steveninck, R. F. M. 1976. Effects of hormones and related substances on ion transport. See Ref. 43, 2 (Part B), pp. 307–42

189. Vignes, D., Calmés, J. 1975. Quelques modifications physico-chimiques et physiologiques liées à la senescence. *Physiol. Plant.* 33:188–93

190. Walker, M. E. 1975. Calcium requirements for peanuts. *Commun. Soil Sci. Plant Anal.* 6:299–313

191. Wiebe, H. J. 1975. Beziehungen zwischen dem Wasserhaushalt der Pflanzen und dem Auftreten der Innenblattnekrosen bei Weißkohl. *Gartenbauwissenschaft* 40:134–38

192. Wiebe, H. J., Schätzler, H. P., Kühn, W. 1977. On the movement and distribution of calcium in white cabbage in dependence of the water status *Plant Soil* 48:409–16

193. Wieneke, J. 1979. Calcium transport and its microautoradiographic localization in the tissue. *Commun. Soil. Sci. Plant Anal.* In press

194. Wieneke, J., Biddulph, O., Woodbridge, C. G. 1971. Influence of growth regulating substances on absorption and translocation of calcium in pea and bean. *J. Am. Soc. Hortic. Sci.* 96:721–24

195. Wieneke, J., Führ, F. 1973. Mikroautoradiographischer Nachweis von ^{45}Ca-Kristallablagerungen im Apfelstiel und Fruchtgewebe. *Angew. Bot.* 47:107–12

196. Wieneke, J., Führ, F. 1973. Untersuchungen zur Translokation von ^{45}Ca im Apfelbaum I. *Gartenbauwissenschaft* 38:91–108

197. Wieneke, J., Führ, F. 1975. Untersuchungen zur Translokation von ^{45}Ca im Apfelbaum IV *Gartenbauwissenschaft* 40:105–12

198. Wiersum, L. K. 1966. Calcium content of fruits and storage tissue in relation to the mode of water supply. *Acta Bot. Neerl.* 15:406–18

199. Wiersum, L. K. 1979. Effects of environmental and cultural practices on calcium nutrition *Commun. Soil Sci. Plant Anal.* In press

200. Wiersum, L. K. 1979. Ca-content of the phloem sap in relation to Ca-status of the plant. *Acta Bot. Neerl.* In press

201. Wilcox, G. E., Hoff, J. E., Jones, C. M. 1973. Ammonium reduction of calcium and magnesium content of tomato and sweet corn leaf tissue and influence on incidence of blossom end rot to tomato fruit. *J. Am. Soc. Hortic. Sci.* 98:86–89

202. Wilkinson, B. G. 1968. Mineral composition of apples IX. Uptake of calcium by the fruit. *J. Sci. Food Agric.* 19:646–47

203. Wilkinson, B. G., Fidler, J. C. 1973. Physiological disorders. In *The Biology of Apple and Pear Storage*, ed. J. C. Fidler, B. G. Wilkinson, K. L. Edney, R. O. Sharples, pp. 63–132. Farnham, England: Commonw. Agric. Bur. 235 pp.

204. Williams, R. J. P. 1976. Calcium chemistry and its relation to biological function. In *Calcium in Biological Systems,* pp. 1–17. Cambridge: Cambridge Univ. Press. 485 pp.

205. Wills, R. B. H., Scott, K. J., Carroll, E. T. 1975. Use of alkaline earth metals to reduce the incidence of storage disorders of apples. *Aust. J. Agric. Res.* 26:169–71

206. Yeh, Y. Y. 1963. Uptake and movement of radioactive sulphur and calcium by tomato plants as influenced by gibberellins. *J. Agric. Assoc. China* 42:52–58 (In Chinese); *Hortic. Abstr.* 1964. 34:3033

207. Ziegler, H. 1963. Verwendung von Ca^{45} zur Analyse der Stoffversorgung wachsender Früchte. *Planta* 60:41–45

208. Ziegler, H. 1975. Nature of transported substances. See Ref. 43, pp. 59–100

209. Zimmermann, M. H. 1969. Translocation of nutrients. In *The physiology of Plant Growth and Development,* ed. M. B. Wilkins, pp. 383–417. London: McGraw

Ann. Rev. Phytopathol. 1979. 17:123–47
Copyright © 1979 by Annual Reviews Inc. All rights reserved

SEROLOGICAL IDENTIFICATION OF PLANT PATHOCENIC BACTERIA

◆3704

N. W. Schaad

Department of Plant Pathology, University of Georgia,
Experiment, Georgia 30212

INTRODUCTION

The correct diagnosis of any disease is a prerequisite of control. The more rapidly and accurately the causal organism is identified, the sooner proper controls can be instituted. Although precise chemotherapy can be applied to most human bacterial infections once the pathogen is identified, specific chemical treatment for controlling plant pathogenic bacteria is in an early stage of development. Antibiotics and various formulations of copper are used to control plant bacterial diseases but with limited success.

Apathy toward research on ecology and control of bacterial diseases is due in large part to difficulties in identifying the pathogens. There is a real need for development of effective and rapid methods for identification of plant pathogenic bacteria. Much of this increased need for rapid identification is a result of a much increased international trade and movement of plant propagative materials such as potato tubers and seeds. Biochemical and physiological tests which are routinely used to identify plant pathogenic bacteria (11, 13) are not entirely satisfactory. The tests are often complicated, difficult to interpret, and require weeks to months to complete. Also, not all strains of each organism always give the same results.

The use of serology for identifying bacteria is almost as old as the science of plant pathology itself. Many medically important bacteria are routinely identified by serological tests. The ability of medical bacteriologists to provide quick and accurate serological identification of bacteria indicates a similar potential for identification of plant pathogenic bacteria.

123

0066-4286/79/0901-0123$01.00

Diagnostic bacteriology as a routine laboratory service is not available to many universities and state and federal departments of agriculture laboratories. By the methods and techniques to be discussed in this review standardized techniques may be chosen and a diganostic service made available to a far greater number of laboratories.

No effort is made in this review to describe in detail the various serological techniques that have been developed or to list all their applications; this has been done elsewhere (14, 26, 49, 62, 141). Also, this review is not concerned with nomenclature or taxonomy. Rather, the purpose of this review is to focus attention upon past problems and future benefits of using serology as a practical tool for identification of plant pathogenic bacteria.

An understanding of the mechanisms of serology (antigen-antibody interaction) is necessary for an adequate interpretion of serological tests. The first part of this paper reviews the principles of serology as a prelude to understanding past results and future possibilities of using this technique to identify plant pathogenic bacteria.

PRINCIPLES OF SEROLOGY

Serology is the science of reactions, preparations, and use of serums. Studies of antigen-antibody reactions constitute the methods of serology. Carpenter (14) defines an antigen as "a substance that elicits a specific immune response when introduced into the tissues of an animal. The response may consist of antibody production, cell mediated immunity, or immunologic tolerance."

Antiserum is the fundamental reagent of serology and provides great versatility and high specificity. Antiserum is produced by an animal's lymphoid system which, recognizing an antigen as a foreign substance and an implied biological threat, manufactures proteins (immunoglobulins) that react selectively with the foreign "body." These proteins carried in body fluids are known collectively as "antibodies."

Serology is extremely sensitive. Precipitating antibody can be induced by a single injection of only 5 μg of egg albumin (20). The dose given may be varied according to the purpose for which the antiserum is needed. Optimal production of antibody in rabbits has been shown to require only 0.5 mg of a hapten-protein conjugate (146). If this dose of antigen per injection is used, a 1% contaminant of the inject antigen could induce precipitating antibodies. Since an amount greater than 0.5 mg is normally used for injection, a contamination of the inject antigen of less than 1% would be required to obtain a monospecific antiserum. However, antisera can often be rendered specific for a given antigen by absorption. The unwanted antibodies can be neutralized by adding the respective antigens to the antiserum. For most immunochemical purposes, however, it is preferable to use isolubilized antigens [protein copolymers (4) or proteins conjugated to insoluble matrices (122)] for the absorption. This will insure the removal of soluble antigen-antibody complexes, which may remain in solution if soluble antigen is used for absorption. Absorption techniques are not generally recommended, however [(27) p. 108].

Antibodies are oblong-shaped molecules with active sites at both ends. The antigen-antibody reaction results in the formation of a three-dimensional network or lattice in which the antigen and antibody alternate. This is termed a *primary antigen-antibody reaction* and cannot be observed in the usual test tube or agar plate procedures; it can be observed by immunofluorescence. Secondary reactions, called *precipitation,* require several hours, or even days, depending upon the conditions. Precipitation is the final visible result of a serologic reaction. Agglutination is the same as precipitation except that the antigen is particulate, e.g. a virus particle or bacterial cell rather than a molecule.

For a detailed discussion of serology and specific methods in serology, including preparation of immunizing materials, injection and bleeding of animals, agglutination, adsorption of agglutinins, precipitation (usually in agar plates), complement fixation, and immunofluorescence staining, see Carpenter (14) and Crowle (26, 27).

HISTORY

Serological Tests

TUBE AGGLUTINATION AND PRECIPITATION   The first report about using serology to identify a plant pathogenic bacterium was published in 1918 (68) when Jensen showed that a strain of *Agrobacterium tumefaciens* from Denmark could be differentiated from a strain of *A. tumefaciens* from the United States by agglutination tests. Agglutination tests with antiserum (AS) that were made by use of live or formalized cell preparations (AS to live or formalized cells) rapidly became very popular during the 1930s among researchers working on the identification of plant pathogenic bacteria and remains so today. Such techniques were thought to be specific and were suggested for differentiating or identifying *Erwinia atroseptica* (52, 53, 103), *E. aroideae* (69, 107), *E. carotovora* (50, 51, 53, 76, 89–92, 108, 156), *E. amylovora* (79), *Agrobacterium tumefaciens* (8, 21, 82, 124), *Rhizobium* spp. (54), *Corynebacterium sepedonicum* (18), *Pseudomonas* spp. (5, 63, 110, 125, 149), *P. solanacearum* (64), yellow-pigmented bacteria (82, 83, 142), *Xanthomonas malvacearum* (39, 172), *X. translucens* (7, 59), and a newly described xanthomonad isolated from filbert (97).

Several investigators did report that bacteria thought to be different from the organisms used for AS production cross-reacted in agglutination tests using AS to live cells (12, 39, 45, 73, 84, 113, 123, 170); however, the specificity of AS to live cells was not openly questioned until 1941 (39). Elrod (39) failed to differentiate between strains of *E. carotovora, E. atroseptica,* and *E. aroideae* and suggested that the cross-reactions might be a result of common flagellar antigens present in the whole cell preparation used for immunization. In 1947 Elrod & Braun (40, 41) made an extensive study of several strains of *Xanthomonas* spp. They found when live cells were used for immunization that cross-agglutination did in fact result between some strains of different species. However, when AS were prepared against cells without the mucoid polysaccharide or against cells from which the polysaccharide had been removed, the cross-agglutination reactions were eliminated. In 1948 Berquist & Elrod (8) used AS to cell suspensions heated at 100°C for 1 hr or to cell suspensions treated with absolute ethanol [somatic ("O") antigens of Edwards & Bruner (37)] and found that *A. tumefaciens, A. radiobacter, A. rubi,* and *A. rhizogenes* were easily differentiated.

Berquist & Elrod were not, however, the first to use heat-killed cells for immunization. In 1929, Stapp (151) used cells (washed from 2–3-day-old potato agar slants in 0.85% NaCl) heated for 1.5 hr at 60°C for immunization. Stapp also used agglutination and precipitin tests to separate 128 strains of *Bacillus phytophthorus* (*E. atroseptica*) and *B. carotovorus* (*E. carotovora*) into five groups. Using the same methods he reported that *P. tabaci, P. angulata,* and *P. mellea* could not be distinguished (152) and that 12 strains of a xanthomonad (*X. begoniae*) isolated from diseased begonia plants were identical but different from *X. campestris* (154). The advantage of using AS to heat-killed cells ("O" antigens) for identifying plant pathogenic bacteria was not generally realized until several workers during the 1960s (73, 74, 96, 101) showed that the tests were specific and useful for identifying pseudomonads.

Another important observation that has gone unnoticed by plant pathologists for many years was a report by Stapp in 1937 (153). He found in agglutination tests (AS to heat-killed cells) that six strains of *P. pisi* and single strains of *P. phaseolicola* and *P. tabaci* agglutinated equally well and could not be distinguished; by contrast in precipitin tests (tube) no reaction occurred between AS to cells of *P. pisi* and cells of *P. phaseolicola* or *P. tabaci* or between AS to *P. phaseolicola* and cells of *P. pisi* (153). Later, referring to serological methods for identifying plant pathogenic bacteria, Stapp (155) made the following statements: "Compared with agglutination the precipitation is not only more sensitive, but also more specific." "Several years experience of the author showed that strains which could be agglutinated by seradilutions of 1:200 or higher, reacted negatively in the precipitation test." Precipitin tests were preferred over agglutination tests by medical bacteriologists during the 1940s but not by plant pathologists.

AGAR PRECIPITIN With the introduction of the agar double diffusion precipitin test by Ouchterlony in 1948 (114), agar precipitin tests became widely used during the 1950s for identification of medically important bacteria. However, Ouchterlony double diffusion (ODD) was not tested for identifying a plant pathogenic bacterium until 1960. That year Lovrekovich & Klement (84) reported that species-specific antigens of *P. tabaci* were detected in ODD tests but not in agglutination tests. Ouchterlony double diffusion tests quickly became popular and have been widely used since then. For example, ODD has been used to identify or differentiate strains of *P. syringae* (86, 93, 112, 113, 165), *P. phaseolicola* (56, 58, 160), *P. lachrymans* (87, 158), *P. savastanoi* (174), *P. solanacearum* (33, 61, 66, 67, 99, 120, 138), *P. pisi* (162), *P. avenae*, *P. setariae*, *P. glumae*, and *P. andropogonis* (165), and *Pseudomonas* spp. (116, 118, 127, 128). Strains of *X. vesicatoria* (16, 85, 95, 106, 115, 133, 151), *X. oryzae* (88, 105), *X. ampelina* (42), and *X. cyamopsidis* (111) were differentiated by ODD tests.

Ouchterlony double diffusion tests were used to confirm the identity of a strain of *Rhizobium* isolated from galls as the original inoculum strain (36). This technique was used also to identify noninfective strains of *Rhizobium* (31) and to identify and group strains of *Agrobacterium* into two serovars that correlated completely with the biovars (71). They have been used to identify *E. atroseptica* isolated from potato plants with black-leg symptoms (70, 75, 159); to differentiate between an unidentified *Erwinia* spp. isolated from alfalfa and other erwiniae (144); to identify strains of *E. rubrifaciens* and *E. carotovora* (132); to identify a strain of *E. carotovora* responsible for blackleg of potato which was serologically unrelated to strains of *E. carotovora* that did not cause blackleg symptoms (157); and to differentiate *E. aroideae* and *E. carotovora* isolated from callas and hyacinth (69).

IMMUNOFLUORESCENCE Immunofluorescent (IF) staining was suggested for identification of bacteria in plants in 1943 (119). However, no further interest was shown until 1965 when Morton (98) reported the superiority of direct IF over agglutination or bentonite flocculation tests for rapid identification of *X. vesicatoria* in extracts of diseased tissue. In 1967 indirect IF was used to identify *E. aroideae* in extracts of host tissues and soil (72). It has been used to identify *P. phaseolicola* in preparations of bean seeds (22, 166) and leaves (160), *P. solanacearum* in soil (67), *P. tabaci* in leaves and agar culture (109), *E. atroseptica* in culture (3, 109) and in preparations of tubers, leaves, insects, and soil (3), *X. albilineans* in host tissues (80), *X. campestris* in culture (135) and in preparations of soil (35), *C. michiganense* in seeds (2, 166), *Spiroplasma citri* in host tissues (126), and the phony peach bacterium extracted from peach roots (44).

Immunization Preparations

NONPURIFIED IMMUNOGENS   The specificity of serological tests is improved when AS to heat-killed rather than live cells are used (8, 73, 84, 88, 152, 162). In spite of this, few preparations other than live or killed cells are used for immunization.

Perlasca (121) used a soluble extract from acetone-precipitated cells of *P. syringae;* however, he observed a considerable degree of heterogeneity among strains isolated from stone fruits in agglutination tests. Perez (120) used a formamide heat-extracted (150°C for 15 min) polysaccharide preparation. Antisera were prepared by use of extracts from nine strains of *P. solanacearum* and single strains of *P. tabaci* and *E. carotovora.* All 59 strains of *P. solanacearum* tested were positive with the AS to *P. solanacearum* but not with the other AS in ring precipitin tests. An AS to a soluble extract of sonicated cells of a strain of *E. nigrifluens* was used by Zeitoun & Wilson (177) to differentiate eight strains of *E. nigrifluens* from single strains of *E. rubrifaciens, E. amylovora, E. aroideae, E. atroseptica, E. carotovora, E. tracheiphila,* and *Escherichia coli* in ODD tests. Of six to eight bands observed in the homologous reaction, several were common to all the bacteria tested. Antisera to preparations of sonicated cells were specific for *P. lachrymans* (87), *P. syringae* (113), and *X. vesicatoria* (31) only when heat-killed cells were used for test antigens. In like manner, Tanii & Baba (158) reported that AS to soluble preparations of *P. lachrymans* reacted either identically or cross-reacted with 16 other pseudomonads including *P. coronafaciens, P. glycinea, P. phaseolicola, P. syringae,* and *P. cichorii* as well as *E. aroideae* and *C. michiganense* when untreated soluble preparations were used as test antigens. However, when heat-treated cell preparations were used as test antigens *P. lachrymans* cross-reacted only with *P. tabaci.*

PURIFIED IMMUNOGENS   Pastushenko & Simonovich (117, 118) used an unidentified nucleoprotein fraction from cells of *P. cerasi* extracted with sodium deoxycholate. They concluded that the AS was specific at the generic level in agglutination and ODD (two to seven precipitin bands observed) tests, but they failed to present adequate data for such a conclusion. I have reported that AS to purified ribosome preparations were useful

for differentiating several enterobacteria (133). Antisera to two strains of
E. carotovora and *E. rubrifaciens,* and single strains of *Escherichia coli,*
Salmonella typhimurium, Klebsiella pneumonia, and *X. campestris* were
tested against test antigens (crudely extracted ribosomes) of 25 strains of
enterobacteria and four other bacteria (133). Two precipitin bands were
observed in ODD tests; one was always present in homologous systems and
specific at the subspecies level. Moreover, AS to ribosome preparations of
X. vesicatoria were specific at the subspecies level and used to type 25 strains
into three serovars (134). Similarly, Digat & Cambra (33) presented evidence that AS to a glycoprotein extracted from cells by ammonium sulfate
precipitation was specific at the subspecies level (33). Antiserum to four
strains of *P. solanacearum* reacted in agglutination and IF tests with all 14
strains of *P. solanacearum* but not with any of 30 other plant pathogenic
bacteria. Agglutination-absorption tests were used to divide the 14 strains
into several serovars.

Antiserum to glutaraldehyde-fixed cells has been shown to be specific at
the species level. Such AS were used by Allen & Kelman (3) to identify
E. atroseptica from potato, Stanghellini and co-workers (150) to identify
E. carotovora from potato and sugar beet, and Slack and co-workers (148)
to identify *C. sepedonicum* from potato. Results similar to that found with
AS to ribosome preparations were observed with a membrane protein complex (MPC) extracted with lithium chloride at 45°C from intact cells of
E. chrysanthemi (176) and *P. solanacearum* (138). The MPC was highly
immunogenic and a single precipitin band was specific at the subspecies
level in ODD tests. Antisera to five strains of *E. chrysanthemi* were used
to type 27 strains of *E. chrysanthemi* (from 18 hosts) into four serovars.
Antiserum to MPC of *S. typhimurium* reacted identically with homologous
MPC but not with MPC of *E. chrysanthemi* and AS to *E. carotovora*
cross-reacted very weakly with *E. chrysanthemi* after the AS was concentrated four times. The AS to *E. chrysanthemi* failed to react with MPC of
*E. carotovora, E. atroseptica, S. typhimurium, E. coli. Enterobacter agglom-
erans (Erwinia herbicola), P. solanacearum,* or *X. campestris.* Antisera to
P. solanacearum were also specific at the subspecies level and used to type
19 strains of *P. solanacearum* from Brazil into five serovars using AS to six
strains (138).

PURIFIED AND IDENTIFIED IMMUNOGENS In contrast to the above
uses of unidentified immunogens, Mazzucchi and associates (94) used AS
to pectic lyases to differentiate soft-rotting *Erwinia* spp. They tested AS to
two strains of *E. chrysanthemi* against partially purified test antigens of 78
strains of *E. chrysanthemi,* 21 strains of *E. carotovora,* and 14 strains of

E. atroseptica in ODD tests. No extracts from *E. carotovora* or *E. atroseptica* produced a band of precipitin whereas all extracts of *E. chrysanthemi*
produced a reaction of identity or partial identity with both AS. Another
novel approach is the use of AS to L-asparaginase to differentiate *E.
carotovora* from other enterobacteria (6, 15).

APPRAISAL AND FUTURE OUTLOOK

Serology is an extremely valuable tool in molecular biology. Serological
techniques are presently used for purification, determination of structure
and function, and identification of specific compounds. Although serological techniques are very useful, they are also very exacting. Every reagent
employed must be carefully tested for nonspecific activity, and the degree
of specific activity must be accurately assessed before they can be used for
identification or research investigations. There are no short cuts to obtaining pure reagents with specific properties, and those who attempt to use
short cuts will obtain results with inadequately controlled reagents and tests
that are totally unreliable.

Contrasting Results

Indeed, considerable data has accumulated in the last 40 years on the
serology of plant pathogenic bacteria, but much of it does not agree.
Pseudomonas tabaci, P. angulata, and *P. mellea* are reported to be serologically identical (9, 84, 123) but distinct (102). In like manner, *P. syringae* and
P. morsprunorum are reported to be identical by some workers (25, 43, 86,
113) but distinct by others (46, 110, 147). Similarly, *P. lachrymans* is not
distinguishable from *P. tabaci* (161) but is easily distinguishable from *P.
mori, P. tomato,* and *P. phaseolicola* (87) and from several other pseudomonads (25, 158). *Pseudomonas phaseolicola* cannot be differentiated from *P.
glycinea* or *P. mori* in agglutination tests or IF tests but is differentiated in
ODD tests (160). However, *P. phaseolicola* is easily differentiated from most
other pseudomonads in agglutination tests when AS prepared with heat-
killed cells are used (56, 84, 132, 152). Strains of *P. syringae* from stone
fruits are too heterogeneous in agglutination tests to be differentiated (121)
but 450 strains of *P. syringae* from several hosts where differentiated into
10 serovars in ODD tests (112). Furthermore, both positive (79, 81, 85, 106,
131, 145) and negative (1, 16, 23, 87, 113, 134) correlation between host of
origin or virulence and serological reactions are reported for the same
organism. Finally, cross-reactions do not occur in agglutination tests between *Corynebacterium* spp. according to Claflin & Shepard (18) and Lazar
(77) but do according to Rosenthal & Cox (128) and Slack and co-workers
(148).

Standardization Needed

It is evident from the above contrasting results that there is a great need for some standardized procedures such as number of strains, source of strains, methods of preparing antigens for immunization, immunization schedule, and type of test used. Because serological tests are very sensitive, small changes in methods can result in significant changes in the results. This makes it difficult to decide from published results whether a certain serological method can be used successfully for identification.

Routine Usage

Currently, most commonly isolated bacteria are easily identified to a genus by doing a simple Gram stain and observing growth and colony characteristics on certain agar media (N. W. Schaad, in preparation). With some additional knowledge of the host and disease symptoms and results from one or two biochemical tests it is quite easy to narrow the identification to one or two species. It is in this final confirmation of the presumptive identification that serology is of most value. In fact, several workers (U. Mazzucchi, A. Kelman, H. Vruggink, and A. Takatsu, personal communication) are presently using serology for that purpose. Otta (112) identifies *P. syringae* isolated from wheat by testing all oxidase negative, fluorescent strains against AS to six serovars of *P. syringae* in ODD tests. In contrast, U. Mazzucchi (personal communication) routinely uses agglutination tests and AS to heat-killed cells for rapid confirmation of bacteria presumptively identified on agar media. Identification is finally completed by one or two key biochemical tests. He uses the ring precipitin test for direct identification of *E. atroseptica* from diseased tissues and finds a high correlation with presumptive identification on pectate gel plates. On the other hand, A. Kelman in Wisconsin and H. Vruggink in the Netherlands (personal communication) use direct IF to identify colonies of *E. atroseptica,* which are isolated from rotted tubers onto crystal violet pectate medium (28). They also confirm the identification by doing a couple of biochemical tests. In like manner, I routinely use indirect IF with a highly specific antiserum (135) to confirm the presumptive identification of *X. campestris* isolated onto SX (139) and nutrient-starch-cycloheximide (137) agars. Normally between 100 and 150 samples of 10,000 crucifer seeds are assayed for *X. campestris* during a three to four month period each year. In Brazil, ODD tests with AS to MPC (138) are used to confirm the identity of the banana race (race 2) of *P. solanacearum* isolated from samples of banana trees from the Amazon Basin (A. Takatsu, personal communication).

Several authors use agglutination (5, 7, 18, 47, 48, 52, 59, 60, 65, 66, 79, 98, 99, 103–105), precipitation (57, 70, 99, 105), immunofluorescence staining (2, 22, 44, 98, 99, 164, 169), and serological specific microscopy (20, 29,

32) for identifying bacteria obtained directly from diseased tissues. Some nonconventional tests such as growth inhibition, metabolic inhibition, and organism deformation are used routinely to identify *Spiroplasma* spp. (17, 30, 167, 173) even though ODD and IF would be quicker and more specific.

Factors Affecting Antisera Characterization

Serological identification is clearly moving from an era of establishing the usefulness and value of serology for rapid identification to an era of establishing the antigenic structure of plant pathogenic bacteria. To accomplish this more quickly and accurately a basic understanding of the factors that can affect the characteristics of an AS is needed so that proper tests and methods can be adopted. Factors such as the species of animal injected and the presence of contaminating substances affect the specificity of an AS. Certain nonspecific bacterial polysaccharides and dextrans are not antigenic in rabbit but are antigenic in swine. Immunization of the rabbit with whole cells, however, induces formation of antibodies that will precipitate purified polysaccharides. It is clear that the species of animal injected and the presence of certain substances affect the formation of precipitating antibodies. Precipitin antibody formation parallels the ability of the various animals to produce antibodies in general and is usually better in rabbits, horses, and chickens than in guinea pigs, rats, and mice.

ANIMALS USED FOR IMMUNIZATION   Antisera from the different species of animals have their own predictable antigen-precipitating characteristics. Rabbits are most often used because they respond well to many antigens, yield reasonably large volumes of AS, are easy to handle, and are inexpensive. Proteins are often antigenic in rabbits but not in mice or guinea pigs, and purified polysaccharides are antigenic in mice and guinea pigs but not in rabbits. Rabbit AS will produce more stable precipitate bands in agar gel under more adverse conditions than AS produced in horses, but the bands will be broader and more diffuse than horse AS with a consequent possible loss of resolution among closely spaced precipitin bands. Horse AS are widely used but have the disadvantage of forming more unstable precipitins with antigen over a much narrower range of antigen-antibody proportions. Goat AS rivals horse AS because it can be produced in comparatively large volumes and yet contains the more desired R-type precipitins of rabbits. Like horse AS, sheep AS precipitates antigen only over a narrow optimum proportion ratio with antigen, but it will still produce precipitates in antibody excesses as do rabbit AS. Mice and rat AS precipitate antigen in a narrow optimal proportion like H-type antibody, but the resulting precipitates are stable in antibody excess. Chickens are excellent alternates for rabbits, easily producing good volumes of potent precipitating AS against many different kinds of antigens. Frequently they precipitate well in liquid media at high salt concentration (8% NaCl) but precipitate not at all at physiologic strength. Furthermore, AS that do not precipitate well in liquid media may or may not be effective in agar gels.

METHOD OF IMMUNIZATION   The method of immunization determines how the capacity of an antigen to induce AS formation will be exploited, and controls whether AS will be produced and what type and quantity. Sera drawn early, after modest immunization will contain low titers of AS, often nonprecipitating, with high specificity; intermediate immunization and later bleedings provide moderate titers of varied antibodies, some of which will contain precipitin antibodies of good specificity; immunization followed by bleedings carefully timed for peaking of an animal's precipitin response will produce AS with very high titers of several different antibodies, including potent precipitins of mediocre specificity. In addition to the inherent adjuvancy of an antigen, the "strength" of immunization depends on its schedule and the method of administering the antigen. Within limits, which may differ slightly for each antigen and animal used for immunization, the longer and more repeatedly the animal is stimulated with antigen, the greater variety and quantity of precipitins it will make, and the stronger but less specific will be its AS. Although high titered AS have been shown to be less specific (74), many continue to choose a higher titered AS over a lower titered one.

METHOD OF INJECTION   Several factors should be considered when choosing a method of administering antigens. The weakest AS responses develop when the injected antigen is disposed of quickly (e.g. intravenous injections); intermediate responses are obtained when the antigen is protected by route administration (e.g. subcutaneous versus intravenous injection) or especially by the form in which it is injected (e.g. insolubilized by heat or chemicals as opposed to merely being dissolved in saline or buffers); strongest responses result from injecting the antigen together with a potent adjuvant which in addition to protecting the antigen also releases small amounts of it repeatedly over a long time for a more prolonged immunizing stimulus, and distributes the antigen throughout the antibody-producing reticuloendothelial system. Generally the more foreign an antigen is to an animal the better it will induce precipitin production.

CHOICE OF TEST   Determining the species of animal and immunization schedule is an important consideration, but of even greater importance is the choice of serological test and its proper use. The failure in the past of many researchers to determine adequately the specificity of their serological test employed is partly responsible for the reluctance of plant pathologists to use serological techniques for routine identification of plant pathogenic bacteria. Often the specificity of a certain test is ascertained by using test antigens to 10 or fewer strains (7, 18, 24, 25, 55, 61, 67, 86, 95, 99, 100, 111, 144, 175, 177) or AS to only one or two strains (3, 10, 25, 36, 60, 66, 79, 86, 125, 129, 157, 160, 170). Cells for immunization are commonly harvested from agar slants (3, 84, 101, 127, 148), which can result in cross-precipitins, due to the agar (27). Results of agglutination and ODD tests are sometimes misinterpreted or not reported correctly. Although sera-dilutions of 1:200 or higher in agglutination tests are often due to non-specific agglutination (155), sera-dilutions of 1:800 and 1:320 are interpreted as being different (131). Ouchterlony double diffusion precipitin bands are correctly reported as reactions of (a) identity (fusion), (b) non-identity (intersection), (c) partial identity (partial intersection or "spur"), or (d) no reaction [p. 259 in (27)]. However, results are reported as "antigens reacted" (143) and as "a greater number of bands" or "more intense bands" (77, 78).

Such tests as agglutination, IF, and ODD are well adapted for identifying bacteria when they are properly used. Agglutination tests are easily performed and rapid but are inadequate for distinguishing between closely related immunogens and are not very sensitive. They can be useful for identifying closely related bacteria (33, 39, 130), if time-consuming cross-absorption tests (14) are included. Immunofluorescent staining is also easily performed, very sensitive, and rapid but requires expensive equipment and is only moderately specific.

The high sensitivity of IF (163) makes the test useful for rapid diagnosis. The fact that IF is less specific than ODD tests can also be used to advantage. For example, P. solanacearum is easily differentiated into seven serovars in ODD tests, but strains of two of the serovars do not form precipitin bands with all the AS (138). By using a pooled AS (equal volume of each), however, strains of P. solanacearum are immunofluorescent positive whereas other pseudomonads are negative (N. W. Schaad and co-workers, unpublished data). Pseudomonas solanacearum can quickly be identified with a single AS (pool) by IF and then the serovar determined by using seven separate AS in ODD tests.

The best one of the three tests for differentiating among strains of the same organism is clearly the ODD test. This test is a commonplace analytic tool for molecular biologists because closely related immunogens are easily distinguished. If there are several antigen-antibody systems that precipitate, they cannot be distinguished from each other in tube tests. But in agar diffusion tests each system of several coexisting ones can be observed to precipitate in a different position in the agar medium. Agar diffusion tests are routinely used to detect fraudulent products, standardize biologicals, monitor human physiology and pathology, identify and classify plants and animals, and study the epidemiology of animal and plant diseases. Also, there is good evidence that ODD is the best test available for identifying and differentiating strains of plant pathogenic bacteria (33, 84, 87, 160, 176, 177).

When developing a serological test for identification, one should first have a clear purpose and then choose a test with the specificity to meet that purpose. As pointed out earlier, the most specific test is not always the best test for a particular need. Tanii and associates (160) compared agglutination, ODD, and IF using AS to heat-killed cells for identification of *P. phaseolicola* isolated from bean and kudzu (*Pueraria thunbergiana*.) Several of the 25 *Pseudomonas* spp. cross-reacted with *P. phaseolicola* in agglutination tests and four species, *P. glycinea, P. mori, P. adzukicola,* and *P. coronafaciens* cross-reacted with *P. phaseolicola* in IF. No cross-reactions between *P. phaseolicola* and the other species occurred in ODD tests. Still, the IF test was recommended for rapid initial identification of *P. phaseolicola* isolated from diseased tissues and the ODD test for final confirming identification. Allen & Kelman (3) recommend IF for identification of *E. atroseptica* because IF is specific and rapid. The highly sensitive enzyme linked immunosorbent assay (ELISA) has recently been reported to be useful for serodiagnosis of *E. atroseptica, X. pelargonii* (168), and *C. sepedonicum* (19) but not *P. phaseolicola* (171).

CHOICE OF INJECT ANTIGEN   Although the inherent specificity of the test used is very important, of even more importance is the specificity of the AS. When deciding which antigen should be used for immunization, it is most important to have a clear purpose for which the resulting AS will be used. Will it be employed to (*a*) detect differences in antigenic determinants or similar molecules of antigen, (*b*) to demonstrate relationships between different species, or (*c*) to diagnose disease? For the first purpose the AS should have maximum specificity, for the second maximum cross-reactivity, and for the third a combination of high titer and high specificity. Antisera usually provides one of the following three broad functions: to detect a maximum number of antigens in a mixture, to detect qualitative or quantitative differences in complex mixtures of antigens, or to reveal the immunologic anatomy of individual populations of antigens.

Identifying a particular bacterial species would most likely require an AS of less specificity than the identification of a bacterium at the subspecies level would require. However, when in doubt about the needed specificity, an AS of high specificity should be chosen. If the AS turns out to be too specific, one could use a less specific test such as agglutination or IF. For example, IF with a highly specific AS is an ideal test for rapid confirmation of the identity of *X. campestris* (152).

Determination of Antigenic Structure

I subscribe to the view of Lucus & Grogan (87) that the antigenic structure and variation within the organism must be known when using serological methods for identification. The antigenic structure of an organism consists of the sum of its various antigenic components. The following are three major groups of surface antigens: (*a*) capsular, which are generally polysaccharide, (*b*) flagellar, which are protein, and (*c*) somatic, which are protein-polysaccharide-phospholipid complexes. The antigenic structure of these surface antigens is not, however, always constant. Both somatic (O) and flagellar (H) constituents are subject to variation. The smooth to rough transformation affects bacterial O antigens. Antisera for one rough form of *Salmonella* may agglutinate the rough form of other *Salmonella* spp. or even *Shigella* spp. and other more distantly related bacteria (38). The flagella of certain species exist in two alternative antigenic forms. H antigen of one form appears to be characteristic of the species and is called *specific antigen,* and those of the other forms are found in many species and are designated *group antigens.* In 1947 Berquist & Elrod (8) suggested that results that were reported earlier as specific (21, 124) could not be so because of the presence of nonspecific flagellar antibodies present in AS to live cells. The value of using a more specific AS such as one prepared to to heat-killed cells (152) becomes clearer when one realizes that cells of the same culture may have both specific and nonspecific O and H antigens.

The failure to determine adequately the antigenic variation within an organism has probably contributed more than any other factor to the deterrence of the widespread use of serology to identify plant pathogenic bacteria. It is difficult to have confidence in a technique that was based upon as few as 10 strains of the organism. It is, however, difficult to say what specific number of strains should be tested because the variability of each organism is different. An organism such as *E. rubrifaciens,* which has a very narrow host range, infects a single walnut cultivar, and is found only in California (140), would require the testing of fewer strains than an organism such as *E. carotovora* which has a very wide host range and is found worldwide. Clearly, an understanding of the ecology of the organism can be very useful in choosing how many and which strains of the organism and other organisms should be studied. Stapp's 1929 paper (151) on the serology of *Bacillus phytophorus* (*E. carotovora*) and *B. atrosepticus* (*E. atroseptica*) is of interest in illustrating the above need. A total of 129 strains were studied. He isolated 113 in Germany, 2 in Sweden, 3 in Switzerland, 3 in

England, and he received 7 from other collections, including *E. atroseptica* from American Type Culture Collection and *B. carotovorus* from L. R. Jones in Wisconsin. Agglutination and precipitin tests with AS to saline-washed cells heated for 1.5 hr at 60°C were used to group the 129 strains into five serovars. Group 1 contained 107 of his strains and two Dutch strains received as *B. melanogenes* (*E. aroideae*) and *B. atrosepticus;* group 2 contained 13 of his strains; group 3 contained *E. atroseptica* from ATCC; group 4 contained four strains of *B. carotovorus* and one strain of *B. solanisarpus* (*E. atroseptica*); group 5 contained a *B. carotovorus* strain received from J. G. Leach. Had Stapp used only his strains such heterogeneity would not have been observed. The presence of four serovars among 133 strains of *E. carotovora* isolated from potato (34) are consistent with the above results.

Realizing the variability within *P. solanacearum,* Perez (120) used AS homologous to nine strains and test antigen from 59 strains in ring precipitin tests to identify *P. solanacearum* in Puerto Rico. All 59 strains were positive with the nine AS whereas no reaction occurred with AS to single strains of *P. tabaci* or *E. carotovora.* His failure to differentiate any groups within the 59 strains was probably due to his use of the ring precipitin test instead of ODD. His purpose for identifying *P. solanacearum* was accomplished, however. As I have said earlier, Lucus & Grogan (87) were aware of the need to study a large number of strains. They found that 300 strains of *P. lachrymans* typed into three serovars using ODD tests and AS to nine strains. No reaction occurred between AS to one serovar and antigen of another. Strains of *P. phaseolicola, P. mori, P. tomato, X. campestris, C. sepedonicum,* and *E. carotovora* did not react with *P. lachrymans* antisera. In like manner, Otta & English (113) studied the antigenic structure of 450 strains of *P. syringae.* They found that strains of *P. syringae, P. aptata,* and *P. morsprunorum* were indistinguishable and grouped them into 10 serovars. Although cross-reactions were not observed in reciprocal tests with six of the serovars, they were observed with the other four serovars. Of 21 *Pseudomonas* spp. studied all strains of *P. aptata, P. morsprunorum,* and *P. pisi* which were pathogenic to peach were identical with one of the 10 serovars. However, several strains of seven different species that were not pathogenic to peach were also identical with one of the serovars. The latter results suggest that either the AS used (to broken cells) was not specific enough or that many pseudomonads they had received had been misidentified.

There is good evidence using IF that *E. atroseptica* is a very homogenous species with respect to AS to glutaraldehyde-fixed cells of a strain of *E. atroseptica* from Wisconsin (3). Fifty-seven of 58 strains from the United States and 15 to 25 strains from eight other countries were positive. Only one of 86 strains of *E. carotovora,* none of 15 strains of other *Erwinia* spp., and none of 60 other bacteria including several isolated from soil and decayed potato tubers were positive. Additional strains of *E. atroseptica* from the United States have been tested since and all have been positive (A. Kelman, personal communication).

In contrast to this, AS to several strains are needed to identify *E. chrysanthemi* (176) and *P. solanacearum* (138) using MPC and ODD. For example, AS to MPC of six strains of *P. solanacearum* are needed to type 19 strains from five states and one territory of Brazil into five serovars (138). Furthermore, when we extended the study to include strains of *P. solanacearum* from outside Brazil the results were even more complex. Antisera to six additional strains (one each from United States, Kenya, Rhodesia, Costa Rica, Colombia, and Peru) are needed to type the 19 strains from Brazil and 35 additional strains from outside Brazil into 9 serovars. Contrary to our initial results showing that serovars correlate with host of origin, no correlation was observed when strains outside Brazil were used (N. W. Schaad and co-workers, unpublished data).

CONCLUSION

Serology has advanced a long way since Jensen first used agglutination tests to identify a strain of *Agrobacterium* (68). The reliability of serology in molecular biology is well known, but only recently has serology been accepted as a reliable tool for identifying plant pathogenic bacteria. Furthermore, serology will never replace the more conventional and time-consuming bacteriological techniques as the primary tool for identification unless its reliability can be clearly established.

Much of the literature on serological identification does not instill such confidence. Not until the best methods currently available are used for characterization of the organism will those working with the identification of bacteria begin to use serology on a routine basis. Presently, several researchers are using a particular method they have developed to identify a specific organism, but few are using a method developed by someone else.

Ideally, we should have standardized methods for identifying all plant pathogenic bacteria. Perhaps having standardized methods for identifying the most commonly isolated species is more realistic. Once standardized methods are agreed upon, AS could be made commercially for each pathogen. However, much research is still needed before standardized methods can be chosen.

One very important question needs to be answered. What should be used for immunization? Do we use (*a*) whole cells, (*b*) heat-killed cells, (*c*) glutaraldehyde-fixed cells, (*d*) glycoprotein extracts, (*e*) ribosome preparations, (*f*) membrane protein complexes, or (*g*) purified enzymes. Additional time and purification techniques are required to obtain some of these preparations; however, the resulting AS are more specific and need no further attention. As stated earlier, a relatively nonspecific AS may be made more specific by absorbing out unwanted antibodies, but there are many technical problems in doing so and absorption is not generally recommended [p. 108 in (27)]. A major advantage of specific AS is that crude extracts of cells or untreated cells can be used as test antigens.

I believe the use of monospecific AS such as AS to pectic lyases (94) and L-asparaginases (8, 15) should be investigated further. Various protein-specific cross-linking compounds (4) such as the glutaraldehyde used by Allen & Kelman (3) should be tested further, also. Perhaps membrane proteins that are highly immunogenic (136) and that can be obtained by sucrose density gradient centrifugation could be used. Finally, a specific immunogen present in a complex perhaps could be purified by polyacrylamide gel electrophoresis. That portion of the gel containing the antigen could be cut out and then injected into the rabbit with or without the gel.

The following suggestions for developing a serological test can be made from what is currently known about serology:

1. Use ODD tests to determine the specificity of the AS and the serological variability of the organism.
2. Use AS homologous to at least two or three strains of each nomenspecies.
3. Use AS homologous to several other closely related bacteria in the same genus.
4. Use test antigens from a relatively large number of strains (20 or more, depending upon the organism). Choose carefully from several different geographical areas and hosts, if applicable.
5. Use test antigens from several other bacteria in different genera.

For routine identification, IF with a specific AS such as to membrane protein complex or glutaraldehyde-fixed cells is probably the best test available. I prefer the indirect method over the direct method because it is simpler (although the staining takes more time) and is more standardized. In the direct method the AS must be conjugated to the fluorescein isothiocyanate (FITC) dye correctly [molecular relationship between FITC and globulin protein (F/P ratio) must be determined (62)]. In the indirect method the dye is conjugated to antirabbit globulin rather than the AS to the organism to be identified and is available commercially with a proper F/P ratio. A block test (49) is easily performed to determine the optimum dilution of AS and antirabbit globulin.

Indirect IF requires very little AS and whole cells taken from agar cultures are used as test antigens. For example, using AS to ribosome preparations 160 suspected cultures of *X. campestris* can be tested with 1.0 ml of AS (0.05 ml of a 1 : 8 dilution per sample).

I hope that this review will stimulate further interest in using serological tests for identification and that the increased interest will eventually result in the adoption of standardized methods.

ACKNOWLEDGMENTS

I wish to express my sincere appreciation to Drs. R. A. Lelliott, H. P. Maas Geesteranus, H. Vruggink, and K. Rudolph for allowing me access to their card files and to A. Trigalet and R. Davis for sending me lists of references. I also wish to express my appreciation to Dr. Z. Klement, U. Mazzucchi, T. Tominaga, and A. Tanii for supplying reprints. I wish to thank Dr. A. Kelman and H. Vruggink for sending me copies of unpublished manuscripts. I am endebted to Dr. K. Rudolph for providing reprints and English translations of certain parts of Professor Stapp's work and to Dr. T. Tominaga for translating parts of several Japanese papers. I am most grateful to Dr. B. Cunfer, Dr. J. Demski, and my wife Terry for suggestions concerning the manuscript.

Literature Cited

1. Addy, S. K., Dhal, N. K. 1977. Serology of *Xanthomonas oryzae*. *Indian Phytopathol.* 30:64–69
2. Akerman, A., Zutra, D., Zafrira, V., Henis, Y. 1973. Application of an immunofluorescent technique for detecting *Corynebacterium michiganense* and estimating its extent in tomato seed lots. *Phytoparasitica* 1:128 (Abstr.)
3. Allen, E., Kelman, A. 1977. Immunofluorescent stain procedures for detection and identification of *Erwinia carotovora* var. *atroseptica*. *Phytopathology* 67:1305–12
4. Arrameas, S., Ternynck, T. 1969. The cross-linking of proteins with glutaralydehyde and its use for the preparation of immunoabsorbents. *Immunochemistry* 6:53–76
5. Barbakadze, N. A. 1974. Sero diagnosis of legume bacteriosis caused by the genus *Pseudomonas*. *Soobsch. Akad. Nauk Gruz. SSR* 74:193–95
6. Bascomb, S., Bettelheim, K. A. 1976. Immunological relationships of bacterial L-asparaginases. *J. Gen. Microbiol.* 92:175–82
7. Belenkii, D. E., Popova, N. N. 1939. The drop method of agglutination and

its application to the diagnosis of black bacteriosis of wheat. *Proc. Lenin. Acad. Agric. Sci. USSR* 14:26–29
8. Berquist, K. R., Elrod, R. P. 1948. The somatic antigens of the genus *Agrobacterium*. *Proc. SD Acad. Sci.* 27:104–11
9. Braun, A. C. 1937. A comparative study of *Bacterium tabacum* Wolf and Foster and *Bacterium angulatum* Fromme and Murray. *Phytopathology* 27:283–304
10. Braun, A. C., McNew, G. L. 1940. Agglutinon-absorption by different strains of *Phytomonas stewartii*. *Bot. Gaz.* 102:78–88
11. Breed, R. S., Murray, E. G. D., Smith, N. R. 1957. *Bergey's Manual of Determinative Bacteriology*. Baltimore: Williams & Wilkins. 1094 pp. 7th ed.
12. Brooks, R. S., Nain, K., Rhodes, M. 1925. The investigation of phytopathogenic bacteria by serological and biochemical methods. *J. Pathol. Bacteriol.* 28:203–9
13. Buchanan, R. E., Gibbons, N. E. 1974. *Bergey's Manual of Determinative Bacteriology*. Baltimore: Williams & Wilkins. 1268 pp. 8th ed.

14. Carpenter, P. L. 1975. *Immunology and Serology.* Philadelphia: Saunders. 346 pp. 2nd ed.

15. Chalkovskaya, S. M., Makarova, R. A., Tochenaya, N. P., Frolova, M. A. 1976. Antigenic properties of L-asparaginase isolates from different microbiol species and strains. *Antibiotiki Moscow* 21: 255–58

16. Charudattan, R., Stall, R. E., Batchelor, D. L. 1973. Serotypes of *Xanthomonas vesicatoria* unrelated to its pathotypes. *Phytopathology* 63:1260–65

17. Chen, T. A., Liao, C. H. 1973. Corn stunt spiroplasma: Isolation, cultivation, and proof of pathogenicity. *Science* 188:1015–17

18. Claflin, L. E., Shepard, J. E. 1977. An agglutination test for the serodiagnosis of *Corynebacterium sepedonicum. Am. Potato J.* 54:331 38

19. Claflin, L. E., Uyemoto, J. K. 1978. Serodiagnosis of *Corynebacterium sepedonicum* by enzyme-linked immunosorbent assay. *Phytopathol. News* Vol. 12, p. 156 (Abstr.)

20. Coe, J. E., Salvin, S. B. 1964. The immune response in the process of delayed hypersensitivity of circulating antibody. *J. Immunol.* 93:495–510

21. Coleman, M. F., Reid, J. J. 1945. The serological study of strains of *Aklaligenes radiobacter* and *Phytomonas tumefaciens* in "M" and "S" phases. *J. Bacteriol.* 49:187–92

22. Coleno, A. 1968. Utilisation de la technique d'immunofluorescence pour le depistage de *Pseudomonas phaseolicola* (Burkh) Dowson dans les lots de semences contamines. *C. R. Seances Acad. Agric. Fr.* 54:1016–20

23. Coleno, A., Hingand, L., Rat, B. 1976. Some aspects of the serology of *Pseudomonas solanacearum* E. E. Smith. In *Proc. 1st Planning Conf. and Workshop on the Ecology and Control of Bacterial Wilt Caused by Pseudomonas solanacearum,* ed. L. Sequeira, A. Kelman, pp. 110–19. Raleigh, NC: North Carolina State Univ. 166 pp.

24. Coleno, A., Hingand, L., Renee, B. M. 1970. Contribution a letude serologique de *Pseudomonas phaseolicola* (Burk) Dowson. *Ann. Phytopathol.* 2:199–207

25. Coleno, A., LeNormand, M., Bariz, M. R. 1972. A qualitative study on the antigens involved in the complement-fixation among some phytopathogenic pseudomonads. In *Proc. 3rd Int. Conf. Plant Pathog. Bact., Wageningen,* ed. H. P. Maas Geesteranus, pp. 143–50. Toronto: Univ. Toronto Press. 365 pp.

26. Crowle, A. J. 1960. Interpretation of immunodiffusion tests. *Ann. Rev. Microbiol.* 14:161–76

27. Crowle, A. J. 1973. *Immunodiffusion.* New York: Academic. 545 pp. 2nd ed.

28. Cupples, D., Kelman, A. 1974. Evaluation of selective media for isolation of soft-rot bacteria from soil and plant tissue. *Phytopathology* 64:468–75

29. Damann, K. E., Tejada, J., Derrick, K. S. 1979. Serologically specific microscopy of the bacterium associated with ratoon stunting disease of sugarcane. In *Proc. 4th Int. Conf. Plant Pathog. Bact., Angers,* ed M. Ride. In press

30. Davis, R. E. 1978. Spiroplasma associated with flowers of the tulip tree (*Liriodandron tulipifera* L.). *Can. J. Microbiol.* 24:954–59

31. Dazzo, F. B., Hubbell, D. H. 1975. Antigenic differences between infective and non-infective strains of *Rhizobium trifoli. Appl. Microbiol.* 30:172–77

32. Derrick, K. S., Brlansky, R. H. 1976. Assay of virus and mycoplasmas using serologically specific electron microscopy. *Phytopathology* 66:815–20

33. Digat, B., Cambra, M., 1976. Specificity of antigens in *Pseudomonas solanacearum* E. F. Smith and application of serology for studying bacterial wilt. See Ref. 23, pp. 38–57

34. Dobias, K. 1973. The serological relationship of strains of *Erwinia carotovora* (Jones) Holland isolated from potato. *Rostl. Vyroba* 19:277–84

35. Domen, H. Y., Alvarez, A. 1979. Detection of *Xanthomonas campestris* in soil using a direct immunofluorescence techniques. See Ref. 29

36. Dudman, W. F., Brockwell, J. 1968. Ecological study of root-nodule bacteria introduced into field environments. I. A survey of field performance of clover inoculants by gel immunodiffusion serology. *Aust. J. Agric. Res.* 19:739–47

37. Edwards, P. R., Bruner, D. W. 1942. Serological identification of *Salmonella* cultures. *Ky. Agric. Exp. Stn. Circ.* 154:1–35

38. Edwards, P. R., Ewing, W. H. 1962. *Identification of Enterobacteriaceae.* Minneapolis, Minn: Burgess. 258 pp. 2nd ed.

39. Elrod, R. P. 1941. Serological studies of the *Erwinieae.* I. *Erwinia amylovora. Bot. Gaz.* 103:123–31

40. Elrod, R. P., Braun, A. C. 1947. Serological studies of the genus *Xanthomonas.* I. Cross-agglutination relationships. *J. Bacteriol.* 53:509–18

41. Elrod, R. P., Braun, A. C. 1947. Serological studies of the genus *Xanthomonas*. II. *Xanthomonas translucens* groups. *J. Bacteriol.* 53:519–24

42. Erasmus, H. O., Matthee, F. N., Louw, H. A, 1974. A comparison between plant pathogenic species of *Pseudomonas, Xanthomonas,* and *Erwinia* with special reference to the bacterium responsible for bacterial blight of vines. *Phytophytactica* 6:11–18

43. Fliege, H. F. 1971. Beitrag zur atiologie einer bakteriellen erkrankung an sauerkirschen und zur taxonomie paraenchympathogener pseudomonaden. *Zentrabl. Bakteriol. Parasitenkd. Infektionskr. Hyg. Erste Abt. Orig. Reihe A* 126:171–225

44. French, W. J., Stassi, D. L., Schaad, N. W. 1978. The use of immunofluorescence for the identification of phony peach bacterium. *Phytopathology* 68:1106–8

45. Friedman, B. A. 1953. Serological tests with some phytopathogenic species of *Pseudomonas. Phytopathology* 43:412–14

46. Gehring, F. 1961. Investigations on the occurrence of plant pathogenic bacteria of the genus *Pseudomonas* on woody plants in 1960 in the Federal Republic. *Nachrichtenbl. Dtsch. Pflanzenschutzdienstes Stuttgart* 13:172–82

47. Gehring, F. 1962. On an occurrence of *Pseudomonas solanacearum* in Egyptian imported potatoes and on a simple serological demonstration method for this bacterium in heavily infected tuber material. *Nachrichtenbl. Dtsch. Pflanzenschutzdienstes Stuttgart* 14:27–29

48. Gillaspie, A. G. 1978. Ratoon stunting disease of sugarcane: Serology. *Phytopathology* 68:529–32

49. Goldman, M. 1968. *Fluorescent Antibody Methods,* pp. 157–58. New York: Academic 303 pp.

50. Goto, M., Okabe, N. 1957. Studies on strains of *Erwinia carotovora.* (Jones) Holland. IV. Antigenic variations. *Bull. Fac. Agric. Shizuoka Univ.* 7:11–20

51. Goto, M., Okabe, N. 1958. Studies on strains of *Erwinia carotovora* (Jones) Holland. V. Antigenic structures of soma, their relation to biochemical properties, and heat inactivation of agglutinins. *Bull. Fac. Agric. Shizuoka Univ.* 8:1–31

52. Graham, D. C. 1963. Serological diagnosis of potato blackleg and tuber soft rot. *Plant Pathol.* 12:142–44

53. Graham, D. C. 1964. Taxonomy of the soft rot coliform bacteria. *Ann. Rev. Phytopathol.* 2:13–42

54. Graham, P. H. 1963. Antigenic affinities of the root-nodule bacteria of legumes. *Antonie van Leeuwenhoek J. Microbiol. Serol.* 29:281–91

55. Graham, P. H. 1971. Serological studies with *Agrobacterium radiobacter, A. tumefaciens* and *Rhizobium* strains. *Arch. Mikrobiol.* 78:70–75

56. Guthrie, J. W. 1968. The serological relationship of races of *Pseudomonas phaseolicola. Phytopathology* 58:716–17

57. Guthrie, J. W. 1970. The detection and eradication of *Pseudomonas phaseolicola* in *Phaseolus vulgaris. Proc. Int. Seed Testing Assoc.* 35:89–93

58. Guthrie, J. W., Huber, D. M., Fenwick, H. S. 1965. Serological detection of halo blight. *Plant Dis. Reptr.* 49:297–99

59. Hagborg, W. A. F. 1946. The diagnosis of bacterial black chaft of wheat. *Sci. Agric.* 26:140–46

60. Hale, C. N. 1972. Rapid identification methods for *Corynebacterium insidiosum* (McCulloch, 1925) Jensen 1934. *NZ J Agric. Res.* 15:149–54

61. Harrison, D. E., Freeman, H. 1961. Bacterial wilt of potato. II. Serological relationship of two strains of *Pseudomonas solanacearum* and a culture of *Corynebacterium sepedonicum. Aust. J. Agric. Res.* 12:872–77

62. Herbert, G. A., Pittman, B., McKinney, R. M., Cherry, W. B. 1972. *The Preparation and Physiochemical Characterization of Fluorescent Antibody Reagents.* Atlanta, Ga.: US Dep. Health Educ. Welfare, Cent. Dis. Control. 41 pp.

63. Hevesi, M. 1968. The occurrence of a bacterium sp. belonging to the *Pseudomonas syringae* van Hall group on *Sorghum* spp. *Acta Phytopathol. Acad. Sci. Hung.* 3:321–29

64. Horgan, E. S. 1931. The value of serological tests for the identification of *Pseudomonas malvacearum. J. Bacteriol.* 22:1181–84

65. Israilskii, V. P., Shkylar, S. N., Orlova, G. I. 1967. Use of the serological method of diagnosing *E. amylovora* without isolating the culture from diseased plant. *Mikrobiologiya* 36:412–16

66. Jenkins, S. F., Morton, D. J., Dukes, P. D. 1966. Distinguishing *Pseudomonas solanacearum* infections from other peanut wilt diseases by the use of serological techniques. *Plant Dis. Reptr.* 50:836–38

67. Jenkins, S. F., Morton, D. J., Dukes, P. D. 1967. Comparison of techniques for detection of *Pseudomonas solanacearum* in artificially infected soil. *Phytopathology* 57:25–27

68. Jensen, C. O. 1918. Undersogelser vedrorende nogle svulstilignende dannelser hos planter. *Ser umlaboratorium Meddelelser fra den Kgl. veterinaer-og landbohojskoles aarsskrift.* Copenhagen: Nielsen & Lydiche. 143 pp.

69. Kabashnaja, L. V. 1977. Serological study of *Erwinia aroideae* (Townsend) Holland and *E. carotovora* (Jones) Holland isolated from hyacinths and Callas affected by soft rot. *Microbiol. Z. Kiev* 39:450–57

70. Kawakami, K., Kobayashi, T. 1975. Diagnosis of *Erwinia atrospetica* infection of potatoes by antisera. *Ann. Phytopathol. Soc. Jpn.* 41:120 (Abstr.)

71. Keane, P. J., Kerr, A., New, P. B. 1970. Crown gall of stone fruit. II. Identification and nomenclature of *Agrobacterium* isolates. *Aust. J. Biol. Sci.* 25:585–95

72. Kikumoto, T., Sakamoto, M. 1967. Ecological studies on the soft rot bacteria of vegetables. III. Application of immunofluorescent staining for the detection and the counting of *Erwinia aroideae* in soil. *Ann. Phytopathol. Soc. Jpn.* 33:181–86

73. Klement, Z., Lovrekovich, L. 1960. Comparative study of *Pseudomonas* species affecting hemp and the mulberry tree. *Acta Microbiol.* 7:113–19

74. Klement, Z., Lovrekovich, L., Hevesi, M. 1960. Studies on the biochemistry phase sensitivity and serological properties of *Pseudomonas* pathogenic to mulberry. *Phytopathol. Z.* 38:18–32

75. Kotylyarova, L. O., Pasheva, V. A. 1975. Protection of potato agent latent forms of bacterial and virus diseases. *Nauchno-Issled. Tr. Tsentr. Nauchno-Issled. Inst. Shelkovoi Promsti.* 23:143–45

76. Lacey, M. S. 1926. Studies in bacteriosis. XIII. A soft rot of potato tubers due to *Bacillus carotovora* and a comparison of the cultural, pathological, and serological behavior or various organisms causing soft rots. *Ann. Appl. Biol.* 13:1–11

77. Lazar, I. 1968. Serological relationships of corynebacteria. *J. Gen. Microbiol.* 52:77–88

78. Lazar, I. 1971. Serological relationships between the 'amylovora', 'carotovora' and 'herbicola' groups of the genus Erwinia. In *Proc. 3rd Int. Conf. Plant Pathog. Bact.,* ed. H. P. Maas Geesteranus, pp. 131–141. Toronto: Univ. Toronto Press. 365 pp.

79. Lelliott, R. A. 1968. The diagnosis of fire blight (*Erwinia amylovora*) and some diseases caused by *Pseudomonas syringae*. Report of the International Conference on fire blight. *Eur. Mediterr. Plant Prot. Organ. EPPD Publ.* 45:27–34

80. Leoville, F., Coleno, A. 1976. Detection de *Xanthomonas albilineans* (Ashby) Dowson, agent delechaudure da La Lanne a sucre dans des boutures contaminees. *Ann. Phytopathol.* 8:233–36

81. Lin, D. O., Li, O. C., Kuo, T. T. 1969. Bacterial leaf blight of rice. I. Serological relationship between virulent and weakly virulent strains of *Xanthomonas oryzae. Bot. Bull. Acad. S. Taipei* 10:130–33

82. Link, G. K. K., Link, A. D. 1928. Further agglutination tests with bacterial plant pathogens. I. *Bacterium campestre—Bact. phaseoli* group; *Bact. medicaginis* var. *phaseolicola; Bact. tumefaciens. Bot. Gaz.* 85:178–97

83. Link, G. K. K., Sharp, C. G. 1927. Correlation of host and serological specificity of *Bact. campestris, Bact. flaccumfaciens, Bact. phaseoli,* and *Bact. phaseoli* sojense. *Bot. Gaz.* 83:145–60

84. Lovrekovich, L., Klement, Z. 1961. Species specific antigens of *Pseudomonas tabaci. Acta Microbiol. Acad. Sci. Hung.* 8:303–10

85. Lovrekovich, L., Klement, Z. 1965. Serological and bacteriophage sensitivity studies on *Xanthomonas vesicatoria* strains isolated from tomato and pepper. *Phytopathol. Z.* 52:222–28

86. Lovrekovich, L., Klement, Z., Dowson, W. J. 1963. Serological investigation of *Pseudomonas syringae* and *Pseudomonas morsporunorum* strains. *Phytopathol. Z.* 47:19–24

87. Lucus, L. T., Grogan, R. G. 1969. Serological variation and identification of *Pseudomonas lachrymans* and other phytopathogenic *Pseudomonas* nomenspecies. *Phytopathology* 59:1908–12

88. Mahanta, I. C., Addy, S. K. 1977. Serological specificity of *Xanthomonas oryzae* incitant of bacterial blight of rice. *Int. J. Syst. Bacteriol.* 27:383–85

89. Matsumoto, T. 1929. On the diagnosis of certain plant infectious diseases by means of serological reactions. *J. Soc. Trop. Agric. Formosa* 1:14–22

90. Matsumoto, T. 1930. Further studies on some putrefactive phytopathogenic bac-

teria by agglutinin absorption. *J. Soc. Trop. Agric. Formosa* 2:16–25

91. Matsumoto, T., Okabe, N. 1931. On the causal organisms of the bacterial soft rot of Kotyo-ran, *Phalaenopsis aphrodite* Reich. *J. Soc. Trop. Agric. Formosa* 3:117–34

92. Matsumoto, T., Somazawa, K. 1931. On the relationship between the serological reaction and other biological characters of some putrefactive phytopathogenic bacteria. *J. Soc. Trop. Agric. Formosa* 3:317–36

93. Mazzucchi, U. 1975. Vascular blackening in sugarbeet toproots caused by *Pseudomonas syringae* vanHall. *Phytopathol. Z.* 84:289–99

94. Mazzucchi, U., Alberghina, A., Garibaldi, A. 1974. Comparative immunological study of pectic lyases produced by soft rot coliform bacteria. *Phytopathol. Mediterr.* 13:27–35

95. McNew, G. L., Braun, A. C. 1946. Agglutination test applied to strains of *Phytomonas stewartii. Bot. Gaz.* 102:64–77

96. Medeiros, A. G. 1961. Estudos serologicos de *Pseudomonas solanacearum* (Smith) Smith. In *Estudo Sobre Pseudomonas solanacearum (Smith) Smith,* ed. A. G. Medeiros, pp. 5–18. Rio de Janeiro: Escritorio Tech. Agri. 54 pp.

97. Miller, P. W., Bollen, W. B., Simmons, J. E., Cross, H. N., Brass, H. P. 1940. The pathogen of filbert bacteriosis compared with *Phytomonas juglandis,* the causal agent of walnut blight. *Phytopathology* 30:731–33

98. Morton, D. J. 1965. Comparison of three serological procedures for identifying *Xanthomonas vesicatoria* in pepper leaves. *Phytopathology* 55:421–24

99. Morton, D. J., Dukes, P. D., Jenkins, S. F. Jr. 1965. Serological identification of *Pseudomonas solanacearum* in four solanaceous hosts. *Phytopathology* 55:1191–93

100. Morton, D. J., Dukes, P. D., Jenkins, S. F. Jr. 1966. Serological relationships of races 1, 2 and 3 of *Pseudomonas solanacearum. Plant Dis. Reptr.* 50:275–77

101. Mushin, R., Naylor, J., Lahovary, N. 1959. Studies on plant pathogenic bacteria. II. Serology. *Aust. J. Biol. Sci.* 12:233–46

102. Novakova, N. 1936. *Mikrobiologichnii Zhurnal (Akad. Nauk URSR)* 3:139

103. Novakova, J. 1957. A new method of isolation of blackleg pathogen from diseased plants. *Phytopathol. Z.* 29:72–74

104. Obata, T. 1968. Studies on the hyacinth yellows and the use of antiserum as an aid to its diagnosis in the field. *Res. Bull. Plant Prot. Serv. Jpn.* 5:7–16

105. Obata, T., Tsuboi, F. 1972. Studies on the serological diagnosis of bacterial plant diseases. I. Cross-agglutination and gel-diffusion tests with some *Xanthomonas nomen* species. *Res. Bull. Plant Prot. Ser. Jpn.* 10:8–16

106. O'Brien, L. M., Morton, D. J., Manning, W. J., Scheetz, R. W. 1967. Serological differences between apparently typical pepper and tomato isolates of *Xanthomonas vesicatoria. Nature* 215:532–33

107. Okabe, N., Goto, M. 1955. Bacterial plant diseases in Japan. II. Studies on soft rots due to *Erwinia aroideae* (Townsend) Holland, with special reference to the antigenic structures of flagella. *Bull. Fac. Agric. Shizuoka Univ.* 5:72–86

108. Okabe, N., Goto, M. 1956. Studies on the strains of *Erwinia carotovora* (Jones) Holland. I. Antigenic structure of flagella and their relations to pathogenicity and maltose fermentation. *Bull. Fac. Agric. Shizuoka Univ.* 6:16–32

109. Ono, K. 1976. Ecological studies on the wildfire disease of tobacco. *Bull. Morioka Tob. Exp. Stn.* 11:1–52

110. Oprea, F. 1971. Serological investigations on bacteria *Pseudomonas morsprunorum* and *Pseudomonas syringae. Rev. Roum. Biol. Ser. Bot.* 16:213–20

111. Orellana, R. G., Weber, D. F. 1971. Immunodiffusion analysis of isolates of *Xanthomonas cyamopsidis. Appl. Microbiol.* 22:622–24

112. Otta, J. D. 1977. Occurrence and characteristics of isolates of *Pseudomonas syringae* on winter wheat. *Phytopathology* 67:22–26

113. Otta, J. D., English, H. 1971. Serology and pathogenicity of *Pseudomonas syringae. Phytopathology* 61:443–52

114. Ouchterlony, O. 1948. In vitro method for testing the toxin-producing capacity of diptheria bacteria. *Acta Pathol. Microbiol. Scand.* 25:186–91

115. Pastushenko, L. T., Koroleva, I. B., Sydorenico, S. S. 1975. Identification of *Pseudomonas atrofaciens* by a serological method. *Mikrobiol. Z. Kiev* 37:373–74

116. Pastushenko, L. T., Koroleva, I. B., Sidorenko, S. S., Simonovich, I. D. 1976. Serological study of the pathogen

of base bacteriosis in wheat. *Sk. Biol.* 11:582–86

117. Pastushenko, L. T., Simonovich, I. D. 1971. Obtaining specific sera to phytopathogenic bacteria of the genus *Pseudomonas Mikrobiol. Z. Kiev* 33: 37–40

118. Pastushenko, L. T., Simonovich, I. D. 1971. Study of methods for obtaining antigens of pathogenic bacteria of the genus *Pseudomonas. Mikrobiol. Z. Kiev* 33:389–95

119. Paton, A. M. 1943. The adaption of the immunofluorescence technique for use in bacteriological investigations of plant tissues. *J. Appl. Bacteriol.* 27:337

120. Perez, J. E. 1964. The use of fuller's foramide method in the serological identification of *Pseudomonas solanacearum. J. Agric. Univ. PR* 46:144–53

121. Perlasca, C. 1960. Relationships among isolates of *Pseudomonas syringae* pathogenic on stone fruits. *Phytopathology* 50:889–99

122. Porath, J., Axon, R., Ernback, S. 1967. Chemical coupling of proteins to agrose. *Nature* 215:149–50

123. Reid, J. J., Naghski, J., Farell, M. A., Huley, D. E. 1942. Bacterial leafspots of Pennsylvania tobacco. I. Occurrence and nature of the microorganism associated with wildfire. *Penn. State Coll. Bull.* 422:1–36

124. Riker, A. J., Barfield, W. M., Wright, W. J., Keitt, G. W., Sagen, H. E. 1930. Studies on infections, hairy root on nursery apple trees. *J. Agric. Res.* 41: 507–40

125. Romero, P., Callao, V. 1969. Tecnica Estandardizada para el aislamiento e identificación del agente etiologico de la tuberculosis del olivo. *Microbiol. Esp.* 22:219–31

126. Rosen, H. R., Bleecker, W. L. 1933. Comparative serological and pathological investigations of the fire-blight organism and a pathogenic, fluorescent group of bacteria. *J. Agric. Res.* 46:95–119

127. Rosenthal, S. A., Cox, C. D. 1953. The somatic antigens of *Corynebacterium michiganense* and *Corynebacterium insidiosum. J. Bacteriol.* 65:532–37

128. Rosenthal, S. A., Cox, C. D. 1954. An antigenic analysis of some plant and soil *Corynebacteria. Phytopathology* 44: 603–4

129. Saaltink, G. J., van Slogteren, D. H. M., Kamerman, W. 1972. Serological identification of bacteria in bulbs. See Ref. 25, pp. 153–56

130. Samson, R. 1972. Heterogenicity of heat stable antigens in *Erwinia amylovora. Ann. Phytopathol.* 4:157–63

131. Samson, R. 1973. The pectinolytic *Erwinia* II. Studies on somatic antigens of *Erwinia carotovora* var. *chrysanthemi. Ann. Phytopathol.* 5:377–88

132. Sato, M., Takahashi, K., Wakimato, S. 1971. Properties of the causal bacterium of bacterial blight of mulberry, *Pseudomonas mori* (Boyer et Lambert). Stevens and its phages. *Ann. Phytopathol. Soc. Jpn.* 37:128–35

133. Schaad, N. W. 1974. Comparative immunology of ribosomes and disc gel electrophoresis of ribosomal proteins from erwiniae, pectobacteria, and other members of the family Enterobacteriaceae. *Int. J. Syst. Bacteriol.* 24: 42–53

134. Schaad, N. W. 1976. Immunological comparison and characterization of ribosomes of *Xanthomonas vesicatoria. Phytopathology* 66:770–76

135. Schaad, N. W. 1978. Use of direct and indirect immunofluorescence tests for identification of *Xanthomonas campestris. Phytopathology* 68:249–52

136. Schaad, N. W., Donaldson, R. C. 1978. Identification of nonribosomal immunogens in ribosomal preparations of *Salmonella typhimurium. Abstr. Ann. Meet. Am. Soc. Microbiol.,* p. 55 (Abstr.)

137. Schaad, N. W., Kendrick, R. 1975. A qualitative method of detecting *Xanthomonas campestris* in crucifer seed. *Phytopathology* 65:1034–36

138. Schaad, N. W., Takatsu, A., Dianese, J. C. 1979. Serological identification of *Pseudomonas solanacearum.* See Ref. 29

139. Schaad, N. W., White, W. C. 1974. A selective medium for soil isolation and enumeration of *Xanthomonas campestris. Phytopathology* 64:876–80

140. Schaad, N. W., Wilson, E. E. 1971. The ecology of *Erwinia rubrifaciens* and the development of phloem canker of Persian walnut. *Ann. Appl. Biol.* 69:125–26

141. Sever, J. L., Madden, D. L. 1977. Enzyme-linked immunosorbent assay (ELISA) for infectious agents. *J. Infect. Dis.* 136:5257–5340 (Oct Suppl.)

142. Sharp, C. G. 1927. Virulence, serological, and other physiological studies of *Bact. flaccumfaciens, Bact. phaseoli,* and *Bact. phaseoli sojense. Bot. Gaz.* 83:113–44

143. Shekhawat, P. S., Chakravarti, B. P. 1977. Serological tests to find out antigenic differences among isolates of *Xan-*

thomonas vesicatoria (Doidge) Dowson and some other phytopathogenic bacteria. Cur. Sci. 46:46–47

144. Shinde, P. A., Lukezic, F. L. 1974. Characterization and serological comparisons of bacteria of the genus Erwinia associated with discolored alfalfa roots. Phytopathology 64:871–76

145. Singh, N., Chohan, J. S., Khatri, H. L. 1978. Immunogenicity of different isolates of Xanthomonas oryzae the causal organism of bacterial leaf blight of rice. Indian J. Microbiol. 17:98–99

146. Deleted in proof

147. Skripal, I. G. 1969. Serologichui vlastyvosli Pseudomonad zbudnykiv bakterioziv pladovgkh derev. Mikrobiol. Z. Kiev 31:588–95

148. Slack, S. A., Kelman, A., Perry, J. B. 1979. Comparison of three serodiagnostic assays for the detection of Corynebacterium sepedonicum. Phytopathology. In press

149. Soda, J. A., Cleveyodn, R. C. 1960. Some serological studies of the genera Corynebacterium, Flavobacterium and Xanthomonas. Antonie van Leeuwenhoek J. Microbiol. Serol. 26:98–102

150. Stanghellini, M. E., Sands, D. C., Kronland, W. C., Mendonca, M. M. 1977. Serological and physiological differentiation among isolates of Erwinia carotovora from potato and sugar beet. Phytopathology 67:1178–82

151. Stapp, C. 1929. Die schwarzbeingkeit and knollennassfaule der kartoffel. Arb. Biol. Reichsans. Land Forstwirtsch. Berlin-Dahlem. 16:643–703

152. Stapp, C. 1930. Bakterielle tabakk rankheiten und ihre erreger. Angew. Bot. 12:241–74

153. Stapp, C. 1937. Der bakterielle stengelband der erbsen. Zentralbl. Bakteriol. Parasitenkd. Infectionskr. Hyg. Abt. 2 96:1–17

154. Stapp, C. 1939. Der bakterielle Erreger einer Blattfleckenkrankheit von Begonien und seine Verwandtschaft mit Pseudomonas campestris, dem Erreger der Adernschwarze des Khols. Arb. Biol. Reichsanst. Land Forstwirtsch. Berlin-Dahlem 22:379–97

155. Stapp, C. 1958. Pflanzanpathogene Bakterien. Berlin & Hamburg: Parey. 259 pp.

156. Takimoto, S. 1925. Studies on the putrefaction of vegetables. J. Plant Prot. 8:344–55

157. Tanii, A., Akai, J. 1975. Blackleg of potato plant caused by a serologically specific strain of Erwinia carotovora var.

carotovora (Jones) Dye. Ann. Phytopathol. Soc. Jpn. 41:513–17

158. Tanii, A., Baba, T. 1973. Bacterial plant diseases in Mokkaido. III. Angular leaf spot of cucumber caused by Pseudomonas lachrymans (Smith et Byran) Carsner. Bull. Hokkaido, Prefect. Agric. Exp. Stn. 28:70–78

159. Tanii, A., Ozaki, M., Baba, T. 1973. Blackleg of potato plant caused by Erwinia atroseptica (van Hall) Jennison (Erwinia carotovora var. atroseptica van Hall) in Japan. Ann. Phytopathol. Soc. Jpn. 39:351–60

160. Tanii, A., Takakuwa, M., Baba, T., Takita, T. 1976. Studies on halo blight of beans (Phaseolus vulgaris) caused by Pseudomonas phaseolicola (Burkholder) Dowson. Misc. Publ. Hokkaido Prefect. Takachi Agric. Exp. Stn. 6:1–60

161. Taylor, J. D. 1970. Bacteriophage and serological methods for the identification of Pseudomonas phaseolicola (Burkh.) Dowson. Ann. Appl. Biol. 66:387–95

162. Taylor, J. D. 1972. Specificity of bacteriophages and antiserum for Pseudomonas pisi. NZ J. Agric. Res. 15: 421–31

163. Thomason, B. M., Moody, M. D., Goldman, M. 1956. Staining bacterial smears with fluorescent antibody. II. Rapid detection of varying numbers of Malleomyces pseudomallei in contaminated materials and infected animals. J. Bacteriol. 72:302–67

164. Thomson, S. V., Schroth, M. N. 1976. The use of immunofluorescent staining for rapid detection of epiphytic Erwinia amylovora on healthy pear blossoms. Proc. Am. Phytopathol. Soc. 3:321 (Abstr.)

165. Tominaga, T. 1971. Studies on the diseases of forage crops in Japan. Bull. Natl. Inst. Agric. Sci. Jpn. Ser. C, pp. 205–306

166. Trigalet, A., Rat, B. 1976. Immunofluorescence as a tool for detecting the internally borne bacterial diseases: Corynebacterium michiganense (E. F. Smith) Jensen and Pseudomonas phaseolicola (Burkholder) Dowson. Rept. 15th Int. Workshop Seed Pathol. 8–14 Sept. 1975, Paris, France, p. 16. 46 pp.

167. Tully, J. G., Whitcomb, R. F., Bove, J. M., Saglio, P. 1973. Plant myocoplasmas: Serological relation between agents associated with citrus stubborn and corn stunt diseases. Science 182: 827–29

168. Vruggink, H. 1979. Enzyme linked immunosorbent assay (ELISA) in the serodiagnosis of plant pathogenic bacteria. See Ref. 29

169. Vruggink, H., DeBoer, S. H., 1979. Detection of *Erwinia carotovora* var. *atroseptica* in potato tubers with immunofluorescence following induction of decay. *Potato Res.* 21:225–29

170. Vruggink, H., Maas Geesteranus, H. P. 1975. Serological recognition of *Erwinia carotovora* var. *atroseptica.* The causal organism of potato blackleg. *Potato Res.* 18:546–55

171. Weaver, W. M., Guthrie, J. W. 1978. Enzyme linked immunospecific assay: Application to the detection of seed borne bacteria. *Phytopathol. News 12:*, p. 156 (Abstr.)

172. Williams, O. B., Glass, H. B. 1931. Agglutination studies of *Phytomonas malvacearum. Phytopathology* 21:1181–84

173. Williamson, D. L., Whitcomb, R. F. 1975. Plant mycoplasma: A cultivible spiroplasma causes corn stunt disease. *Science* 188:1018–20

174. Wilson, E. E., Magie, A. R. 1963. Physiological, serological and pathological evidence that *Pseudomonas tolelliana* is identical with *Pseudomonas savastanoi. Phytopathology* 53:653–59

175. Wu, H. M., Wang, C. K., Chiu, K. Y. 1977. Studies on *Xanthomonas vasculorum,* the causal organism of gummosis of sugar cane. Part 1. Serology. *Plant Prot. Bull.* 19:162–67

176. Yakrus, M., Schaad, N. W. 1979. Serological relationships among strains of *Erwinia chrysanthemi. Phytopathology.* In press

177. Zeitoun, F. M., Wilson, E. E. 1966. Serological comparisons of *Erwinia nigrifluens* with certain other *Erwinia* species. *Phytopathology* 56:1381–84

Ann. Rev. Phytopathol. 1979. 17:149–61
Copyright © 1979 by Annual Reviews Inc. All rights reserved

INSURANCE, INFORMATION, AND ORGANIZATIONAL OPTIONS IN PEST MANAGEMENT

♦3705

Gerald A. Carlson

Department of Economics and Business, North Carolina State University, Raleigh, North Carolina 27650

Insurance markets have not developed in the crop damage area as they have in the human health and real property areas. Unavailability of insurance contracts at any price can result in the reduction of farm output (2). All-risk crop insurance and disaster payments have been provided in the United States with some federal subsidy to help offset the lack of private market insurance. Other countries have also developed various crop-loss insurance schemes.

Self-insurance or self-protection (such as by the use of pesticides or integrated pest management methods) from crop pests are substitutes for market insurance. Optimal decisions about market insurance purchases for a farm operator who is averse to risks would dictate that he attempt to obtain equal amounts of protection per dollar expended on various market and self-protection methods (9). If use of pesticide products is an important form of providing self-protection, then relatively low private costs (as distinct from social costs) of these forms of protection may discourage development of insurance markets. As a result, insufficient levels of crop insurance might exist in these cases.

Several organizations are currently proposing changes in crop insurance, with reduction in pesticide use and an increase in the availability of crop damage insurance as major considerations (23, 24). Despite additional financial support by the US government for various pest management activities, there still are unresolved issues on organizational and delivery mechanisms.

149

0066-4286/79/0901-0149$01.00

Several changes from farmer-directed pest management are considered in addition to insurance in this paper. These alternative approaches might be called group or collective approaches to pest management. They focus on the management and information aspects of pest management. Often, biological features of pests may suggest that economically efficient control systems should involve groups of farmers and pest control specialists.

This paper examines how transaction costs, pest control costs, information costs, and organizational structure set some constraints on future institutional approaches to pest management. The first section briefly surveys the difficulties with insurance related to pest control. The next section reviews existing and proposed insurance schemes for overcoming these problems. The section entitled "Information" examines ways that public and private provision of pest control information assists both market insurance and self-protection. Discussion of the advantages and disadvantages of various organizational forms for pest control follows. Finally, there is an assessment of the most striking gaps in our knowledge in the collective or group approaches to pest management.

MARKET INSURANCE OBSTACLES

The problems of development of and trade in insurance contracts have received some attention by economists and others. Arrow (2), writing on health insurance, drew attention to the "adverse selection" and "moral hazard" difficulties. "Adverse selection" refers to the purchase of insurance by those who know that their probability of payoff on the contract is higher than the average for the group upon which their premiums are based. It also includes the nonpurchase of insurance contracts by those who know that their expected payoffs on the contract plus the value of risk aversion is less than the price of the insurance. "Moral hazard" refers to the ability of the insured (or insurer) to selectively use protective or other resources so as to influence the probability of the insured event occurring or affecting the magnitude of indemnities paid.

An example of moral hazard for a private pest damage insurance contract would be for a farmer to underfertilize (saving him money) and then file an insurance claim for the reduced yield due supposedly to a pest attack that could not be controlled. Adverse selection for this insurance contract would be the nonpurchase decisions by those farmers who typically have very little pest damage compared with their neighbors, but who are in the same yield or rate class. It can be seen that a private insurance firm would not offer to sell a contract which has high costs of preventing moral hazard and adverse selection. This is sometimes referred to as the *high enforcement cost problem.*

A third type of difficulty for private insurance contracts is the high cost of claim adjusting and establishing actuarial tables for premium setting. Claim adjusting costs are similar to moral hazard actions in that it is difficult to specify and verify if and when an insured event has occurred. Crop physiology and pest damage are sufficiently complex in most cases that it is difficult to separate insured from noninsured plant damage (for example, drought stress and insect or disease stress may have similar symptoms). Establishing the probabilities of insured pest damage events is confounded by the many forms of pest protection that may be used, such as resistant crop varieties, adjusting planting dates, pesticides, crop rotation, and others. Damage probabilities would have to be based on the assumption that some standard protection measures are utilized. Also, the pest population information services are seriously lacking in ability to provide area-wide information on a consistent basis for long time periods (6, 17).

The risk spreading potential of pest damage insurance contracts may be quite limited in some cases. Risk pooling by insurance requires that large pest damages in one area be offset by low pest damages in other insured areas, or in other time periods. Crop diseases, nematodes, and insects that have similar amounts of damage over wide areas do not permit insurance risk pooling at low costs. This may apply to many pests which are usually present at constant, low levels. This seems to be a particular problem with weed pests and some insect pests of extensive acreage crops such as soybeans, corn, and wheat. Sale of insurance contracts to other (insurance) firms that have other offsetting risks (known as reinsurance) is one way to overcome this problem.

Self-insurance mechanisms such as credit, crop storage, futures contracts, and forward price contracts can substitute for crop loss insurance contracts. In addition, government-subsidized all-risk crop insurance, disaster payments, and flood insurance limit crop loss exposure of farmers. Lower costs of self-insurance or government-subsidized insurance will lead to less purchase of market insurance.

The above difficulties that insurance companies face, plus the relatively low cost of chemical and other methods of pest control, undoubtedly, have limited the development of pest damage insurance. Yet, a few pest damage insurance contracts are available, and there may be methods of overcoming some of the difficulties mentioned above. All-risk crop insurance does cover pest damage if appropriate protective measures are taken by farmers. There is private insurance for stored product damage including that of some pests, and termite damage is routinely protected against by homeowners with various prepaid inspection, protection, and insurance programs. Disease losses of farm animals can be insured against in private markets.

INNOVATIONS IN INSURANCE CONTRACTS

Both the adverse selection problem and moral hazard problem arise in most insurance markets. The use of *deductibles* and *co-insurance* are the most common pricing mechanisms designed to decrease these problems. Having the first x dollars of damage uninsured (a deductible) will result in policy holders attempting to use resources to prevent occurrence of the insured event. Co-insurance involves a proportional sharing of damage by insurer and insured over all or some range of damage. Combinations of deductibles, co-insurance, and premium changes can reduce false damage claims and increase the options for efficient resource allocation (2, 15, 16). In the broad substitution set outlined by Ehrlich & Becker (9), one can see that there can be an optimum combination of self-insurance (e.g. credit status or crop rotation), self-protection (pesticide and other pest controls), and market insurance with various deductibles and co-insurance. However, these models need to be modified to include the social costs (off-farm damages) associated with use of some of the self-protection options, especially pesticide use.

Prevention of moral hazard is also enhanced by peer associations. These take the form of oaths of allegiance and actions to maintain "good will" by some professions (doctors, lawyers, mechanics) who are sometimes asked to act as third party judges in establishing whether an insured event has occurred. Similar actions might be expected for veterinarians, pest management consultants, or university personnel in the plant protection sciences who may be asked to verify pest-induced agricultural losses.

Group insurance plans are an important organizational innovation in insurance. Although group plans enable some cost savings because of increased size, they may be more important as a device for selecting a low-risk clientele. A health insurance contract offered through an employer may be using conditions for employment as a way to eliminate moral hazards and adverse selection. Barzel (3) has shown how group plans provide information on expected risks (e.g., health status) so that the information on probable damages is equalized between insurer and insured. Group plans can also elicit pressures among peers against "cheating" on levels of damage. Group plans of crop insurance might use cooperative membership of farmers in a given location as a means to establish expected crop losses and premiums. Farmer organizations such as the Farm Bureau have played important roles in other forms of insurance (auto, farm liability); perhaps they have a role in crop-loss insurance. Also, group plans might be written for limiting the liability of various occupational groups involved in pest control such as aerial applicators, pest management consultants, and extension personnel. Group certification if it signifies competency, can lower the cost to insurance companies attempting to estimate probable claims from liability contracts on members of the group.

In a more general way, inexpensive *signals* of expected damages are aids in low cost insurance provision. Zeckhauser (26) has examined the advantages and disadvantages of monitoring one or more features of yield, use of productive resources, or crop losses of insurance participants. The idea he pursues is that if the insurer has a thorough knowledge of the production process he may only need to measure indirectly the insured event to efficiently operate an insurance contract. An example in pest management may be that a farmer may only need to report his pest control expenditures, and the insurance firm can use trap data or surveys of pest populations to infer the amount of damage and indemnities that are paid in an area.

There are wide differences in insurance plans that are offered in various countries. Table 1 lists some of the characteristics of these crop insurance schemes. Some of these programs have been operating for several decades while others are recent or have expired. Most natural hazard and all "all-risk" policies insure against pest damage. The only programs that include a large proportion of crops and acreages are the compulsory insurance schemes. The compulsory programs have crop storage and farm incomes support elements as well. There is a wide range in the proportion of the cost of the programs paid by growers.

Because the current crop insurance program in the United States has not provided disaster insurance or protection for large areas, there are several proposals to revise it (14, 23). Table 2 gives features of current and proposed crop insurance and disaster programs. The disaster payments to growers of only six major crops have provided protection to a large proportion of

Table 1 Characteristics of major crop insurance programs

Country	All risk or specific risks	Compulsory or voluntary	Crops covered	Premium classes	Premium paid by farmer	Eligible area in the program	Source of information
Soviet Union	all risk	compulsory and voluntary	all crops	area	subsidized	all area	Rogers (20)
Sweden	all risk	compulsory	23 crops	area	30%	99%	Swed. Min. Agric. (21)
Australia	specific risks (hail, flood, etc)	voluntary	4 crops	area	subsidized	—[a]	Battese & Francisco (4)
Canada	natural hazards	voluntary	—[a]	area	80%	—[a]	Ray (19)
Japan	natural hazards	compulsory	5 crops	area	40%	all area	Ray (19)
Ceylon	natural hazards	compulsory	rice	area	60%	rice	USDA (22)
Puerto Rico	hurricane loss	voluntary	coffee and shade trees	individual	100%	—[a]	Camacho (5)
Brazil	natural hazards	voluntary	5 crops	area	subsidized	—[a]	Pombo (18)
USA	all risk	voluntary	26 crops	area	100%	13%	USDA (23)

[a] Unknown.

Table 2 Current and proposed crop insurance and disaster programs in the United States

	Number of crops included	Percentage of eligible acreage in program	Total risk coverage ($ billions)	Govern- ment costs ($ millions)	Premium subsidy rates (percentage)	Area (A) or individual farm (I) premiums
Current programs						
Government						
FCIC (1938)[a]	26	13	2.3	30	0	A
Disaster payments (1974)	6	60–100	—[b]	450	100	—
Disaster loans[a]						
FHA (1918)						
SBA (1976)	—[c]	—[b]	1.4[d]	75	50–75[e]	—
Private Hail	—[b]	25	8	0	0	A
Proposed programs[f]						
I. All Federal Crop Insurance	18–all	68	14.9	542	50, 20, 0[g]	I
II. Joint Federal-Private Crop Insurance	18–all	68	14.9	857	50, 20, 0[g]	I
III. Subsidized Reinsurance[h]	18–all	55	10.8	671	50, 20, 0[g]	A

[a] FCIC, Federal Crop Insurance Corporation; FHA, Farmers Home Administration; SBA, Small Business Association.
[b] Unknown.
[c] All designated disaster crops are eligible; this has been about two thirds of all crops in recent years.
[d] Not risk exposure, but average total loan amounts of recent years.
[e] Interest charges are 5% on all loans beginning in 1979; this is an estimated subsidy of 50–75%.
[f] Proposed program estimates are not adjusted for inflation; they represent 3rd–5th year operating characteristics [US Dep. Agric. (23)].
[g] Premium subsidies for three levels of insured crop prices, low to high.
[h] Reinsurance is the sale of insurance contracts to other firms that have offsetting risks.

eligible acreage. However, disaster payments have been fully subsidized and represent a large cost to taxpayers for the acreage covered. The only major private insurance available has been for hail damage.

The recently proposed crop insurance programs listed in Table 2 represent a decrease in disaster loans and payments, and a major expansion in crop insurance. The crop insurance will include many more crops, beginning with 18 and eventually covering all crops. The premium subsidies are expected to attract subscribers on a much larger proportion of eligible acreage (55–88% as compared with 13% for the current crop insurance program). Government costs of the proposed insurance programs will be slightly higher than the current Federal Crop Insurance Corporation program plus disaster program.

The three proposed options (I, II, and III in Table 2) represent different uses of public and private insurance resources. In each case major improvements are needed for information on probabilities of crop destruction. Premium schedules for two of the proposed crop insurance options will be assigned to individual farms rather than to groups of farmers over wide areas. This will require much more crop damage data collection, but it will increase the incentives for better managed and higher yielding farms to participate. Halcrow (12) has discussed the feasibility of individual farm rate classes for the case of weather-damage insurance. The same procedures can be applied to compute pest loss insurance premiums for individual farms.

INFORMATION

Decisions made with uncertainty will usually be improved with information. Information often applies to more than one producer (farmer). Its use by one producer does not seriously diminish its usefulness for another producer. Also, it may not be profitable for a private firm to compile and sell information if it is available to nonpaying producers. In addition, there are often economies of size in producing and distributing information. Thus, information and pest control recommendations produced by the private sector may be inadequate.

Information can assist in the preparation and enforcement of insurance contracts. Could the public provision of information help establish pest control insurance? This may be the case, but the same information will also help improve self-insurance (for example, direct pest controls by pesticides) and self-protection (crop diversification).

Miranowski (15) concluded that information for improved pest control decisions would reduce pesticide use more than would insurance support. This was based on a simulation of farmer pest control decisions with unknown pest populations and assumptions about how much farmers value avoiding crop losses. Pest population information is more likely to save crop losses and reduce pesticide use, thereby smoothing farm income, than is pest damage insurance.

The forms of information that would be most helpful to private and public insurance firms in the pest control area are those associated with frequency and extent of pest damages. Currently, insurance firms do not know infestation probabilities by areas for determining expected losses and setting premiums. Also, area-wide assessments of actual pest damage and pest densities could reduce moral hazards and claim adjusting costs. In addition, research on low cost methods of distinguishing pest damage in plants from other stresses could assist insurance firms in claim adjustments.

Information currently provided by USDA laboratories, land grant universities, and the Extension Service is directed at lowering costs of pest control. These include new pesticide application methods, pest population and damage thresholds, recommendations on how and when to use chemical pesticides, and how to comply with pesticide use regulations. However, this information from public sources reduces the profitability of privately provided pest control information.

One of the most important sources of information is field scouting by growers, hired scouts, and pest control advisors (1, 11, 13). Hall (13) found that those California citrus and cotton growers who used pest control advisors used fewer insecticide applications with no reduction in profits. Grube & Carlson (10) found that the use of scouting for pests increased if expected levels of pest damage were higher and if farm managers were better trained. Returns from scouting were more than its cost indicating that greater utilization would increase profits. Most of the analysis of extension pest management programs has indicated a high return from these programs, which involve scouting and advisory activities (25).

The importance of information disseminated by chemical firms in pest control decisions should not be minimized. It is frequently quite highly ranked relative to scouts and extension personnel in surveys on sources of information for pest control decision making. For example, several chemical advertisements indicate that their product "saves beneficial insects" or "does not have crop maturity delaying effects."

It seems that use of pest management information and insurance schemes is lagging behind direct chemical pest controls. This may be just the result of past public funding in support of chemical intensive methods. Or it could reflect the true state-of-the-art in technology and information for pest management. However, institutional and organizational innovations are areas that seem to offer untapped potential in pest management.

ORGANIZATIONAL INNOVATIONS

Organization of pest control has been highly decentralized in the United States. There are some federal quarantine and control actions, some local government districts for public pest control, and a small number of grower cooperatives. But most agricultural pest control resources are used or directed by individual farm operators using purchased inputs for individual decisions. Individual farmers may not have specialized pest management skills and they may not have large enough acreages to manage mobile pests. Some organizational innovations that may have possibilities are: (*a*) credit-pest control associations, (*b*) prepaid pest control organizations, and (*c*) procedures for determining pest control actions and grower fees in a centralized pest control organization.

Credit-Pest Control Associations

In many risky operations, credit firms, such as farm Production Credit Associations, or commercial banks ask for and receive guarantees of performance. A livestock herd might not be eligible as loan collateral unless vaccinations are used and a veterinarian is on call. Liability and collision insurance is needed for equipment loans and, perhaps, termite insurance must be purchased before housing and building loans are approved. It seems reasonable that the credit organizations could expand their capacity in the pest management area if loans were contingent on pest control guarantees or plans.

An effective crop pest management plan could serve as a basis for establishing the use of a growing crop for a farm operation loan. The pest management plan for a crop could be based on university or other standard recommendations. Usually, the credit firm would want the plan biased toward more certainty of pest control over the duration of the loan (a high use of chemical pest control plan in many cases). However, if the federal government (to reduce environmental and health costs of pesticides) had a credit subsidy program or a loan guarantee program for plans with low chemical use, farm operators and credit-pest management specialists would be induced to select these plans. Use of these plans could lower the default risk and interest costs compared with loans not contingent on effective pest management. Government costs of interest rate subsidies of this type might be very low.

Farm credit firms would be able to tailor pest management plans to fit the environment of a given farm. More important, they know the management skills of their customer farmers and could provide a pest management plan that would use known operator skills. Credit firms would be able to induce community-wide pest control efforts for mobile (common property) pests. Of course, this organizational change would need a group of pest management specialists, and it would need a reliable set of recommended controls.

Prepaid Pest Control Organizations

Prepaid pest control implies advanced payment for monitoring and pest suppression activities carried out by a firm or organization other than the farmer himself. Such organizations are present today primarily as public abatement districts and pest management cooperatives. Urban pest control such as for structures is also often prepaid. Mosquito abatement districts are a prime example of compulsory, prepaid pest control for a local area. In the Eastern United States they have had good success in lowering pest populations. However, they tend to reduce pest populations to levels lower than tax payers may desire (7). Also, the districts tend to overspend in the use of permanent controls (water drainage and empoundments) relative to pesticide controls. Nevertheless, they offer an organizational form that may be workable for agricultural pests and needs further attention.

Compulsory pest control by government agencies has usually taken the form of federal quarantine or eradication efforts. However, the lowering of costs by increasing the size of a control area sought by such programs might also be obtained by local pest management districts. Irrigation districts sometimes have weed control activities.

Grower cooperatives are another form of prepaid pest control. The voluntary cotton protection groups that have been operating in North Carolina for up to 12 years have been quite successful in providing quantity discounts, pest scouting, and use of economic thresholds (11, 25). One striking feature of the cooperatives is that they monitor pests, purchase and apply chemical pesticides, but they have not influenced long-run pest control measures such as use of resistant host crops, planting dates, or crop rotation. One area needing further development and research is the type of year-round recommending services that are applicable for multiple pests and crops. How can pest monitoring, resistant variety selection, and use of cultural controls be coordinated on a community basis? Community cooperatives have encouraged the use of area-wide pest management methods such as diapause control for boll weevils.

The long-run record of grower cooperatives in reducing pesticide use per unit of crop output is not clear. They often seem to induce more treatments per acre. But crop yields seem to be increased even more with a net reduction of chemicals per dollar of crop output (11, 25).

Centralized Pest Control Decisions

The presence of mobile pests, pesticide drift, or other biological interrelationships can sometimes lead to more effective pest management on an area-wide basis. Because pest populations vary from farm to farm, it is unprofitable to provide equal pest suppression on all farms; not all farmers may be willing to take equal pest damage risks. Therefore, one of the major problems of an area-wide pest management program is how to determine what levels of pest suppression to provide and how to charge the member farmers for this service.

Figure 1 shows a hypothetical example of this problem. Levels of pest suppression effort per unit area are measured on the horizontal axis. Costs and benefits (demand) are in dollars on the vertical axis. The downward sloping lines represent demand for pest control by three subgroups in the area organization. The high (H) subgroup has more severe pest problems and does not wish to take many pest damage risks. At the area-wide pest control costs (MC_1) this group desires level Q_H of effort. Likewise, median and low pest threat growers demand median (Q_M) and low (Q_L) levels of effort, respectively. If the organization must choose one level of effort for all growers it will tend to choose the median level (Q_M). The high effort farmers will tend to join the group and accept the median level of group effort; they will supplement it with individual pest control up to level Q_H'. The median group of farmers will accept the median level of effort (Q_M). The low level pest control farmers will usually be worse off if they join the group. The only exception to this is if the costs of pest control are much less expensive per additional unit of control to the group (MC_1) than they are to individual growers (MC_2). Since effectiveness of the group effort will often fall rapidly as farmers fail to join (causing MC_1 to rise), it may be necessary to require participation by all growers. This results in a direct tax on low pest control demanding growers.

An alternative approach suggested for problems of this type by Clark (8) is to charge each group member based on their contribution to group costs. Members can be encouraged to correctly report their pest control demand levels (H, M, L) by a complex tax and rebate system. This system will encourage low pest control members to join, but it may have high administrative costs.

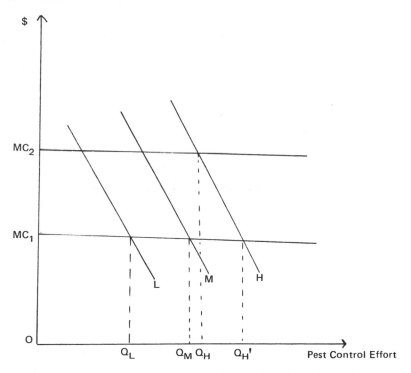

Figure 1   Group pest control with various levels of demand for effort.

GAPS IN INSTITUTIONAL PEST MANAGEMENT RESEARCH

Very little research is being conducted on the institutions and organizations for pest management mentioned above. Most research on pest control is in the biological sciences. Farm cooperatives, diagnostic clinics, public districts, all-risk crop insurance, and various information sources are evolving around the world. Agriculturalists have not closely examined the benefits and costs of various pest control methods. There are a few economic studies completed on field scouting (1, 10, 25) and one on pest consultants (13).

Very little is known about the role of farm operator pest management skills in pest control. These people are making most pest management decisions, but we know little about how their pest control knowledge is changing over time (10). Typically, the level of skills has greatly influenced rates of adoption of new technologies in agriculture.

Lastly, government policies are becoming more important in shaping the materials and methods available for pest control. We need to know how cancellations of chemical pesticide registrations and lengthy approval periods influence rates of discovery of new compounds and nonpesticide techniques. We should investigate whether subsidies for crop insurance or field scouting will lower crop losses and pest control costs. Before large sums are expended, program options need careful analysis by biologists, economists, and others.

Literature Cited

1. Allen, W. A., Roberts, J. F. 1974. Economic feasibility of scouting soybean insects in late summer in Virginia. *J. Econ. Entomol.* 67(6):644–47
2. Arrow, K. J. 1963. Uncertainty and the welfare economics of medical care. *Am. Econ. Rev.* 53:941–73
3. Barzel, Y. 1977. *Insurance Screening Mechanisms.* Dep. Econ., Univ. Washington, Seattle, Wash. (Mimeo)
4. Battese, G. A., Francisco, E. M. 1977. Distributions of indemnities for crop insurance plans with application to grain crops in New South Wales. *Aust. J. Agric. Econ.* 21(2):67–79
5. Camacho, P. A. 1957. Puerto Rico coffee insurance. *Agric. Finance Rev.* 19:40–43
6. Carlson, G. A. 1970. A decision theoretic approach to crop disease prediction and control. *Am. J. Agric. Econ.* 52(2):216–23
7. Carlson, G. A., DeBord, D. V. 1976. Public mosquito abatement. *J. Environ. Econ. Manage.* 3:142–53
8. Clark, E. H. 1971. Multipart pricing of public goods. *Public Choice* 11:17–33
9. Ehrlich, I., Becker, G. 1972. Market insurance, self insurance and self protection. *J. Polit. Econ.* 80(4):623–48
10. Grube, A., Carlson, G. 1978. *Analysis of Field Scouting and Insecticide Use.* NC State Univ., Raleigh, NC (Mimeo)
11. Grube, A., Carlson, G. 1978. *Economic Analysis of Cotton Insect Control in North Carolina.* EIR No. 52, Dep. Econ. Bus., NC State Univ., Raleigh, NC
12. Halcrow, H. G. 1978. A new proposal for federal crop insurance. *Ill. Agric. Econ.* 18(2):20–29
13. Hall, D. C. 1977. *An Economic and Institutional Evaluation of Integrated Pest Management with an Empirical Investigation of Two California Crops.* EPA Rep., Washington DC
14. Miller, T. A., Walter, A. S. 1977. *Options for improving government programs that cover crop losses caused by*

national hazards. ERS No. 654, ERS, USDA, Washington DC

15. Miranowski, J. 1974. *Crop Insurance and Information Services to Control Use of Pesticides.* EPA 600/5-74-018, EPA, Washington DC

16. Mossin, J. 1968. Aspects of rational insurance purchasing. *J. Polit. Econ.* 76(4):553–68

17. Newsom, D. 1977. *Relationship Between Pest Management and Survey and Detection Programs.* Dep. of Entomol., Louisiana State Univ., Baton Rouge, La. (Mimeo)

18. Pombo, R. 1959. Crop and livestock insurance in Brazil. *Agric. Finance Rev.* 21:86–88

19. Ray, P. K. 1967. *Agricultural Insurance.* New York: Pergamon. 240 pp.

20. Rogers, P. P. 1964. Soviet agricultural insurance. *Agric. Finance Rev.* 25:1–12

21. Swed. Minist. Agric. 1972. *Cost-Benefits of Crop Insurance Programmes.*

Stockholm: Swedish Minist. Agric. 13 pp.

22. US Dep. Agric. 1958. All-risk paddy rice crop insurance in Ceylon. *Agric. Fin. Rev.* 20:71

23. US Dep. Agric. 1978. *Comprehensive Agricultural Insurance Options.* Food Agric. Policy Work. Group, Memor. for the Secre., Washington DC. 14 pp.

24. US Environ. Prot. Agency 1978. *Integrated Pest Management and Crop Insurance Issues Paper.* Off. Pestic. Programs, Washington DC. 31 pp.

25. Von Rumker, R., Carlson, G. A., Lacewell, R. D., Norgaard, R. B., Parvin, D. W. Jr. 1975. *Evaluation of Pest Management Programs for Cotton, Peanuts, and Tobacco in the United States.* EPA 540/9-75-031. Washington DC: EPA

26. Zeckhauser, R. 1970. Medical insurance: A case study of the trade-off between risk spreading and appropriate incentives. *J. Econ. Theory* 2(1):10–26

Ann. Rev. Phytopathol. 1979. 17:163–79
Copyright © 1979 by Annual Reviews Inc. All rights reserved

AGROBACTERIUM RADIOBACTER STRAIN 84 AND BIOLOGICAL CONTROL OF CROWN GALL

♦3706

Larry W. Moore

Department of Botany and Plant Pathology, Oregon State University, Corvallis, Oregon 97331

Guylyn Warren

Department of Chemistry, Montana State University, Bozeman, Montana 59715

INTRODUCTION

Crown gall caused by *Agrobacterium tumefaciens* is of particular economic concern to nurseries growing rosaceous plants, *Rubus* species, grapevines, and various nut-bearing trees. Epidemics of 80–100% galled nursery stock have occurred in the Pacific Northwest, but they are usually sporadic and infrequent. Economic losses to nurseries occur primarily because most states prohibit interstate shipment or receipt of galled plants, which must then be destroyed. Losses are not confined to nurseries but can be severe in some orchards and landscape plantings.

Since E. F. Smith's (57) demonstration that a bacterium was the causal agent of crown gall, there have been many studies on the host range of *A. tumefaciens,* and major attempts have been made to control the disease with bactericides, which for the most part are ineffective.

A novel approach to crown gall control was suggested in 1972 by Kerr (26) and colleagues (46). A nonpathogenic *Agrobacterium* strain, *A. radiobacter* strain 84, was inoculated onto peach seeds planted in soil infested with *A. tumefaciens.* Three months later, 31% of the plants from seeds inoculated with strain 84 were galled compared to 79% for the uninocu-

163

0066-4286/79/0901-0163$01.00

lated treatment. Since 1972, *A. radiobacter* strain 84 has been used world-wide (41) (Table 1) on thousands of plants, including species of *Prunus, Rubus, Malus, Salix, Vitis, Libocedrus, Chrysanthemum, Crategus, Carya, Rosa, Pyrus,* and *Humulus.* It is remarkable that a single strain of bacteria would be disseminated so widely in the space of 4–6 years and be used successfully by numerous investigators to control crown gall. The spectacular success of this biological control probably is unparalleled in the history of plant pathology.

This review examines the use of strain 84 for biological control of crown gall and considers some factors affecting its antagonistic action to *A. tumefaciens.*

FACTORS AFFECTING SUCCESS OF BIOCONTROL

Strain 84 has reduced the incidence of galled plants markedly in most instances (Table 1), even when tested against high populations of the pathogen or a mixture containing some pathogenic strains insensitive to strain 84 (39, 54). However, the ratio of strain 84 to pathogenic bacteria should be $\geq$ 1:1 (46). Control generally was better against naturally occurring pathogenic strains in field soil than against inoculum grown on synthetic media, which often contains 10^6–10^8 bacteria per ml. The latter population is 10 to 1000 times greater than *Agrobacterium* populations normally present in soil (37, 46), and most of the agrobacteria in soil are nonpathogenic.

Prevention of gall formation by strain 84 has been predictable under the following conditions:

The Absence of Latent Infections

Latent infections have been shown to occur in wounds of cherry and plum rootstocks (38). Although these plants were symptomless at planting time, the Ti plasmid from *A. tumefaciens* may have been transferred to host cells at the wound site and treatment of these plants with strain 84 would have been ineffective in preventing gall formation. To test the effectiveness of strain 84 against latent infections, dormant myrobalan plum seedlings were wounded and inoculated with *A. tumefaciens* and stored for 90 days at 6–9°C before inoculation with strain 84. These plants were heavily galled when harvested 10 months after planting (40). Conversely, seedlings wounded and inoculated with *A. tumefaciens* and strain 84 at planting time remained disease-free for 10 months.

Planting stock can carry latent infections when harvested from layering beds containing galled mother block plants, even when the incidence of galled plants is low. Garrett (19) reported that three sprays of strain 84 on plants in an infected layering bed of F12-1 mazzard cherry did not control galling. Similarly, when F12-1 rootstocks from these beds were dipped in

Table 1 Effectiveness of *Agrobacterium radiobacter* strain 84 to prevent crown gall on a variety of crop plants produced in different countries

Country[a]	Crop	Number of plants treated[b]	Galled plants (%)[c]	
			Greenhouse	Field
Australia (26)	peach	99	31/79	
Canada (12)	peach	100, 2,668	7/46	0/17.2
Greece (48)	almond	not reported	40/100	79/100
Hungary (63)	raspberry	102		4.9/65.9
		146		28.7/58.3
Italy (6)	cherry	2,000		4.7/88.3
	plum	2,000		3.7/93.2
New Zealand (14)	rose	4,750		0.9/23.4
USA				
California (54)	almond	37		30/82
	peach	46		46/97
	plum	87		22/98
Michigan[d]	apple	50		12/32
	cherry	42, 42	3/98	26/98
	plum	48		17/98
Ohio (47)	euonymus	50	4/79	
Oregon (39–41)	boysenberry	73		14/62
	cherry	40, 7,845	8/90	2.5/32
Texas[e]	rose	276		51/41
Washington (41)	apple	11,200		0.5/8.5

[a] The number in parentheses are literature references.

[b] The number of plants that were inoculated with *A. radiobacter* strain 84.

[c] The numerator is the percentage of galled plants that were inoculated with *A. radiobacter* strain 84 (often mixed with *A. tumefaciens*); the denominator is the percentage of galled plants inoculated only with *A. tumefaciens* or infected by naturally occurring soilborne inoculum.

[d] A. Jones, personal communication.

[e] E. W. Lyle, personal communication.

strain 84 after harvesting and again before planting, the incidence of galling was comparable to the untreated controls. Biological control was achieved only when strain 84 was applied as a preventative preplanting dip on healthy cuttings. Such stock obviously should be used to establish new layer beds; however, the new wounds made on the mother block plants at harvest should be treated with strain 84.

Sensitivity of A. tumefaciens Strains to Agrocin 84

Kerr & Htay (28) showed that *A. tumefaciens* strains sensitive to agrocin 84 (see section on mode of action for explanation of "agrocin 84") in vitro are prevented by strain 84 from infecting stems of tomato seedlings, but insensitive strains produced tumors in the presence of strain 84. Thus, a high positive correlation existed between sensitivity of pathogenic strains to agrocin 84 and their inability to produce galls on tomato stems when coinoculated with strain 84. Nevertheless, a pathogenic strain may be insensitive to strain 84 in laboratory and glasshouse studies, but be controlled by the antagonist in the field. Schroth & Moller (54) inoculated *Prunus* seedlings with strain 84 and a mixture of three pathogenic strains, one of which was insensitive to agrocin 84 in vitro, and reduced the incidence of galling about 60% below that of the control. Moore (39) showed that a mixture of six pathogenic strains, two of which were agrocin-insensitive, was not inhibited by strain 84 on tomato seedlings in a glasshouse but was inhibited on mazzard cherry seedlings grown in the field. Conversely, Panagopoulos et al (48) reported that strain 84 did not prevent galling of field-grown almond trees when coinoculated with an agrocin 84–insensitive strain of *A. tumefaciens.*

In later field experiments, biological control of crown gall was obtained on pear seedlings, but not on apple, using four agrocin-insensitive pathogenic strains combined with strain 84 (41). The incidence of galled apple seedlings was reduced when the seedlings were dipped for 5 to 10 sec in 10^{-3} M indole butyric acid (IBA) before inoculation with the insensitive pathogens and strain 84. Eighty-one percent of the mazzard cherry seedlings inoculated at planting time in 1978 with these same four pathogens were galled after 6 months, compared to 7.6% when seedlings were inoculated additionally with strain 84 (L. W. Moore, unpublished data). Apparently, the host species influences the interaction between strain 84 and the pathogens.

The mechanism of this host influence is unknown. At the below-ground wound site, the interaction of strain 84 with the pathogen is only one of several complex sets of factors (7, 8, 53, 59, 67) that impinge on each other (Figure 1). Other competing microbes and fauna are present in the soil and on the root; soil salts, moisture, gases, temperature, pH, etc vary from site to site; and root metabolism, uptake, and exudates contribute a pervasive influence over all these factors.

Recent evidence demonstrates that nutrition can affect the strain 84-pathogen interaction in vitro. D. Cooksey and L. W. Moore (unpublished data) showed that growth of *A. tumefaciens* B_6 was inhibited around a colony of strain 84 growing on potato dextrose agar (PDA), but no inhibition zone was observed on a defined medium routinely used to assay agrocin sensitivity (39). The effect of dextrose, a component of PDA, was tested on agrocin-resistant B_6 and agrocin-sensitive B_{234} using the defined medium. Zones of inhibition were produced against B_6 when 5, 10, 15, or 20 g/liter of glucose were added to the defined medium. Conversely, these increasing concentrations of glucose caused a progressive decrease in the diameter of inhibition zones of *A. tumefaciens* B_{234} around strain 84. Apparently, glucose differentially affects sensitivity of the test strains to agrocin 84 rather than the production of agrocin 84 by strain 84.

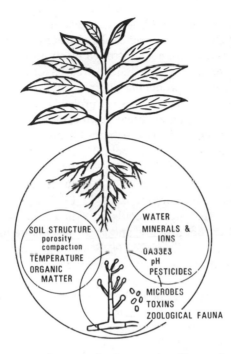

Figure 1   The environment at the root surface is a complex of interacting physical, chemical, and biological factors, here depicted as subsets within the influence of the root. Alteration of any of these factors could affect the antagonism of strain 84 against *A. tumefaciens.* Thus, biological control of crown gall below-ground is more demanding than at an aerial stem wound and should be given proper consideration when using stems of herbaceous plants to test the effectiveness of a bacterium as a biological control agent.

INSTANCES OF INEFFECTIVE BIOCONTROL

The number of cases of successful crown gall control using strain 84 is impressive, but some applications have been unsuccessful. The usual diversification of biological systems from mutation, recombination, and selection makes it unlikely that all *A. tumefaciens* strains will be controlled in every instance. For example, strain 84 failed to control 7 of 14 *A. tumefaciens* strains from Europe (28), was ineffective against strains of the pathogen from galled grapes in Greece (29), and failed to protect field-grown apple seedlings in Washington and rose plants in Texas and Pennsylvania (41). In contrast, crown gall has been reduced to less than 1% on commercially produced roses in Australia (A. Kerr, personal communication) and New Zealand (14). One possible reason for the high success in Australia is that the population of *A. tumefaciens* appears to be quite homogeneous (27). Most of the Australian pathogens apparently are biotype 2 (24, 45), utilize nopaline, and are sensitive to agrocin 84 (30). More heterogeneity is found elsewhere in the world (3, 29, 41, 62), and in some instances, virulent biotype 1 and 2 strains have been isolated from the same tumor (4).

Pathogenic agrobacteria not controlled by strain 84 can occur naturally (29, 41, 62) and might arise by genetic transfer of the Ti plasmid (25) to strain 84 or the agrocin-producing plasmid from strain 84 to a pathogenic strain (15) (Figure 2). Genetic exchange probably is the basis of Panagopoulus's report (48) of an apparent breakdown of biological control on almond. A new class of agrobacteria was recovered from the almond tumors; these agrobacteria were resistant to agrocin 84, pathogenic, and producers of a bacteriocin with a bacterial host range similar to agrocin 84.

In commercial plantings, the lack of uninoculated control treatments is a major problem in interpreting data. In 1978, one West Coast nursery growing seedlings that were inoculated with strain 84 immediately after harvest, again after root pruning, and lastly, just before planting had crown gall in 40% of the seedlings (M. N. Schroth, personal communication). Historically, crown gall on pear had never been observed in this nursery, but unfortunately no untreated seedlings were available to assess whether strain 84 increased galling above the level of natural background infection. Garrett (19) reported a higher incidence of crown gall on young layered plants and rootstocks of F12-1 mazzard cherry that were treated with strain 84 than on the untreated.

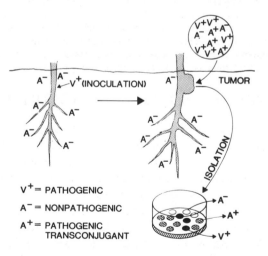

Figure 2 A diagrammatic representation of the Kerr Cross (25). A virulent bacterium, V^+, initiates a tumor and serves as a donor of Ti plasmid DNA to avirulent recipient agrobacteria, A^-, naturally present in the soil or inoculated with the pathogen. The progeny of the mating represents a new pathogenic strain, A^+. If the tumor synthesizes octopine, the frequency of conjugation can be increased 100-fold (49).

On the other hand, the sudden appearance of crown gall or hairy root epidemics in a crop is not without precedent. Munnecke et al (44) reported a severe outbreak of hairy root on roses that persisted only a few years. In Oregon, a grower plowed under a new planting of boysenberries because most of the plants were galled. This grower remarked that he had never observed such an outbreak in over 30 years of farming, especially during the first growing season (L. W. Moore, personal communication). These observations emphasize the importance of growers maintaining blocks of uninoculated control plants. Pathologists can then determine whether an agrocin-resistant pathogen is present, whether the stock bore latent infections prior to treatment, or whether new *Agrobacterium* genotypes have arisen.

MODE OF ACTION

Kerr & Htay (28) showed that the active agent produced by *A. radiobacter* strain 84 diffused in agar and was similar to the bacteriocin ("agrocin") described by Stonier (60). Agrocin 84 has not been demonstrated in a wound inoculated with strain 84 despite the high correlation of biological control with agrocin sensitivity in vitro. Kerr & Htay (28) showed a tenfold reduction of the sensitive pathogen population in plant wounds inoculated with a 1 : 1 ratio of strain 84, but agrocin-resistant strains were unaffected. M. Staver, R. Morris, and L. W. Moore (unpublished data) have shown that several applications of a highly purified fraction of agrocin 84 to wounded tomato stems prevented tumor formation by a sensitive strain or allowed only small galls to develop.

Physiochemical Nature of the Agrocin Molecule
Liquid cultures of *A. radiobacter* strain 84 contain small quantities of agrocin 84 which are very active biologically. The molecule was initially described as a peptide (35), but this was later retracted. Heip et al (21) characterized agrocin 84 as acidic, ethanol-soluble, heat stable, less than 1000 mol wt, and with reduced biological activity at pH below 4 and above 9. Their report of heat stability is contrary to the observations of M. Staver, R. Morris, and L. W. Moore (unpublished data) and A. Kerr (personal communication) that biological activity of highly purifed agrocin is lost at 21 to 23° C. Roberts et al (52) identified agrocin 84 as a phosphoramidate of an adenine deoxyarabinoside, and the highly reactive phosphoramidate linkage probably accounts for the lability of the molecule. If the molecule has two phosphoramidate linkages, and if the phosphoramidate linkage is necessary for biological activity, then two possible breakdown products could be generated depending upon which linkage is hydrolyzed. M. Staver,

R. Morris, and L. W. Moore (unpublished data) have evidence for the presence of two active components, but their identity is unknown at this time. Production of more than one bacteriocin by a bacterial strain has been shown for four species of *Corynebacterium* (19a), and Brock (7) states that many organisms produce more than one antibiotic, sometimes as many as five or six.

Das et al (9) proposed that agrocin 84 inhibits DNA synthesis in sensitive strains, with inhibition occurring after 30 min of treatment in a log phase culture. Since the full response of the sensitive strain requires 30 min, the toxic action of agrocin 84 may be similar to blockage of purine and pyrimidine synthesis by adenosine and its analogues (22).

Smith & Hindley (58) suggested that agrocin 84 blocks attachment of the pathogen to plant tissues by interfering with the synthesis of the outer cell envelope of the bacteria, thus impairing binding of the pathogen to the plant cell and transfer of tumorigenic DNA. This report also presents electron micrographs of thin sections of *A. tumefaciens* C58 cells preincubated in agrocin 84 that contain electron dense granules which are considered possible cell envelope precursors. These granules also look like induced phage particles.

The impaired binding of pathogenic cells to host tissues hypothesized by Smith & Hindley (58) apparently occurs because, instead of the agrocin molecules physically blocking the host receptor site, the pathogenic bacteria are damaged by agrocin 84. Competitive binding experiments are needed to confirm this proposal. It has been shown that heat-killed (18) and chloroform-killed (39) cells of strain 84 do not prevent galling. Only living cells provided biological control, suggesting that a metabolic product from strain 84 was necessary rather than strain 84 physically blocking pathogenic cells from attaching to the host cell (32).

Factors Affecting Sensitivity to Agrocin 84

Agrobacterium radiobacter strain 84 is an avirulent biotype 2 bacterium which carries a small plasmid (30×10^6 mol wt) that codes for agrocin production and a larger plasmid (124×10^6 mol wt) that codes for nopaline utilization (15, 36, 55).

Kerr & Roberts' data (30) confirmed the close association between pathogenicity, octopine/nopaline utilization, and between agrocin sensitivity and nopaline utilization. They also found several exceptions to the correlation between pathogenicity and the other characteristics, some of which we present in Table 2. Although agrocin 84 sensitivity is often associated with the presence of a Ti plasmid (16, 55), the interaction between strain 84 and pathogenic agrobacteria is not restricted to the presence or absence of the

Table 2 Some phenotypic characteristics of *Agrobacterium* strains differentially sensitive to agrocin 84

Strain[a]	Agrocin 84 sensitivity	Pathogenicity	Biotype	Octopine/nopaline utilization
590(L), 22, 23 (30)	+	−	1	nopaline
C58, T37 (30)	+	+	1	nopaline
IIBV7 (36)	−	+	1	nopaline
A6 (30)	−	+	1	octopine
B6, (30)	−[b]	+	1	octopine
TT133, 223 (30)	+	+	?	both
Ag 43 (30)	−	+	2	both
57 (Ag 43 transc)[c] (30)	+	+	1	nopaline
LBA57 (23)	−	+	1	octopine
LBA601 (LBA57 transc) (23)	−	+	1	octopine
542[d]	+	+	1	neither
A281 (542 transc)[d]	−	+	1	neither
AT1[d]	−	+	1	neither
AT4[d]	slight	+	1	neither
A543 (AT4 transc)[d]	−	+	1	neither

[a] The number in parentheses is the literature reference.
[b] B6 is sensitive to agrocin 84 when tested in vitro on potato dextrose agar (cf text).
[c] The abbreviation "transc" means transconjugant.
[d] M-D. Chilton, personal communication.

Ti plasmid (Table 2). For example, strain Ag43 is agrocin insensitive, yet its transconjugate (strain 57) is sensitive. Conversely, strain 542 is agrocin sensitive but its transconjugate (strain A281) is insensitive. Furthermore, the plasmids from agrocin insensitive AT1 have the same fingerprint cleavage patterns as those plasmids from 542 (M. D. Chilton, personal communication), thus suggesting that agrocin sensitivity also can be affected by the bacterial chromosome.

Mutant colonies from a sensitive strain often develop in the inhibition zone surrounding growth of strain 84 on solid media. The mutants are resistant to agrocin 84 and some have lost the Ti plasmid (16), rendering them avirulent. Kerr & Htay (28) found that the majority (33 of 35 colonies) of mutants from 7 sensitive strains were avirulent. In contrast, 18 of 25 resistant colonies of strain B234 were still pathogenic (D. Cooksey and L. W. Moore, unpublished data). Süle (62) also found that 70% of the resistant mutants from 12 sensitive stains were still virulent. On the other hand, strain 23 is sensitive to agrocin 84 but is not pathogenic on tomato plants (30). The inability of strain 23 to infect tomato may result from its host specificity (4). Alternatively, strain 23 may be similar to strain 5 GlyFeAvir, a glycine attenuated (64) derivative of A_6 whose plasmid restriction endonuclease fingerprint is exactly like that of B6-806 (55). Although attenuated, strain 5 GlyFeAvir can transfer its Ti plasmid to a recipient, resulting in oncogenic transconjugants.

Other mechanisms that possibly affect agrocin sensitivity include changes in cell permeability. Plasmids are known to produce alterations in the cell surface such as pili which alter permeability. Since the agrocin molecule is near the molecular cutoff size (ca 900 mol wt) (13) for passive diffusion through the outer membrane, a small change in permeability due to alteration of the outer membrane protein or capsule production could change sensitivity. Such a change might explain D. Cooksey and L. W. Moore's results (unpublished) with glucose causing a resistant strain B6 to become sensitive. High levels of glucose are known to switch off the production of cAMP (50), and cAMP levels control the function of various catabolic pathways in most bacteria. Conversely, strain B6's resistance may be due to the presence of a degradative enzyme active on agrocin 84 or production of an agrocin inhibitor similar to those produced by certain strains of *Pseudomonas syringae* against the bacteriocin of *P. syringae* PS14 (65a).

A general cellular defect in the cell surface could cause resistance to agrocin 84 and mask the presence of Ti plasmid by blocking plasmid transfer and inhibiting oncogenicity. For example, there is a genetic defect in *Escherichia coli* at the *qme* locus that provides tolerance to high levels of glycine (66). A defect at this locus has pleiotrophic effects on cell surface membrane protein content. Mutants at the *tsx* locus also have defective cell surfaces for nucleoside transport (20).

Classification of agrobacteria sensitivity to agrocins by in vitro studies can be misleading relative to biological control in the field. Strains resistant in vitro have been controlled in the field (39, 41, 54), and strains producing agrocins in vitro have failed to provide field control (29). Clearly, all agrocin producers are not fit organisms for biological control, as also demonstrated by genetic transfer of the agrocin 84 plasmid to different recipients. One recipient strain became an effective biological control agent but the other did not (15). Although the latter recipient produced an antibiotic in vitro, its ability to grow and produce the antibiotic at the site of control apparently was limited.

The interaction of pathogens with strain 84 may be one portion of an entire spectrum of effects including those observed by the Lippincotts (33) showing complementation of pathogens and nonpathogens (other than strain 84) leading to an increase in host range or change in tumor type. This spectrum may result from control mechanisms regulating synthesis of agrocin 84 or from interactions between chromosomal backgrounds and other products of these species.

Alternate Explanations for Tumor-Inhibiting Effects

It is possible that phosphodiesterase inhibitors or cAMP could inhibit tumors, and both could be isolated and concentrated from bacteria-free culture extracts. Cyclic AMP, 5'-ADP, and IAA theophylline will inhibit the production of tumors by pathogenic strains (5, 17). In addition, Ames (2) has shown that genetic induced tumors also are reduced by increasing cAMP levels.

In plants, the control of *A. tumefaciens* by strain 84 could be related to lethal zygosis, especially when strain 84 cells are used much in excess of the pathogen. Lethal zygosis occurs with excess attachment of donor cells to a recipient during conjugation, causing inhibition, leakage, and death (56). Interactions between plasmids such as exclusion and compatibility effects would explain much of this as suggested by Sciaky et al (55).

OTHER ANTAGONISTS

To date, *A. radiobacter* 84 has been superior to hundreds of other *Agrobacterium* strains tested for biological control (19, 29, 39; S. Süle, personal communication). Moore (39) found that 3 of 32 avirulent agrobacteria prevented crown gall when mixed with laboratory-grown *A. tumefaciens* and inoculated to mazzard cherry seedlings growing in the field. These three potential antagonists did not produce an antibiotic against the pathogens on a defined medium, and coinoculations have not been effective as a preplanting treatment (L. W. Moore, unpublished data). The antagonistic action in situ may have been via exclusion of the pathogen at the infection site (32) or complementation with antagonistic microflora already present on the root.

Crown gall disease usually occurs sporadically in the nursery, suggesting the presence of other natural defenses. Since soil fumigation can increase crown gall (10), some of the natural enemies of *A. tumefaciens* probably were eradicated. Deep & Young (11) used unidentified fungi isolated from cherry roots to reduce the incidence of galling on mazzard cherry seedlings. D. Cooksey and L. W. Moore (unpublished data) isolated from plant roots species of *Penicillium, Aspergillus, Trichoderma,* and *Bacillus* which produce potent antibiotics in vitro against species of *Agrobacterium,* including strain 84. Some of these antagonists mixed with *A. tumefaciens* also prevented infection of radish and tomato seedlings in greenhouse experiments. In 1978, these antagonists were mixed with *A. tumefaciens* and field-tested on mazzard seedlings. Some of the antagonists reduced the incidence of galling, depending upon the group of pathogens used, but none was as effective as strain 84. However, the trials showed that there are microorganisms other than agrobacteria that can influence crown gall disease. Their success as biocontrol agents will be governed by their survival and activity in the rhizosphere, in which the agrobacteria seem well suited (27, 59). In turn, microbial antibiosis to strain 84 may render it ineffective when introduced into a new plant-soil regime colonized by *A. tumefaciens* strains that have been selected over time for resistance to the endemic antagonistic microflora.

ATTRIBUTES OF STRAIN 84 AS A BIOCONTROL AGENT

In addition to selectivity, there are a number of other attributes considered desirable for biological control (65). *A. radiobacter* strain 84 fulfills most of these criteria, and they are considered individually below.

Antagonist Persistence

Although wounds for *A. tumefaciens* can occur on the crown or root throughout the life of host plants, damage by crown gall is greatest when the plants are young (43). Consequently, protection of wounds is most critical during the early stages of plant development. Treatment of seeds, cuttings, and seedling liners with strain 84 offers a means of continuing protection [up to two years (12)] probably because the rhizoplane-rhizosphere region is a natural habitat for *Agrobacterium* (27, 37, 59).

Cool soil or low ambient air temperatures can delay wound healing and extend the period of host susceptibility (38). Wounds on the roots of mazzard cherry trees inoculated with *A. tumefaciens* failed to develop galls when kept at or below 10°C for 80 days, but galls appeared within 2–3 weeks after these trees were placed at 24–27°C (L. W. Moore, unpublished data). Mean soil temperatures usually are about 10°C at spring planting time in Oregon and Washington, and pruning wounds callus slowly on roots planted to these soils. Inoculation of these roots with strain 84 prior to planting can provide protection against soilborne *A. tumefaciens.*

Safety

Toxicity studies on *A. radiobacter* suggest that it is harmless to mammals. No deaths occurred when 20 mice were entubulated with 6×10^{10} colony forming units (CFU)/ml each of *A. radiobacter* 84. Similarly, no mice died when entubulated with 5×10^{10} cells of *A. radiobacter* or a closely related bacterium, *Rhizobium phaseoli,* a commonly used inoculant of legume seeds for nitrogen fixation (41; L. W. Moore, unpublished data). Eye, nasal, and skin tests for toxicity of *A. radiobacter* 84 to mice and rabbits were negative. These results agree with conclusions from clinical studies that showed agrobacteria were harmless to humans (31, 51).

Neither a partially purified preparation of cell-free agrocin 84 nor living cells of *A. radiobacter* 84 induced mutagenesis of *Salmonella typhimurium* strains used in the Ames test (42). The strains tested included TA1537 and TA 98 (detectors of frameshift mutagenesis) and TA 100 (detects basepair substitution activity), both in the presence and absence of liver microsomes (1). These tests for detection of carcinogenic activity of agrocin 84 were presumptive only, but McCann & Ames (34) showed a 90% correlation between 175 known carcinogens and their mutagenicity to strains of *S. typhimurium*. Thus, it appears there is little risk to the health and safety of workers handling *A. radiobacter* 84.

Aesthetic Acceptibility

Plants inoculated with strain 84 exhibit no unsightly appearance, disagreeable odor, or unpleasant handling characteristics. Once the bacteria are distributed over the plant surfaces inoculated areas cannot be distinguished from the surrounding parts.

Production and Storage

A. radiobacter 84 is cultured readily in liquid or on solid nutrient media. The bacteria can be washed off solid media or centrifuged out from liquid culture prior to inoculation of plant tissues. To prepare a plant inoculum, collected bacteria can be resuspended in water, carboxymethyl cellulose (CMC), or in a peat preparation similar to *Rhizobium* inoculant. The half-life of a peat inoculum of strain 84 is about 6 months at room temperature (61). Viability of strain 84 in CMC also remained high during 6 month's storage at 4–6°C (40), and refrigerated agar cultures can be stored 6–8 weeks without loss of viability. The peat preparation is lightweight for shipment in contrast to agar or CMC preparations.

Cost

Cost of the treatment varies with the size of the plant propagule being treated. In the Pacific Northwest, the cost of treating barerooted *Prunus* seedlings is less than $0.01 per plant. This cost is below or comparable to standard chemical treatments such as streptomycin sulfate and gives superior protection (39).

Application

Strain 84 has been applied successfully to seeds, cuttings, barerooted seedlings, aerial grafts, and plants growing in situ. The antagonist usually is applied by dipping or spraying the plant tissues with an aqueous suspension containing 10^8–10^9 colony forming units/ml. Treated plants must be protected after inoculation to prevent desiccation of the antagonist cells.

Avoiding undue desiccation of small seeds that are inoculated is an inconvenience to nurserymen if the seeds are to be planted mechanically. Wet seeds feed poorly through the planting machine. No experiments have been reported on the survival of strain 84 cells on roots or seed under field conditions while awaiting planting, but common sense dictates shading, cooling, and screening the inoculated tissues from adverse meterological conditions to protect the living bacteria.

CONCLUSION

Few if any biological control measures have ever been tested and used to the extent that *Agrobacterium radiobacter* strain 84 has been used for control of crown gall. In the space of about 4 years, biological control of crown gall has been demonstrated with Kerr's strain 84 in most areas of the world where susceptible plants are grown. It is remarkable that a single strain of bacteria could be so effective in such diverse geographic habitats. The effectiveness of this biological control probably occurs because of the unique infection process of *A. tumefaciens.* Pathogenic strains sensitive to strain 84 are prevented from transferring their Ti plasmid to the wounded host, apparently because of a bacteriocin (agrocin 84) produced by the antagonist which either kills or prevents attachment of the pathogen to the host receptor site.

Strain 84 has not prevented crown gall in some instances because either the pathogens are insensitive to agrocin 84 or produce a bacteriocin against strain 84 or an inhibitor against agrocin 84. We know of only a few reports of ineffective biocontrol, but the negative nature of such reports precludes their being published. To avoid failures when coinoculating plants with a mixture of strain 84 and *A. tumefaciens,* it is imperative that their ratio be at least 1:1 or greater. Other factors that influence the in situ interaction of pathogens insensitive to strain 84 include the host species being treated, planting site, and associated soil microflora. Different media also can alter the in vitro response of pathogens to the agrocin from strain 84. Other antagonists have been found that inhibit some of these insensitive pathogenic strains in vitro and prevent infection in greenhouse tests, but they have not been as effective in field tests. In general, no other antagonist has been found as effective as Kerr's strain 84.

It should be emphasized that biological control with strain 84 is preventative, not curative, and will not stop latent infections or preclude use of proper sanitation and nursery management. Planting stock treated with the antagonist should be protected from undue desiccation prior to planting. Commercial growers who are just beginning to use strain 84 should always include untreated control plants in their planting to assess the effectiveness of the treatment. Strain 84 can be applied to trees as a dip or spray; the latter is recommended if part of the planting stock might be contaminated with pathogenic fungi such as *Phytophthora* that could be spread to other plants during dipping. Given these precautions, biological control with strain 84 is the most effective means of controlling crown gall available to the grower.

Literature Cited

1. Ames, B. N., McCann, J., Yamasaki, E. 1975. Methods for detecting carcinogens and mutagens with the Salmonella/mammalian-microsome mutagenicity test. *Mutat. Res.* 31:347–64

2. Ames, I. H. 1976. The possible role of cyclic AMP in the control of genetic tumor induction. *Plant Cell Physiol.* 17:1059–66

3. Anderson, A. R. 1977. *Taxonomy and host specificity of the genus Agrobacterium.* PhD thesis. Oregon State Univ., Corvallis. 69 pp.

4. Anderson, A. R., Moore, L. W. 1979. Host specificity in the genus *Agrobacterium. Phytopathology* 69: In press

5. Babula, M. J., Galsky, A. G. 1975. Effects of cyclic-AMP on the formation of crown-gall tumors on the primary leaves of *Phaseolus vulgaris* var. Pinto. *Plant Cell Physiol.* 16:357–60

6. Bazzi, C., Mazzucchi, U. 1978. *Biological control of crown gall in cherry, myrobalan, and peach rootstocks in Italy.* Presented at IVth Int. Conf. on Plant Pathogenic Bacteria, Angers, France

7. Brock, T. D. 1966. *Principles of Microbial Ecology.* Englewood Cliffs, New Jersey: Prentice-Hall. 300 pp.

8. Chan, E. C. S., Katznelson, H., Rouatt, J. W. 1963. The influence of soil and root extracts on the associative growth of selected soil bacteria. *Can. J. Microbiol.* 9:187–97

9. Das, P. K., Basu, M., Chatterjee, G. C. 1978. Studies on the mode of action of agrocin 84. *J. Antibiot.* 31:490–92

10. Deep, I. W., McNeilan, R. A., MacSwan, I. C. 1968. Soil fumigants tested for control of crown gall. *Plant Dis. Reptr.* 52:102–5

11. Deep, I. W., Young, R. A. 1965. The role of preplanting treatments with chemicals in increasing the incidence of crown gall. *Phytopathology* 55:212–16

12. Dhanvantari, B. N. 1976. Biological control of crown gall on peach in Southwestern Ontario. *Plant Dis. Reptr.* 60:549–51

13. DiRienzo, J. M., Nakamura, K., Inouye, M. 1978. The outer membrane proteins of Gram-negative bacteria: Biosynthesis, assembly and functions. *Ann. Rev. Biochem.* 47:481–532

14. Dye, E. W., Kemp, W. J., Amos, M. J., Parker, W. C. 1975. Crown gall in roses can be controlled. *Commer. Hortic.* 7:5, 7

15. Ellis, J. G., Kerr, A. 1978. *Developing biological control agents for soil borne pathogens.* Presented at IVth Int. Conf. on Plant Pathogenic Bacteria, Angers, France

16. Engler, G., Holsters, M., Van Montagu, M., Schell, J., Hernalsteens, J. P., Schilperoort, R. 1975. Agrocin 84 sensitivity: A plasmid determined property in *Agrobacterium tumefaciens. Mol. Gen. Genet.* 138:345–49

17. Favus, S., Gonzalez, O., Bowman, P., Galsky, A. 1977. Inhibition of crown-gall tumor formation on potato discs by cyclic-AMP and prostaglandins E_1 and E_2. *Plant Cell Physiol.* 18:469–72

18. Garrett, C. M. E. 1978. Biological control of crown gall, *Agrobacterium tumefaciens. Proc. Ann. Appl. Biol.* 89:96–97

19. Garrett, C. M. E. 1979. Biological control of crown gall in cherry rootstock propagation. *Ann. Appl. Biol.* 91: In press

19a. Gross, D. C., Vidaver, A. K. 1979. Bacteriocins of phytopathogenic *Corynebacterium* species. *Can. J. Microbiol.* In press

20. Hantke, K. 1976. Phage T_6—colicin K receptor and nucleoside transport in *Escherichia coli. FEBS Lett.* 70:109–12

21. Heip, J., Chatterjee, G. C., Vandekerckhove, J., Van Montagu, M., Schell, J. 1975. Purification of the *Agrobacterium radiobacter* 84 agrocin. *Arch. Int. Physiol. Biochim.* 83:974–76

22. Henderson, J. F., Smith, C. M., Snyder, F. F., Zombor, G. 1975. Effects of nucleoside analogs on purine nucleotide metabolism. *Ann. NY Acad. Sci.* 255:489–99

23. Hooykaas, P. J. J., Klapwijk, P. M., Nuti, M. P., Schilperoort, R. A.,

Rorsch, A. 1977. Transfer of the *Agrobacterium tumefaciens* Ti plasmid to avirulent Agrobacteria and to *Rhizobium* ex planta. *J. Gen. Microbiol.* 98:477–84

24. Keane, P. J., Kerr, A., New, P. B. 1970. Crown gall of stone fruit. II. Identification and nomenclature of *Agrobacterium* isolates. *Aust. J. Biol. Sci.* 23:585–95

25. Kerr, A. 1971. Acquisition of virulence by non-pathogenic isolates of *Agrobacterium radiobacter. Physiol. Plant Pathol.* 1:241–46

26. Kerr, A. 1972. Biological control of crown gall: Seed inoculation. *J. Appl. Bacteriol.* 35:493–97

27. Kerr, A. 1974. Soil microbiological studies on *Agrobacterium radiobacter* and biological control of crown gall. *Soil Sci.* 118:168–72

28. Kerr, A., Htay, K. 1974. Biological control of crown gall through bacteriocin production. *Physiol. Plant Pathol.* 4:37–44

29. Kerr, A., Panagopoulos, C. G. 1977. Biotypes of *Agrobacterium radiobacter* var. tumefaciens and their biological control. *Phytopathol. Z.* 90:172–79

30. Kerr, A., Roberts, W. P. 1976. *Agrobacterium:* Correlations between and transfer of pathogenicity, octopine and nopaline metabolism and bacteriocin 84 sensitivity. *Physiol. Plant Pathol.* 9:205–11

31. Lautrop, H. 1967. *Agrobacterium* Spp. isolated from clinical specimens. *Acta Pathol. Microbiol. Scand.* 187(Suppl.): 63–64

32. Lippincott, B. B., Lippincott, J. A. 1969. Bacterial attachment to a specific wound site as an essential stage in tumor initiation by *Agrobacterium tumefaciens. J. Bacteriol.* 97:620–28

33. Lippincott, J. A., Lippincott, B. B. 1978. Tumor initiation complementation on bean leaves by mixtures of tumorigenic and nontumorigenic *Agrobacterium rhizogenes. Phytopathology* 68:365–70

34. McCann, J., Ames, B. N. 1976. Detection of carcinogens as mutagens in the *Salmonella*/microsome test:Assay of 300 chemicals: Discussion. *Proc. Natl. Acad. Sci. USA* 73:950–54

35. McCardell, B. A., Pootjes, C. F. 1976. Chemical nature of agrocin 84 and its effect on a virulent strain of *Agrobacterium tumefaciens. Antimicrob. Agents Chemother.* 10:498–502

36. Merlo, D. J., Nester, E. W. 1977. Plasmids in avirulent strains of *Agrobacterium. J. Bacteriol.* 129:76–80

37. Moore, L. W. 1973. *Colonization and recovery of Agrobacterium tumefaciens from roots of Prunus seedlings and adjacent nursery soil.* Presented at 2nd Int. Congr. Plant Pathol, Minneapolis, Minn.

38. Moore, L. W. 1976. Latent infections and seasonal variability of crown gall development in seedlings of three *Prunus* species. *Phytopathology* 66: 1097–1101

39. Moore, L. W. 1977. Prevention of crown gall on *Prunus* roots by bacterial antagonists. *Phytopathology* 67:139–44

40. Moore, L. W. 1978. Biological control of crown gall. *Proc. 2nd Woody Ornamental Disease Workshop, Univ. Missouri, Columbia.* In press

41. Moore, L. W. 1978. Practical use and success of *Agrobacterium radiobacter* strain 84 for crown gall control. In *Biology and Control of Soil Borne Plant Pathogens,* ed. B. Schippers, W. Gams. New York: Academic. In press

42. Moore, L. W., Tindall, K., Warren, G., Staver, M. 1978. *Nonmutagenicity of agrocin 84 and Agrobacterium radiobacter strain 84 in the Ames test.* Presented at Ann. Meet. Am. Phytopathol. Soc., 70th, Tucson, Ariz.

43. Moore, L. W., Tingey, D. T. 1976. Effect of temperature, plant age, and infection site on the severity of crown gall disease in radish. *Phytopathology* 66: 1328–33

44. Munnecke, D. E., Chandler, P. A., Starr, M. P. 1963. Hairy root (*Agrobacterium rhizogenes*) of field roses. *Phytopathology* 53:788–99

45. New, P. B., Kerr, A. 1971. A selective medium for *Agrobacterium radiobacter* biotype 2. *J. Appl. Bacteriol.* 34:233–36

46. New, P. B., Kerr, A. 1972. Biological control of crown gall: Field measurements and glasshouse experiments. *J. Appl. Bacteriol.* 35:279–87

47. Nuesry, S. M. 1975. *Biological control of crown gall on Euonymus.* MS thesis. Ohio State Univ., Columbus. 47 pp.

48. Panagopoulos, C. G., Psallidas, P. G., Alivizatos, A. S. 1978. *Evidence of a breakdown in the effectiveness of biological control of crown gall.* Presented at 3rd Int. Congr. Plant Pathol., Munich, Germany

49. Petit, A., Tempe, J., Kerr, A., Holsters, M., Van Montagu, M., Schell, J. 1978. Substrate induction of conjugative activity of *Agrobacterium tumefaciens* Ti plasmids. *Nature* 271:570–71

50. Rickenberg, H. V. 1974. Cyclic AMP in prokaryotes. *Ann. Rev. Microbiol.* 28:139–66

51. Riley, P. S., Weaver, R. E. 1977. Comparisons of thirty-seven strains of VD-3 bacteria with *Agrobacterium radiobacter*: Morphological and physiological observations. *J. Clin. Microbiol.* 5:172–77

52. Roberts, W. P., Tate, M. E., Kerr, A. 1977. Agrocin 84 is a 6-N-phosphoramidate of an adenine nucleotide analogue. *Nature* 265:379–81

53. Rouatt, J. W., Katznelson, H. 1957. The comparative growth of bacterial isolates from rhizosphere and nonrhizosphere soils. *Can. J. Microbiol.* 3:271–75

54. Schroth, M. N., Moller, W. J. 1976. Crown gall controlled in the field with a nonpathogenic bacterium. *Plant Dis. Reptr.* 60:275–78

55. Sciaky, D., Montoya, A. L., Chilton, M. D. 1978. Fingerprints of *Agrobacterium* Ti plasmids. *Plasmid* 1:238–53

56. Skurray, R. A., Reeves, P. 1954. F factor-mediated immunity to lethal zygosis in *Escherichia coli* K-12. *J. Bacteriol.* 117:100–6

57. Smith, E. F., Brown, N. A., Townsend, C. O. 1911. Crown gall of plants: Its cause and remedy. *USDA Plant Ind. Bull. 213*

58. Smith, V. A., Hindley, J. 1978. Effect of agrocin 84 on attachment of *Agrobacterium tumefaciens* to cultured tobacco cells. *Nature* 276:498–500

59. Starkey, R. L. 1931. Some influences of the development of higher plants upon the microorganisms in the soil. IV. Influence of proximity to roots on abundance and activity of microorganisms. *Soil Sci.* 32:367–93

60. Stonier, T. 1960. *Agrobacterium tumefaciens* Conn. II. Production of an antibiotic substance. *J. Bacteriol.* 79:889–98

61. Süle, S. 1978. Biological control of crown gall with a peat cultured antagonist. *Phytopathol. Z.* In press

62. Süle, S. 1978. Pathogenicity and agrocin 84 resistance of *Agrobacterium tumefaciens*. Presented at IVth Int. Conf. on Plant Pathogenic Bacteria, Angers, France

63. Süle, S., Kollanyi, L. 1977. Biological control of crown gall of raspberry. *Novenyvedelem* 13:241–44 (Engl. summ.)

64. Van Lanen, J. M., Baldwin, I. L., Riker, A. J. 1952. Attenuation of crown gall bacteria by cultivation in media containing glycine. *J. Bacteriol.* 63:715–34

65. Vidaver, A. K. 1976. Prospects for control of phytopathogenic bacteria by bacteriophages and bacteriocins. *Ann. Rev. Phytopathol.* 14:451–65

65a. Vidaver, A. K., Mathys, M. L., Thomas, M. E., Schuster, M. L. 1972. Bacteriocins of the phytopathogens *Pseudomonas syringae*, *P. glycinea*, and *P. phaseolicola*. *Can. J. Microbiol.* 18:705–13

66. Wijsman, H. J. W., Pafort, H. C. 1974. Pleiotropic mutations in *Escherichia coli* conferring tolerance to glycine and sensitivity to penicillin. *Mol. Gen. Genet.* 128:349–57

67. Weinhold, A. R., Bowman, T. 1963. Selective inhibition of the potato scab pathogen by antagonistic bacteria and substrate influence on antibiotic production. *Plant Soil* 28:12–24

Ann. Rev. Phytopathol. 1979. 17:181–202
Copyright © 1979 by Annual Reviews Inc. All rights reserved

GENETIC SYSTEMS IN ◆3707
PHYTOPATHOGENIC BACTERIA

G. H. Lacy

Department of Plant Pathology and Botany, The Connecticut Agricultural
Experiment Station, New Haven, Connecticut 06504

J. V. Leary

Department of Plant Pathology, University of California,
Riverside, California 92521

INTRODUCTION

Studies of bacterial genetics have made a major contribution to the knowledge of the nature, function, and regulation of the genetic determinants of phenotype. Although the genetics of bacteria has been intensively investigated, the scope of these studies remained so narrow that Millard Susman, with tongue in cheek, stated that "bacteria" means *Escherichia coli* K12 (93). This viewpoint is apparent when surveying the literature for information on the role of the bacterial genome in biological interactions between organisms, especially in phytopathogenicity. It is evident that this area has received little attention.

Phytopathologists are interested in defining mechanisms of pathogenicity among diverse bacterial pathogens (leaf-spotters, wilt-inducers, toxin-producers, blighters, gall-inducers, and soft-rotters). One obvious advantage in using genetics to study pathogenicity would be employing a different point of view in approaching these studies. Hitherto, classical studies of phytopathogenesis have considered how the environment affects the expression of pathogenicity, how the host defends against the pathogen, or how the chemical products produced in host-pathogen interactions might contribute to pathogenesis. The critical question of what is in the genome of the pathogen that makes it capable of causing pathogenesis has been largely left unstudied.

0066-4286/79/0901-0181$01.00

However, some remarkable recent advances would indicate that the situation is changing. These include mapping of pathogenicity genes in *Erwinia chrysanthemi* (14), elucidation of the molecular basis for oncogenicity in the *Agrobacterium tumefaciens*-crown gall system (15), syringomycin production in the *Pseudomonas syringae*-holcus spot system (39), and construction of a chromosomal map in *Pseudomonas glycinea* (34). These advances appear to be the first green fruits of a new growth resulting from application of the techniques of classical and molecular genetics to phytopathology.

Development of genetic systems among phytopathogenic bacteria is a prerequisite to studying the genetics of phytopathogenicity. Since development has been meager and has occurred only in sporadic bursts to date, it is necessarily the task of this first comprehensive review to collate the existing information. To this end, the material here is divided into three sections. The first considers extrachromosomal inheritance, the second deals with substitutive recombinational phenomena such as transformation and conjugation, and a third section considers the additive recombinational systems of transduction, plasmid integration, and transposable elements.

EXTRACHROMOSOMAL INHERITANCE

Extrachromosomal inheritance in bacteria is mediated by plasmids. Since the involvement of plasmids in conjugative chromosomal transfer is covered in a later section, this discussion of extrachromosomal inheritance considers only plasmid genes and their contribution to the host bacterial genotype and phenotype.

Basic to an understanding of bacterial extrachromosomal inheritance is a knowledge of the nature of plasmids. Two kinds of plasmids exist: conjugative plasmids capable of mediating their own transfer from bacterium to bacterium and nonconjugative plasmids incapable of self-transfer.

Bacterial plasmids consist of circular double-stranded deoxyribonucleic acid (DNA) molecules resident in the cytoplasm of the host. Conjugative plasmids are often associated with pili, which mediate cell-to-cell contact for transfer of DNA from one bacterium to another. Occasionally, nonconjugative plasmid or chromosomal DNA may be cotransferred with conjugative plasmids.

The critical difference between a bacteriophage and a plasmid is that the plasmid genetic material normally does not occur outside the host bacterium in a bacteriophage-like virion, but remains within a bacterium until conjugal transfer occurs. Even during transfer, the plasmid DNA may be "protected" by the pilus extracellularly. Single-stranded plasmid DNA passes from the donor to the recipient, and is used as a template for a new double-stranded molecule in the recipient.

Plasmids often carry many genes not required for plasmid maintenance. The resultant genetic flexibility allows plasmids to exist in symbiotic associations with their bacterial hosts rather than a parasitic one. Most plasmids studied in detail have been found to carry genes beneficial to their bacterial hosts. Examples are those for antibiotic resistance, heavy metal resistance, unique biochemical pathways, bacteriocin synthesis, bacteriocin resistance, and toxin production. Therefore, it seems reasonable to expect that plasmids also play a role in pathogenicity.

This section of the review is devoted to plasmids detected in, plasmids transferred into, or phenotypic expression of plasmids in phytopathogenic bacteria. Phytopathogenic bacteria are found in the Gram-negative genera *Agrobacterium* (Rhizobiaceae), *Erwinia* (Enterobacteriaceae), *Pseudomonas* and *Xanthomonas* (both in the Pseudomonadaceae), as well as in the Gram-positive genera *Corynebacterium* (coryneform group of bacteria) and *Streptomyces* (Streptomycetaceae).[1]

Oncogenicity of Agrobacterium

Some of this information has been reviewed (15, 74). Kerr (55) detected transfer of oncogenicity from donor *Agrobacterium tumefaciens* strains inoculated onto plant surfaces to recipient, nontumor-inducing *A. radiobacter* inoculated at later times onto the same tissue. Serological and biochemical analyses of genetically marked recipients confirmed transfer of oncogenicity (54).

A positive correlation between the presence of large plasmids (112 to 156 megadaltons) and pathogenesis was established in strains of *A. tumefaciens* (101). Pathogenic strains of *A. tumefaciens* were made nonpathogenic when cured of the large plasmid (49, 97, 98, 99). Analyses showed that the plasmid was not integrated into the bacterial chromosome (99). Pathogenicity was restored by conjugative transfer of a large plasmid from *A. tumefaciens* strains back to the cured strains (48, 98, 99). In a similar manner, the ability to induce root proliferation, a character of *Agrobacterium rhizogenes*, was transferred in planta[2] to *A. tumefaciens*. The transconjugant had the biochemical characters of *A. tumefaciens* (1).

Not all plasmids in agrobacteria have a role in tumor induction. Diverse kinds of large plasmids may occur in strains of *Agrobacterium* (29, 75, 90).

---

[1]Several of the species names used in this review are *species incertae sedis* according to Bergey's *Manual of Determinative Bacteriology* (ed. R. E. Buchanan, N. E. Gibbons. Baltimore: Williams & Wilkins. 8th ed.); however, they have been retained here to indicate pathotypes or biotypes important in phytopathology without intending any taxonomic clarification.

[2]"In planta" conjugation refers to conjugative transfer among bacteria on or in plants (62).

In planta conjugal transfer or "Kerr-transfer" of the tumor-inducing (Ti) plasmid in *Agrobacterium* was time consuming (two or more weeks were required for tumor growth), required greenhouse facilities, and was liable to extraneous contamination by other nonpathogenic agrobacteria. Liao & Heberlein (70) developed a carrot disk system for in planta transfers under more stringent aseptic conditions. Other laboratories developed conjugative transfer systems with IncP-1 plasmids in *Agrobacterium* (65, 86).

The Ti plasmid was also cotransferred by IncP-1 plasmids in ex planta[3] matings (17, 69, 73). The host range of a limited host range strain was extended by cotransfer of a Ti plasmid from a broad host range strain of *Agrobacterium tumefaciens* (73).

Direct transfer of Ti in ex planta matings was facilitated by the presence of octopine or nopaline (36) or their precursors. Analogues of these compounds did not have the same effect and sulfhydryl-containing amino acids were inhibitory (P. J. J. Hooykaas, C. Roobol, and R. A. Schilperoort, manuscript in preparation).

Further, Ti was transferred directly to both nonpathogenic agrobacteria and *Rhizobium trifolii* during ex planta matings. Oncogenicity and the presence of a large plasmid (120 megadaltons) were conferred on the *R. trifolii* Ti[+] transconjugant (51). Significantly, the *R. trifolii* Ti[+] transconjugant was still able to nodulate *Trifolium pratense* but not *T. parviflorum*. The inability to nodulate *T. parviflorum* was correlated with the loss of an even larger plasmid (250 to 350 megadaltons) which had been present in the recipient *R. trifolii* strain prior to mating! Other rhizobia have large plasmids and their role in nodulation and host specificity is under investigation (82).

Australian workers (56, 81) reported that some nonpathogenic strains of *Agrobacterium radiobacter* carried a bacteriocin that could limit the spread of crown gall in field and greenhouse experiments with infested soil. These findings have been confirmed and extended (79) (see the review by L. W. Moore in this volume). Sensitivity to agrocin 84 was shown to be a property of Ti plasmids (33). Other properties associated with the Ti plasmids were ability of the bacterium to utilize the guanidoamino acids octopine and/or nopaline (6, 78), and the induction of synthesis of the same amino acids by the tumor tissue.

Renaturation kinetic analysis of radioactively labeled, tumor-inducing plasmid DNA, in the presence of crown gall tumor DNA suggested that about 5% of the plasmid genes were present in tumor DNA (16). The plasmid was digested with a restriction endonuclease (*Sma*I) and the frag-

---

[3]"Ex planta" conjugation refers to conjugative transfer among plant-related bacteria in culture vessels rather than on or in plants (51).

ments were separated by electrophoresis. The fragments, used separately as renaturation probes, revealed homologies to fragments 3b and 10c in tumor DNA. Since these fragments are proximal on the circular map of Ti (B. P. Koekman, G. Ooms, P. M. Klapwijk, and R. A. Schilperoort, manuscript in preparation), it is possible that a single region is transferred to the plant cell during its conversion to an incipient tumor. Further evidence for DNA transfer to the plant was the detection of ribonucleic acid transcription products of Ti in aseptic tumor tissue (32). The mechanisms whereby DNA can be removed from a plasmid, transferred across a membrane, two cell walls, another membrane into a nucleus (across another membrane), possibly integrated into a eucaryotic genome, and expressed in a higher plant cell are of great interest. Evidently, the evolutionary and genetic chasm between procaryotes and eucaryotes is bridged easily by a "simple" phytopathogenic bacterium.

Plasmids in Corynebacterium

Plasmid DNA was isolated from *C. fascians*, *C. insidiosum*, *C. michiganense*, *C. nebraskense*, *C. oortii*, *C. rathayi*, *C. sepedonicum*, and *C. tritici*. The molecular weights of these plasmids ranged from 20 to 75 megadaltons (44). Since no phenotypic traits have been correlated with these plasmids, they are "cryptic." Transfer of plasmids between Gram-positive and -negative bacteria has not been reported.

Plasmids in Erwinia

Plasmids involved in extrachromosomal inheritance may be introduced from bacterial genera other than *Erwinia* (nonindigenous) or they may be indigenous in *Erwinia* species.

NONINDIGENOUS PLASMIDS   A list of plasmids that have been transferred into *Erwinia* is presented in Table 1. Further, *E. amylovora* (10, 11, 100), *E. carotovora* var. *atroseptica* (11, 18), *E. carotovora* var. *carotovora* (18), *E. dissolvens* (11), *E. herbicola* (10, 11, 37; G. H. Lacy, unpublished), *E. nigrifluens* (11), *E. nimipressuralis* (11), *E. chrysanthemi* (10, 18, 60), and *E. stewartii* (21, 22, 39) are capable of intergeneric transfer of one or more of the plasmids $F'lac^+$ (11, 21, 100), pKR210 (22), R100drd-56 (11), and RP1 (18, 60; G. H. Lacy, unpublished) into *E. coli* (10, 11, 21, 25, 37, 60), *P. aeruginosa* (18, 37), *Salmonella typhimurium* (11, 100), and *Shigella dysentariae* (10, 11). Interspecific transfer (between different *Erwinia* species) has been accomplished with R100drd-56 (11), $F'lac^+$ (10), RP1 (G. H. Lacy, unpublished). Intraspecific plasmid transfer (between strains of the same species) has been detected in *E. amylovora*, *E. chrysanthemi*, and *E. herbicola* with $F'lac^+$ (10), *E. chrysanthemi* (60), *E. herbicola* (37;

Table 1 Plasmids transferred by conjugation into *Erwinia* species and phytopathogenic *Pseudomonas* species

Recipient	Plasmids	References
Erwinia amylovora	F'*lac*$^+$	10
	N3, R446b, R478, RK2, RP1–1, Rts1	86
	R1*drd*19	3
	R100*drd*–56	11
	RP1	18*[a]
E. aroideae	R18–1	18
	R100*drd*–56, SR–1	11
E. carotovora var. *atroseptica*	N3, R391, R446b, R471a, R478, RK2,	
	RS–a	86
	R18–1, RP1	18
	R68.45	87
	R100*drd*–56	11
E. carotovora var. *carotovora*	N3, R18–1, R446b, R471a, R478,	
	RS–a	86
	R68, RP4, RP4::Mu–1*cts*62	87
	R91, RP1	18
	RK2	45, 46, 86
	Unnamed plasmid	5
E. chrysanthemi	F'*lac*$^+$	10
	F$_{ts_{lac}}$	45, 86
	R18–1	18
	R100*drd*–56, SR–1	11
	RK2, RP11–42, Rts1	45, 46
	RP1	18, 60
	RP4	45, 46, 65
E. cytolytica	R100*drd*–56	11
E. dissolvens	R100*drd*–56	11
E. herbicola	F'*lac*$^+$	10
	R18–1	18
	R100*drd*–56, SR–1	11
	RP1	37*[a]
E. nigrifluens	R18–1, RP1	18
	R100*drd*–56, SR–1	11
E. nimipressuralis	R100*drd*–56	11
E. stewartii	pRK210	22
	pRK212.1	23
	R68.45, R751, RP1	21
E. uredovora	R68, R91, RP1	18
Erwinia spp. B–4	N3, R387, R446b, R478, RK2	86
SH–2	N3, R446b, R471a, R478, RK2, RP1–1	86
USB	N3, R387, R446b, R471a, R478, RK2	86

Table 1 *(Continued)*

Recipient	Plasmids	References
Pseudomonas glycinea	FP2, R68.45, R91–5	68
	R18–1, RK2	85
	R68	63
	RP1	62, 85
P. lachrymans	R18–1	85
	RK2, RP4	86
P. marginalis	R18–1	85
	RK2, RP4	86
P. phaseolicola	N3, R391	86
	R18–1, R91–1, RK2, RP1–1	85
	R68.45	47
	RP1	62, 85
P. pisi	R18–1	85
	RK2	86
P. solanacearum	RP4	76
	RP4::Mu, RP4::Mu–1cts62	7
P. syringae	R18–1, R91–1, RK2, RP1, RP1–1	85

a* indicates G. H. Lacy, unpublished data.

G. H. Lacy, unpublished), and *E. stewartii* (21) with RP1 (10, 21, 60) and R68.45 (21). Intrastrain transfer of plasmids (among genetically marked members of the same *Erwinia* strain) is described later. The nonconjugative plasmid pML2, important as a cloning vehicle for recombinant DNA, was cotransferred from *E. coli* by the conjugative plasmid $F_{ts}lac$ into *E. amylovora* and *E. chrysanthemi* (86).

In summary, plasmids of the *E. coli* incompatibility groups[4] F (R100drd-56, F'*lac*+), J (R391), K (R387), L (R471a), M (R446b), N (N3), P-1 (RP1, RP1-42, R18-1, R68, R68.45, R9, RP4::Mu-cts62, RP4, RK2), P-2 (RP1-1), S (R478), T (Rts1), and two unclassified plasmids (SR-1 and the unnamed plasmid) have been transferred directly into *Erwinia* species by conjugation. In one survey, plasmids of the incompatibility groups A, C, FII, FIV, H, I, and O failed to transfer into *Erwinia* species (86).

Genetic exchange in planta could offer phytopathogenic bacteria an opportunity to acquire traits useful for plant colonization, overcoming host resistance, and establishment of infection. Unfortunately, this subject has been practically ignored among the genera of phytopathogens with the exception of *Agrobacterium*. Only three *Erwinia* species have been examined

[4]The members of the plasmid incompatibility group are determined by the failure of genetically similar plasmids to superinfect a bacterium already containing a plasmid of the same incompatibility group.

for ability to exchange plasmids in planta. Transfer of RP1 between *E. chrysanthemi* strains was readily detected in seedling maize plants (60). The frequency of transfer was almost 100-fold greater in planta than ex planta. In a second maize line, the presence of 2,4-dihydroxy-7-methoxy-(2H)-1,4-benzoxazin-3(4*H*)-one (DIMBOA), a cyclic hydroxamate involved in resistance of maize to bacterial stalk rot (61), did not decrease the frequency of plasmid transfer. Transconjugants (RP1$^+$) of *E. chrysanthemi* were able to cause stalk rot symptoms on maize seedlings.

Transfer of RP1 from *E. coli* into *E. amylovora* and *E. herbicola* in pear blossoms occurred at high frequencies. The plasmid could be transferred from *E. herbicola* to epiphytic *P. syringae* and pathogenic *E. amylovora* at low frequencies (G. H. Lacy and L. Hankin, unpublished results). This study suggests that epiphytes may act as reservoirs of antibiotic resistance plasmids for a phytopathogen whose control often depends on antibiotic chemotherapy.

INDIGENOUS PLASMIDS   Lactose-utilizing (Lac$^+$) strains of *Erwinia herbicola* from human clinical material were able to donate the ability to utilize lactose to Lac$^-$ strains of *E. amylovora* and *E. herbicola* of both plant and human origins at high frequencies (12). Donor ability (E) was found to be proximal to *lac* by transductional analysis. The E-*lac* factor did not confer susceptibility to infection by the F plasmid-specific bacteriophage M13 and coexisted with F-like plasmids without any reduction in the stability. This information suggests that E-*lac* may be extrachromosomal on a plasmid unrelated to F.

Other indigenous plasmids exist among *Erwinia* species. One strain of *E. stewartii* has been shown to contain ten or more cryptic plasmids (25). Two *E. stewartii* plasmids have been found to mediate conjugative transfer of nonconjugative plasmids (23).

Some strains of *E. chrysanthemi* contain plasmids of about 50 megadaltons or larger. Characterization of supercoiled plasmid DNA purified from cell lysates of one strain of *E. chrysanthemi* by gel electrophoresis, contour measurement, and restriction endonuclease analysis revealed a large plasmid (pEC1, 50.4 megadaltons) and a small plasmid (pEC2, 4.8 megadaltons) (93). Similar analyses of strains of *E. carotovora* var. *carotovora* have indicated that many have plasmids of various sizes (30).

Naturally occurring streptomycin-resistant strains of *Erwinia amylovora* are often encountered in nature. This resistance may arise by spontaneous mutation or conjugal transfer of antibiotic resistance. Streptomycin-resistant strains have been found to contain plasmids, but no correlation between their presence and antibiotic resistance has been discovered (86). However, plasmid-borne streptomycin resistance has been introduced on nonindigenous plasmids in ex planta matings (3).

Plasmids in Phytopathogenic Pseudomonads

NONINDIGENOUS PLASMIDS   A list of plasmids transferred into phytopathogenic *Pseudomonas* sp. may be found in Table 1. Plasmids FP2 (68), RP1 (47, 62, 85), R18-1 (85), RP4::Mu (7), RP4::Mu-1*cts*62 (7), and RK2 (85) have been transferred from *P. glycinea* (68, 85), *P. solanacearum* (7), and *P. phaseolicola* (85) into *E. coli* (7, 85) and *P. aeruginosa* (68, 85). Interspecific transfers of RP1 (62, 85) and R18-1 (85) have been detected from *P. glycinea* to *P. phaesolicola* (62) and from *P. phaseolicola* to *P. glycinea* (62) and *P. syringae* (85). Intraspecific transfers of FP2 (68), R68 (63), R68.45 (68), RP1 (85), and R18-1 (85) have been shown to occur in *P. glycinea* (63, 68) and *P. phaseolicola* (85).

Plasmid RP1 was transferred readily in planta between *P. glycinea* and *P. phaseolicola* (62). However, RP1-mediated transfer of chromosomal genes in planta in *P. glycinea* intrastrain matings could not be reliably detected (59). Plasmid pML2 was contransferred by plasmid pRK212.1 into *P. glycinea, P. lachrymans, P. marginalis, P. phaseolicola,* and *P. pisi* (86).

In summary, plasmids of the *E. coli* incompatibility groups J (R391), N (N3), P-1 (RP1, RP4, RP4::Mu, RP4::Mu-1*cts*62, R68, R68.45, R91-1, R91-5, R18-1, RK2), P-2 (RP-1), and the unclassified *P. aeruginosa* plasmid FP-2 have been moved into phytopathogenic pseudomonads. In one survey (86), plasmids of the *E. coli* incompatibility groups A, C, FII, FIV, H, I, K, L, M, O, S, T, and W failed to transfer by conjugation into *P. phaseolicola.* Evidently, the host range of many *E. coli* plasmids does not include phytopathogenic pseudomonads.

INDIGENOUS PLASMIDS   Cryptic plasmids have been found in *P. cepacia* (40), *P. glycinea* (28), *P. lachrymans* (86), *P. phaseolicola* (35), *P. savastanoi* (86), *P. syringae* (39), and *P. tonelliana* (86). Phenotypic properties have not been associated with these plasmids (28, 86) except in *P. syringae* (39) and *P. phaseolicola* (35).

Syringomycin, a phytotoxic peptide, is produced by strains of *P. syringae.* Some strains became nonpathogenic and lost the ability to produce syringomycin (Syr⁻). The loss of syringomycin production could also be induced by acridine orange. Electrophoresis of cleared lysates indicated a plasmid (pCG101) of about 35 megadaltons in Syr$^+$ but not in Syr$^-$ strains (39).

Tyrosinase Inheritance in Streptomyces

Gregory & Shyu (43) demonstrated that tyrosinase inheritance in *Streptomyces scabies* was extrachromosomal. They found that 0.2% of the colonies plated on a tyrosine-containing medium were deficient in tyrosinase (Tye⁻) and, therefore, in the ability to form a dark brown pigment (probably melanin). However, if two complementary diauxotrophic mutants, one Tye$^+$ and the other Tye$^-$, were grown together on a nutritionally deficient medium, greater than 99% of colonies of both parental nutritional types gave positive tyrosinase reactions. Genetic studies (41) indicated that *tye* was not linked to chromosomal genes. Acridine dyes increased the frequency of loss of *tye*$^+$ among strains of *S. scabies* (42).

The factor conferring fertility in *S. scabies* (41) may also be extrachromosomal in nature. Whether or not it is identical with the factor mediating the Tye$^+$ phenotype is unknown, but another *Streptomyces* fertility factor has physically characterized as a plasmid (4).

Plasmids in Xanthomonas

NONINDIGENOUS PLASMIDS   Plasmids RP4 and RK2 were transferred from *E. coli* and plasmid R18-1 was transferred from *Pseudomonas aeruginosa* to *Xanthomonas vesicatoria* ex planta. In turn, *X. vesicatoria* transconjugants transferred plasmids RP4 and RK2 to other strains of *X. vesicatoria* as well as *X. campestris, X. corylina, X. hedarae, X. incanae, X. juglandis, X. malvacearum, X. pelargonii,* and possibly, *X. dieffenbachia.* Transfer of both plasmids was also detected from *X. vesicatoria* to *Agrobacterium tumefaciens, Erwinia chrysanthemi,* and *Pseudomonas phaseolicola,* but not to the Gram-positive species *Corynebacterium michiganese* (65). The β-lactmase (penicillinase) activity of plasmids RP4 and RK2 was more stable in colonies isolated from foliar lesions or air-dried plant tissue infected with *X. vesicatoria* transconjugants than in clones stored at 5°C on an agar medium not fortified with antibiotics (66).

INDIGENOUS PLASMIDS   Several plasmids are present in *X. manihotis,* the causal agent of cassava blight. No phenotypic functions have been associated with these plasmids (71).

SUBSTITUTIVE RECOMBINATION

Substitutive recombination requires that some DNA sequences on homologous segments be exchanged, i.e. substituted by an even number of crossover events. The resultant composite molecule is usually about the same size as the original recipient molecule. In this section recombination resulting from DNA transformation and conjugal transfer of chromosomal DNA is discussed.

Transformation

Purified DNA may be taken up by competent recipient bacterial cells and incorporated through recombination into the recipient genome. This process is called *transformation.* Transformation of plasmid or bacteriophage

DNA into a recipient cell does not require recombination events with homologous DNA segments of the chromosome. However, since the method of DNA introduction into the cell may be similar, transformation of plasmid and bacteriophage DNA is also included under this heading.

Transformation is useful for mapping the bacterial chromosome. It is possible to determine the linear relationship of loci near selected markers from the frequency of coinheritance of unselected markers. Further, transformation of plasmid DNA makes it possible to extract cryptic plasmids from phytopathogenic bacteria of relatively unknown genetic backgrounds and insert them into a well-known system (*Escherichia coli* K12) to study their functions.

TRANSFORMATION OF AGROBACTERIUM   Early reports of transformation of oncogenicity from *Agrobacterium tumefaciens* to nonpathogenic agrobacteria, related bacteria, and plant tissue were unreliable because DNA sterility tests (58), deoxyribonuclease (DNase) controls (2), or genetic and physical demonstration of the genetic determinant were lacking (52). However, later demonstrations of transformation (50, 53) and transfection (transformation of bacteriophage DNA) of DNA into *A. tumefaciens* (50, 77) are more complete.

TRANSFORMATION OF ERWINIA   Purified DNA of plasmid pBR322 was transformed into *E. herbicola* (Nal^r) using the calcium-chloride method developed for *E. coli*. Primary selection for both Amp^r and Tet^r and counterselection for Nal^r precluded any reasonable possibility that the phenotype was due to spontaneous mutation. Physical characterization of supercoiled DNA recovered from the *E. herbicola* transformants showed the presence of a plasmid that was similar to pBR322 as determined by gel electrophoresis, endonuclease restriction patterns, and contour measurement (G. H. Lacy and R. B. Sparks, manuscript in preparation).

TRANSFORMATION OF PHYTOPATHOGENIC PSEUDOMONADS Transformation systems have been reported for *Pseudomonas solanacearum*. A nonpathogenic mutant deficient in tryptophan biosynthesis (Try$^-$) was transformed with DNA from a prototrophic donor and pathogenic prototrophs were recovered at low frequencies (20–24).

Also, an indigenous chromosomal exchange system was detected among auxotrophic, antibiotic resistant mutants of *P. solanacearum* (8). Since DNase destroyed the transfer ability and crude DNA preparations could replace living donor cells, transformation is now considered to be the method of genetic exchange (67).

Restoration of prototrophy by transformation in auxotrophic mutants of *Pseudomonas syringae* has also been reported (72, 96). Double auxotrophic mutants with complementary genotypes were used as donors and recipients.

TRANSFORMATION OF *XANTHOMONAS* Recombinant phenotypes were recovered after transformation of *Xanthomonas phaseoli* with DNA from strains resistant to streptomycin or of different colonial morphology. The transforming activity of the DNA preparation was eliminated by treatment with DNase (26, 27).

Conjugative Transfer of Chromosomal Loci

Conjugative transfer may be exclusively mediated by plasmids in bacteria. Conjugal transfer of chromosomal loci may be detected in bacteria if the DNA sequences have physically transferred into a recipient cell, have recombined in some stable manner with a DNA molecule capable of replication, and have expressed the gene products of the transferred loci in the recipient genetic background.

Plasmids may acquire insertions of chromosomal material or they themselves may become inserted into the bacterial chromosome. Therefore, this discussion reviews the transfer and expression of loci of chromosomal origin without prejudice to their replicating molecular host, whether it be chromosome or plasmid, in the recipient bacterium.

CHROMOSOMAL TRANSFER IN *STREPTOMYCES* The first mention of chromosomal transfer in a phytopathogenic bacterial species was made in 1964 from studies of *Streptomyces scabies* (41). Complementary diauxotrophic mutants were grown together and their spores harvested, washed, and plated on selective media. Stable recombinant clones were isolated on minimal media at frequencies of 10^{-5} to 10^{-4} recombinants per parental type. The mechanism of transfer was not defined, but, drawing upon the well-studied genetics of *S. coelicolor* (4) and the presence of an extrachromosomal determinant of tyrosinase activity in *S. scabies* strains (42, 43), it was probably due to plasmid-mediated conjugal transfer of chromosomal markers.

F'*lac*+-MEDIATED CONJUGATIVE TRANSFER *Escherichia coli* plasmid F'*lac*+, containing the base sequences for the lactose operon, was introduced into strains of *Erwinia amylovora, E. chrysanthemi,* and *E. herbicola.* The plasmid was unstable in *E. chrysanthemi* (10).

Some F'*lac*+ transconjugants of *E. amylovora* resisted acridine orange-induced loss of the plasmid. This suggested that the plasmid existed in a chromosomally integrated state in these strains. These donors were used in interrupted matings with multiply marked, auxotrophic recipients. The results indicated an efficient and oriented chromosomal transfer (*cys, met,* and *trp* were proximal; *ser* and *gua* intermediate; and *arg, ilv,* and *pro* were distal markers) (13).

Pathogenic, acridine orange-stable, F'*lac*$^+$ transconjugants of *E. amylovora* could transfer the gene or genes that determine the ability to produce fireblight symptoms on immature pear fruit or *Pyracantha* twigs to auxotrophic, nonpathogenic, streptomycin-resistant strains. Interrupted matings suggested that *ser* and *vir* (the gene or genes conferring pathogenicity) entered recipient cells during the first 15 min of a 3 hr mating period. At 75 min, *pro* entered and *lac* (originating on the F'*lac*$^+$ insertion) entered toward the end of the mating period (89).

Supplementation of inoculum, consisting of recipient cells, with serine did not allow development of symptoms of fireblight. Further, among prototrophic Ser$^+$ revertants recovered from nonpathogenic Ser$^-$ strains, five were pathogenic and nine remained nonpathogenic. These data would suggest that the locus required to produce fireblight symptoms, is close to, but not identical with *ser* (89).

Donor strains of *E. chrysanthemi* were developed by isolating stable F' *lac*$^+$ transconjugants by repeated selection on media with lactose as the sole carbon source. Interrupted matings using these donors have indicated the following linear chromosomal transfer sequence: origin . . . *leu* . . . *thr* . . . *ade* . . . *lys* . . . *mcu* . . . *pat* . . . *his* . . . *trp* . . . *gal* . . . *lac* . . . F (14). The locus *pat* designates genetic sequences required for the degradation of polygalacturonic acid (at pH 8) by polygalacturonic acid *trans*-eliminase. Mutants with lesions at this locus were not capable of macerating plant tissue even though they formed hydrolytic polygalacturonase (13).

Conjugal transfer of *trp* genes from *E. amylovora*/F'*lac*$^+$ to *Salmonella typhimurium* Trp$^-$ strains provides additional evidence for transfer of chromosomal genes from phytopathogenic to zoopathogenic species (100). Further analyses of the stable recombinant Trp$^+$ *S. typhimurium* clones by cotransduction of markers proximal to the tryptophan region detected no cryptic *trp* alleles and indicated that a considerable degree of genetic homology exists between *E. amylovora* and *S. typhimurium trp* genes. The LD$_{50}$ in mice for the recombinant *S. typhimurium* clones was only slightly lower than the wild type. *Erwinia amylovora* produced fireblight symptoms on pear fruit, but was nonpathogenic in mice.

CONJUGATIVE TRANSFER MEDIATED BY P-1 PLASMIDS   One disadvantage of F'*lac*$^+$ for developing general chromosomal transfer systems in phytopathogenic bacteria is that its host range is limited to the enterobacteriaceae (86). Fortunately, plasmids of the P-1 incompatibility group (including the antibiotic resistance plasmids RK2, RP1, RP4, R68, and R91) have wide host ranges including *Agrobacterium, Erwinia, Pseudomonas,* and *Xanthomonas* (84).

These plasmids have been used to mobilize chromosomal genes among phytopathogenic pseudomonads and erwiniae. Chromosomal transfer mediated by P-1 plasmids (RP1 and R68) was described in *P. glycinea* (63). Complementing diauxotrophic mutants of the pathogen were mated in combinations in which none, one, or both mutants contained a plasmid. The transfer of chromosomal markers was scored by the appearance of prototrophy at one or more of the unselected auxotrophic loci. No recombinants to prototrophy were recovered when a P-1 plasmid was not present in either parental strain. In matings with a P-1 plasmid in only one parental strain, recombinants at two loci were recovered at frequencies of 10^{-8} to 10^{-6} recombinants per donor. Recombinants at one locus were recovered 8 to 18 times more frequently than recombinants at both loci. When both strains participating in a mating had P-1 plasmids, the frequency of recombination was depressed by plasmid entry-exclusion since both plasmids belong to the same incompatibility group.

Pseudomonas glycinea donors more efficient in chromosomal donor ability were recovered from ultraviolet irradiated R68$^+$ transconjugants (64). Recombinants to prototrophy at two auxotrophic loci were recovered at frequencies of 10^{-3} from matings with one irradiated donor strain. However, the plasmid was very unstable in this donor.

The plasmids R68 and R68.45 have been used to map a segment of the chromosome of *P. glycinea* by testing for frequency of coinheritance of prototrophy at unselected auxotrophic loci after intrastrain matings (34). Furthermore, plasmid R68.45 promoted interspecific transfer of *P. glycinea* chromosomal genes into *P. aeruginosa* at frequencies equal to or greater than those in the *P. glycinea* intrastrain crosss (68). Thus, it may be possible to compare the genetic map of a phytopathogenic pseudomonad directly with that of a genetically well characterized human pathogenic pseudomonad. These studies are interesting since *P. aeruginosa* is known to exist in agricultural soils, on the surface of vegetables, fruits and flowers, and has occasionally been implicated in soft-rot diseases of plants as well as in human disease (19, 59).

Evidence for chromosomal mobilization by RP1 in *Erwinia chrysanthemi* was based on recovery of prototrophs at frequencies significantly higher when RP1 was present than in matings without RP1 between complementary auxotrophic mutants (60). Recombinants were recovered from matings with an unstable donor strain at frequencies of 10^{-4} to 10^{-3} per donor.

Guimaraes & Panopoulos (46, 47) reported that P-1 plasmids RPI, RP11-42, RP4, and RK2 promoted chromosomal transmission among auxotrophic mutants of *E. chrysanthemi*. Plasmid R68.45 was effective in transferring genes from prototrophic wild-type *Erwinia carotovora* var. *carotovora* strains to auxotrophic, antibiotic-resistant mutants (87, 88).

CONJUGATIVE TRANSFER MEDIATED BY OTHER PLASMIDS   Two plasmids other than F (F'lac^+) or P-1 incompatibility group plasmids (RP1, RP4, R68, etc) have been used to mobilize chromosomal genes in phytopathogenic bacteria. Leary (68) used the *P. aeruginosa* chromosome-mobilizing plasmid FP2 to mobilize genes in intraspecific crosses with *P. glycinea* and interspecific crosses between *P. glycinea* and *P. aeruginosa*. Transductional and conjugational analyses of recombinants showed that the *P. glycinea* recombinant fragments were stably integrated into the recipient *P. aeruginosa* chromosome. Unfortunately, FP2 is too unstable in *P. glycinea* to be used for constructing time-of-entry maps.

Guimaraes & Panopoulos (46, 47) reported chromosomal mobilization in *Erwinia chrysanthemi* with the T incompatibility group plasmid Rts1. The chromosomal mobilizing-ability (cma) of donors carrying thermosensitive Rts1 plasmids was increased 25- to 100-fold by repeated selection for plasmid retention in donors incubated at high temperatures.

ADDITIVE RECOMBINATION

In additive recombination, genetic elements recombine with one another to form a composite molecule which is the sum of the recombining molecules. Several mechanisms of additive recombination, including integration of bacteriophages, plasmids, and transposons, may be explained by the expanded Campbell model as reviewed by Schwesinger (92). This model requires that (*a*) the DNA sequences be circularized prior to insertion, (*b*) there be a single reciprocal cross-over between specific regions (attachment sites) on the genetic element to be inserted and the recipient DNA molecule, and (*c*) the process of excision reverse the events leading to insertion and regenerate the original attachment sites and the individual components of the composite molecule.

Transduction

Transduction is recombination mediated by bacteriophage. Temperate bacteriophages are most effective since they often are inserted as prophages into the host chromosome. A prophage may be induced to replicate autonomously and cause lysis of its host bacterium. Imprecise excision of prophage DNA results in acquisition of adjacent bacterial DNA sequences. Upon virion contact with a recipient bacterium, the donor DNA sequences packaged with the bacteriophage DNA enter the recipient via the same processes that accomplish bacteriophage infection. Once in the recipient cell, additive or substitutive recombination may occur with indigenous DNA molecules.

Transduction allows great precision in genetic mapping (fine-structure mapping) which is not possible with techniques employing fertility plasmids. Two kinds of transducing bacteriophages exist: (*a*) the generalized transducing bacteriophages that may move any genetic locus and (*b*) specialized transducing bacteriophages that move certain regions of the bacterial chromosome at high frequencies.

Bacteriophages for phytopathogenic and epiphytic bacteria are easily isolated. Logically, among these bacteriophages, one would expect to find bacteriophages capable of transduction. Several attempts to demonstrate that temperate bacteriophages were capable of transducing chromosomal markers in phytopathogenic bacteria have failed. However, in 1955, Okabe & Goto reported that a temperate bacteriophage, T-$_c$200, mediated host specificity changes in tobacco and sesame strains of *Pseudomonas solanacearum* (83). However, no further reports have appeared on the subject.

Expansion of the host range of a known transducing bacteriophage is a second strategy that may be followed to develop a transduction system for phytopathogenic bacteria. A heat-inducible, kanamycin-resistance-mediating derivative of bacteriophage P1 was used to detect P1-sensitive mutants in *Erwinia carotovora*. Thermal induction of the transducing bacteriophage from *E. carotovora* yielded T1 virions (38). However, no markers of chromosomal origin were transduced.

At this point, then, we must state that no confirmed reports of transduction in phytopathogenic bacteria have appeared in the literature surveyed for this review. This is unfortunate considering the value of transduction systems in the study of the genetics of bacteria.

Plasmid Integration

Integration of plasmids (such as F) into bacterial chromosomes occurs at a low frequency at a limited number of specific sites. Since F' elements, like F'*lac*$^+$, are F plasmids imperfectly excised from the bacterial chromosome, heteroduplex studies of hybrids between various F' elements have indicated that preferred sites for plasmid integration may actually represent loci where insertion sequences are available on the chromosome.

Insertion sequences are small regions of DNA associated with genetic loci capable of insertion and transposition (92). These sequences cause polar mutations and often include inverted repeat sequences detectable by intrastrand DNA hybridization.

Evidence for insertion of plasmids into the bacterial chromosome is scarce among phytopathogenic bacteria. Indirect evidence for chromosomal insertion of F'*lac*$^+$ into *Erwinia amylovora* and *E. chrysanthemi* can be assumed (13, 14). The significance of integration of F'*lac*$^+$ in phytopathogens is that it indicates homology among insertion sequences shared between zoopathogens and phytopathogens. This, in turn, may mean that recombinant molecules have been formed in vivo between phytopathogens and other bacteria sharing similar insertion sequences without recourse to in vitro recombinant DNA techniques.

Laboratory synthesis of sites for integration of plasmids in phytopathogenic bacteria is a logical course of action to develop conjugative donors with high chromosomal donor ability. The temperate bacteriophage Mu is a prime candidate for creation of synthetic insertion sequences in phytopathogenic bacteria. Mu may integrate at random into the chromosome of its host, cause polar mutations, cause deletions, cause transposition of host DNA, cause attachment of unrelated DNA sequences, and promote integration of circular DNA (31). Unfortunately, Mu has a narrow host range, limiting natural infections to *E. coli* and *Shigella dysenteriae*. To transfer Mu to phytopathogenic bacteria, Boucher et al (7) developed RP4::Mu cointegrate molecules in vivo. Conjugative transfer of Mu vectored by RP4 was possible to *Erwinia carotovora* var. *carotovora* (87, 88), *Pseudomonas solanacearum* (7), and *Rhizobium meliloti* (7). Similar techniques were used to move Mu onto the chromosome of *Erwinia stewartii* (22).

Transposons

Transposons are genes associated with insertion sequences that transpose from one location to another on genetic elements. They differ from plasmids or bacteriophages in that they are not capable of replication unless integrated in a plasmid, bacteriophage, or chromosome.

A transposon is implied in the transfer of genes conferring oncogenicity from their location on the Ti plasmid into incipient tumor cells (16). The significance of transposons, capable of moving between procaryotes and eucaryotes, has not been lost to plant geneticists. Proposals to introduce genes of agronomic importance (such as nitrogen fixation or *nif*) into higher plants have been made (16, 32, 80, 91).

CONCLUDING REMARKS

A review of the development of genetic systems in phytopathogenic bacteria prior to 1977 would have been unnecessary. Almost 71% of the literature cited in this article was published after 1974 with about 50% appearing after 1977.

Much of the recent development of genetic systems in phytopathogenic bacteria is a direct result of the overwhelming interest in plasmids generated from work with *Escherichia coli* and *Pseudomonas aeruginosa*. The rapid extension of these studies to phytopathogens has led to recent demonstrations that (*a*) a variety of plasmids can be detected in phytopathogenic bacteria, (*b*) at least some of these plasmids contribute to host phenotype, and (*c*) plasmids can be transferred to phytopathogenic bacteria where they alter the phenotype of the new host.

The most widely studied and best understood example of the association of pathogenicity with the presence of a particular plasmid in bacterial cells is the *Agrobacterium tumefaciens*/Ti plasmid system. Since the burst of research on this system has been reviewed extensively elsewhere, we restricted our review to a brief discussion of the background literature and concentrated on the development of the genetic system which clearly and convincingly defined the role of the plasmid in determining the pathogenic phenotype. We did this, admittedly, for reasons reflecting a strong personal commitment to the viewpoint that without the means to examine correlative biochemical evidence genetically to demonstrate the relationship of genotype to phenotype, such evidence may be meaningless.

It is gratifying that researchers concerned with the genetics of phytopathogenic bacteria very quickly recognized the potential of plasmids and plasmid-mediated genetic systems from other organisms and applied them successfully to phytopathogens. It is in the area of simply developing useful genetic systems for the study of the inheritance of the pathogenic phenotype that plasmids have proven most useful to the plant pathologist. Without the demonstration that such plasmids as $F'lac^+$ and those of the IncP-1 group could be transferred to, be expressed in, and mobilize chromosomal genes in phytopathogens, much of the research we have discussed here would have been impossible.

It would be tempting to suggest that since the evidence is so convincing that the pathogenicity of *Agrobacterium tumefaciens* appears to be coded for by plasmid-borne genes, we should ignore the chromosome of phytopathogenic bacteria and concentrate only on plasmids to understand the genetics of phytopathogenicity. We feel that this would be a serious mistake, and such a "band-wagon" effect should be avoided. Although considerable evidence is being accumulated that extrachromosomal elements are involved in expressions of pathogenicity such as toxin production and tumor-induction, there is another body of data that points to the involvement of chromosomal genes as determinants of pathogenicity. Both possibilities should be investigated as intensively as possible now that the systems for such studies are available.

We have described genetic systems developed in every major category of phytopathogenic bacteria: leaf-spotters, wilt-inducers, toxin-producers, blighters, gall-inducers, and soft-rotters. It is important that the research on the genetics of phytopathogenic bacteria begin to concentrate on specific problems rather than to simply proliferate to other related bacteria.

The areas that should receive the greatest attention are transduction so that the fine-structure of chromosomal genes for pathogenicity can be determined, transformation in order to move genes for pathogenicity into other organisms where they can be studied under ideal conditions, and recombinational analysis of pathogenicity determinants by in vitro techniques. The possibilities for research in the last area are overwhelming. Once the location of a gene involved in the expression of pathogenicity is known, it can be isolated and transferred to experimental strains specifically designed to permit identification of the particular gene product. However, cloning of genes for pathogenicity in phytopathogenic bacteria depends on the development of efficient transformation systems. Therefore, it is essential that we concentrate our efforts. The advances in microbial genetics owe much to the intensively concentrated research on *E. coli*. We should learn from that.

This review has dealt with the systems available for studying the genetics of phytopathogenic bacteria. We look forward to the next review—on the genetics of phytopathogenicity in bacteria.

ACKNOWLEDGMENTS

The authors gratefully acknowledge the assistance of D. F. Voss and D. C. Savo in preparing the manuscript; P. R. Day, N. T. Keen, S. Rich, R. B. Sparks, Jr., and G. A. Zentmyer for their critical reviews; and to all those persons who shared their unpublished data with us.

Literature Cited

1. Albinger, G., Beiderbeck, R. 1977. Übertragung der Fahigkeit zur Wurzelinduktion von *Agrobacterium rhizogenes* auf *A. tumefaciens. Phytopathol. Z.* 90:306–10

2. Beltra, R., Rodriguez de Lecea, J. 1971. Aseptic induction of crown gall tumors by the nucleic acid fraction from *Agrobacterium tumefaciens. Phytopathol. Z.* 70:351–58

3. Bennett, R. A., Billing, E. 1975. Development and properties of streptomycin resistant cultures of *Erwinia amylovora* derived from English isolates. *J. Appl. Bacteriol.* 39:307–15

4. Bibb, M. J., Freeman, R. F., Hopwood, D. A. 1977. Physical and genetical characterization of a second sex factor, SCP2, for *Streptomyces coelicolor* A3(2). *Mol. Gen. Genet.* 154:155–66

5. Bogoroditskaya, S. V., Shenderov, V. A., Shevyakova, N. N. 1973. Transfer of extrachromosome resistance to antibiotics from *Escherichia coli* to *Erwinia carotovora f. citrulis* by conjugation. *Biol. Nauki Moscow* 16:110–12

6. Bomhoff, G. H., Klapwijk, P. M., Kester, H. C. M., Schilperoort, R. A., Hernalsteens, J. P., Schell, J. 1976. Octopine and nopaline: Synthesis and breakdown genetically controlled by a plasmid of *Agrobacterium tumefaciens. Mol. Gen. Genet.* 145:177–81

7. Boucher, C., Bergeron, B., Barate de Bertalmio, M., Dénarié, J. 1977. Introduction of bacteriophage Mu into *Pseudomonas solanacearum* and *Rhizobium meliloti* using the R factor RP4. *J. Gen. Microbiol.* 98:253–63

8. Boucher, C. A., Sequeira, L. 1978. Evidence for the cotransfer of genetic markers in *Pseudomonas solanacearum* strain K60. *Can. J. Microbiol.* 24:69–72

9. Deleted in proof

10. Chatterjee, A. K., Starr, M. P. 1972. Genetic transfer of episomic elements among *Erwinia* species and other enterobacteria: F'*lac*+. *J. Bacteriol.* 111:169–76

11. Chatterjee, A. K., Starr, M. P. 1972. Transfer among *Erwinia* spp. and other enterobacteria of antibiotic resistance carried on R factors. *J. Bacteriol.* 112:576–84

12. Chatterjee, A. K., Starr, M. P. 1973. Transmission of *lac* by the sex factor E in *Erwinia* strains from human clinical sources. *Infect. Immun.* 8:563–72

13. Chatterjee, A. K., Starr, M. P. 1973. Gene transmission among strains of *Erwinia amylovora. J. Bacteriol.* 116:1100–6

14. Chatterjee, A. K., Starr, M. P. 1977. Donor strains of the soft-rot bacterium *Erwinia chrysanthemi* and conjugational transfer of the pectolytic capacity. *J. Bacteriol.* 132:862–69

15. Chilton, M.-D., Drummond, M. H., Gordon, M. P., Merlo, D. J., Montoya, A. L., Sciaky, D., Nutter, R., Nester, E. W. 1978. Foreign genes of plasmid

origin detected in crown gall tumor. *Microbiology* 1978:136–38

16. Chilton, M.-D., Drummond, M. H., Merlo, D. J., Sciaky, D., Montoya, A. L., Gordon, M. P., Nester, E. W. 1977. Stable incorporation of plasmid DNA into higher plant cells: The molecular basis of crown gall tumorigenesis. *Cell* 11:263–71

17. Chilton, M.-D., Farrand, S. K., Levin, R., Nester, E. W. 1976. RP4 promotion of transfer of a large *Agrobacterium* plasmid which confers virulence. *Genetics* 83:609–18

18. Cho, J. J., Panopoulos, N. J., Schroth, M. N. 1975. Genetic transfer of *Pseudomonas aeruginosa* R factors to plant pathogenic *Erwinia* species. *J. Bacteriol.* 122:192–98

19. Cho, J. J., Schroth, M. N., Kominos, S. D., Green, S. K. 1975. Ornamental plants as carriers of *Pseudomonas aeruginosa. Phytopathology* 65:425–31

20. Coplin, D. L. 1972. *Pseudomonas solanacearum: Genetic transformation and virulence of biochemical mutants.* PhD thesis. Univ. Wisconsin, Madison. 98 pp.

21. Coplin, D. L. 1978. Properties of F and P group plasmids in *Erwinia stewartii. Phytopathology* 68:1637–43

22. Coplin, D. L. 1978. *Phytopathol. News* 12:155 (Abstr.)

23. Coplin, D. L., Rowan, R. G. 1979. Conjugative plasmids in *Erwinia stewartii. Proc. 4th Int. Conf. on Plant Pathog. Bact., Angers, France. 27 August to 2 September, 1978,* pp. 67–73

24. Coplin, D. L., Sequeira, L., Hanson, R. S. 1974. *Pseudomonas solanacearum:* Virulence of biochemical mutants. *Can. J. Microbiol.* 20:519–29

25. Coplin, D. L., Stetak, T. A. 1976. *Proc. Am. Phytopathol. Soc.* 3:22 (Abstr.)

26. Corey, R. R., Starr, M. P. 1957. Genetic transformation of colony type in *Xanthomonas phaseoli. J. Bacteriol.* 74:141–45

27. Corey, R. R., Starr, M. P. 1957. Genetic transformation of streptomycin resistance in *Xanthomonas phaseoli. J. Bacteriol.* 74:146–50

28. Curiale, M. S., Mills, D. 1977. Detection and characterization of plasmids in *Pseudomonas glycinea. J. Bacteriol.* 131:224–28

29. Currier, T. C., Nester, E. W. 1976. Evidence for diverse types of large plasmids in tumor-inducing strains of *Agrobacterium. J. Bacteriol.* 126:157–65

30. Daughtery, M. L. 1978. *Native extrachromosomal deoxyribonucleic acid in Erwinia carotovora var. carotovora.* MS thesis. Univ. Massachusetts, Amherst. 64 pp.

31. Dénarié, J., Rosenberg, C., Bergeron, B., Boucher, C., Michel, M., Barate de Bertalmio, M. 1977. In *DNA Insertions,* ed. A. I. Bukhari, J. Shapiro, S. Adhya, pp. 507–20. New York: Cold Spring Harbor Lab.

32. Drummond, M. H., Gordon, M. P. Nester, E. W., Chilton, M.-D. 1977. Foreign DNA of bacterial plasmid origin is transcribed in crown gall tumors. *Nature* 269:535–36

33. Engler, G., Holsters, M., Van Montagu, M., Schell, J., Hernalsteens, J. P., Schilperoort, T. 1975. Agrocin 84 sensitivity: A plasmid determined property in *Agrobacterium tumefaciens. Mol. Gen. Genet.* 138:345–54

34. Fulbright, D. W., Leary, J. V. 1978. Linkage analysis of *Pseudomonas glycinea. J. Bacteriol.* 136:497–500

35. Gantotti, B. V., Patil, S. S., Mandel M. 1978. *Phytopathol. News* 12:154 (Abstr)

36. Genetello, C., Van Larebeke, N., Holsters, M., De Picker, A., Van Montagu, M., Schell, J. 1977. Ti plasmids of *Agrobacterium* as conjugative plasmids. *Nature* 265:561–63

37. Gibbins, L. N., Bennett, P. M., Saunders, J. R., Grinsted, J., Connolly, J. C. 1976. Acceptance and transfer of R-factor RP1 by members of the "Herbicola" group of the genus *Erwinia. J. Bacteriol.* 128:309–16

38. Goldberg, R. B., Bender, R. A., Streicher, S. L. 1974. Direct selection for P1-sensitive mutants of enteric bacteria. *J. Bacteriol.* 118:810–14

39. Gonzalez, C. F., Vidaver, A. K. 1977. *Proc. Am. Phytopathol. Soc.* 4:10 (Abstr.)

40. Gonzalez, C. F., Vidaver, A. K. 1978. *Phytopathol. News* 12:154 (Abstr.)

41. Gregory, K. F., Huang, J. C. C. 1964. Tyrosinase inheritance in *Streptomyces scabies.* I. Genetic recombination. *J. Bacteriol.* 87:1281–86

42. Gregory, K. F., Huang, J. C. C. 1964. Tyrosinase inheritance in *Streptomyces scabies.* II. Induction of tyrosinase deficiency by acridine dyes. *J. Bacteriol.* 87:1287–94

43. Gregory, K. F., Shyu, W.-J. 1961. Apparent cytoplasmic inheritance of tyrosinase competence in *Streptomyces scabies. Nature* 191:465–67

44. Gross, D. C., Vidaver, A. K. 1978. *3rd Int. Congr. Plant Pathol. Munchen, Germany. 16 to 23 August 1978* (Abstr.)

45. Guimaraes, W. V. 1976. *Studies of plasmids in plant pathogenic bacteria. I. Conjugational plasmid transmission and plasmid-promoted chromosomal mobilization in Pseudomonas phaseolicola and Erwinia chrysanthemi. II. Effect of plasmids on the physiology and pathogenicity of Erwinia chrysanthemi.* PhD thesis. Univ. California, Berkeley. 109 pp.

46. Guimaraes, W. V., Panopoulos, N. J. 1977. *Proc. Am. Phytopathol. Soc.* 4: 170–71 (Abstr.)

47. Guimaraes, W. V., Panopoulos, N. J., Schroth, M. N. 1979. Conjugative properties of *inc*P, *inc*T, and *inc*F plasmids in *Erwinia chrysanthemi* and *Pseudomonas phaseolicola.* See Ref. 23, pp. 53–65

48. Hamilton, R. H., Chopan, M. N. 1975. Transfer of the tumor-inducing factor in *Agrobacterium tumefaciens. Biochem. Biophy. Res. Commun.* 63: 349–54

49. Hamilton, R. H., Fall, M. Z. 1971. The loss of tumor-initiating ability in *Agrobacterium tumefaciens* by incubation at high temperature. *Experientia* 27:229–30

50. Holsters, M., deWaele, D., De Picker, A., Messens, E., Van Montagu, M., Schell, J. 1978. Transfection and transformation of *Agrobacterium tumefaciens. Mol. Gen. Genet.* 63:181–87

51. Hooykaas, P. J. J., Klapwijk, P. M., Nuti, M. P., Schilperoort, R. A., Rorsch, A. 1977. Transfer of *Agrobacterium tumefaciens* TI plasmid to avirulent agrobacteria and to *Rhizobium ex planta. J. Gen. Microbiol.* 98: 477–84

52. Kern, H. 1965. Untersuchungen zur genetischen transformation zwischen *Agrobacterium tumefaciens* und *Rhizobium* spec. I. Ubertragung der fahigkeit zur induktion Pflanzlicher tumoren auf *Rhizobium* spec. *Archiv. Mikrobiol.* 51:140–55

53. Kern, H. 1969. Interspezifische transformationen zwischen *Agrobacterium tumefaciens* und *Rhizobium leguminosarum. Arch. Mikrobiol.* 66:63–68

54. Kerr, A. 1969. Transfer of virulence between isolates of *Agrobacterium. Nature* 223:1175–76

55. Kerr, A. 1971. Acquisition of virulence by non-pathogenic isolates of *Agrobacterium radiobacter. Physiol. Plant Pathol.* 1:241–46

56. Kerr, A., Htay, K. 1974. Biological control of crown gall through bacteriocin production. *Physiol. Plant Pathol.* 4:37–44

57. Kerr, A., Manigault, P., Tempe, J.

1977. Transfer of virulence *in vivo* and *in vitro* in *Agrobacterium. Nature* 265:560–61

58. Klein, R. M., Braun, A. C. 1960. On the presumed sterile induction of plant tumors. *Science* 131:1612

59. Lacy, G. H. 1975. *Genetics of Pseudomonas glycinea Coerper.* PhD thesis. Univ. California, Riverside. 96 pp.

60. Lacy, G. H. 1978. Genetic studies with plasmid RP1 in *Erwinia chrysanthemi* strains pathogenic on maize. *Phytopathology* 68:1323–30

61. Lacy, G. H., Hirano, S. S., Victoria, J. I., Kelman, A., Upper, C. D. 1979. Inhibition of soft-rotting *Erwinia* strains by 2,4-dihydroxy-7-methoxy-(2H)-1,4-benzoxazin-3(4H)-one (DIMBOA) in relation to their pathogenicity on *Zea mays. Phytopathology* 69: In press

62. Lacy, G. H., Leary, J. V. 1975. Transfer of antibiotic resistance plasmid RP1 into *Pseudomonas glycinea* and *Pseudomonas phaseolicola in vitro* and *in planta. J. Gen. Microbiol.* 88:49–57

63. Lacy, G. H., Leary, J. V. 1976. Plasmid-mediated transmission of chromosomal genes in *Pseudomonas glycinea. Genet. Res.* 27:363–68

64. Lacy, G. H., Leary, J. V., Grantham, G. L. 1976. *Proc. Am. Phytopathol. Soc.* 3:223 (Abstr.)

65. Lai, M., Panopoulos, N. J., Shaffer, S. 1977. Transmission of R plasmids among *Xanthomonas* spp. and other plant pathogenic bacteria. *Phytopathology* 67:1044–50

66. Lai, M., Shaffer, S. Panopoulos, N. J. 1977. Stability of plasmid-borne antibiotic resistance in *Xanthomonas vesicatoria* in infected tomato leaves. *Phytopathology* 67:1527–30

67. Le, T. K. T., Leccas, D., Boucher, C. 1979. Transformation of *Pseudomonas solanacearum* strain K60. See Ref. 23, pp. 819–22

68. Leary, J. V. 1979. Transfer and integration of chromosomal genes from *Pseudomonas glycinea* into *Pseudomonas aeruginosa. Can. J. Microbiol.* In press

69. Levin, R. A., Farrand, S. K., Gordon, M. P., Nester, E. W. 1976. Conjugation in *Agrobacterium tumefaciens* in the absence of plant tissue. *J. Bacteriol.* 127:1331–36

70. Liao, C. H., Heberlein, G. T. 1978. A method for the transfer of tumorigenicity between strains of *Agrobacterium tumefaciens* in carrot root disks. *Phytopathology* 68:135–37

71. Lin, B.-C., Chen, S. J. 1978. *3rd Int.*

Congr. Plant Pathol., Munchen, Germany, 16 to 23 August, 1978 (Abstr.)

72. Liu, S. C. Y., 1973. *2nd Int. Congr. Plant Pathol., Minnesota* 5 to 12 September, 1973 (Abstr.)

73. Loper, J. E., Kado, C. I. 1978. *Phytopathol. News* 12:155 (Abstr.)

74. Merlo, D. J. 1979. In *Plant Disease,* Vol. III, ed. J. G. Horsfall, E. B. Cowling, pp. 201–13. New York: Academic

75. Merlo, D. J., Nester, E. W. 1977. Plasmids in avirulent strains of *Agrobacterium. J. Bacteriol.* 129:76–80

76. Message, B., Boucher, C., Boistard, P. 1975. Transfert d'un facteur R, RP₄ dans une souche de *Pseudomonas solanacearum. Ann. Phytopathol.* 7:95–103

77. Milani, V. J., Heberlein, G. T. 1972. Transfection in *Agrobacterium tumefaciens. J. Virol.* 10:17–22

78. Montoya, A. L., Chilton, M.-D., Gordon, M. P., Sciaky, D., Nester, E. W. 1977. Octopine and nopaline metabolism in *Agrobacterium tumefaciens* and crown gall tumor cells: Role of plasmid genes. *J. Bacteriol.* 129:101–7

79. Moore, L. W. 1977. Prevention of crown gall on *Prunus* roots by bacterial antagonists. *Phytopathology* 67:139–44

80. Nester, E. W., Chilton, M.-D., Drummond, M., Merlo, D., Montoya, A., Sciaky, D., Gordon, M. P. 1977. In *Recombinant Molecules: Impact in Science and Society,* ed. R. F. Beers, E. G. Basset, pp. 178–88. New York: Raven

81. New, P. B., Kerr, A. 1972. Biological control of crown gall: Field measurements and glasshouse experiments. *J. Appl. Bacteriol.* 35:279–87

82. Nuti, M. P., Ledeboer, A. M., Lepidi, A. A., Schilperoort, R. A. 1977. Large plasmids in different *Rhizobium* species. *J. Gen. Microbiol.* 100:241–48

83. Okabe, N., Goto, M. 1955. Studies of *Pseud. solanacearum* X. Genetic change of the bacterial strains induced by the temperate phage T-₂200. *Rept. Fac. Agric., Schizuoka Univ.* 5:57–62

84. Olsen, R. H., Shipley, P. 1973. Host range and properties of the *Pseudomonas aeruginosa* R factor R1822. *J. Bacteriol.* 113:772–80

85. Panopoulos, N. J., Guimaraes, W. V., Cho, J. J., Schroth, M. N. 1975. Conjugative transfer of *Pseudomonas aeruginosa* R factors to plant pathogenic *Pseudomonas* spp. *Phytopathology* 65:380–88

86. Panopoulos, N. J., Guimaraes, W. V. Hua, S.-S., Sabersky-Lehman, C., Resnik, S., Lai, M., Shaffer, S. 1978. Plasmids in phytopathogenic bacteria. *Microbiology* 1978:238–41

87. Perombelon, M. C. M. 1977. *Ann. Rept. Scott. Hortic. Res. Inst.* 02027 (Abstr.)

88. Perombelon, M. C. M., Boucher, C. 1978. See Ref. 23

89. Pugashetti, B. K., Starr, M. P. 1975. Conjugational transfer of genes determining plant virulence in *Erwinia amylovora. J. Bacteriol.* 122:485–91

90. Rapp, B., Kemp, J. D., White, F. 1979. Isolation of a non-tumor inducing mutant of the Ti plasmid of *Agrobacterium tumefaciens* strain B6. *Can. J. Microbiol.* In press

91. Schell, J., Van Montagu, M. 1977. In *Genetic Engineering for Nitrogen Fixation,* ed. A. Hollander, pp. 159–79. New York: Plenum

92. Schwesinger, M. D. 1977. Additive recombination in bacteria. *Bacteriol. Rev.* 41:872–902

93. Sparks, R. B., Lacy, G. H. 1977. *Proc. Am. Phytopathol. Soc.* 4:198 (Abstr.)

94. Susman, M. 1970. General bacterial genetics. *Ann. Rev. Genet.* 4:135–76

95. Deleted in proof

96. Twiddy, W., Liu, S. C. Y. 1972. *Phytopathology* 62:794 (Abstr.)

97. Van Larebeke, N., Engler, G., Holsters, M., Van Den Elsacher, S., Zaenen, I., Schilperoort, R. A., Schell, J. 1974. Large plasmid in *Agrobacterium tumefaciens* essential for crown gall inducing ability. *Nature* 252:169–70

98. Van Larebeke, N., Genetello, C., Schell, J., Schilperoort, R. A., Hermans, A. K., Hernalsteens, J. P., Van Montagu, M. 1975. Acquisition of tumor-inducing ability by non-oncogenic agrobacteria as a result of plasmid transfer. *Nature* 255:742–43

99. Watson, B., Currier, T. C., Gordon, M. P., Chilton, M.-D., Nester, E. W. 1975. Plasmid required for virulence of *Agrobacterium tumefaciens. J. Bacteriol.* 123:255–64

100. Wu, W. C., Middleton, R. B., Hsu, W. H. 1977. Transfer of episome F'lac⁺ and chromosomal trp⁺ genes from *Erwinia amylovora* to *Salmonella typhimurium. Chin. J. Microbiol.* 10:37–47

101. Zaenen, I., Van Larebeke, N., Teuchy, H., Van Montagu, M., Schell, J. 1974. Supercoiled circular DNA in crown-gall inducing *Agrobacterium* strains. *J. Mol. Biol.* 86:109–27

Ann. Rev. Phytopathol. 1979. 17:203–22
Copyright © 1979 by Annual Reviews Inc. All rights reserved

COMPONENTS OF RESISTANCE ❖3708
THAT REDUCE THE RATE
OF EPIDEMIC DEVELOPMENT

J. E. Parlevliet

Institute of Plant Breeding, Agricultural University, PO Box 386,
6700 AJ Wageningen, the Netherlands

INTRODUCTION

Pathogens vary greatly in the way and rate they multiply and spread. Resistance generally affects the multiplication of the pathogen rather than its spread. Some pathogens have only one reproductive cycle per growing season of the host (smuts and bunts of small cereals, potato cyst nematode); others have several to many cycles (rusts and mildew of small cereals, rice blast). The latter group of diseases was classified by Van der Plank (53) as "compound interest diseases." The epidemic development of such diseases is determined by the amount of disease present at the start of the epidemic, x_o, and by its multiplication rate, described by the apparent infection rate r. Resistance may reduce x_o and/or r. If x_o is reduced the epidemic is delayed; if r is reduced the epidemic is slowed down.

In discussing resistance to such compound interest diseases the concepts of horizontal (HR) and vertical resistance (VR) must also be dealt with because Van der Plank (54) confounded these concepts, which he defined in terms of population kinetics (see section on terminology), with the epidemiological parameters r and x_o. He stated that VR in the host delays the start of an epidemic, whereas HR slows it down after it has started (54, p. 24). In order to reduce the confusion, Nelson (31) suggested that Van der Plank's definitions of HR and VR, which indeed are not very consistent, be dropped and that HR and VR be defined solely in terms of their effects on disease increase and disease onset.

203

0066-4286/79/0901-0203$01.00

The aim of this chapter is to discuss the quantitative aspects of r-reducing and x_o-reducing resistances as well as their race-specific and race non-specific aspects.

Since the meaning of the terms used in this field of research is far from consistent it is necessary to explain in a short paragraph the terms used in this chapter.

TERMINOLOGY

In general the terminology of Robinson (42) has been followed. Host resistance is defined as the ability of the host to hinder the growth and/or development of the pathogen. Nonhost resistance has been kept out of the discussion here. The term *complete resistance* is used when the multiplication of the pathogen is totally prevented, that is, when the spore production (SP) is zero. *Incomplete resistance* refers to all resistances that allow some SP. *Partial resistance* is a form of incomplete resistance in which the SP is reduced even though the host plants are susceptible to infection (susceptible infection type).

Van der Plank (53, 54) classified resistance as horizontal and vertical. His definitions, however, were not consistent, as Nelson (31) pointed out. Robinson's definitions (42), reflecting Van der Plank's ideas, are clearer and I follow his definitions, and use HR in the sense of race-nonspecific resistance, characterized by the absence of genetic interactions between host genotypes and pathogen genotypes. VR or race-specific resistance then is characterized by the presence of genetic interactions between host and pathogen genotypes. Complementary to the terms HR and VR, which refer to the host's reaction, a set of terms to describe the pathogen's reaction is needed. *Aggressiveness* and *virulence* have been used for this purpose but since they have also been used in other senses, especially the former one, it is better to avoid these terms and to follow Robinson's more consistent terminology. He proposed horizontal and vertical pathogenicity which are equivalent to cultivar-nonspecific and cultivar-specific pathogenicity.

Another term in need of some delimitation is *tolerance.* This is the ability of the host to endure the presence of the pathogen, and can be expressed by less severe (reduced) disease symptoms and/or less (reduced) damage. If one wants to measure tolerance experimentally, both the severity of symptoms and the amounts of damage to host genotypes should be compared at equal amounts of pathogen at the same stage of host development. This is exceedingly difficult. In several publications, less severe disease symptoms or smaller amounts of damage have been ascribed to tolerance even though the pathogen population was not measured. In many cases, such observations are caused by a mixture of incomplete resistance and tolerance or by the former alone.

ASSESSMENT OF RESISTANCE

To assess resistance, the growth and development of the parasite should be measured. With animal parasites this is the usual procedure; with plant pathogens this is possible only in the case of ectopathogens like powdery mildews and to a limited extent with some other biotrophic organisms like rusts. In most cases the disease symptoms are assessed with the assumption that they reflect quantitatively the growth of the pathogen in the host.

The assessment is made in various ways. One may measure the *disease incidence,* defined as the number of plant units infected, and expressed as a percentage of the total number of units assessed (e.g. percentage of diseased plants or ears), or the *disease severity,* defined as the area of plant tissue affected by disease, and expressed as a percentage of the total area assessed (19). Differences in the rate of increase in disease severity are reflected by differences in apparent infection rates (53). It is also possible to assess the amount of disease in more detail by measuring the number of successful infections often expressed in number of lesions or as disease incidence, and by measuring the size or extent of the lesions. With compound interest diseases (53) the disease severity is the cumulative result of several factors or components like *infection frequency,* defined as the proportion of spores that result in sporulating lesions, *latent period* measured as the time from infection to spore production, *spore production* expressed as spores produced per lesion or per unit area of affected tissue and/or per unit of time, and *infectious period* being the period over which the diseased tissue sporulates.

A different method of assessment is widely used with many biotrophic leaf pathogens like rusts and powdery mildews. The host may react to infection by the pathogen in a hypersensitive way. Around the infection court the host tissue becomes necrotic or chlorotic. The growth of the pathogen is hindered either qualitatively (no sporulation) or quantitatively (restricted sporulation). This is described by means of *infection types* (IT), where an IT 0 indicates a necrotic or chlorotic fleck without sporulation and an IT 4 means a fully susceptible reaction, a sporulating pustule without chlorosis or necrosis. Infection types 1, 2, and 3 describe pustules surrounded by necrotic or chlorotic tissue with increasing intensities of sporulation. An IT 3 or 4 or 4 only, depending on the situation, is generally indicated as a susceptible or high IT, the others as a resistant or low IT. Low IT resistances can be expressed from the seedling stage onward (seedling resistance) or in the adult plant stage only (adult plant resistance). Low IT resistances are often race-specific and simply inherited (38). Infection types assess the resistance partly in a qualitative and partly in a quantitative sense. The IT is in principle measured on individual lesions; infection frequency, latent period, and infectious period are not measured at all, but the spore production per pustule is assessed in a quantitative sense because pustule size and spore production tend to be correlated.

The assessment of resistance to compound interest diseases can be accomplished in various ways. Basically one measures the disease severity either once at the peak of epidemic development or several times from the beginning to the end of the epidemic. The former is assumed to represent the cumulative result of all resistance factors operating during the progress of the epidemic (40), the latter data are used to measure the apparent infection rate r (53) or the area under the disease progress curve (58). Each method has its merits. One problem all these methods have in common is interplot interference, which can be very considerable with airborne leaf pathogens (40) as shown below.

r-REDUCING RESISTANCES

Resistances of this type are very common, but it is not easy to assess them accurately. Especially interplot interferences may strongly reduce the magnitude of the resistance. Table 1 shows the progress of a leaf rust epidemic in six barley cultivars over several weeks. When not disturbed by plot interference, the most susceptible and most resistant cultivar differed by a factor of over 2500. The same cultivars were grown in small adjacent plots of four rows and of one row wide. Here the cultivars L94 and Vada differed by only a factor of 30 and 17, respectively, a reduction in cultivar differences in the order of 80 to 150 times. Apparently the resistances measured in small adjacent plots can be underestimated very considerably. Nevertheless, the ranking order of the cultivars remained the same; only Monte Christo ranked somewhat more susceptible in the adjacent plots.

The r-value of an epidemic varies not only with the cultivar but also with such factors as the stage of development of the plants and the progress of

Table 1 Infection types and estimated number of *Puccinia hordei* uredosori per tiller on six spring barley cultivars grown on plots isolated by winter wheat to avoid interplot interference (38, 40).

Cultivar	Infection type		Number of uredosori per tiller in 1973 on sampling date			
	Seedling	Adult plant	14/6	27/6	5/7	12/7
L94	4	4	5.0	440	2,800	—[a]
Sultan	4	4	1.5	20	750	—[a]
Volla	4	4	0.7	5.4	115	750
Monte Christo	3	2	0.9	8.0	28	—[a]
Julia	4	3⁺–4	0.7	1.9	17	100
Vada	4	3⁺–4	0.3	0.5	1.1	4.5

[a] No data as leaves were dead because of earliness and/or too much rust.

the epidemic. The data in Table 1 show this. L94 and Monte Christo reached their highest rates of rust increase in the first period, Vada in the third, and the others in the second period. The first two cultivars are considerably earlier in heading than the other four, which probably explains their early peak in r. Vada reached its peak in r later, most likely because its rust epidemic was still in an early phase.

What types of resistance reduce r? Van der Plank (53, 54) concluded that r-reducing resistances are of a race-nonspecific or horizontal nature (see Introduction.) This assumption has been widely accepted but is certainly not true. In fact all resistances result in a reduced r, whether they are race-nonspecific or race-specific. Complete and nearly complete resistances with infection types of 0 or 1 do not allow any epidemic development; r is zero. But all other incomplete resistances give values for r which are smaller than the highly susceptible control as shown in Table 1. The resistance of Monte Christo is of a low IT: it is probably monogenically inherited and race-specific (38), but it has a smaller value of r like the other cultivars with resistance of a high IT, polygenically inherited (37) and largely race-non-specific in nature. Resistances similar to that of Monte Christo, which are race-specific and reduce r, because they allow moderate sporulation, occur fairly frequently. Several of the Sr genes, conferring stem rust resistance to wheat, have an IT 2 like $Sr7b$, $Sr8$, and $Sr14$. The *Hordeum laevegatum* resistance of barley to powdery mildew is inherited monogenically, is race-specific, and has an IT 3 (22). The powdery mildew builds up every season in the barley cultivars having this resistance but never so strongly as in truly susceptible cultivars. Another reported example is formed by the rice culti-var St-1. This cultivar appeared to have a high level of partial resistance (called *field resistance* in rice) to rice blast, *Pyricularia oryzae*, which is assumed to be horizontal. Several years later races were found, which could overcome this partial resistance of St-1, and genetic studies revealed that its resistance is governed by one dominant gene, Pi-f (49).

Components of r-Reducing Resistances

Factors that may reduce r can broadly be classified as avoidance mecha-nisms or as resistance. Little is known about the former, but they should not be underrated. The primitive cultivars of small grains tend to have a less dense leaf canopy, allowing spores to escape into the air or to fall onto the ground more easily, thus reducing the rate of epidemic development (30). Also the growth habit of the cultivar may influence the deposition of spores onto the plant parts (43). Stomatal behavior may affect the chance of penetration as Hart (14) showed long ago. Such avoidance mechanisms are difficult to select for, especially when the conditions of testing deviate from field conditions.

The components of r-reducing resistances have drawn considerably more attention recently. One might subdivide the r-reducing resistances into resistance to infection, resistance to colonization, and resistance to reproduction. They form a convenient classification but their evaluation, however, is not easy. To reduce r, the rate of spore production should be reduced, this results from fewer lesions which start to produce spores later at a lower rate. These latter components are measured by infection frequency, latent period, and spore production.

INFECTION FREQUENCY   In the strictest sense of the word, there often is no resistance to infection; gerimination, appressorium formation, and penetration by biotrophic pathogens are in general nonspecific processes. Host-pathogen interactions start only after the first contact between the host cell and the pathogen has been made. Measuring the infection frequency (IF), by measuring the number of successful infections in terms of sporulating lesions, indicates not only that resistance to first contact but also resistance to colonization is involved. A good example is reported by Clifford (4). He followed the course of infection and colonization of leaf rust in seedlings of two barley cultivars. Up to the formation of the substomatal vesicle, 24 hr after inoculation, the two cultivars did not respond different ͵. After 48 hr Vada, however, a resistant cultivar (see Table 1), had two times as many infections arrested in the substomatal vesicle phase than Midas, a highly susceptible cultivar. The infections not arrested grew less vigorously in Vada than in Midas. The accumulated result was that the pustules in Midas ruptured the epidermis 8 to 9 days after inoculation and on Vada after 9 to 14 days. The pustules on Vada were significantly smaller and fewer (22%) in number than those on Midas.

Differences in IF reflect differences accumulated over various development stages, from the establishment phase just after penetration to the late phases of colonization just prior to spore formation. Infections may become arrested at various stages from substomal vesicle formation onward. This is indicated by some observations on mature barley plants inoculated with leaf rust. The first visible symptoms of the infections are small pale flecks in the center of which the spores become visible one to several days after the flecks become visible. In cultivars with long latent periods some to many of these flecks become arrested before the spore-formation phase. Other infections have been arrested even earlier, before this fleck stage. This can be deduced from the appearance of teleutospore formation on sites carrying no visible symptoms of infections. Normally these teleutospores are formed around the uredosori when the leaves mature (J. E. Parlevliet, unpublished observations).

When looked for, differences in IF or lesion number arc observed in most host-pathogen systems (4, 16, 17, 30, 32, 39, 45, 52, 56). It should be realized, however, that IF not only varies with host genotypes, but also with the developmental stage of the host and the environmental conditions. This is well known for the late blight disease induced by *Phytophthora infestans* on potato (50). The IF of barley cultivars to leaf rust depends on the stage of plant development and the age of the leaves (39). Table 2 shows some interactions between cultivars and stage of plant development. In the adult plant stage the IF tended to become higher as the leaves became older. Similar results were obtained with oats inoculated with the powdery mildew fungus—*Erysiphe graminis* f. sp. *avenae* (2). The percentage of germinated spores that formed haustoria in the epidermal cells was determined on leaf 1 and leaf 5 of the susceptible cultivar Manod and the partially resistant cultivar Maldwyn. The penetration percentages for the two cultivars were 24.8 and 23.1% on leaf 1 and 18.2 and 10.7% on leaf 5. Plant stage and cultivar both appeared important.

LATENT PERIOD   In some host-pathogen relationships the incubation period (ICP), the time between inoculation and first visible symptoms, is measured rather than the latent period (LP). When measuring the ICP one assumes that the ICP and LP vary in a parallel way. In most cases this is probably a fair assumption. Considerable differences in ICP and LP may exist between cultivars as demonstrated in the potato—*Phytophthora infestans* (44, 56), the wheat—*Puccinia striiformis* (8), the wheat–*P. recondita* f. sp. *tritici* (32), the rye–*P. recondita* f. sp. *recondita* (35), and the barley–*P. hordei* (4, 33) relationships.

Differences in ICP and LP are often assumed to reflect differences in growth rate of the pathogen in the host (30, 51). This, however, is not always so. Five *Avena sterilis* accessions and the oats cultivar Fulghum were tested with five isolates of *Puccinia graminis* f. sp. *avenae* (48). The host genotypes did not differ for LP, despite differences in colony size.

Table 2   Number of uredosori per cm^2 leaf area of six spring barley cultivars inoculated with leaf rust, *Puccinia hordei*, in the seedling and adult plant stage, relative to those of L94 (39)

Cultivar	Relative infection frequency*	
	Primary leaf	Flag leaf
L94	100[a]	100[a]
Berac	63[b]	51[cd]
Sultan	54[b]	75[abc]
Julia	52[b]	78[ab]
Pauline	33[c]	73[bc]
Vada	32[c]	47[d]

*IF with different letters are significantly different (5% level) according to Duncan's multiple range test.

When measuring LP the plants compared should be in the same stage since the development stage of the plant and age of the leaf can be important factors (33). Table 3 shows four barley cultivars inoculated at various plant stages. The LP increases from the primary leaf to the young flag leaf stage for all cultivars after which it decreases again. The differences between cultivars are small in the seedling stage and large at the adult plant stage. A similar pattern was observed in rye for *P. recondita* f. sp. *recondita* (35).

COLONY AND LESION SIZE   Although not mentioned before as a separate component of *r*-reducing resistances, colony or lesion size is discussed here because several publications report on this topic. "Lesion size" refers to the area showing disease symptoms; "colony" measures the area actually invaded by the pathogen. Both colony and lesion size have been measured in various ways like area, diameter, length, or assessed using a scale devised for that purpose. In many host-pathogen systems, differences in colony size, pustule size, or lesion size have been mentioned or reported. A few of these are wheat–*Erysiphe graminis* f. sp. *tritici* (45), oats–*Puccinia graminis* f. sp. *avenae* (48), barley–*Puccinia hordei* (4), wheat–*Puccinia recondita* f. sp. *tritici* (32), potato–*Phytophthora infestans* (52), and rice–*Xanthomonas oryzae* (59).

As with LP, lesion size is assumed to reflect the growth rate of the pathogen in the host and therefore its spore production. Although most data confirm this, in some cases it does not hold. Habgood (13) studied the partial resistance of five barley cultivars to *Rhynchosporium secalis.* The lesion size and LP on the second leaf did not vary with the cultivars; however, the spore production did. The partially resistant cultivars Proctor and Ruby produced about half the number of spores compared to the susceptible Cambrinus.

SPORE PRODUCTION   Measuring spore production (SP) accurately is notoriously difficult because collecting the spores unavoidably means interfer-

Table 3 Latent periods of four spring barley cultivars inoculated with leaf rust, *Puccinia hordei*, at various stages relative to the latent period of L94 in the seedling stage (33).

| Cultivar | Relative latent period of leaf number | | | | |
| | 1 | 4 | 7 | 9, the flag leaf | |
				Young	Old
L94	100	106	113	117	109
Volla	104	113	122	142	135
Julia	110	125	141	182	166
Vada	123	140	157	233	201

ing with the plant in one way or another. There is no need to go into this in detail as some excellent papers have dealt with SP comprehensively (21, 28, 29, 60). Nevertheless, some general aspects and those that have a bearing on the other components should be discussed.

SP has been expressed in various ways, such as SP per unit leaf area, SP per lesion or pustule, SP per unit area of lesion, or SP per unit area of sporulating surface. These unit areas can be measured in units of time or over the whole infectious period. These various ways of expressing SP reflect its dependence on many different factors. At high pustule density the car-bohydrate production of the leaf is limiting and one can express the SP per unit leaf area. Under such conditions one may not measure differences in host resistance but rather differences in host productivity under stress. The pustules compete for the nutrients, which are scarce. This results in a negative correlation between SP per pustule and pustule size on the one hand and pustule density or IF on the other (29, 60). Such negative interfer-ences operate not only at high pustule densities, but also at moderate pustule densities, when they interfere with the accurate measurements of components of partial resistance (30, 60). To compare the SP of various genotypes, one should do so at similar IF. This is exceedingly difficult to accomplish as differences in IF are the rule rather than the exception.

Because of all these difficulties most studies concerning SP have been carried out on seedlings; this is certainly not the most representative stage of the plant. It is therefore risky to extrapolate observations of cultivar effects on SP in the seedling state to the adult plant stage in the field. One of the very few studies done on both seedlings and adult plants is the comparison of the *Avenae sterilis* accessions mentioned in the section on latent period (48). The six host genotypes tested differed considerably in SP, with an adult plant pattern similar to the seedling pattern. The differences in the adult plant stage, however, were somewhat larger.

INFECTIOUS PERIOD   The pustules or lesions of many leaf pathogens often sporulate over extended periods although the bulk of the spores tends to be produced in the early phase of the infectious period (30, 48). Sporula-tion may be terminated because of exhaustion or dying off of the infected plant parts. In mature plants the uredosori of rusts and the pustules of powdery mildew may stop producing uredospores and conidia because of the formation of telia and cleistothecia, respectively. The infectious period (IP) like SP shows a negative correlation with IF. At high IF, the leaves are apparently exhausted sooner. This negative interference again makes the evaluation of the IP difficult. Despite this negative association, clear differ-ences in IP appeared to exist in the barley-leaf rust relationship (30). The cultivar Vada, despite a lower IF and a reduced SP, showed a shorter IP compared with the very susceptible L94. This was partly because of a longer LP and partly because of an earlier termination of SP.

ASSOCIATED VARIATION OF THE COMPONENTS When host genotypes were studied for some or all of the components described here, the components often varied in association with each other. This is shown clearly in the barley-leaf rust relationship. Table 4 shows cultivar effects on the various components as well as on the total spore production per unit leaf area, which is obtained by multiplying the relative IF with SP and IP. The data shown are representative of many trials (30, 33, 39, 40; J. E. Parlevliet, unpublished observations). The cultivars vary for all four components and the variations are clearly associated. Partial resistance tends to go together with a lower IF, a longer LP, a reduced SP, and a shorter IP. This association, although fairly strong, is not complete. The SP on L98 is much higher than on any other cultivar tested although it does not score as most susceptible for the other components. The cultivars Pauline, Berac, and Julia have a similar LP (39, 40) but they differ in IF (see Table 2). It should be realized, however, that only a restricted part of the variation among cultivars has been studied here—only those cultivars that react to leaf rust with a susceptible infection type (IT). All the cultivars with some SP but a lower IT (ranging from 1 to 3) may behave differently. Only one cultivar, Monte Christo, with such a lower IT was studied in detail; it suggests a different pattern. It has a high IF, probably higher than Sultan, its LP is similar to the one of Sultan, and its SP is below that of Vada (J. E. Parlevliet, unpublished). The observations suggest that the hypersensitive or low-IT reaction mainly reduces the production of spores and affects the other components only to a limited extent. Data reported by Clifford (5) support this.

Table 4 Components of partial resistance to *Puccinia hordei* in barley[a]

Cultivar	IF	LP	SP	IP	TSP	Res
L94	100	100	100	100	100	2,000
L98	70	120	160	90	100	1,000
Sultan	65	130	80	100	52	230
Volla	70	130	110	90	69	35
Julia	65	165	50	85	28	6
Vada	40	195	50	70	14	1

[a] IF, infection frequency in uredosori per unit leaf area; LP, latent period; SP, spore production per uredosorus per day; IP, infectious period; TSP, total spore production per unit leaf area; RES, partial resistance in the field expressed by the number of uredosori per tiller. IF, LP, SP, IP, and TSP all on flag leaves and relative to Lg_4, which is set at 100%.

Partial or field resistance in potato to *Phytophthora infestans* behaves in a similar way. IF, lesion growth, and SP tend to be associated when different cultivars are compared (51, 52). LP, lesion size, lesion growth, and SP are more strongly associated with each other than with IF. The latter varies in a more independent way (51, 52, 56). Similar results were reported for the disease of wheat caused by *Puccinia recondita* f. sp. *tritici* (32). Although only four cultivars were studied, the association between LP and pustule size was very clear. The IF varied to a lesser extent in association with these two variables.

Dehne (8) made a comprehensive study of the yellow rust disease of wheat caused by *Puccinia striiformis.* He measured the growth rate of the pathogen in the leaf, the LP, SP, and infection type (IT) at various developmental stages of the plant. The variation in LP, SP, and growth rate were clearly but not completely associated. The variation in SP was almost completely associated with IT. The lower the SP the lower the IT. This is of course not surprising because the IT classification is to a large extent based on size and number of the visible pustules. A wide variation in SP among cultivars with a susceptible infection such as is found in the barley–leaf rust relationship does not seem to exist here or the cultivars chosen did not register it.

In most of the examples discussed above, all components of disease varied with the host genotype. This is not necessarily so, however. The six oat genotypes inoculated with oat stem rust (48), as already mentioned, varied for colony size and SP but not for LP. The barley cultivars studied by Habgood (13) did not vary for lesion size and LP, but did so for SP when infected with *Rhynchosporium secalis.*

EFFECT OF THE COMPONENTS ON "r"-REDUCING RESISTANCES
Johnson & Taylor (21) concluded that the measurement of the total spore production of pathogens growing on infected host cultivars provides an accurate method of measuring the pathogenicity of the pathogen and the resistance of the host—the resistance being the sum of the effects of all components. This may be true for monocyclic tests, but the *r*-reduced resistances result from a concerted effect of the various components over several reproduction cycles of the pathogen, where the contribution of the individual components to the epidemic development may differ considerably. Zadoks (61) demonstrated this clearly by simulating epidemics with parameters that approximate the range found in potato late blight and the cereal rusts and cereal powdery mildews. Only one component at a time was allowed to vary. The components IF and SP were combined into one value; the daily multiplication factor, which in view of the strong negative interfer-

ence between the two components, makes good sense. The response of r to a variation of the three components studied varied greatly. It was large when the LP varied, somewhat smaller when the daily multiplication factor varied, and very small when the IP varied.

The data collected in the barley–leaf rust relationship corroborate these results. The partial resistance in the field and the LP measured on the young flag leaves were highly correlated; in 1973 with 9 cultivars r was 0.94 and in 1974 with 16 cultivars r was 0.91 (40). The IF was less well correlated with partial resistance; r with 15 cultivars was –0.72 (39). The multiple regression equation, predicting the partial resistance from both the LP and the IF, suggested that the differences in partial resistance in the field are largely explained by differences in LP and hardly at all by differences in IF. The fairly high correlation between partial resistance and IF is most likely caused by the high correlation between IF and LP (r = –0.80). This is confirmed by a study on all components in the flag leaf stage of eight barley cultivars (30). The correlation coefficient, r, between the total spore production and partial resistance was –0.85; between LP alone and partial resistance r also was 0.85. Thus, LP appears as good a parameter for estimating the partial resistance as the combined result of all four parameters, total spore production. This is partly due to the association of the components; a longer LP tends to go together with a lower IF, a reduced SP, and a shorter IP. This is true partly because LP is a measure of the first spores produced by a lesion. The spores produced over the life time of a lesion are not equally important in the development of an epidemic. Those produced early have a much larger effect on the epidemic than those produced at the end of the infectious period of the lesion (30, 54, 61). In several host-pathogen relationships the epidemic developed in some four to six weeks. A pustule that started to produce spores in the early phase of an epidemic may still sporulate when the epidemic is nearly completed, but these spores produced at the last minute no longer have an impact on the epidemic.

The LP therefore is the crucial component determining the apparent infection rate r when a large number of reproductive cycles of the pathogen are required to complete the epidemic. The fewer the number of reproductive cycles the more important the effect of the other components will become. In the ultimate situation, where only one reproductive cycle of the pathogen occurs per reproductive cycle of the host plant, the IF and the SP are the determining factors. This is shown, for example, in the case of the smuts and bunts in small cereals.

In other host-pathogen relations partial resistance also could be explained reasonably well from the components studied. In 1951, Schaper (44) mentioned the importance of LP in the selection for partial resistance in potato to *Phytophthora infestans*. Umaerus & Lihnell (52) also conclude that screening for partial resistance in potato to late blight can be done by measuring the cultivar effects on the individual components. In wheat the partial resistance to leaf rust, *Puccinia recondita* f. sp. *tritici,* was well explained from the LP and pustule size. The number of cultivars, however, was restricted—only four (32).

Not all variation in partial resistance can be explained by the components discussed here, however; the barley cultivar Volla has been shown to be more resistant to leaf rust than Sultan (Table 4), although this is not expected from the component analysis. Which other factor(s) causes Volla to be more resistant is not yet known, but it might be found among avoidance mechanisms as mentioned in the beginning of the section on components of r-reducing resistances.

When trying to evaluate partial resistance from its components one must be careful to measure the components at the correct stage of plant or leaf development (51). Dehne (8) observed that the wheat cultivars he studied were all highly susceptible to yellow rust in the seedling stage, showing at most small variations for the components LP, growth rate, and SP. The variation in resistance in the field correlated much better with the variation in these components measured in the adult plant stage. With leaf rust in barley a similar situation exists. The correlation coefficient r of partial resistance with the LP in the flag leaf stage was on average 0.9 (30, 40); with the LP in the seedling stage it was less than 0.6 (40).

INHERITANCE OF THE COMPONENTS   Nelson (31) comprehensively discusses the inheritance of *r*-reducing resistances. He concluded that many but not all *r*-reducing resistances are of a polygenic nature. My objective in this discussion is also to point out that monogenic or oligogenic control of *r*-reducing resistances is quite common.

All the resistance genes in small grains that give moderately resistant infection types (IT 2 or x) to rusts or powdery mildew are examples of monogenic *r*-reducing resistances. Some examples have already been discussed in the section on *r*-reducing resistances. The resistance genes in tomato to *Cladosporium fulvum,* Cf1 and Cf3, allow flecks to develop with some sporulation (25). In the field or greenhouse such genes permit some, albeit very slow, development of the epidemic. In maize the rate of epidemic development of northern leaf blight is severely limited by either gene *Ht*1 or gene *Ht*2. Resistance of these genes is expressed as a restriction of the growth and sporulation of the fungus (15, 18). A similar resistance gene *rhm* is known in maize to restrict growth and sporulation of the southern leaf blight fungus (46). In barley, *r*-reduced resistances to *Rhynchosporium secalis* can be under monogenic control, like in the cultivar Vulcan (10) or under polygenic control as in the cultivars Proctor and Ruby (11).

Nelson (31) therefore concluded quite rightly that the inheritance of *r*-reducing resistances ranges from a single gene to many genes.

Race Specificity of r-Reducing Resistances

Several cases of monogenic *r*-reducing resistances have been discussed in the section on *r*-reducing resistances; they were all of a race-specific type, like the genes *Cf*1, *Cf*3 (25), and *Ht*1 (27) discussed in the preceding paragraph. Those *r*-reducing resistances that are inherited monogenically appear to react race-specifically when exposed to a range of pathogen genotypes. In some cases, the matching vertical gene for pathogenicity has not yet been observed as in the cases of the Ht2 and rhm resistance genes. In the sugar beet *Cercospora beticola* system, Solel & Wahl (47) observed large differences in number but not in type of lesions. This variation in resistance appeared race-specific and the Israel isolates were differentiated into three races. Two races could be discerned in the USA. They were differentiated by a single resistance gene, *Cb*. This gene confers resistance to race *C*2, but not to *C*1, causing a low IT in the seedling stage (26, 57).

What about the polygenically inherited resistances. Van der Plank is quite definite that they are race-nonspecific and cannot operate on a gene-for-gene basis with the pathogen (55; p. 167). Parlevliet & Zadoks (41), on the other hand, concluded that polygenic host-pathogen systems operating on a minor gene–for–minor gene basis also behave largely as race-nonspecific and cultivar-nonspecific systems. The major part of the genetic variation in resistance is independent of the pathogen genotype; that is, it is race-nonspecific or horizontal. The remaining minor part is dependent of the pathogen genotype, i.e. race-specific or vertical. These race-specific effects are so small that only in more refined experiments one may hope to discern them from the unavoidable experimental error.

The question of whether nature has employed the system envisaged by Van der Plank or the one described by Parlevliet & Zadoks, or both, will have to be resolved by more detailed research. In a few polygenic host-pathogen systems small race-specific effects have been reported, suggesting that at least for those systems a minor gene–for–minor gene relationship may exist. These systems are potato–*Phytophthora infestans* (3, 24); barley–*Puccinia hordei* (7, 36), and barley–*Rhynchosporium secalis* (12). In other systems, of which the inheritance of the quantitatively expressed resistance is not yet known, small race-specific effects have been reported. Ziv & Eyal (62) observed small differential variations in disease severity of wheat cultivars infected with *Septoria tritici*. Kuhn et al (23) studied the IF, LP, and postule size on the flag leaves of four wheat cultivars for 21 *Puccinia recondita* isolates. As in the other studies mentioned above most variation was race-nonspecific, but small race-specific effects on IF and LP were present as well.

In general, race-specificity is considered to indicate instability of the resistance (54, 55). This may be true for the majority of monogenetically inherited resistances, but not necessarily so for resistances of the polygenic type (41). In fact there is no evidence yet that resistances, assumed to be inherited polygenically, tend to erode in time, not even for those where small race-specific effects have been reported.

If it is true that polygenic resistances are durable, it would be convenient when simply inherited resistances could be discerned easily from polygenically inherited ones. This unfortunately is not always the case. When screening for resistance to rusts and powdery mildews in small cereals, Parlevliet (38) advised to select for relatively low levels of disease severity at as susceptible an infection type as possible. Although it increases considerably the chances of selecting polygenic resistances, it does not safeguard completely against simply inherited resistances. In some host-pathogen systems it seems fairly easy to discern partial resistance from low infection type resistances [as with barley-leaf rust (40) and rice-bacterial leaf blight (59)] but in others it is less easy [as in the wheat-yellow rust system (8, 21)].

x_o-REDUCING RESISTANCES AND THEIR RACE SPECIFICITY

Although one might wish to describe the onset of an epidemic in terms of the time and amount of inoculum arriving at a healthy crop, this is not easy. When the time and amount of inoculum are considered as the onset of an epidemic, one has a parameter that is inconvenient, inaccurate, and difficult to measure. After the arrival of the inoculum the disease may spread and increase almost directly or it may simmer at a low level for a long time, waiting for more favorable conditions to occur. The arrival of rust spores in the spring on spring sown cereals and in the late autumn on autumn sown cereals represents these two situations, respectively. Another point is that the parameters describing the onset and the increase of a disease must be derived from comparable observations. Van der Plank (53, 54) recognized these problems. He considered an epidemic to have begun when the disease enters the logarithmic phase of increase (54), and the parameters x_o and r are derived from disease severity observations.

x_o describes the proportion or amount of disease present at the onset of an epidemic. It depends on the amount of inoculum that comes into contact with the host tissue to be diseased and the proportion of that inoculum forming reproducing lesions (the IF). The former factor cannot be classified as resistance and falls outside the scope of this discussion. The latter is a component of resistance already discussed within the context of r-reducing resistances (see section on infection frequency). Evidently the variation in

IF between cultivars not only affects x_o; but it also affects r. To assume that the disease onset and disease increase are independent elements as Van der Plank (54) did is simply not correct; they are linked together by the common component, IF. Partial resistance, due to a decreased IF, a longer LP, and a reduced SP, as shown by Vada against leaf rust (Table 4), not only results in a slowing down of the epidemic development, but also causes a delay in the onset of the epidemic (Table 1); the delay is caused by the lower IF, the slowing down by the combined effect of the lower IF, the longer LP, and the reduced SP.

Resistances that delay the onset of an epidemic are the resistances that reduce IF. These can be of different kinds. Monogenic, low-infection-type (IT) resistances do reduce the IF, when the infections result in necrotic, nonsporulating (IT 0;) lesions. This was the type of resistance Van der Plank was considering when he stated that VR in the host delays the onset of an epidemic. A decreased IF can also result from polygenic, partial resistance, as has been shown for the potato–*Phytophthora infestans* (1, 51, 52), the barley–*Puccinia hordei* (37, 39), the maize–*Cochliobolus heterostrophus* (16), and the maize–*Trichometasphaeria* (Septosphaeria) *turcica* (17) relationships. The former type of resistance, the low IT, is often race-specific; the latter is generally considered to be of a largely race-nonspecific type.

x_o-Reducing resistances therefore can be, like the r-reducing resistances, race-specific as well as race-nonspecific.

CONCLUSIONS

The epidemic development of diseases within the growth cycle of the host is determined by the initial amount of disease, x_o, and the rate at which the disease increases, described by the apparent infection rate r (53, 54). To reduce the severity of disease produced by such compound interest diseases, the reproductive rate of the pathogen must be decreased. The major components of resistance which affect the reproduction rate of the pathogen, are a reduction in the *infection frequency* (IF) or lesion number, a lengthening of the *latent period* (LP), and a decrease in *spore production* (SP). Components, like incubation period and lesion size are not components of resistance as such since they do not affect the reproductive rate of the pathogen directly; they tend, however, to be strongly correlated with LP and SP respectively.

The resistances that affect the components IF, LP, and SP may differ in type of lesion as reported for the two leaf blights in maize (16, 17) and for the rusts and mildew in small grains. One type is more of a qualitative nature, in the sense that its inheritance is simple; the different genotypes are fairly easily to discern. The chlorotic-lesion type of resistance to northern leaf blight in maize conditioned by single genes (15, 18, 27) and the low IT resistances in many crops to biotrophic leaf pathogens, like rusts and downy and powdery mildews, belong to this type. The other type is more of a quantitative nature. The various host genotypes show a continuous rather than a discontinuous distribution in the expression of resistance; in general, this is assumed to be polygenically controlled. In both types, the colonization process of the individual infections often is hampered from a very early phase of colonization onward. If the hindrance is only moderate, the resistance is incomplete; this usually will result in a reduced SP in the case of the monogenic types (Table 1, Monte Christo). In the case of polygenic types of resistance, this will result in a reduced IF, a longer LP and/or a reduced SP (Table 4). The latter resistances rarely result in complete resistances; monogenic resistances, however, often do so. When the growth of the pathogen is restricted severely, either none (or only a part) of the lesions may reach the stage of SP. The IF and SP both are either strongly reduced or have become zero. Resistances characterized by large race-specific effects appear to be inherited in a simple way, while with polygenic resistances the race-nonspecific effects dominate (34, 41). Incomplete resistance, expressed by a reduced IF and/or SP, can therefore be race-specific as well as race-nonspecific as Johnson reported (20) for yellow rust resistance in wheat; durable resistance usually is incomplete; incomplete resistance, however, is not always durable. In many cases incomplete resistance appeared race-specific and unstable (20, 21).

The rapid adaptation of the pathogen population to newly introduced resistances seems restricted to monogenic resistances, especially those of the low infection or hypersensitive type. The need to differentiate between monogenic and polygenic resistances in a simple way led to the classification of resistances into two classes: the vertical, unstable, monogenic, and qualitative type, and the horizontal, stable, polygenic, and quantitative type. The former is supposed to delay the onset of an epidemic (i.e. decrease x_o); the latter is supposed to slow it down once it has started [i.e. reduce r (54)]. Slow-rusting and slow-mildewing refer to the latter, horizontal type. Parlevliet & Zadoks (41) warned against this confounding of meanings. This can be illustrated by the barley cultivar Vada, which carries a gene for incomplete resistance to mildew (22) and a series of minor genes for partial resistance to leaf rust (37). The slow-mildewing character of this cultivar is race-specific (22) the slow-rusting character to leaf rust is largely race-nonspecific (36). It is even possible that a VR does not affect x_o, but does reduce r when the gene for incomplete resistance only reduces SP. The resistance to leaf rust of the barley cultivar Monte Christo (38) illustrates this. Vada, on the other hand, with a resistance of the horizontal type, not only reduces r but also reduces x_o through a strongly reduced IF (Tables

1 and 4). The epidemic parameters x_o and r, therefore, in no way indicate VR and HR, respectively.

Nelson (31), trying to clarify the concepts and meanings around HR and VR, proposed to change Van der Plank's definitions. Van der Plank first defined HR and VR in population kinetic terms, i.e. as race-nonspecific and race-specific, but followed this definition with an interpretation in epidemiological sense, i.e. as r-reducing or x_o-decreasing resistance (53, 54). Nelson proposed to use the terms horizontal and vertical strictly in the epidemiological sense as r-reducing and x_o-decreasing resistances, respectively. This, however, is not a good solution either, because x_o and r are not independent. Variation in the component of resistance, IF, does affect the variation in both x_o and r. In Nelson's proposal HR and VR would not be independent, nor would it classify the resistances in any better way than Van der Plank's concepts do.

Clifford worded this problem of trying to classify the endless variation in resistance very well (6). He wrote, "In common with other workers, the author accepts the convenience of cataloguing resistance in two types. Nature, I am sure, never intended this division." With this last sentence Clifford is right. Resistance cannot be classified unambiguously into two groups nor is there an easy way to discern stable from unstable resistances (9, 20, 34, 41). There is no simple characteristic by which one can always discern polygenic from monogenic resistance without fail. Only a sound knowledge of host-pathogen systems in general and of the host-pathogen system implicated in particular is a good guarantee that one uses the resistance that is most appropriate for the situation concerned.

Literature Cited

1. Black, W. 1970. The nature and inheritance of field resistance to late blight (*Phytophthora infestans*) in potatoes. *Am. Potato J.* 47:279–88
2. Carver, T. L. W., Carr, A. J. H. 1977. Race non-specific resistance of oats to primary infection by mildew. *Ann. Appl. Biol.* 86:29–36
3. Caten, C. E. 1974. Inter-racial variation in *Phytophthora infestans* and adaptation to field resistance for potato blight. *Ann. Appl. Biol.* 77:259–70
4. Clifford, B. C. 1972. The histology of race non-specific resistance to *Puccinia hordei* Otth. in barley. *Proc. Eur. Mediterr. Cereal Rusts Conf. Prague, 1972* I:75–78
5. Clifford, B. C. 1974. Relation between compatible and incompatible infection sites of *Puccinia hordei* on barley. *Trans. Br. Mycol. Soc.* 63:215–20
6. Clifford, B. C. 1975. Stable resistance to cereal diseases: Problems and progress. *Rep. Welsh Plant Breed. Stn. 1974,* pp. 107–13
7. Clifford, B. C., Clothier, R. B. 1974. Physiologic specialization of *Puccinia hordei* on barley. *Trans. Br. Mycol. Soc.* 63:421–30
8. Dehne, D. 1977. *Untersuchungen zur Resistenz von Sommerweizen gegenüber Gelbrost (Puccinia striiformis West.) im Feld und unter kontrollierten Bedingungen während der Ontogenese der Wirtspflanze.* PhD Diss. Georg-August-Univ. Göttingen, Germany. 200 pp.
9. Eenink, A. H. 1976. Genetics of host-parasite relationships and uniform and differential resistance. *Neth. J. Plant Pathol.* 82:133–45
10. Habgood, R. M. 1972. Resistance to *Rhynchosporium secalis* in the winter

barley cultivar Vulcan. *Ann. Appl. Biol.* 72:265–71

11. Habgood, R. M. 1974. The inheritance of resistance to *Rhynchosporium secalis* in some European spring barley cultivars. *Ann. Appl. Biol.* 77:191–200

12. Habgood, R. M. 1976. Differential aggressiveness of *Rhynchosporium secalis* isolates towards specified barley genotypes. *Trans. Br. Mycol. Soc.* 66:201–4

13. Habgood, R. M. 1977. Resistance of barley cultivars to *Rhynchosporium secalis. Trans. Br. Mycol. Soc.* 69:281–86

14. Hart, H. 1929. Relation of stomatal behaviour to stem rust resistance in wheat. *J. Agric. Res.* 39:929–48

15. Hooker, A. L. 1963. Monogenic resistance in *Zea mays* L. to *Helminthosporium turcicum. Crop Sci.* 3:381–83

16. Hooker, A. L. 1972. Southern leaf blight of corn—Present status and future prospects. *J. Environ. Qual.* 1:244–49

17. Hooker, A. L. 1973. Maize. In *Breeding Plants for Disease Resistance: Concepts and Applications,* ed. R. R. Nelson, pp. 132–52. Univ. Park & London: Penn. State Univ. Press. 401 pp.

18. Hooker, A. L. 1977. A second major gene locus in corn for chlorotic-lesion resistance to *Helminthosporium turcicum. Crop Sci.* 17:132–35

19. James, W. C. 1974. Assessment of plant diseases and losses. *Ann. Rev. Phytopathol.* 12:27–48

20. Johnson, R. 1978. Practical breeding for durable resistance to rust diseases in self-pollinating cereals. *Euphytica* 27:529–40

21. Johnson, R., Taylor, A. J. 1976. Spore yield of pathogens in investigations of the race-specificity of host resistance. *Ann. Rev. Phytopathol.* 14:97–119

22. Jørgensen, J. H., Torp, J. 1978. The distribution of spring barley varieties with different powdery mildew resistances in Denmark from 1960 to 1976. *R. Vet. Agric. Univ. Yearb.* 1978:27–44

23. Kuhn, R. C., Ohm, H. W., Shaner, G. E. 1978. Slow leaf-rusting resistance in wheat against twenty-two isolates of *Puccinia recondita. Phytopathology* 68:651–56

24. Latin, R. X., Mackenzie, D. R., Cole, H. 1978. A significant host/pathogen interaction determined among apparent infection rates. *Proc. 35th. Ann. Meet. Potomac Div. Am. Phytopathol. Soc. 1978.* In *Phytopathol. News* 12:70–71 (Abstr.)

25. Lazarovits, G., Higgings, V. J. 1976. Histological comparison of *Clados-porium fulvum,* race 1 on immune, resistant and susceptible tomato varieties. *Can. J. Bot.* 54:224–34

26. Lewellen, R. T., Whitney, E. D. 1976. Inheritance of resistance to race C2 of *Cercospora beticola* in sugar beet. *Crop Sci.* 16:558–61

27. Lim, S. M., Kinsey, J. G., Hooker, A. L. 1974. Inheritance of virulence of *Helminthosporium turcicum* to monogenic resistant corn. *Phytopathology* 64:1150–51

28. Manners, J. G. 1971. Spore formation by certain pathogens in infected leaves. In *Ecology of Leaf Surface Micro-organisms,* ed. T. F. Preece, C. H. Dickinson, pp. 339–52. London & New York: Academic

29. Mehta, Y. R., Zadoks, J. C. 1970. Uredospore production and sporulation period of *Puccinia recondita* f. sp. *triticina* on primary leaves of wheat. *Neth. J. Plant Pathol.* 76:267–76

30. Neervoort, W. J., Parlevliet, J. E. 1978. Partial resistance of barley to leaf rust, *Puccinia hordei.* V. Analysis of the components of partial resistance in eight barley cultivars. *Euphytica* 27:33–39

31. Nelson, R. R. 1978. Genetics of horizontal resistance to plant disease. *Ann. Rev. Phytopathol.* 16:359–78

32. Ohm, H. W., Shaner, G. E. 1976. Three components of slow leafrusting at different growth stages in wheat. *Phytopathology* 66:1356–60

33. Parlevliet, J. E. 1975. Partial resistance of barley to leaf rust, *Puccinia hordei.* I. Effect of cultivar and development stage on latent period. *Euphytica* 24:21–27

34. Parlevliet, J. E. 1977. Plant pathosystems: An attempt to elucidate horizontal resistance. *Euphytica* 26:553–56

35. Parlevliet, J. E. 1977. Variation for partial resistance in a cultivar of rye, *Secale cereale* to brown rust, *Puccinia recondita* f.sp. *recondita. Cereal Rusts Bull.* 5:13–16

36. Parlevliet, J. E. 1978. Race-specific aspects of polygenic resistance of barley to leaf rust, *Puccinia hordei. Neth. J. Plant Pathol.* 84:121–26

37. Parlevliet, J. E. 1978. Further evidence of polygenic inheritance of partial resistance in barley to leaf rust, *Puccinia hordei. Euphytica* 27:369–79

38. Parlevliet, J. E. 1978. Aspects of and problems with horizontal resistance. *Crop Improv.* 5:In press

39. Parlevliet, J. E., Kuiper, H. J. 1977. Partial resistance of barley to leaf rust, *Puccinia hordei.* IV. Effect of cultivar

and development stage on infection frequency. *Euphytica* 26:249–55

40. Parlevliet, J. E., van Ommeren, A. 1975. Partial resistance of barley to leaf rust, *Puccinia hordei.* II. Relationship between field trials, micro-plot tests and latent period. *Euphytica* 24:293–303

41. Parlevliet, J. E., Zadoks, J. C. 1977. The integrated concept of disease resistance; a new view including horizontal and vertical resistance in plants. *Euphytica* 26:5–21

42. Robinson, R. A. 1969. Disease resistance terminology. *Rev. Appl. Mycol.* 48:593–606

43. Russell, G. E. 1975. Deposition of *Erysiphe graminis* f.sp. *hordei* conidia on barley varieties of differing growth habit. *Phytopathol. Z.* 80:316–21

44. Schaper, P. 1951. Die Bedeutung der Inkubationszeit für die Züchtung Krautfaule resistenter Kartoffelsorten. *Z. Pflanzenzücht.* 30:292–99

45. Shaner, G. 1973. Reduced infectability and inoculum production as factors of slow mildewing in Knox wheat. *Phytopathology* 63:1307–11

46. Smith, D. R., Hooker, A. L. 1973. Monogenic chlorotic-lesion resistance in corn to *Helminthosporium maydis. Crop Sci.* 13:330–31

47. Solel, Z., Wahl, I. 1971. Pathogenic specialization of *Cercospora beticola. Phytopathology* 61:1081–83

48. Sztejnberg, A., Wahl, I. 1977. Mechanisms and stability of slow rusting resistance in *Avenae Sterilis. Phytopathology* 67:74–80

49. Toriyama, K. 1975. Recent progress of studies on horizontal resistance in rice breeding for blast resistance in Japan. *Proc. Semin. Hortic. Res. Blast Dis. Rice, 1971. CIAT Ser.* CE9:65–100

50. Ullrich, J. 1976. Epidemiologische Aspekte bei der Krankheitsresistenz von Kulturpflanzen. *Adv. Plant Breed.,* Vol. 6, Suppl. to *J. Plant Breed.* Berlin & Hamburg: Parey. 87 pp.

51. Umaerus, V. 1970. Studies on field resistance to *Phytophthora infestans.* 5. Mechanisms of resistance and application to potato breeding. *Z. Pflanzenzücht.* 63:1–23

52. Umaerus, V., Lihnell, D. 1976. A laboratory method for measuring the degree of attack by *Phytophthora infestans. Potato Res.* 19:91–107

53. Van der Plank, J. E. 1963. *Plant Diseases: Epidemics and Control.* New York & London. Academic. 349 pp.

54. Van der Plank, J. E. 1968. *Disease Resistance in Plants.* New York & London. Academic. 206 pp.

55. Van der Plank, J. E. 1975. *Principles of Plant Infection.* New York, San Francisco & London: Academic. 216 pp.

56. Van der Zaag, D. D. 1959. Some observations on breeding for resistance to *Phytophthora infestans. Eur. Potato J.* 2:278–87

57. Whitney, E. D., Lewellen, R. T. 1976. Identification and distribution of races C1 and C2 of *Cercospora beticola* from sugar beet. *Phytopathology* 66:1158–60

58. Wilcoxson, R. D., Skovmand, B., Atif, A. H. 1975. Evaluation of wheat cultivars for ability to retard development of stem rust. *Ann. Appl. Biol.* 80:275–81

59. Yamamoto, T., Hifni, H. R., Machmud, M., Nishizawa, T., Tantera, D. M. 1977. Variation in pathogenicity of *Xanthomonas oryzae* (Uyeda et Ishiyama) Dowson, and resistance of rice varieties to the pathogen. *Contrib. Centr. Res. Inst. Agric. Bogor,* No. 28. 22 pp.

60. Yarwood, C. E. 1961. Uredospore production by *Uromyces phaseoli. Phytopathology* 51:22–27

61. Zadoks, J. C. 1971. Systems analysis and the dynamics of epidemics. *Phytopathology* 61:600–10

62. Ziv, O., Eyal, Z. 1978. Assessment of yield component losses caused in plants of spring wheat cultivars by selected isolates of *Septoria tritici. Phytopathology* 68:791–96

Ann. Rev. Phytopathol. 1979. 17:223–52
Copyright © 1979 by Annual Reviews Inc. All rights reserved

SCIENTIFIC PROPORTION AND ECONOMIC DECISIONS FOR FARMERS

❖3709

John Grainger[1]

Former Head, Plant Pathology Department, West of Scotland Agricultural College, Auchincruive, Ayr, Scotland

The science of crop pathology exists to play a vital part in making a farmer's practical venture as economically productive as possible, but it does not at present make full impact. It is very easy to concentrate on our desire for scientific understanding and to neglect our duty to be useful to farmers. It is, however, our responsibility so to direct our energies as to make proper scientific proportion the basis of all calculation involved in decision making in the farming industry, thus securing an economic increase in the abundance of food. This chapter is designed to assist in the development of this approach.

EQUAL STUDY OF HOST, PARASITE, AND ENVIRONMENT

The Host Plant as a Habitat for Fungal and Bacterial Pathogens

A piece of scientific *dis*proportion in phytopathology is the tacit assumption that host plants are invariably receptive to fungal and bacterial pathogens. This assumption is not correct, but a host can be *made* receptive at any time by various items of bad human management. For this reason, there is great need to test the scientific proportion of our approach in research. We can only do this effectively by new measurements of new parameters, particularly of the physiological "disease potential" or "receptivity" of the host.

---

[1]Dr. Grainger died after completing this review. An obituary is located on page 251.

223

0066-4286/79/0901-0223$01.00

In temperate regions, well-tended commercial crops are usually healthy in midgrowth and least healthy during earlier and later stages of development. This differentiation in general health status is induced by physiological changes that override whatever genetic resistance or susceptibility is possessed by a particular crop or cultivar. These changes can be monitored by careful subjective observation of well-grown hosts and by objective measurement of the Cp/Rs ratio.

THE Cp/Rs RATIO AS AN INDICATION OF DISEASE POTENTIAL   A parasitic fungus or bacterium must obtain all the materials needed for its growth and reproduction from the host plant attacked. Although the amount required varies, energy is always a major need of the pathogen and the suitability of the host as a habitat for the parasitic fungi or bacteria which are capable of attacking it (see below) can be gauged very largely by the amount of carbohydrate it has available for breakdown by the parasite.

The carbohydrate fractions in plant leaves or shoots vary with every passing cloud; for this reason we need a measure of "disease potential" that changes far more deliberately. Healthy growth in midseason is generally associated with low carbohydrate in the whole plant coupled with rapid growth. Since the host uses its carbohydrate "from hand to mouth" for its own growth there is usually none that an invading pathogen can plunder, and hence no disease (12).

Cp/Rs is a specialized measure of available carbohydrate. It is obtained by dividing the weight in grams of total carbohydrate in the whole plant (Cp) by the residual (= carbohydrate-free) dry weight (also in grams) of the shoot (Rs). It has sometimes been miscalculated, but Table 1 shows the proper way.

As I have found in some 40 host-parasite relations, the higher the calculated Cp/Rs ratio the greater the disease potential, and vice versa. Many of these relationships have been checked by inoculation with suitable pathogens together with the provision of favorable environment. In other instances plots grown side by side and established at intervals through growth gave sustained inoculum and mutually suitable conditions, but had different disease potentials, according to their Cp/Rs values at any one time. Disease appeared with increasing severity the higher the ratio, whenever a pathogen was also present (8), but hosts having suitably low Cp/Rs were not attacked though growing along side.

Disease phases   Table 1 shows the disease phases likely to be found with most productive crop plants, but there are variations. Low carbohydrate seeds like cocksfoot grass or red clover do not have supersensitive or permissive phases, and seedlings with unrestricted root growth should remain

Table 1 Calculation of Cp/Rs of potato, and some disease phases indicated by it (averages per plant, simplified from actual values for ease of following)

		Dry weight, grams* (a)	Percentage total carbo hydrate (b)	Weight of total carbo- hydrate, grams (c) = (a × b)	Rs, grams (a − c)	Cp/Rs ratio	Disease phases
Before growth	Seed tuber	20	70	14 Cp	6 Rs	2.33	A. Dormant (tuber only)
Planted; first growth	Old tuber	10	70	7			B. Supersensitive (sprouts only)
	Roots	0.001	30	—			
	Sprouts	0.06	40	0.024	0.036	195.1	
	Whole plant	10.06		7.024 Cp	Rs		
Early mid growth							C. Permissive†
Mid growth	Old tuber	2	25	0.5			D. Physiological barrier (no disease)
	Roots, stolons	2	30	0.6			
	Shoots	6	20	1.2	4.8	0.47	
	Whole plant			2.3 Cp	Rs		
Maximum growth	Roots, stolons	6	25	1.5			E. Epidemic; Cp/Rs over 1.0 and rising
	Tubers	350	70	245	105‡		
	Shoots	110	20	22	88	3.05	
	Whole plant			268.5 Cp	Rs		

*All dry weights are obtained by the dehydration of fresh material to constant weight in an aspirated oven at 43 to 45° C.

†Cp/Rs is falling between 10 and 1. Disease may occur but the plant "tends to grow away from it"—a much needed explanation of this statement.

‡Cp/Rs of the new tubers as distinct from all other organs of the plant is obtained by dividing 245 by 105 to give 2.33.

below Cp/Rs 1.0 for at least 50 days after sowing. The unreceptive physiological barrier phase has to be defined as below Cp/Rs 1.0 for epidemic pathogens, below 0.5 for "normal" pathogens, and below 0.4 for "low carbohydrate" pathogens (12).

Some mild pathogens (e.g. *Penicillium* spp.) do not attack until the host is supersensitive. Kamoen (16) has shown, however, that tuberous begonia still maintains its changing disease potential toward *Botrytis cinerea* according to the magnitude of the Cp/Rs ratio even though this is high enough right through growth to come into the supersensitive category.

We may now use Cp/Rs measurements along with factors of the environment to elucidate situations where disease is less than might be expected.

Costless Controls: Worldwide Factors That Keep Disease to Manageable Levels

We know from fossil forms that the first land plants (*Rhynia, Hornea,* and *Asteroxylon*) were attacked by fungi and bacteria in Devonian times, some 400 million years ago. Since we still have their descendants—club mosses and ferns—on earth today, the diseases never drove these primitive hosts to extinction. More modern-looking crops (e.g. barley, wheat) have been grown for about 12,000 years in SW Asia. There have indeed been blights, famines, and pestilences with very local effects, but overall occurrence has been rare enough to suggest that suffering from disease and pest attack is abnormal, while health is the normal state.

This generalization is supported by data from four long-term studies of disease losses. Our own detailed survey (9) gave 11.4% loss in temperate west Scotland (excluding rough grazing). Padwick (22) found 11.6% in the former British colonies, mostly in warm or very warm regions. LeClerg (17) published figures from which the overall loss in the USA can be calculated as 13.3%, and Cramer (6) collated world results to give no more than 11.8%. These figures are all in keeping with the "norm of equal chance" in the disease triangle (below). Warm and very warm regions have as low amounts of disease as cooler ones, so, though details must inevitably be different, we are likely to find restraining influences all over the agricultural world.

Costless "disease escape" controls are even possible in some situations, while disease is so naturally curtailed in a few instances that no extraneous control is justified.

"DISEASE ESCAPE" DURING A TEMPERATE SPRING AND SUMMER
Most fungi have narrower ranges of activity in relation to temperature than farm plants. For this reason, crops begin to grow in a temperate spring long before most parasites can attack. At this time, many hosts are extremely receptive ("supersensitive" to pathogens; B in Table 1), but low temperature provides an effective restraint on disease development (12). Later when temperatures are high enough to permit fungal activity, Cp/Rs measurements show that the crops are entering the midgrowth healthy phase when they are not receptive to pathogens (D in Table 1). This joint action thus postpones disease until so late in the growing season that there is not enough time left for a severe disease to develop on the plants which by then have grown quite large. This relationship shows up in the calculation of Cp/Rs ratio—the value Rs of the Cp/Rs ratio is a good measure of relative size of plant.

"DISEASE ESCAPE" FROM AN EELWORM ATTACK BY EXPLOITING DIFFERENTIAL ENVIRONMENTAL EFFECTS AND A SPECIAL CULTIVAR   The potato cyst eelworm, *Heterodera rostochiensis,* does not become active until soil temperature rises above 70°C (45°F), typically about mid April in west Scotland. The potato cultivar 'Epicure', planted in February, starts growth at 2°C (36°F)—thus there are about 6 weeks of initial eelworm-free host activity, which itself halves the economic loss. 'Epicure', moreover, devotes a very large proportion of its dry weight to the production of useful produce (tubers), as measured by the crop production efficiency ratio Du/Drp. Du is the dry weight of useful produce per plant, and this is divided by the dry weight of the rest of the Plant (Drp) that actually makes the produce.

'Kerr's Pink', a maincrop potato, has a Du/Drp ratio of about 3.2 at maximum production after a long growth time, whereas early 'Epicure' has a Du/Drp ratio of about 5.5 at the end of a much shorter period. It is thus possible to market an economically high-priced yield of about 12,500 kg/ha (5 ton/acre) by June 15, before the eelworm's life cycle has been completed.

Any new population that does develop can be still further depleted in the farmer's sowing a catch crop of Italian ryegrass immediately after the first early potatoes are harvested. 'Groundkeeper' potatoes appearing through this growth have their roots attacked by eelworms, but again these cannot complete a life cycle before they and their host plants are destroyed by winter frosts. This is an effective use of a trap crop: An earlier-established one would tend to *increase* the soil population of potato cyst eelworms and only makes matters worse.

"Disease escape" in this instance is a costless control which, if applied with full knowledge of host, parasite, and environment, leads to virtual elimination of one of our most serious diseases (13). The method, however, only works in temperate areas with a dormant winter: it does not work in regions (e.g. the Channel Isles) where the soil temperature remains above 7°C throughout the year [Figure 71 in (13)].

"DISEASE ESCAPE" FROM A FUNGAL ATTACK BY EXPLOITING DIFFERENTIAL ENVIRONMENTAL EFFECTS AND LOW RECEPTIVITY
Potato late blight does not appear at all on first early 'Epicure' in southwest Scotland. The fungus *Phytophthora infestans* is inactive under field conditions until the temperature rises above 10°C (50°F), typically in early May. The Cp/Rs of the potato host is then below 1.0 and remains at unreceptive levels until the end of that month, when periods of humidity suitable for spore germination (e.g. 'Beaumont' periods) are rare. The crop remains healthy right up to its early harvest.

Second early 'Epicure' potatoes planted in mid March and harvested in July do not escape late blight. Low initial temperature and then low host receptivity both favor the host and not the fungus, but Cp/Rs begins to rise to receptive levels from mid June onwards in west Scotland. Blight then appears shortly after a period of favorable humid weather occurs, and develops rapidly on the foliage. Maincrops are typically in the nonreceptive physiological barrier phase at the time and are not attacked until after a later period of suitable weather—a good test of the Cp/Rs concept when crops are contiguous.

"DISEASE ESCAPE" IN WARM REGIONS   Warm parts of the world also have low overall amounts of disease (above). Low receptivity apparently occurs in mid growth, but is often followed by later temperatures *above* the range of parasite activity. Late blight of potato, for example, presents no practical problem in many warm areas. Although the overall yield of potatoes in such regions may be far from the world optimum, yields may be good enough to make the crop economically competitive in local markets because long-distance transport is avoided. In many places with a dry climate lack of very high humidity periods (for many pathogens) or of free water deposition (e.g. for potato late blight) may also limit disease.

Details of "disease escape" thus vary from time to time within a growing season and from place to place according to climate. Full practical use of general worldwide principles can only be made by considering local conditions in relation to them. Much more local work is necessary to take full advantage of these possibilities.

DISEASE CONTROL BY GROWING HEALTHY PLANTS   The tomato—a greenhouse crop in Britain—produces flowers and fruit sympodially and continuously over a long growing period. This is equivalent to an extension in time of the mid growth healthy period of farm crops (above). Rs keeps ahead of Cp, so Cp/Rs is low enough (below 0.5) to maintain the crop in good health for a long time; fruit production does not claim relatively large overall amounts of carbohydrate until well through the season. During the period of low Cp/Rs of any plant, various items of bad management can cause that ratio to rise untypically and if a pathogen is present (as it usually is), disease appears.

Specification of good management   A survey of management bad enough to make Cp/Rs rise untypically in tomato revealed five items which, when specified per contra, make a schedule for growing healthy plants (11–13). To avoid disease there must be (*a*) no restriction of the root system; (*b*) no shortage of water; (*c*) no lack of essential nutrients (which must be available both in proper amount and in proper balance); (*d*) no large change of temperature; and (*e*) no phytotoxic pollutants. None of these costs anything more than the usual provisions for a healthy crop. The same specifications apply to all crops.

"Logarithm growth"   Along with low Cp/Rs of healthy growth there is a smooth rise in log dry weight of the whole plant; but if any item of bad management occurs (or natural senescence sets in!), the smooth rise falls away markedly and disease usually appears. The term "logarithm growth"

is used to denote unrestricted (but not unpruned) growth where no inimical factor curtails the steady rate of progression in optimal growth. It provides a most useful confirmatory check that Cp/Rs is remaining low.

The greenhouse tomato maintained in a temperate region probably has a longer period of potentially healthy growth than any other crop. Having the concepts of Cp/Rs and "logarithm growth," however, we can now assess the maximum period of healthy growth in any crop—a highly economic parameter.

SOIL CONDITIONS SIMULTANEOUSLY BEST FOR GROWTH AND WORST FOR DISEASE

Soil pH We can specify the best soil pH for each crop's optimum yield, but disease severity is likely to be minimal at rather different values of relative soil acidity. Equal study of host and parasite would here seek to find a mutually suitable soil pH where host yield will be very near its optimum and, simultaneously, parasite activity will be very near the minimum. As shown below, this approach has proved practical and consistent over several years (unpublished).

A soil pH between 5.0 and 5.5 permitted high potato yield, also low common tuber scab (*Streptomyces*) and low russeting (a nonparasitic blemish), while even late blight and *Rhizoctonia* on the tubers were not excessive. Scab severity was minimal as pH fell below 6.2, but at that value yield was, on the average, 12% less than the yield of about 37,000 kg/ha (15 tons/acre) at soil pH 5.0 to 5.5.

A soil pH of 6.5 permitted the highest yield of turnips, about 59,500 kg/ha (23.7 tons/acre), with club root disease averaging only 1.6% (0 to 6%) over 5 years. At pH 6.7 this minute amount of disease was halved, but the average yield was down by about 4,770 kg/ha (1.9 tons per acre) —8% below the average of about 59,500 kg/ha (23.7 tons/acre) at pH 6.5.

Tomato gave greatest yield when grown on soil between pH 6.0 and 6.5. Botrytis diseases on stem and leaves, root rots, and root knot eelworm (*Meloidogyne hapla*) were collectively at a minimum at pH 6.4. If, however, the scab organism *Spongospora* was also present on the roots, all four were minimal at pH 6.1 while the yield stayed maximal. We thus have the possibility of mutual relations with crop yield according to parasite *groups* —a welcome economic approach.

Most agricultural soils naturally drift toward acid pH conditions; liming provides a relatively inexpensive correction toward neutrality which must be made for each crop. Then, if we are able to find a mutually suitable soil pH value for host and parasite, disease management can be achieved with little or no cost.

Soil fertility Fertilizer added to a soil affects crop yield in well-known ways, but may also influence differentially the amount of disease that appears. This action is subject to yield variation, to precropping amounts of inorganic nutrients in the soil, to weather, and doubtless to other factors. In west Scotland cereal mildew seems to be lower when any deficiency of soil phosphate is corrected, and when overall fertility is not too high. Normal dressings of nitrogen alone make the disease slightly worse. Details vary for particular diseases, but cereal mildew is usually so low in amount in west Scotland as scarcely to affect farm productivity except possibly on highly susceptible cultivars such as Golden Promise barley in certain years.

Genetics and Disease: Prohibition and Partial Resistance

THE SMALL NUMBER OF DISEASE SITUATIONS   Stapley & Gayner (23) describe 337 major parasites of all kinds attacking 22 world crops, but only 60 attack more than one crop. If all parasites attacked all crops there would be at least $(337 - 60) \times 22 = 6094$ disease situations instead of the 337 (= 5.5%). The first part of Miller & Pollard (19) gives a comparable figure of 2.7%.

The productive crops of the world, while accepting a few parasites, are evidently equipped to reject totally a large number of harmful organisms. A few mechanisms are worthy of mention in this connection.

Hypersensitivity   A parasite attacks but the host responds so vigorously that cells are killed round the invader, which is thus completely inactivated, often without any visible symptoms. Müller (20) thus explained why spores of *Phytophthora infestans,* for example, did not attack cabbage or lettuce, for hypersensitivity is, indeed, not very specific.

Phytoalexins   These are particular antibiotic chemicals, many of them polyphenols, which are produced by hosts in response to injury, to poisons, or to invasion by infecting organisms. The materials are absorbed by an invading fungus and inhibit its growth. This work was also originated by Müller [with Börger (21)], and notably continued by Cruickshank and co-workers [e.g. with Perrin (7)]. The comparative activity of saprophytes and parasites and also the degree of genetic host resistance seem related to the relative amounts of phytoalexin present. Accelerator genes may increase the rate of its production.

Order of attack: antagonism of fungi   Clark (3) has shown that if *Penicillium cyclopium* attacks a hyacinth leaf before *Pseudomonas marginalis* the resulting joint damage is more severe than if *P. marginalis* were the first invader.

Millard & Taylor (18) explained the reduction of potato common scab *Actinomyces* (= *Streptomyces*) *scabies* by the addition of organic matter to soil as an antagonistic overgrowing of parasitic *A. scabies* by saprophytic *A. praecox*. More research is needed into whether such results are related to phytoalexin concepts.

Specific chemical and physical barriers   Link & Walker (17a) found the substance catechol and/or related compounds in onions resistant to *Colletotrichum circinans*. Catechol differs from phytoalexins in that it is a normal host metabolite.

An uninterrupted cork layer over a potato tuber well nigh prevents the entrance of the gangrene fungus, *Phoma foveata*, for which wounding is necessary; but late blight and skin spot (*Oospora pustulans*) can enter without wounding. Glandular or sticky hairs on the host are likely to inactivate virus vectors.

There are thus certain instances where genetic factors apparently prohibit attack by disease or pests. Genetic resistance itself is, however, variable in both performance and permanence.

Genetics: polygenically controlled characters necessary   Two resistant turnip cultivars 'Bruce' and 'Wallace' were both affected by club root *Plasmodiophora brassicae* to an extent of about 13% over 18 years at Auchincruive during which time attack of susceptible cultivars on the same soil averaged 45%. 'Bruce' and 'Wallace' have polygenic resistance to club root disease, maintained at a similar level for over 130 years, and controlled by the combined action of many genes brought together by "field" or "horizontal" selection.

This contrasts with many hosts having oligogenic or "vertical" resistance, e.g. of new potato cultivars to late blight on the foliage, where the resistance is controlled by one or a very few potent genes bred into the system. Such cultivars typically lose resistance within a few years after their introduction, as the parasite changes its genetic constitution to overcome the host resistance; such impermanence restricts productivity very insidiously. Long-term polygenic resistance of potato *tubers* to blight is available in several cultivars, e.g. 'Golden Wonder' and 'Dunbar Rover'. Tubers are always very receptive to late blight (Cp/Rs about 2.0 to 2.5), so as the physiological disease potential is fairly standard, the genetic value of polygenic "field" or "horizontal" resistance is emphasized.

The introduction of one particular gene for disease resistance in host genetics is "scientific" but may be impermanent. Field selection for resistance demands less "scientific" thought but is in better proportion, is surer, more permanent, and is thus better science in the long run. Many polygenic characters selected in the field may not even be concerned directly with disease resistance. A host having high crop production efficiency (above), for instance, may well show less effect from disease than would be conferred by single gene resistance.

Geneticists have begun to deplore the inadequacy of gene banks in crop breeding—there are not enough new gene sources. They need have no alarm if they concentrate on polygenic inheritance by field selection. There will still be need for cross breeding to combine populations having favorable genes, but the selection of desired cultivars will take longer and be more practically searching than with oligogenic material.

COMPREHENSIVE POLYGENIC CHARACTERS   The resistant turnip cultivars 'Bruce' and 'Wallace' (above) are resistant to club root disease but are fairly prone to soft rot (*Pectobacterium carotovorum*). Their produce has only about 9% solids. Susceptible turnips are swedes, with 11% dry matter and also are susceptible to soft rot. The swede 'Wilhelmsburger', with 11% dry matter in the produce, is moderately resistant to both club root and soft rot diseases. Performances and calculated dry weight yields are listed in Table 2.

The best all-round cultivar for west Scotland is not one of the most disease-resistant turnips but rather 'Wilhelmsburger', with the highest dry matter yield whether the diseases are present or not.

The tomato 'Curabel' is resistant to several diseases and if its resistance is sustained it will indeed be an economic asset.

We are equipped to help a farmer's economic decisions when we give host and parasite equal study, integrate as many factors as possible, and maintain a comprehensive interest in all crop-husbandry practices.

Table 2 Performance and total dry weight yield per hectare of some turnip cultivars in conditions of health and disease

		Percentage of club root	Percentage of soft rot	Farmer's yield kg/ha	Dry weight yield kg/ha
Average susceptible	healthy	0	0	52,705	5,798
(swedes)	diseased	45	16	21,082	2,259
Average resistant	healthy	0	0	62,744	5,647
(turnips)	diseased	13	13	44,423	3,915
'Wilhelmsburger'	healthy	0	0	60,234	6,626
	diseased	26	9	46,682	5,145

ECONOMICS OF CROP MANAGEMENT

The science of economics compares dissimilar items of worth to the human race (e.g. knowledge, skill, rent, fuel, useful produce), by measuring them in a common value—money, and "maximum profitability" is an obvious standard of reference for evaluation of farming systems.

A farmer often must incur production costs months or even years before he reaps a harvest and hopefully gains a profit. His expenditures are at risk all this time and our science of plant pathology must minimize such risk within the larger concept of crop protection, which is concerned with losses from all causes.

Different Economics of Different Crop Values

Agricultural income is derived from crops of very different annual value (13), here expressed in relative 1965 prices. Unimproved permanent grazing gives a very small although fairly consistent profit of about £12 per hectare per annum, but it will not sustain any recognizable expenditure on crop protection. Cereals and improved grass usually make a profit about £148 per hectare per annum, though in years of low yield there may be a loss from cereals; maximum spending on plant protection is one twentieth of crop value (£7.4). Potatoes (£500 per hectare per annum) offer a surer and larger profit than cereals and can sustain an expenditure of about one tenth of crop value (£50). Fruit crops (£990 per hectare per annum) offer the largest potential profits among outdoor crops and can justify about one fifth of crop value (£200) on crop protection. These proportionate spendings reflect the relative amounts of loss expected which, for diseases caused by fungi and bacteria, vary according to the average Cp/Rs values of the hosts throughout growth, which are higher for higher value crops.[2]

One Provision for All Loss Agents

The amounts mentioned above for crop protection must cover all losses, whether from viruses, bacteria, fungi, insects, eelworms, mites, weeds, or even from nonparasitic causes. Crops do not recognize our human specializations, and we must adopt a multidisciplinary approach to see how much expenditure is available for control of all pathogens. The amount often will not be large, and the control of weeds usually takes the highest priority.

Ratio of Benefit to Cost

The introduction of any contingent value into crop production costs (e.g. expenditure on crop protection) might still be justified if it conferred benefit

---

[2]Although crop values and control costs have risen sharply since these figures were derived, *relative proportions* of economic spending remain broadly similar.

worth very much more than its cost. The minimum sustainable ratio of benefit to cost also varies with relative annual crop value—e.g. 7 to 1 for cereals, 3.5 to 1 for potatoes, and only with fruit crops can it be as little as 2 to 1. There is often an extremely restricted budget for crop protection, so the methods used must be of a rather uncommonly high economic performance. An important part of our work is to find costless or low-cost controls (above). A second part is to make sure that disease losses are correctly interpreted, and yet another part is to ensure that there is a sufficiently high ratio of benefit to cost in any chemical control deemed necessary.

Host and Parasite in Economic Relations

The first measure we need with which to design a disease management system is always of the amount of disease or damage ($\%D$). This is then corrected by multiplying constants for the time the damage appears (Kt), its effect on yield (Ky), on market value (Km), and on the necessity for replacement or restitution (Kr). The first two most often reduce the presumed economic effect, but the last two almost always increase loss. There are sometimes interactions between crop price, amount of disease, and time, which generally tend toward lower economic loss (11, 13). The use of constants implies that relations are fairly exact and consistent enough to be expressed as graphs; this is largely true, particularly if results are to be integrated (below).

The Estimation of Economic Loss

It is desirable to define the relative amount of disease on a plant as "severity," disease in a crop as "intensity," and disease over an area, region, or country as "prevalence." Severity is usually a research measure but can be economically useful with the larger fruit crops. Intensity is largely a combination of severity readings while prevalence combines several intensity estimations, so we must ensure that methods of measurement and of combination are adequate.

MEASUREMENT OF THE AMOUNT OF DISEASE ($\%D$)   There are many methods by key or by standard area diagram used either directly or to denote percentage steps in a disease rating calculation (11, 13). The human eye can compare quantitatively specified values drawn within the shape of an appropriate plant part, with patterns occurring naturally, even if the outlines are not of the same size. Complicated diagrams with, e.g. different leaf sizes, are not necessary.

Inaccuracies of general method   Inaccuracies arise when percentages of disease derived from unequal samples are averaged. A viewing box used in the standard manner can equalize visual samples in field experiments, where almost all readings from keys or standard area diagrams are percentages. Surveys need correct weighting by area. A field of 50 hectares with 4% disease cannot be combined successfully with one of 3 hectares and 60% of disease by directly averaging their percentages. This would give the wrong answer, 32%, whereas the correct position is

Hectares (a)	Percentage of disease (b)	Product (a × b)
50	4	200
3	60	180
53		380

Fifty-three hectares could all have 100% disease, so the correct actual value is

$$\text{Percentage of disease} = \frac{\text{actual}}{\text{possible}} \times 100 = \frac{380}{5300} \times 100 = 7.2,$$

vastly different from the incorrect 32%.

We are given little or no evidence of such correct weighting in the results of surveys, and so may wonder uneasily if there are errors of combination, all of which tend to increase apparent disease prevalence.

Translation of %D into Economic Loss

SIMPLE RELATIONS   Several host-parasite relations can be expressed as a direct and simple link between the number of plants affected ($\%D$) and the percentage or decimal loss of yield (Ky). Any disease causing the death of individual plants or organs has $Ky = 1.0$; for example, every potato tuber attacked by late blight becomes a total loss in store. Take-all disease of cereals, rather surprisingly, also causes a fairly predictable degree of damage, $Ky = 0.5$: It apparently causes damage only after the host has become receptive. A local investigation is needed to see which economic relations are amenable to such simple treatment.

COMPLEX VIRUS RELATIONS   Some virus relations with yield can still be expressed simply, e.g. virus X alone on many potato cultivars has $Ky = 0.03$, and even severe tobacco mosaic virus on 'Ware cross' tomato usually has $Ky = 0.33$ if the infection starts in the seedling stage. Most relations of viruses with crop economics are, however, extremely variable and complex. They are not always consistent over a period of years (13) and vary with host cultivar and virus strain according to whether there are aggressive attacks by two or more virus strains, or relatively slight effects from the protection conferred by previously present mild strains.

Thus, although fairly exact specifications are available for some viruses, there are practical difficulties when we try to explain the formidable variation in effects of the various combinations of viruses on yield in the field. There are also indications of varying host receptivity according to the amount of nitrogen available in the plant (unpublished). This is a further complicating factor, but a logical parallel concept to that of the Cp/Rs ratio using host carbohydrate status to measure "disease potential" for fungi and bacteria.

Virus "protection" in practice   Mild viruses are not always stable and if the virus protected against is not the one to arrive, a serious disease may well result. Chamberlain et al (1) have probably made the most practical use of protection by one against another. Growth from apple scions bearing a mild strain of apple mosaic, when grafted upon trees attacked by a severe strain, is effectively protected against invasion by the severe strain. The overall degree of protection is initially small, but builds up later.

Channon et al (2) established quantitative relations with yields of 'Eurocross BB' tomato which, when preinoculated with attenuated tobacco mosaic virus MII-16, gave 14% more produce when a serious strain of the same virus was also present from an early growth stage. The attenuated virus, however, itself reduced yield by about 5% under certain conditions, and we need more work.

ECONOMIC ASSESSMENT BY FIELD WORK   James (15) has reviewed methods and difficulties of assessing economic effects of diseases. Field assessments are probably the best approach, but healthy controls are necessary and are not always easy to contrive. Genetic resistance does not usually confer a sufficient degree of health and often is not permanent, while chemical control is not complete enough to give any practical approach to "healthy" controls.

Farm yields are, in any case, instances where an unknown value of healthy yield is diminished by an unknown amount of loss from parasites, pathogens, or competitors. This provides only an approximate reference for economic loss, and even high mathematical significance among results will not make it more reliable.

The only way out of such a fragile situation is to try alternative scientific approaches, either to the whole problem, e.g. "disease phenology plots," or to particular parts, e.g. yield compensation, and then to integrate all their effects.

"Crop and disease phenology plots"   My own approaches began in 1944 with the establishment of randomized, replicated plots of cereal and *Vicia* bean cultivars, of potatoes and turnips, all grouped round a weather station.

All humanly controllable factors that affect growth were made as favorable to the farmer and as uniform as possible, but no chemical or other direct control of disease was applied. Apart from disease occurrence this left only the uncontrollable weather factors as causes of variation in yield and loss. Yields of turnip and potato, together with their serious losses from disease, were all explainable when all the relevant weather factors were integrated.

The beans often had spectacular foliage infection (*Botrytis* spp.) on late growth made after productive yield was complete, so its economic effect on the current crop was small. Cereals, however, sustained only small amounts of foliage disease (below 10%), but it was never remotely related to yield. The highest among these small disease amounts were indeed often found among the highest annual yields. This alternative approach confirmed the original observation that economic loss from cereal shoot diseases is likely to be slight.

Yield compensation   Maximum production of useful produce is associated with an optimum plant density per unit area, combined with optimum soil fertility at that density (13). Disease that evenly thins a crop which is established above the optimum density *increases* the yield, while thinning below that optimum always gives more yield than would be expected from the arithmetic degree of thinning. The yield is compensated upwards because individual plants have more room to grow and so make greater individual amounts of produce.

Relations derived from various contrived plant densities show the relative yield compensation by a crop and the remainder, Ky, is the percentage loss in yield for a given $\%D$, but is more significant if integrated with other disease factors (below).

Yield compensation can also take place within plants; for example, various degrees of defoliation maintained from different dates to simulate crop loss always have percentage loss of yield *less* than percentage defoliation.

Migration of cereal nutrients from shoot to grain   A special kind of yield compensation takes place in late growth of cereals as a response to the arrest of shoot growth caused by terminal flower initials on all tillers. Carbohydrate and other internal nutrients are transferred from the shoot to the grain while the overall dry weight of shoot plus grain does not change very much as it would have done with the progressive photosynthesis of indeterminate growth.[3]

[3]Incidentally, the normal channels of translocation are probably too deeply seated in the tissue for the carbohydrate in transit to be affected by comparatively superficial parasites. Such parasites as rust fungi are, moreover, "low carbohydrate" organisms (12): Though they may attack a host at comparatively low Cp/Rs they do not deplete it greatly of carbohydrate.

The value of "no control" economic decisions   Certain diseases cause heavy loss, but others will not be worth any consistent expenditure on control in many areas. A farmer is not equipped to make valid "no control" decisions from his own observations: A scientist *can* advise him, but only after extensive new work and exhaustive cross checking from different approaches. "No control" decisions save much money, but we must be very sure of the facts and interpretations, to have complete confidence and then courage in their application.

Checks of loss decisions via the "disease triangle"   Disease will not occur until, simultaneously, (*a*) a parasite is present; (*b*) environmental conditions are suitable for its development; and (*c*) the host plant is receptive to attack. If all three factors have equal chances of being favorable or unfavorable in space and time (as with a large and varied system of agriculture) we should expect disease in only one out of eight situations, or 12.5% (13). This is the "norm of equal chance" and is very close to four world estimates (6, 9, 17, 22).

If one factor in the disease triangle is always favorable while the other two have equal chances to vary, we should expect disease in 1 out of 4 situations, or 25%. Potato tubers, for example, are always receptive to late blight and our long-term disease amount at Auchincruive is about 28%.

If two factors are always favorable, disease occurrence depends solely on the third. Potato hosts susceptible to the cyst eelworm *Heterodera rostochiensis* also find environmental conditions suitable between April and September in west Scotland. The occurrence of disease then depends on whether the soil parasite is present or not—an equal or 50% chance.

The condition where all three factors are always favorable is, fortunately, a theoretical state found very rarely indeed in practice.

Checks of low loss decisions by crop value: G units   The calculation of *G* units (13) is a system of weighting percentage proportions of crops in five categories of different annual value (above) first by their relative annual values, to give an index (*I*) of the intensity of agricultural production. A second weighting by the relative economic spending on loss control gives another index (*P*) of crop protection opportunity. The product $I \times P = G$ units, which are arbitrary values varying from 0 to 300, and specify several factors effectively (e.g. capacity to feed a given population; opportunity to control loss; potential, though not actual, production). *G* unit values can relate either to the total area of a country or to its agricultural area: The results are often quite different.

G units allow us to adjust the "norm of equal chance" (12.5% disease) upwards or downwards according to the kind of cropping on any scale of farm system, remembering that *P* reflects the proportion of high value crops and that these tend to carry more disease. The "norm" corresponds to about 110 *G* units; 300 *G* has about 15% disease while a value round 66 *G* has about 11%—a strongly exponential relation.

Checks of crop loss by a new kind of integrated truth: Reference to "attainable yield"   Attainable yield is the amount permitted in any year by the collective favorableness (or unfavorableness) of the uncontrollable weather factors operating during that growing season. Correlations over several years are sought between yield and all separate weather or time factors, the most consistent being selected for integration by means of a simple analogue device.

The Auchincruive simple integrating computer is electric (Figure 1, *I*), with several variable resistors each representing an uncontrollable factor of the environment. All the resistors are wired in series, the apparatus is powered by a battery of adjusted standard voltage, and the collective result indicates on a scale of yield as weight per unit of land area. Calibration of each linear resistor is made according to the calculated regression line of the correlation it portrays [Figure 1 for barley; Grainger (13) for potato].

Correlations are derived from long-term results of "crop and disease phenology plots" (see above) in which all growth factors under human control are made as favorable as possible (as should happen in farming practice). The uncontrollable factors that affect yield can then be separated from all others, and while single correlations may have little significance, the integrated result (Figure 1, *II*) is highly significant. Attainable yield is a theoretical value only in that it is never measured by the farmer, but it is a very practical alternative approach when we seek to get our scientific proportions right. It is the only way of assessing yield alone without any modifying factor such as disease, but we can still use it comparatively. If, for example, we obtain a highly significant correlation ($r = + 0.883***$) over 14 years between the farmer's yield of barley (with slight foliage disease) and the respective attainable yields integrated as free from disease, any effects of the disease on yield must be minute.

Potato yield at Auchincruive was not, over 23 years, well correlated with date of first blight nor with that of 100% disease as would be expected from the complete foliage destruction it causes. Yield depended greatly on the suitability (or unsuitability) of weather before and during the times when blight was present. Attainable yield integration, however, used both dates of blight phenology along with measured values of temperature and bright sun occurring in each year before and after the physiological barrier phase (Cp/Rs below 1.0), respectively. It then accounted extremely well ($r = + 0.894***$) for all the vagaries of yield of susceptible cultivars in relation to blight attacks (13).

Work with blight resistant potato cultivars was limited to a few years before they lost resistance: Yield was about twice that of susceptible kinds, but still varied between years according to temperature and bright sun. A special circuit on the analogue device coped with this desirable but temporary condition of blight resistance.

The slight economic effect of cereal leaf diseases and the heavy effect of potato late blight were first postulated from observation and preliminary field experiment. It needed, however, the concepts of attainable yield and of integrated truth to confirm and measure more exactly. We then encounter a new clarity of experimentally verified thought which can be extended to the separate estimation of disease, and further to the effect of an estimated amount of disease on an assessed disease-free yield.

The severity of all parasite attack on foliage   This depends on factors in the disease triangle (above), with parasites reacting differently from host plants to items of weather environment. Factors affecting diseases can again be separated from all others and integrated by a separate bank of resistors (*III* in Figure 1). One variable resistor (*P*) can be calibrated for Cp/Rs changes and another (*Q*) for weekly groupings of high or saturated atmospheric humidity. Two resistors (*R*,*S*) used alternatively interpret temperature for either "cool" pathogens or "warm" ones. The integrated result is led to a percentage scale on the meter. This gives similar amounts of disease to those found in practice and vindicates, by an alternative approach, amounts of disease assessed by other methods.

The effect of all diseases on yield   This can be integrated to read on the yield scale as a diminished value arising from all diseased conditions. Shortcomings of direct assessment are likely [(15) and above] so we freely use the

---

Figure 1   An electric analogue device for integrating uncontrollable factors that affect crop yield, crop disease, and also the effect of disease on yield. The meter unit has an outer percentage scale for disease measurement and the inner yield scale shown is suitable for crops giving up to about 7500 kg/ha (3 tons per acre) of useful produce. Alternative setting units in this range can be connected via the multi core link; that shown is for barley. I, circuit diagram; II, factors affecting yield alone (rainfall, temperature, bright sun) are set on knobs A, B, C, D and the collective result is led to the meter by pressing the "yield" switch; III, a small amount of foliage disease (6% on the outer scale) integrated by setting knobs P, Q, R, S (see text) and pressing the "dis" switch on the meter unit; IV, the effect of this amount of disease on the yield shown in II, using knobs W, X, Y, Z (see text). Only foliage diseases integrated in III are here set on knob Y, with their time of occurrence expressed by knob Z. Pressing the "Dis/Yield" switch shows that the yield with disease is practically the same as the integrated attainable yield (II, to which it refers), so this amount of disease can only have had a minute effect on yield.

Knobs W and X are used to assess, separately if required or together if needed, other barley diseases, e.g. take-all (X) or those causing seedling mortality during a supersensitive phase (W).

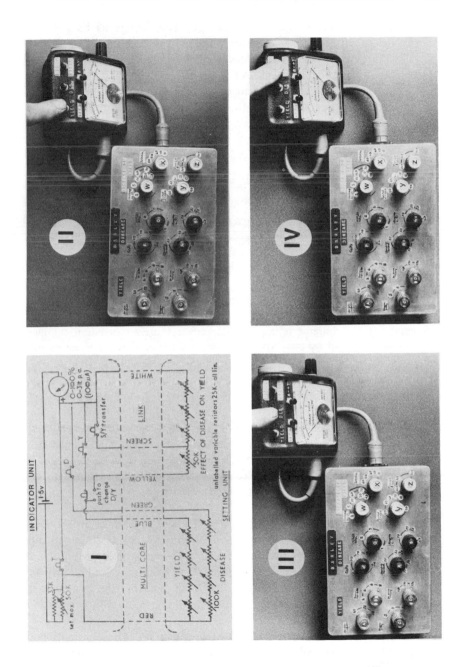

results of quantitatively simulated damage. This brings us back to the farmer's yield, but it is now a *measured* healthy value diminished in known ways by amounts of disease also *measured*. Mathematical significance between actual and calculated values is high e.g., about + 0.9*** for barley with "supersensitive" establishment damage and/or take-all disease.

The barley computer section for this integration (*IV* in Figure 1) includes one resistor (*W*) showing the effect of contrived average crop density on yield. This expresses yield compensation by a crop and thereby permits the assessment of any serious "supersensitive" disease effects during the seedling stage. Another resistor (*X*) is calibrated to read the effects of take-all disease, at the reduced value $Ky = 0.5$ (above). A third resistor (*Y*) expresses the percentage of foliage disease found by integration from III and a fourth (*Z*) corrects for the time of its appearance and also for yield compensation in the individual plant. The last regression is from results of contrived defoliation damage maintained from several given dates: We only get limited information from natural disease occurrence, as hosts are unreceptive in mid growth (above).

Collective results confirm that foliage diseases on barley in west Scotland have minute effect on yield, but take-all disease has a more serious action. "Supersensitive" damage is a potential rather than a consistent menace, adequately assessed by crop density studies, but local work is necessary.

The use in integration of averages from scatter diagrams   Most data from yield and disease studies are shown in the form of scatter diagrams. They must first be averaged—linear correlations as calculated regression lines and nonlinear ones as block averages to make a graph—for calibration as finite values. Scattered results show natural variability and final results after integration do not show any diminution in either scatter or variability. The only difference is that the effects of component variables have been identified and measured during the integration process.

LOSS AVOIDANCE BY CROP CHANGE   Turnips and silage are of about equal annual value for the same purpose of winter stock feed, but turnips may have serious club root disease while silage has no pathological problem. Silage can, however, be rather a wasteful method of storage, and quality hay with equal agricultural value and no disease is probably the best replacement for turnips as it does not diminish the value of overall production.

POTENTIAL DAMAGE TO A SUBSEQUENT CROP   The elimination of pathogen transfer by seed is among the more potent of our present contributions to farming practice, but this depends on an early harvest date. Part of the 1956 Auchincruive oat harvest was delayed from 12 August (normal) to 9 September. The amount of leaf stripe *Helminthosporium* (= *Drechslera*) *avenae* only rose from 7.5% (normal) to 9.9% (delayed), but the amount of seedborne stripe increased from less than 1% (normal) to 9.8% (delayed). Seedborne *Helminthosporium* kills oat seedlings during the supersensitive phase of the next crop, but the economic effect of 9.8% seedling loss is scarcely noticeable because of yield compensation (above). Moreover, the oat yield of 1957 was 4260 kg/ha (1.7 ton/acre), the highest of any during 15 years: Such is the value of cross-checked measurement.

MARKET LOSS   This is perhaps the easiest economic loss to measure and the hardest to interpret. Comparative samples of healthy and diseased produce are submitted to a market or to a grading point, when the percentage change in value of diseased produce is Km. This is usually a positive value when used as a constant to modify $\%D$ (above), and when used to correct other constants must be calculated to give a value above unity in order to express more loss. Rarely, however, it is negative (e.g. to express the greater consumer attractiveness of tulips "broken" by virus over that of the healthy condition).

There is, however, rarely a progressive relation between $\%D$ and price: In most markets, for example, even 5% of venturia scab halves the price of apples and $50\%D$ has no greater effect. Potato tuber scabs exert no effect on price in a market with slight scarcity, but scabby samples will not be sold at all if there is even a minor glut of produce. Therefore, the long-term effect is the frequency of market gluts, which for potatoes is about 6% of the time in west Scotland, so $Km = 0.06$ (but see below).

Analyses of host produce reveal no discernible difference in composition between healthy and scabbed apples or potatoes once the skin is removed (as in human food preparation). The effect of surface blemishes on price is therefore caused only by appearance to humans. Farm livestock are, fortunately, not so particular and readily eat scabbed produce. Although the price is less—often 25% so—the use of produce as food for farm livestock provides a valuable "safety net" for blemished produce, e.g. on potato. The long-term value of Km is 0.06 (above) $\times 0.25 = 0.015$, i.e. a loss of 1.5% as against a short-term one of 25% and a "glut frequency loss" of 6%.

There are different calculations according to whether the market loss is sustained by the diseased produce alone (e.g. club root of crucifers) or by the produce which survives a yield loss Ky (e.g. black currant reversion virus). Complete relations (11) are all arithmetic and amend percentage market losses according to Ky, to give corrected values of Km. They look formidable, but fortunately most of the common ones are among the lower values.

STORAGE LOSSES   Disease in stored products is even more serious than disease in the field. There is no yield compensation and the percentage loss is always at least equal to the percentage of disease and may be very much more. The loss is always one of market value, though yield may also be additionally and adversely affected. A very small attack (e.g. of mold on grain) may decrease the price for both seed and human consumption, but much depends on whether the attacked produce can be marketed separately. Risks of farm storage are rewarded with a higher price at the end, but the operation is always an economically chancy business.

Successful storage demands an early and comprehensive approach [e.g. Clark (4) for potatoes]. If the potato crop is not to be stored, protective sprays to counter phytophthora blight on the foliage may be continued until maximum yield is attained. At Auchincruive, however, blight appears on the tubers 10 to 14 days after it first appears on the foliage. Every tuber with a blight lesion becomes a total loss in storage, so the best yield of stored produce is given by only one protective spray followed by a destructive one (4, 13).

CURING LOSS   The fungus *Alternaria tenuis* develops on the harvested leaves of tobacco during the process of flue curing. C.E. Main (personal communication, 1971) has developed a moving-scale graph (an analogue device) to assess particular economic effects, and approaches control by collective measures to reduce disease during growth.

REPLACEMENT OR RESTITUTION   Fruit trees and bushes that are severely incapacitated or killed by diseases, e.g. black currant by reversion virus, apple by Nectria canker, may be worth replacement in a permanent plantation. Such replacements not only regain full production over a period of years, but give diminished though improving yield in the meantime. Thus a black currant bush takes five years to achieve full production, but a graph of the yield shows that only the equivalent of 3.3 years' full production is actually lost, so $Kr = 3.3$ (11). Comparable figures for bush apples are 7.6 over 10 years. These are heavy losses but would have been worse if replacement had not been made. For example, take 25%D (bushes affected); Ky for reversion virus $= 0.87$:

No replacement   0.25 (%D) $\times$ 0.87 (Ky) $\times$ 5 (years)[4] $= 1.0875$ relative loss.

Replacement   0.25 (%D) $\times$ 3.3 (Kr) $= 0.825$ relative loss over 5 years.

Note that with replacement, the total loss amounts to 3.3 years of full production, so there is no need to make other corrections to %D in this particular case.

---

[4]Km has been omitted here for simplicity.

There would also be great potential spread of reversion virus with no replacement, but very much less from healthy replacements.

A severe soil infestation of potato cyst eelworm *Heterodera rostochiensis* can span 30 years in west Scotland. For 8 yr rotations $Kr = 30/8 = 3.75$ in theory but 4 in practice, so it would take at least 32 yr to starve out the population. During this time the farmer would lose the value of 4 potato crops (each averaging £494 per ha per annum), but he could grow, say, cereals at £148 per ha per annum on the land being rested from potatoes. Kr then becomes $4 \times 148/494 = 1.2$ instead of 4, so there is respite even in this serious situation.

COMBINATIONS AND COMPLETE CONCEPTS

"Joint Damage" or "Disease Complexes" and Control

Virus complexes (above) are matched by joint damage from various other kinds of organisms. Eelworm and fungus (14), sometimes also with bacteria, all work together additively, perhaps synergistically, to cause severe host damage. Insect, eelworm, and fungus vectors of virus diseases often involve close relations and some may be disease complexes.

A plant pathologist must be able to recognize joint damage largely from appearance, and to specify the constituents by the simplest and quickest methods. He must expect fungal and bacterial partners to retire from action when host Cp/Rs is low during midgrowth. The potato host, for example, makes palliative "hunger roots" when the fungus *Rhizoctonia solani* is thus temporarily inactive, to leave cyst eelworm *Heterodera rostochiensis* as the sole but reduced partner.

One partner usually stands out as the major one, to invite control, but as the damage is joint it would seem reasonable to give joint control. There is evidence that a nematicide plus a fungicide is more effective than a nematicide alone (though not yet economic enough) for the control of "potato cyst eelworm" when it is really a complex with *R. solani.*

All this and much more in this review leads to the necessity for a wide, multidisciplinary approach to our subject.

SELECTION AND COMBINATION OF DISEASE CONTROL METHODS
A scientific adviser must be able to specify the best methods of comprehensive control (chemical, biological, or costless) for all the diseases likely to be encountered in the area he serves (13). Palliatives should even be combined with more positive and complete methods. Biological methods are likely to be more widespread and effective in warm regions than in cooler ones, and the important practical work of method selection will usually only be made by independent scientific advisers.

Leadership with Chemical Control

Chemical sprays have long been a preferred control for all kinds of crop parasites. They are not always highly efficient and are hedged about with limitations, dangers, difficulties of precise timing etc. They were historically first in the field and are greatly versatile, but cost money, time, and work. They are percolated with vested interest, not least of the farmer himself who wants to be "up and doing" when parasites might blight his crops. Some costless controls (above) are very efficient—sometimes more so than chemicals—and the scientist will doubtless discover more. We are, however, likely to need chemical controls for some time to come, and a phytopathologist has many ways of making them more effective.

PERFORMANCE AND COST   With any control measure, we have to think in terms of the proportion of the population left for future buildup, and after treatment a method giving 97% control is nearly twice as good as one affording only 83%, so we need a ready measure of performance. The area below a graph of untreated disease development by time from that of zero disease to that of maximum amount, is directly related to the amount of disease to be controlled is taken as 100. A second graph of disease development over the same time interval when control *is* given, is superimposed over the first (vertical scales being equal), to mark out a new and almost invariably smaller area. This area, when referred to that taken as 100, gives the percentage performance of the control material and method. The higher the performance the better, but some quite low values have been recorded.

Performance and cost of a material under test must, however, be considered together, but we have to compare them with the performance and cost of the best standard substance we know:

Performance cost =

$$\text{Actual cost} \times \frac{\text{Percentage performance of standard substance}}{\text{Percentage performance of test substance}}.$$

It is even more realistic if the cost of application is included in the actual cost.

The performance cost relation is inverse—the *lower* it is the better—which most people find rather difficult, for substances with low performance cost usually have high actual performance, and there is no substitute for heavy thinking! This economic value nevertheless tells us the relative price we are paying for the degree of efficiency we select. If we test chemicals with dual action (e.g. to control disease on both potato foliage and tubers) or systemic materials like benomyl, both performances and performance cost are likely to show very large favorable differences from single-effect materials.

It is quite inadequate to show that a substance exerts *some* control: Only *the best* is good enough and the scientist has a clear mandate to lead.

HIGHLY EFFECTIVE COMBINATIONS   The best soil control of meloido-gyne root knot eelworms in British greenhouses is an injection of DD (to treat the soil in depth) followed immediately by a drench of solubilized xylenol (to treat the surface). This gives 97% control while the best single material only inactivates 92% of the population and costs about twice as much as the dual treatment. Only an independent scientist is likely to initiate such a combination, for the two materials are made by different firms, and one is now no longer marketed.

COMPREHENSIVE DISEASE FORECASTING   Prior knowledge of a disease outbreak automatically confers maximum performance on protective foliage sprays and many other treatments. If hosts were treated too long before the parasite appeared, foliage unemerged at that time would have no protective cover later, while sprays after disease had developed could not alleviate damage already caused. All ground-operated spray machinery, moreover, itself causes damage to the treated crop, and it is again important to time the spray so that there will be no redundant applications before disease appears.

A forecast is valid only if it is comprehensive. This is illustrated by an example with potato late blight: (a) Is a forecast needed? Only if local crop growing temperatures lie within the fungus' range (10 to 23°C); (b) Is the parasite present in an area? Use indicator plots of susceptible potatoes established at different times of the year; (c) Has weather suitable for the parasite occurred? Use "Auchincruive" self-calculating recorders calibrated for either Beaumont or Bourke specification of suitable forecast weather; (d) Is the host crop receptive? Use Cp/Rs to define the physiological barrier nonreceptive phase (a rapid method has been found to work in Scotland); (e) Will blight development be rapid or slow? Use Cp/Rs at first blight, also genetic factors and temperature change; (f) Has there been one day of continuous bright sun (over 11 hr)? This can inactivate a late blight epidemic in its early stages; (g) Are there any changes in host/parasite biology?

Many forecasters still use only c, with consequent poor performance. Item f could be forecast by meteorologically synoptic methods which would add precision. The condition is, however, easily recognized by advisers, farmers, and others, but necessitates a rather cumbersome postponement of spraying until the next forecast has occurred. Forecasts have to be made by scientists, but can lead to very important economic decisions as to when to spray. They must however, be fully valid, for their correct application in practice demands great courage, to hold off when other farmers are spraying wastefully.

CORRECT CONTROL MACHINERY   Large insect parasites or weeds with expanded leaves may be controlled sufficiently if a relatively large amount of contact or systemic toxicant falls upon them. Minute fungal spores, however, require minute but numerous drops merging into complete cover. This is supplied by high volume water-diluted spray or by high volume air carriage of well-dispersed minute drops. Any attempt to reduce the overall volume of diluent in this context tends to reduce spray efficiency, though the long-term success of new ultralow volume and fogging methods for greenhouse crops remains to be assessed. Special micronized dusts in a proper machine can be as effective as high volume water sprays, and are much cheaper. A farmer should have his own sprayer and purchase only chemicals with long shelf life.

Soil treatments with nonvolatile materials are made highly effective when mixed very intimately with the top 9 in (23 cm) using the "Auchincruive" soil disease control unit (10). This achieves an incorporation within ± 20% of the required dose, and no other extant field device has so far achieved this.

TREATABLE AND UNTREATABLE SOIL PARASITE POPULATIONS Most field methods of soil disease control do not give a high degree of reduction in parasite population—75% is a common value. Populations of many soil-inhabiting eelworms it is desired to control can, however, be measured quantitatively. With potato cyst eelworm *Heterodera rostochiensis,* a post-treatment population that gives potentially recurrent benefit is 0.3 live cysts per gram of soil. A 75% treatment could reduce to this value an original population of no more than 1.2 live cysts per gram, a level below that of the optimum for increase in a population after treatment.

If the existing population is larger (it often is), a scientific adviser should be able to specify the time in years of fallow with no susceptible host required to bring it down to treatable level. The necessary years of fallow change with district, according to soil temperature and soil water, and this is also necessary knowledge for combinations of chemical treatment with long rotation.

A farmer's economic success or failure in soil disease control here depends entirely on scientific measurement.

LET EXPERIMENTS SIMULATE FARMING PRACTICE

Artificial experimental conditions adopted by many phytopathologists, often even recommended in courses of study, rob our science of its full practical impact. Discard all results from plants growing in those beloved pudding bowls, seed trays, and any containers that restrict root development, and observe all the specifications of good management described above. Plants grown in artificial conditions are often very receptive to disease; I found young tomato plants with Cp/Rs around 1.9 for long periods during which well-grown commercial crops were nonreceptive (Cp/Rs below 0.5). Such artificial experimental conditions usually give neat results with comfortably high mathematical significance, but *they have no practical or economic use whatsoever*. Results of greenhouse experiments conducted in winter are suspect quantitatively: Temperature changes are wrong for summer daylength, while artificial illumination is a poor substitute for full sunshine, and soil water is often inadequate.

Crop pathologists obtain their best information from plots that are as similar as possible to the farmer's own conditions. Differences or correlations revealed by the results are often of low mathematical significance, and it is only when we integrate all the relative factors (e.g. "Attainable yield" above) that we achieve significance—extremely high significance. A scientific result that succeeds in farming practice has to survive a collective test of significance far more rigorous than any single mathematical one! If we extend our farm-simulation plots over several years and the results are still consistent, we have increased the amount of circumstantial evidence arithmetically, but its significance is enhanced exponentially. The concept ceases to be a theory and becomes a truth.

Practical Appreciation of a Farmer's Work

Nothing can give us "scientific proportion" more potently than a year or two of practical work on a farm, preferably before any scientific training. We should then be unlikely to indulge in artificiality of thought or action, while the necessity for a multidisciplinary approach would become clear. A narrow specialist cannot give well-balanced advice: His subject looms too large in his assessment, whereas farming requires a wider outlook and integrated truth.

Much of the work we do at present is of direct value to the farmer, but in other activities the scientific proportion of our approach needs basic change, while many practical results from earlier workers still lie fallow in our literature.

Adjustments for World Environment

It would be very bad scientific *dis*proportion to accept on a worldwide basis without local research, all the examples I have given in this review. Though many of the principles *are* of worldwide application, there is no substitute for the hard work necessary to investigate and interpret local actions.

Part of a Whole Scheme: "Maximum World Crop Production"

Retirement from my responsibilities at the Agricultural College provided the leisure necessary to bring together the factors that affect the physiology, pathology, and agronomy of crops. All these integrate with environment, genetics, soil science, etc in a most satisfying manner and a large book with the title given above is now almost ready for publication. It justifies completely the smaller field of new phytopathology presented in this review.

The Economic Value of a Correctly Proportioned Advisory Service

The ratio of benefit to cost was calculated for my former department in 1968 as about 12.5 to 1, a most suitable return for anyone's money (13). This ratio must inevitably rise as the value of presently controllable loss is enhanced by research and a larger proportion of it *is* controlled, stimulated by the general advisory services.

THE CHALLENGE

I ask every plant pathologist to reequip his or her department so that it will be able to study host, parasite, and environment equally and then to think about and interpret the result within a holistic framework. This may require changes in experimental methods to reveal your own local long-term detail. Throughout history, such invitations to real thought and new action have always been countered by a desire to destroy the author—and that will be the last of him!

No Need to Shoot the Author

Never mind in my case; I am old and Nature will soon do it for you. There are, however, advantages in being old, for I have grappled with the questions that you will sometime ask, and true scientific proportion needs a long time to prove its worth. This has been vouchsafed me and now I pass the results to you.

ACKNOWLEDGMENTS

The author gladly acknowledges help from the West of Scotland Agricultural College Governors, particularly for permission to use long-term results and for the loan of apparatus and equipment to develop, after retirement, the work shown by Figure 1 and its accompanying text.

John Grainger was one of the creative thinkers of plant pathology. He advanced the frontiers of disease loss management and in many other ways made plant pathology useful. These few lines were written by Marjorie Clark to help celebrate his life.

The Editors

JOHN GRAINGER
1904–1978

Dr. John Grainger died on October 20, 1978 at the age of 74 after a long illness. He was Head of the Plant Pathology Department at the West of Scotland Agricultural College, Auchincruive, from 1943 until his retirement in 1969. During this time, he consolidated and added to the work of his predecessor Dr. O'Brien in founding the modern department specializing in research, development, and advisory work in crop protection. He extended his interest in assessment of crop disease losses and control from its detailed approach in West Scotland to an appraisal of the position and opportunities in other countries including those of the Third World. This was recognized in his delivery of a paper at the FAO Conference in Rome in 1967, only one, however, of many such

lectures delivered all over the world. His interest in nematology and practical knowledge of engineering, both rare attributes in general plant pathologists, led to development of machines for intimate mixing of small quantities of pesticides into the soil, and to the award of silver medals at the Highland Show in 1951 and the Royal Agricultural Show in 1959. The machines were also shown to the Seventh Symposium of the Society of European Nematologists when it was held under his auspices at Auchincruive in 1963. His Auchincruive Blight Forecast Recorder is still in use widely. His concept of the physiological receptivity of plants in terms of the "C_p/R_s" ratio elucidated certain apparent anomalies in disease behavior and highlighted the importance of the crop response in host-parasite relationships.

Dr. Grainger's geniality and friendliness, no less than his special and original contribution to his chosen field of plant pathology, were so widely recognized as to merit the rare distinction of an entry in the *World Who's Who in Science* (1968). He was elected an emeritus member of the American Phytopathological Society shortly after his retirement. The editors of the *Annual Review of Phytopathology* requested this article from Dr. Grainger in order to give wider recognition to his practical and scientific contributions to the farming industry. The publication of this article in the seventeenth volume of this Review and Dr. Grainger's soon to be published book *Maximum World Crop Production* will provide enduring reminders of a man of great character and inspired scientific achievement.

Dr. Grainger is survived by his wife to whom he was so happily married for 47 years.

M. R. M. Clark

Literature Cited

1. Chamberlain, E. E., Atkinson, J. D., Hunter, J. A., 1964. Cross-protection between strains of apple mosaic virus. *N. Z. J. Agric. Res.* 7:480–90
2. Channon, A. G., Cheffins, N. J., Hitchon, G. M., Barker, J. 1978. The effect of inoculation with an attenuated mutant strain of tobacco mosaic virus on the growth and yield of early glasshouse tomato crops. *Ann. Appl. Biol.* 88: 121–29
3. Clark, M. R. M. 1967. Joint damage to hyacinths by *Penicillium cyclopium* Westling and *Pseudomonas marginalis* (Brown) Stapp. *Meded. Rijksfac. Landbouw wet. Gent* 32:807–14
4. Clark, M. R. M. 1970. Potato storage diseases. *W. Scotl. Agric. Coll. Res. Bull.* 15 pp.
5. Deleted in proof
6. Cramer, H. H. 1967. *Plant Protection and World Crop Production.* Bayer. *Pflanzenschutz Nachrichten,* p. 488. Leverkusen, Germany. 524 pp.
7. Cruickshank, I. A. M., Perrin, D. R. 1965. VIII. The effect of some further factors on the formation, stability and localization of pisatin in vivo. IX. Pisatin formation by cultivars of *Pisum sativum* L and several other Pisum species. *Aust. J. Biol. Sci.* 18:829–35
8. Grainger, J. 1956. Host nutrition and attack by fungal parasites. *Phytopathology* 46:445–56
9. Grainger, J. 1956. The economic effects of crop disease. *Res. Bull. W. Scotl. Coll. Agric.* 16:80 pp.
10. Grainger, J. 1958. A new soil disease control unit. *Res. Bull. W. Scotl. Coll. Agric.* 25:46 pp.
11. Grainger, J. 1967. Economic aspects of crop losses caused by diseases. *Pap. FAO Symp. Crop Losses.* Rome, Oct 2–6 1967, pp. 55–98
12. Grainger, J. 1968. Cp/Rs and the disease potential of plants. *Hortic. Res.* 8:1–40

13. Grainger, J. 1969. The reduction of crop losses in west Scotland. *Res. Bull. W. Scotl. Coll. Agric.* 43:79 pp.
14. Grainger, J., Clark, M. R. M. 1963. Interactions of *Rhizoctonia* and potato root eelworm. *Eur. Potato J.* 6:31–2
15. James, W. C. 1974. Assessment of plant diseases and losses. *Ann. Rev. Phytopathol.* 12:27–48
16. Kamoen, O. 1972. *Patogenese van Botrytis cinerea op Knolbegonia.* Doctorectsproefschrift Fac. Landbouw. Ghent. 105 pp.
17. LeClerg, E. L. 1964. Crop losses due to plant diseases in the United States. *Phytopathology* 54:1309–13
17a. Link, K. P., Walker, J. C. 1933. The isolation of catechol from pigmented onion scales and its significance in relation to disease in onions. *J. Biol. Chem.* 100:379–83
18. Millard, W. A., Taylor, C. B. 1927. Antagonism of micro-organisms as the controlling factor in the inhibition of [potato] scab by green manuring. *Ann. Appl. Biol.* 14:202–16
19. Miller, P. R., Pollard, H. L. 1976. Multilingual compendium of plant diseases. *Am. Phytopathol. Soc.* 457 pp.
20. Müller, K. O. 1950. Affinity and reactivity of angiosperms to *Phytophthora infestans. Nature* 166:392–94
21. Müller, K. O., Börger, H. 1940. Experimental studies on the Phytophthora resistance of the potato together with a contribution to the problem of 'acquired resistance' in the plant kingdom. *Arb. biol. Reichsanst. Land Forstwirtsch. Berlin Dahlem* 23:189–291
22. Padwick, G. W. 1956. Losses caused by plant diseases in the colonies. *Commonwealth Mycol. Inst. Phytopathol. Pap.* 1:33
23. Stapley, J. H., Gayner, F. C. H. 1969. *World Crop Protection. I. Pests and Diseases.* London: Iliffe

Ann. Rev. Phytopathol. 1979. 17:253–77
Copyright © 1979 by Annual Reviews Inc. All rights reserved

UNIQUE FEATURES
OF THE PATHOLOGY
OF ORNAMENTAL PLANTS

♦3710

Kenneth F. Baker and R. G. Linderman[1, 2]

Ornamental Plants Research Laboratory, US Department of Agriculture, SEA, AR, Oregon State University, Corvallis, Oregon 97330

This paper is concerned with the vast and diverse array of ornamental plants which have high unit value and usually are short-term crops intensively cultivated on small acreages. The first paper (30) in this series on unique features of the pathology of various crops dealt with a relatively few types of forest trees which have comparatively low unit/year value, and are long-term, large-acreage crops receiving limited cultivation. These two papers thus present polar extremes of crop types and of applied plant pathology.

Webster's *Third New International Dictionary* defines an ornamental as "a plant cultivated for its beauty rather than for use." Ornamentals have been referred to as "amenity crops" (22), in the sense that they contribute "to physical or material comfort or convenience or to a pleasant and agreeable life," but this overlooks their usefulness. Ornamental plantings used to control erosion, to provide shade and ameliorate the environment around a home through control of temperature, wind, and traffic noise, to reduce headlight glare on highways, and to help purify the atmosphere of pollutants have become necessities of modern life.

---

[1]M. T. Fossum kindly provided data on the value of the industry. Figures for 1977 were calculated from the 1970 percentages of each crop type (24) and the floriculture crop value for 1977 (25).

[2]Mention of a trademark, proprietary product, or vendor does not constitute a guarantee or warranty of the product by the US Department of Agriculture and does not imply its approval to the exclusion of other products or vendors that may also be suitable.

253

0066-4286/79/0901-0253$01.00

THE ORNAMENTAL PLANT INDUSTRY

The wholesale value of ornamental crops reportedly doubled at seven year intervals for the past two decades; in 1977, US florist and nursery plants were estimated to be worth nearly $1.6 billion (25). It is estimated that these plants generate $4.8 billion in retail sales. This industry includes an extraordinary group of crops of the following six general types.

1. *Floriculture crops* include cut flowers, pot plants, and foliage plants produced in glass-, cloth-, or lathhouses, or outdoors. The flowers and some plants are discarded after use. Commercial production of cuttings for growing-on is also included in this category. In 1977 the wholesale value of US floriculture crops, representing 62.2% of the total industry, was estimated to be $1.0 billion (25). The wholesale value in 1970 was $485 million in nearly 8000 establishments, principally in California, Florida, Ohio, Pennsylvania, New York, Colorado, Michigan, Illinois, New Jersey, and Massachusetts (24). Other than an 881% increase in foliage plants, there was only modest increase in floriculture crops in the period, 1970–1977.

2. *Bedding plants* grown under glass for outdoor planting are largely annuals that are replaced yearly. This sector of the industry is expanding rapidly, and the wholesale value in the United States in 1977 was estimated to be $81.5 million, exclusive of vegetable transplants ($31.5 million) (19). This is an 80% increase over the 1970 value of $44.8 million (49).

3. *Nursery crops*, including trees, shrubs, vines, and perennials, are produced outdoors and transplanted to gardens and parks. Although the nursery life of most of these plants is but 1 to 5 years, they are long-lived and remain for many years once transplanted. Accordingly, they command relatively high prices. Some nursery plants are started as cuttings in glasshouses and then transplanted to the field or outdoor containers. The wholesale value of these crops in the United States in 1977 was estimated to be 36.3% of the industry or $576.8 million (24, 25).

4. *Bulb crops,* including bulbs, rhizomes, tubers, and corms are produced outdoors and sold for long-term planting in gardens and parks, or for short-term planting for cut-flower production of for pot plants (type 1, generally discarded after use). The wholesale value of this group of crops in the United States in 1977 was estimated to be 1.3% of the total industry or $20.8 million (24, 25).

5. *Seed crops,* including annuals, perennials, and grasses, produced in fields or under glass (high-value specialty crops) and sold for producing bedding plants (type 2), flowers (type 1), or turf (type 6), or for direct seeding in gardens or parks. Most of these plants are annuals that are discarded after use. The wholesale value of this group of crops (exclusive of grass seed) in the United States in 1977 was estimated to be 0.2% of the

total industry or $3.2 million (24, 25). However, California, the principal seed-producing state, alone reported $5.3 million in 1975 (43).

6. *Turf grasses* produced in turf or sod farms or in nurseries, are cut as a layer and transplanted to yards, parks, golf courses, and athletic fields where they become perennial lawns. In 1977, 1200 companies in the United States reportedly produced 135,000 acres of sod valued at $225 million (51). Turf was estimated at $11 million wholesale value in California in 1976 (43).

EVOLUTION OF THE ORNAMENTAL PLANT INDUSTRY

The early production of ornamental plants developed as a sideline to growing vegetables, fruit and nut trees, and field crops. With increasing demand, companies were formed that specialized in producing seeds, bulbs, and nursery stock; in the United States this stage was reached by the mid-1700s (11, 29). Training was by the apprentice system, imparting rules-of-thumb arrived at by trial and error. These "secrets" often only represented a practice used on a particularly successful crop, although that practice may have had nothing to do with the success of the crop. For example, the success of such a crop may actually have been due to the fortuitous absence of soilborne pathogens usually present.

The rule-of-thumb system continued until the twentieth century, and still persists among some untrained growers. "Root action," evidenced by new white root tips when the plant was knocked out of the container, was emphasized in growing, particularly in fertilizing and watering practices. This useful concept unfortunately has declined with the increase of technical knowledge and methods.

With the adoption of technological improvements in growing and marketing, particularly since the 1940s, a higher level of skills and training has developed in the industry. The atmosphere of constant improvement in growing practices has resulted in better growers; "how" has been replaced by "why." Agricultural Experiment Stations and Extension Services have played important roles in this improvement through their research, publications, and short courses (3).

Because of relatively limited rapid transportation prior to the 1930s, each grower tended to propagate and grow most of the plants he marketed. Such extensive and generalized production lessened the opportunity for developing specialized knowledge of each crop. Individuals in the industry have now shifted to intensive specialized production of a few crops, with cuttings, seedlings, and lining-out stock obtained from specialist propagators for growing-on or finishing. This shift has given rise to large companies concerned solely with production of propagules. Many crops (chrysanthemum, carnation, poinsettia, geranium, azalea) are now produced to a large extent from propagules from one or a few sources. While this led to improvement in quality, it exposed the industry to the possibility of disease epidemics, some of which are discussed below.

Disease Epidemics and Grower Response

The chrysanthemum, a short-day, fall-flowering plant, has long been grown in gardens and as a florist cut-flower. In the 1930s foliage diseases of this plant were controlled by fungicide applications, and verticillium wilt was minimized by using one of the few available resistant varieties and by careful irrigation to avoid overwatering.

In 1943, Dimock (20) developed the cultured-cutting technique to provide *Verticillium*-free chrysanthemums for research purposes. This method was commercialized by Yoder Brothers Inc., Barberton, Ohio, who parlayed it into a large business. Soil fumigation came into use about that same time and, in conjunction with pathogen-free cuttings, provided *the* method for control of verticillium wilt. Since resistance to *Verticillium* was no longer necessary, the number and types of available varieties increased. Commercial day-length manipulation to control chrysanthemum flowering came in the late 1940s and extended the chrysanthemum season. Precise dependable scheduling of the crop was then a reality, but the buying public regarded the chrysanthemum as a fall flower and was slow to accept it in other seasons. Acceptance was general by 1970, however, and year 'round pot chrysanthemums became big business.

Concentration of propagation in a single company set the stage for, and also the recovery from, the national epidemic of the "stunt disease" in 1947–1950. This disease is a prime example of the rapid rise and decline of an epidemic of a major crop. The disease, first observed in 1945 and still unimportant in 1946 (21), appeared nationally in the United States in 1947, so severely reducing plant size and flower quality that the chrysanthemum industry was seriously threatened (8, 9, 33). From 30 to 100% of the plants were affected; 90% infection was common (12). The highly infectious viroid that caused the disease was spread by contact during routine handling, had a 3 to 8 month incubation period, and conspicuous symptoms appeared only at flowering. Therefore, plants kept vegetative for cutting production became infected during production handling, but remained symptomless. Since a high percentage of the cuttings used commercially in the United States then came from one company, and symptoms could not be detected until the cuttings flowered, the disease appeared more or less simultaneously throughout the United States in 1947–1948, causing widespread concern.

Before the causal agent was identified, the company established a strict system for production of cuttings aimed at eliminating transmission of whatever agent might be involved (12, 44). So carefully was this program carried out that cuttings distributed in late 1949 carried little infection, and by 1950 the disease was reduced to minor importance (10). The staff of the company (Yoder Brothers Inc.) and the research team involved (A. W. Dimock, J. R. Keller, and Kenneth Post at Cornell University, and Philip Brierley and F. F. Smith of the US Department of Agriculture in Beltsville) deserve great credit for this remarkable accomplishment. It should be emphasized that, although this concentration of cutting production in one establishment largely made the epidemic possible, this same concentration made possible the rapid production and distribution of clean stock and the abatement of the disease. In general, the ornamental industry is financially and organizationally able to rapidly take the necessary steps to deal with an epidemic.

Another example of the sudden occurrence and disappearance of a major disease epidemic of an ornamental crop was the cylindrocladium disease of glasshouse azaleas. The occurrence of the disease in the mid-1960s coincided with the rapid expansion of the flowering pot-azalea as a florist item, due largely to development of many new cultivars with horticulturally desirable characteristics (e.g. large flowers, compact growth, and good forcing predictability).

Cuttings were propagated, and liners were shipped primarily from large growers in the southeastern United States to growers in many other states to be forced. Unfortunately, several of the new cultivars were especially susceptible to *Cylindrocladium scoparium,* a pathogen first described on azaleas in 1955 but quiescent until the 1960s. During shipment and subsequent forcing, many of the liners suddenly began to wilt. Since the wilt and root rot symptoms were similar to those caused by *Phytophthora* spp., diagnosis, and therefore control recommendations, often were incorrect. Before long, the industry was faced with a disease that had been spread throughout the United States. Linderman (35) demonstrated in the late 1960s and early 1970s that the leafspot disease, which occurred primarily in the southeastern areas where propagating was done, and the root rot wilt disease, which occurred in other areas where the plants were forced, were two phases of the same disease simply separated geographically. Some 30–60% of the liners being shipped were infected but symptomless. Infections had occurred during propagation, but not to the extent that the infected plants would be detected and discarded. These facts pointed the way to appropriate control measures. Highly susceptible cultivars were eliminated from the trade where possible, no infected cuttings were used in propagation, stock plants were sprayed regularly with foliage fungicides, and softwood cuttings were taken from production plants that had been less exposed to aerial inoculum than had stock plants. The application of the new systemic fungicide, benomyl, also played a key role in checking the dispersal of infected but symptomless liners. By 1973–1974 the cylindro-cladium disease had practically disappeared, but in the process several azalea growers went out of the azalea business.

Growers of rose rootstocks in Washington, Oregon, and California suffered from a disease panic in 1929–1932. Although rose mosaic had been present in commercial stocks without causing serious losses, papers published (42, 54) in 1930 suggested that damage might be far greater than had been thought. A brief survey (42) of all rose understocks on the Pacific Coast reported that 10 to 100% of the plants were virus infected. Michigan rose growers met in 1930 and, after hearing R. Nelson present the results of his survey, adopted a resolution stating that "Grafting stock of Roses grown in North America . . . may be and apparently is largely affected by this disease; and whereas foreign-grown stock seems to be quite free of the disease," they requested "the State Board of Agriculture to investigate the situation and, if deemed necessary, to proclaim a quarantine against Rose stocks coming into the state of Michigan unless certified as clean. . . ." (56). White (55), Milbrath (38, 39), and Weiss & McWhorter (52) immediately expressed doubts about whether symptoms of the virus could have been distinguished from insect and other injuries during the survey, and whether foreign stocks were actually free of the virus.

Eastern growers of glasshouse roses, who were then grafting commercial varieties on these rootstocks, widely canceled orders for material from the West Coast and used unsurveyed stocks from France. The incidence of mosaic in glasshouse roses was little affected, but the western growers of rose rootstocks sustained heavy losses. D. G. Milbrath, pathologist with the California Department of Agriculture, was said to have fittingly summed up the brief inspection tour of West Coast roses as "more of an athletic feat than a scientific accomplishment." This situation arose because insufficient knowledge of symptoms of the disease in the field, as distinct from insect and other injuries, led to an inaccurate survey. Apparently even immaturity of understocks at time of digging, a condition that could lead to plant death, was confused with mosaic (13). The rose mosaic problem reappeared in 1977, when an eastern state inspector ordered a rose shipment destroyed because of mosaic.

It is of interest to note that on October 27, 1948, a meeting of northern California chrysanthemum growers similarly discussed the advisability of imposing an embargo against eastern chrysanthemum cuttings because of stunt disease. In this instance cooler judgment prevailed, and eastern cuttings continued to be used.

Improvements in Culture

The chrysanthemum story illustrates how integrated simultaneous developments in plant pathology, horticulture, soils, and marketing may completely change an agricultural industry. These developments, however, must occur as part of the overall picture, and each must be compatible with the others. For example, light-weight soil mixes such as U. C. Mixes, Jiffy Mix, and Peat-Lite came into general use after 1957 (2), making it economic to ship pot plants, even by air. With the longer flower life of potted chrysanthemums, compared with the cut-flower crop, the increased availability resulting from cheaper transportation, and a wider selection of varieties, the year 'round chrysanthemum became the most popular florist crop, surpassing the rose and the carnation. By 1970 the estimated wholesale value of pot chrysanthemums in the United States reached $35 million (24). It should be noted that when year 'round culture of a crop is achieved, the host-free period is eliminated from the disease cycle, and the hazard of a potential disease epidemic is increased.

It is not worthwhile to introduce pathogen-free propagules into the culture of a crop without the concomitant adoption of soil treatments and sanitary procedures. Some growers at first insist that it is not worthwhile to use healthy cuttings because the soil is infested, or that it is not useful to treat the soil because their propagules are infected. For example, healthy geranium cuttings were developed at the University of California, Los Angeles, in the mid 1950s. These were released to commercial propagators in southern California, but the plants from clean cuttings were quickly contaminated due to inadequate field sanitation. The Carefree geraniums, grown from seed, were released by Pennsylvania State University in the early 1960s in response to the need for pathogen-free propagules; they have since been greatly improved, especially in the rapidity of growth. A market for geraniums from cuttings still persisted, however, and several companies now produce cuttings from culture-indexed stock grown under glass instead of in the field. Apical-meristem culture also is now being used to develop virus-free geraniums for cutting propagation.

The increased cost of labor has made mechanization and labor-saving methods necessary. Modern growing tends to be a production-line operation, particularly in the production of bedding plants and pot chrysanthemums. The time required to produce a marketable plant has been steadily reduced. As with mechanized production lines in other industries, vulnerability of the establishment to labor troubles and to disease losses is increased. Since plant materials must be uniform and production needs to be scheduled in such assembly-line methods, slow growth, small sickly plants, or undependable supply are not acceptable. Mechanization therefore depends upon standardized methods for producing healthy plants (2).

The progress from relatively primitive, apprentice-type production, through technological improvements to the modern, specialized, highly technical operation often leads to large corporate business. However, there is still a place in the business for the grower who carefully operates a small establishment, particularly for specialty items that require more attention than a large operation is willing to give.

DEVELOPMENT OF THE PATHOLOGY OF ORNAMENTAL PLANTS

Studies on the diseases of ornamental plants in the United States began about 1880 (3). B. D. Halsted might justifiably be regarded as the first American pathologist of ornamentals because of his many publications from the New Jersey Agricultural Experiment Station from 1890 to 1905.

Cook (17) nonetheless found in 1916 "a very meager literature on the diseases of ornamental plants" because most people did not consider ornamentals to be economically important, and growers had learned to live with plant diseases without help from pathologists. These conditions have changed completely, particularly since the 1940s. White (53) had reported in 1930 that "Plant pathology as related to ornamental plant materials is the most recent field of the science to be seriously considered."

The Census of Agriculture has provided information on the importance of the ornamental plant industry. Growers have become organized and seek assistance on disease problems of their plants. Pathologists have found, once they establish cooperative relationships with growers, that disease problems of ornamentals are unusually rewarding subjects of study.

The number of pathologists concerned with diseases of ornamentals increased greatly as the industry expanded and became an economically significant segment of agricultural production, but most of them still devote only part time to these crops. Apparently R. P. White (1927, New Jersey), P. E. Tilford (1930, Ohio), A. W. Dimock (1938, New York), and K. F. Baker (1939, California) were the first to devote full time to these crops.

Expressed as full-time equivalents, there were 5.4 pathologists involved in research, extension, and teaching in the pathology of floriculture crops in the United States in 1925, and 14 in 1939. By 1957 this had risen to 61.6, involving 91 people, 41 of whom were full time (3). These data excluded workers dealing with woody nursery crops and turf. Figures on the number of pathologists now working on all ornamentals are unavailable, but the number involved in 1957 certainly has been maintained or possibly increased, although not at a rate comparable to the 1944–1957 period. The 1974 *Directory of the American Phytopathological Society* listed 179 US members who indicated activity and/or interest in diseases of ornamental plants; they were mostly affiliated with state Agricultural Experiment Stations, state Departments of Agriculture, US Department of Agriculture, private companies, or were consultants. States having four or more pathologists of ornamental plants were California, Florida, Pennsylvania, Ohio,

Oregon, North Carolina, Wisconsin, Virginia, Illinois, Maryland, New York, Washington, Georgia, and Alabama. During the period 1957–1974, there probably has been a greater increase in pathologists of nursery crops and turf than of floriculture crops.

An increasing number of journals on ornamental crops have appeared that serve scientists, extension workers, and growers. Numerous trade journals publish semitechnical and popular papers on diseases of ornamentals. There have also been an increasing number of journals, annuals, and year-books published by societies specializing in a given crop or group of related plants. Commercial companies now issue technical publications, and State Agricultural Experiment Stations publish journals that present their re-search on ornamental plants. In several states the Agricultural Extension Service, at both state and county levels publishes small magazines provid-ing current information to growers. Many state Florist Associations, begin-ning with Ohio in 1929, usually in cooperation with the state Extension Service, have published magazines reporting new investigations. There were 75 of these various publications in 1957 that contained papers on floricul-tural pathology (3). Papers on the pathology of all ornamentals probably appear today in well over 100 such restricted publications. Since experimen-tal results often appear first in such journals, and all too frequently nowhere else, it is difficult, even for the specialist pathologist, to keep abreast of new developments on diseases of ornamentals. Probably less than one fourth of the US papers on pathology of ornamentals are published in the journal, *Phytopathology*. Other pathologists rarely or never see these restricted pub-lications, and therefore may not realize the volume of research on diseases of ornamentals. However, the important diseases of the 50 major ornamen-tals are as well understood today as those of comparable vegetable, fruit, or cereal crops.

UNIQUE FEATURES OF ORNAMENTAL PLANTS AND THEIR PATHOLOGY

Ornamental plants differ from other crops in a number of ways that have determined the character and evolution of the industry during recent decades, and have directly or indirectly influenced the pathology of these plants. These unique features have placed unusual demands on, and have presented opportunities to, growers as well as pathologists.

Diversity of Crops, and Transitory Cultivars

The complex of crop types described above is further complicated by the diversity of plants included. It is estimated that at least 1100 genera of plants are grown as ornamentals, most of them infrequently or rarely. There were representatives of about 250 genera sold on the Los Angeles flower market in the 1950s; this probably is typical of most large markets. Some genera of ornamentals include many species, and the number of cultivars may be very high. For example, approximately 20,000 varieties of roses (1), 7000 of gladiolus, and 300 of sweet pea have been developed and named. Because of the large number of ornamentals, the minor diseases of major crops and the major diseases of minor crops often have received minimal attention.

New plant novelties are a prominent aspect of ornamentals, particularly in floriculture crops, bedding plants, and seed crops. Breeders of these crops emphasize the introduction of new varieties to the market each year rather than the improvement of existing lines or the development of varieties resistant to diseases or insects. The All American Trials for annuals and roses, for example, effectively emphasize newness rather than quality improvement. This continuing introduction of new varieties tends to shorten the market life of existing cultivars, and to make uneconomic the time-consuming and expensive development of varieties resistant to disease. It may also deter investigations on resistance of ornamentals by plant pathologists.

Because of the large number of closely similar cultivars, there is a tendency to substitute one for another in filling orders; this has been observed among sweet pea varieties, for example. This varietal uncertainty further lessens interest in developing disease-resistant varieties, and brings into question the practical value of publishing studies evaluating disease resistance or tolerance of cultivars of ornamentals to a given disease. This does not imply, however, that such information is useless.

It was found (K. F. Baker and W. C. Snyder, unpublished data) in several years of observations in California sweet pea seedfields, where pea enation mosaic infects all the plants each year, that cultivars with certain pigments in the petals tended not to show flower breaking. White or cream varieties with white seeds did not show the red color breaking of petals, whereas white or cream varieties with black seeds did; clear blue varities did not show the reddish color breaks common on cultivars with petals of a slight reddish cast. Some cultivars also were severely stunted (shock effect) following aphid transmission of the virus in the seedling stage, but other more tolerant cultivars were almost unaffected. This information was used by breeders, who tended to select away from lines that showed shock effect and conspicuous color breaking. This information thus indirectly benefited the home grower. Had these varietal differences been published, however, they might have confused gardeners because cultivars might be substituted, and would likely become unavailable in a few years. It is not surprising, therefore, that resistant varieties are much less useful and less used in ornamentals than in other annual crops such as vegetables or cereals.

The larger the number of cultivars of a plant grown, the less likely it is that a new variety will have a significant impact on the industry. The emphasis in new releases is on improved horticultural qualities such as plant size and shape, color and size of flowers, and time of flowering, rather than disease resistance. Even when resistance or tolerance to a major disease is mentioned, such cultivars or species may not be chosen by growers unless they can be directly substituted for the most popular cultivars. Thus, the marked tolerance of the rhododendron cultivar Caroline to the widespread severe phytophthora root rot and wilt disease (32) did not greatly change its popularity in the trade. Its resistance to the phytophthora disease has not even been communicated to the consuming public, who may not care about its resistance if its appearance is the major consideration.

It appears that the challenge of breeding or selecting for disease resistance in ornamental plants, where hundreds of commercial cultivars are already grown, is not taken up by breeders or pathologists largely because the returns are too small. This is not to say, however, that researchers have not identified resistance in many ornamental crops, although such resistance was often specific for a given organism in a given localized environment. As examples, resistance has been reported in snapdragons to rust; chrysanthemums to verticillium wilt; roses to powdery mildew, rust, and black spot; gladioli to fusarium yellows; woody plants to armillaria root rot; begonias to pythium root rot; mimosas to fusarium wilt; lilies tolerant of mosaic and fusarium basal rot; sweet peas tolerant of mosaic. Resistance as a type of disease control is destined to assume greater importance in ornamentals as in other crops, but will likely have the greatest impact in crops where relatively few commercial cultivars are grown, where the disease has had a sustained and substantial economic impact, and where a direct substitution of the resistant plant for existing horticulturally desirable, but susceptible, cultivars is possible.

Attempts are being made to increase the reliability of varietal names of ornamentals. Thus, the Royal Horticultural Society in England has attemped to standardize sweet pea cultivars; the American Rose Society, rose varieties; and the Florida Agricultural Experiment Station, camellia varieties. The plant patent laws for vegetatively propagated ornamentals, and the development of F_1 hybrids in ornamentals grown from seed also tend to make cultivars less transitory by protecting the originator. It is doubtful, however, whether any other crop group presents such an extensive array of plants, or is plagued by so many transitory varieties.

Many genera of cultivated ornamentals have species native to or established in some part of this country, often in proximity to commercial plantings. Pathogens on these wild species constitute a reservoir of inoculum for the cultivated crops. The spread of snapdragon rust, *Puccinia antirrhini,* from wild *Antirrhinum* spp. to cultivated snapdragon in California by 1879, and subsequently over much of the world, is a case in point. With the recent strong interest in commercial cultivation of native plants for landscape purposes, the disease hazards are increased. Growers expecting such plants to be more disease tolerant than existing cultivars or species have sometimes been disappointed (36).

High Crop Value in Relation to Disease Control

Perhaps the feature of ornamental plants that has most influenced the approaches to control of their diseases is their comparatively high value. Ornamentals range from those of very high to quite low unit/year value. The annual wholesale value of potted chrysanthemums averaged $3.09 per square foot of production space in 1977 (19), or more than $134,000 per acre. Potted lilies averaged $115,000, and foliage and bedding plants $96,000 per acre. On the other hand, the value of field-grown ornamentals per acre/year may be lower, but still higher than field-grown food and fiber crops. Growers estimate the wholesale acre/year value for rhododendron to be $14,000, shade trees $7500, and bulb crops $5000–$15,000. By comparison, the value for strawberries is $5550, potatoes $860, cotton $300, dry beans $180, and wheat $86 (48).

Several significant features of the pathology of ornamentals, and particularly of floriculture pathology, result from this high unit value per year. The more intensive the cultivation, and the more valuable a crop is, the higher is the tolerable cost of production and of disease control. Thus, low-value field and forest crops perhaps justify only 0.5 to 5.0% of the production cost for disease and pest control. The grower of such a crop usually will hesitate to apply a control procedure unless its economic success has been thoroughly demonstrated. The level of control required is accordingly less than that for high-value crops, and maximum net return, rather than perfect disease control, is sought. Thus, sugar beets given a single application of wettable sulfur to control powdery mildew when it first appears each year give 91.6% of the sugar yield of beets given three additional applications (31).

For median-value vegetable and tree fruit crops, a grower may be willing to spend up to 10% of production cost for plant protection, and to venture into developing or testing a new method of control.

In contrast to the above, growers of high-value crops such as glasshouse plants or field-grown strawberries may willingly spend up to 20% of production cost for disease and pest control. For example, a California flower-seed grower had a field of new tetraploid snapdragon varieties that were to be introduced on the market. Because the fleshy flowers remained moist and did not abscise, and the weather was cool and foggy, infection with *Botrytis cinerea* was common, spreading from the corolla through the pedicel to the main stem, girdling it. To save this valuable seed crop, the grower removed each infected flower by hand, excising lesions of the main stem with a scalpel and painting the wound with a thick Bordeaux mixture.

The more valuable the crop, or the greater its volume of production, the more likely that its disease problems will be investigated. Thus, there have been about 500 papers published on black spot (*Marssonina rosae*) of roses, and more than 175 on snapdragon rust (*Puccinia antirrhini*).

The more difficult a crop is to grow, the higher its value, the greater is the investment risk involved, and the fewer the growers who will successfully produce it. The grower who can produce it has less competition than does one who grows a relatively easy, dependable crop. In the same way, a grower who can dependably produce a crop better than his competitors may have a natural monopoly, but the risk is increased. A grower in Rhinecliff, New York, for many years produced superior F_1 hybrid anemones, and a Washington, DC grower produced exceptional F_1 pansies for cut flowers (45). Disease or changed consumer fancy in such a specialized crop monopoly can be economically ruinous. One needs only to recall that freesias and sweet peas are no longer popular and are now little grown.

The risk factor associated with growing a specialized monopoly crop is reduced as new technology becomes available. Pot chrysanthemum today is such a crop, with diseases largely eliminated and the culture know-how reduced essentially to following published tables (27, 57). This was not always so. Prior to the availability of *Verticillium*-free cuttings and the use of chloropicrin soil fumigation, growing chrysanthemums was a risky business. A grower in the San Francisco area had a local monopoly growing the highly *Verticillium*-susceptible, large-flowered Pocket cultivars of chrysanthemum by critically controlling soil moisture levels. However, the introduction of clean cuttings and soil fumigation nullified the risk, many new local growers produced chrysanthemums, and he lost his competitive advantage.

When culture-indexed chrysanthemum cuttings were placed on the national market, their natural monopoly was not exploited, there was no publicity campaign and only a slight increase in price. The commendable philosophy was that the cost of producing cuttings was also reduced by the disease control effected, and that the old price still made the venture economical. Since no special claims were made for the cuttings, grower complaints and lawsuites were lessened. The subsequent rise of the pot chrysanthemum business indicates the correctness of the analysis.

PIONEERING NEW METHODS OF DISEASE CONTROL   The high unit/ year value of ornamental crops has required and attracted growers with technical training and experience. Such growers are likely to seek and use disease-control information. It is not surprising, therefore, that several important techniques of disease control have come from studies on pathology of ornamental crops. Dimock's (20) method for obtaining culture-indexed cuttings of chrysanthemum free of *V. albo-atrum* was later adapted to sweet potato and rose. This remained the standard commercial method until the apical meristem technique came into use, following the work of F. Quak (46) in 1954 for obtaining virus-free cuttings of carnation. This technique has since been widely adapted to other commercial crops such as strawberry, sweet potato, and fruit trees, sometimes in conjunction with heat treatment. A spin-off from meristem culturing is the tissue-culture method of rapid propagation and development of pathogen-free propagules of ferns, lilies, orchids, and potato. The novelty "Plants in Vitro" of Oakdell Inc., Apopka, Florida has applied these techniques; a variety of plants are widely sold as cultured meristems in test tubes and, after several months, can be planted out by the purchaser.

The virus indexing of cuttings by grafting onto an indicator variety that dependably exhibits marked symptoms (12, 33, 44) came into extensive, continuing, commercial use following the development of the chrysanthemum-stunt control program, and has since been applied to many other crops.

Although thermotherapy of plants had been used much earlier by J. L. Jensen for control of late blight of potato (*Phytophthora infestans*), and by G. Wilbrink for sereh disease of sugarcane, its application against aster yellows of *Impatiens* by Kunkel in 1941 (34) stimulated studies on its use against other plant pathogens.

Steam treatment of commercial glasshouse soil was pioneered by the Rudd Brothers in the Chicago area in 1893 (47). Over 60 years later, Baker & Olsen developed the aerated steam treatment of soil for floriculture crops (7).

The concept that, in the final analysis, the sources of plant pathogens are the soil (including water and nonliving organic matter) and living plants, and that disease control can be achieved by using pathogen-free soil and propagules, coupled with careful sanitation, was first commercially exploited in management of ornamentals (2), and has been emphasized in the pathology of ornamentals.

The important point is that the above methods were developed for high-value crops that would justify the expense, and were used by progressive growers likely to be found in this business. The methods were then adapted for other less valuable agricultural crops.

High Capital Investment

There is high capital investment in commercial glasshouses and in equipment for mechanizing production of ornamentals. In the United States in 1970 there were 49,000 acres of glass and plastic houses producing floriculture crops, more than 325 acres of cloth houses, and over 24,000 acres of outdoor-grown floriculture crops (24).

The ornamental industry has become highly mechanized. Glasshouses, soil handling equipment, moving belts and rollers, boilers for heating glasshouses and for soil steaming, air conditioning equipment, automatic seed planters and transplanting devices, refrigerated storage rooms, controlled environment growth rooms, spray equipment, irrigation equipment, fertilizer injectors, tissue-culture facilities, and large paved areas to control mud and avoid dust contamination, plus many types of mobile carts and trucks, are considered to be standard in the trade. Often a repair shop is maintained on the premises.

The high capital investment and high risk in many segments of the ornamental industry have made disease loss intolerable; a crop failure may jeopardize the whole enterprise. Only intelligent, knowledgeable, and conscientious growers generally enter the ornamental plant business and remain in it. The competition is intense in cheap, low-quality, easily grown crops, but somewhat less in high-value, high-quality, difficult crops.

Controlled Environment

Ornamental plants are grown under stricter environmental control than almost any other crops. Light intensity, daylength, temperature, humidity, water quality (ion content), quantity, and application method (subirrigation, drip, mist, constant level, capillary), soil type, time of planting, fertilizer application (liquid, dry, slow-release), atmospheric pollution (carbon filtration to remove smog), CO_2 amendment, and freedom from pathogens, inter alia, can be controlled to best suit the crop. The decision to provide such control hinges largely on the cost-benefit ratio and on the grower's aggressiveness.

The benefit changes with time, however, necessitating continuous adaptation. Early glasshouse rose growers had to spray with fungicides to control black spot (*Marssonina rosae*) because the plants were syringed with water to control red spider mites, and the water spread the spores and provided favorable conditions for infection. With the development of adequate miticides, syringing ceased, and fungicide spraying became unnecessary under glass. With current restrictions on the use of chemicals it is uncertain whether a reversion to syringing may become necessary. Gray mold (*Botrytis cinerea*) on a series of glasshouse crops, such as snapdragon, has been controlled effectively by applying heat at night, with the ventilators "cracked" to permit moist warm air to escape, thus avoiding condensation on the plants. With the current emphasis on energy conservation, heating is reduced and the gray-mold problem is returning. Will this lead to attempted fungicidal control, with its attendant restriction? Very few control methods long remain satisfactory, but where it is economical, disease control in glasshouses by environmental manipulation has proved reliable, and generally is the method of choice. Certainly pathologists working on glasshouse crops are blessed with a greater variety of potential disease-control methods than are available for other groups of crops.

An outbreak of downy mildew (*Peronospora sparsa*) on glasshouse roses in San Leandro, California in 1951 illustrated the effectiveness of properly applied environmental control. Heavy rainfall flooded the glasshouse area and quenched the boilers in the basement. To conserve the sun's heat in the glasshouses, the vents mistakenly were kept closed, and the humidity became very high. Downy mildew rapidly developed and defoliated the plants. With the boilers back in operation, the ventilators were kept open and the houses dried out. The roses were severely pruned, and the new growth, developed under drier conditions, produced an excellent crop.

Control of relative humidity has played a key role in minimizing plant disease of ornamentals. Glasshouses can be kept drier during rainy weather than can lathhouses. Azalea flower blight (*Ovulinia azaleae*) and flower blight of camellia (*Sclerotinia camelliae*), often destructive in southern California under lath, are less common outdoors, and are rare under glass, due largely to the different humidity levels maintained.

Because space is valuable, both floriculture and nursery crops are grown closely spaced, often in blocks of a single variety. They are "forced" by controlling temperature, light, nutrients, and irrigation. These conditions may favor pathogen spread through the tops (rusts, powdery mildews, foliar nematode) or at the soil surface (*Rhizoctonia solani*). However, because the soil and plant propagules usually are pathogen-free, such epidemics are no longer common.

The glasshouse roof and walls provide partial barriers to pathogen and insect spread, but may favor their increase once they are introduced. Moistened pads, through which air is pulled into the house, are commonly used for cooling in warm months. The effectiveness of these pads in filtering out insects may be enhanced by spraying them with a nonvolatile insecticide. However, these pads may increase the relative humidity, and harbor and support pathogens, as in fusarium stub dieback of carnation (*F. roseum* 'Graminearum') (41). The glasshouse, when the ventilators are closed, also serves as a fumigation chamber for insect control, and for control of powdery mildew (*Sphaerotheca pannosa* var. *rosae*) on rose (18). Furthermore, the tightness of glasshouses makes possible charcoal filtration banks for smog control for sensitive plants (28).

The controlled environment involved in growing many ornamentals offers unusual opportunities for biological control (5). For example, treating soil with aerated steam eliminates pathogens, while leaving the soil biologically buffered against pathogen reinvasion (4). Soil that has been steam-treated or fumigated can be inoculated with beneficial microorganisms capable of protecting host plants (15) or increasing their growth (14). Soil mixes can readily be amended with substrates that favor the antagonists, and can be held at soil temperature or moisture favorable to them. The potential has been demonstrated for inoculating seed to control damping-off (15), and for inoculating roots at transplanting with antagonists to control crowngall (*Agrobacterium tumefaciens*) of nursery stock (40).

Emphasis on Elimination of Disease

The emphasis in disease control in ornamentals is on elimination of the causal agent, more than with most other crop plants. This is accomplished by planting healthy propagules in treated soil, and maintaining this pathogen-free condition by careful sanitation. These points have been discussed in other sections of this paper.

High Standards for Crop Quality

ACCEPTABLE LEVEL OF DISEASE CONTROL   The acceptable level of disease control in ornamentals varies with the crop and its intended use. Those ornamental plants grown primarily for their appearance require essentially perfect disease control because blemishes are unacceptable on a plant whose raison d'être is its beauty. However, certain strictures on methods of disease control are imposed. Visible spray residues on flowers or foliage are unacceptable, particularly for floriculture crops. Spotting of foliage by residue from the water sometimes requires either deionizing the water used in misting, or washing the plants before sale. Any material that injures or discolors flowers cannot be used. This dilemma is resolved whenever possible in commercial growing by avoiding application of chemicals to the tops, relying on disease avoidance through use of pathogen-free planting material and soil, and by environmental manipulation.

In gardens a different problem is faced. So few plants of one kind may be grown that to obtain and use the preferred chemical for disease control may not be worthwhile. The solution often is to use a less effective but wider-spectrum material that can be used on several kinds of plants.

The performance and beauty of many ornamentals depend on the quality of the planting stock. The amount of infection tolerated on that stock sometimes depends on the intended use of the crop. Slight fusarium basal rot (*F. oxysporum* f. sp. *lilii*) can be tolerated in potted Easter lilies because the plants usually are discarded after Easter anyway. For garden lilies, where good flowering performance is required year after year, any infection is intolerable. That the product of ornamentals is the plant itself, rather than its fruit or nuts, emphasizes the requirement of an unblemished crop, particularly for long-term nursery crops such as trees and shrubs. Consumer expectations generally are higher for plants they buy, despite subsequent mistreatment, than for mechanical products.

ECONOMICS OF DISEASES OF ORNAMENTAL PLANTS   There are several types of disease losses of ornamental plants, some obvious and immediate, others obscure and delayed.

Diseases have played a significant role in the development of the ornamentals industry. These diseases have extracted a significant toll in terms of direct losses and costs to control them. According to 1965 estimates in California (16), disease losses on ornamentals ranged from 3.2% for bedding plants to 14.5% for bulb crops grown as pot plants, exclusive of control costs. The average loss for ornamentals was 7.1%. Considering the estimated wholesale value of the national industry to be approximately $1.6 billion (25), these direct annual disease losses could amount to $50—200 million, or $100 million on the average, despite excellent control of the major diseases.

Obviously decreased yield or lowered grade of the product generally are the reasons for applying disease control. However, these factors may be less important in the total picture than are other less obvious results of disease. Disease may cause a significant decline in plant productivity. A relatively small loss of leaf area may cause a disproportionate diminution of root area or effectiveness (37). *Pythium* and *Phytophthora* spp. usually infect tips of rootlets, and thus reduce plant growth through loss of the sites of amino acid and hormone synthesis, as well as of absorption area (5). Such root "nibblers" may cause greater aggregate crop loss than pathogens that kill some of the plants.

The reputation of a grower or area as a dependable, continuing source of high-quality material is an important intangible asset for highly specialized intensive crops such as ornamentals. Once this reputation is lost it may not be regained. Thus, Easter lily bulbs in the United States were largely supplied by Bermuda up to 1920, and by Japan prior to the 1950s. However, the stock was so badly infected by viruses that, when bulbs with less virus infection became available from Oregon, northern California, and Washington, they largely displaced Japanese bulbs from the market. Similarly, geranium cuttings were produced in the field in southern California and shipped to eastern growers until the 1950s. Because of low quality of the cuttings and high incidence of bacterial blight (*Xanthomonas pelargonii*), gray mold (*Botrytis cinerea*), and other diseases, this market was largely lost when geraniums were developed that bred true from seed, and pathogen-free cuttings were produced under glass. Freesias were grown from corms produced in the field in California until varieties that bred true from seed became available from the Netherlands in the 1950s. Because of poor flower quality due to virus infection of the corm-grown freesias, the flowers declined in popularity after 1930, and despite the improved quality of the seed-propagated freesias the market has not been regained.

If an area gets a reputation for an inferior product, it can market only at uneconomically low prices, or may be forced to dump the crop. Thus, aster growers near Redondo Beach, California in 1946 were unable to sell their crop to wholesalers who refused to risk shipment because *Stemphylium callistephi* caused petal and leaf spots during transit. Garden stocks in southern California in 1952–1953 were similarly downgraded and rejected because of *Botrytis cinerea* that spread during shipment.

The cost of the disease control procedure obviously must be included in the economic loss from disease, although that cost is usually not recognized because it is built into production costs. The immediate loss from a disease epidemic, such as that of chrysanthemum stunt, is too obvious to require discussion, although the continuing cost of controlling the disease is enormous, yet often forgotten.

Before the 1950s field growers of crops susceptible to soilborne pathogens moved to new land when the soil became infested. Aster growers in southern California moved their crop every 3–4 years because of fusarium wilt (*F. oxysporum* f. sp. *callistephi*). Some who produced bulb crops, such as gladiolus and lilies, did the same thing. Growers really do not face up to a disease problem by running away. With the increasing scarcity of land, and the advent of economic soil treatments in the 1950s, this migratory production largely ceased, and the quality of both the growers and crops improved. The poorer growers simply changed crops, usually to a less profitable one, or sold the land for real estate development. Neither of these solutions is very satisfactory for plant pathologists or, in the long run, for the industry.

There were interesting illustrations of the Law of Lesser Concessions during the period of acceptance of field soil treatment. Growers first accepted the relatively inexpensive treatments with D-D or EDB and found that their use was economically profitable. They were then willing to try more expensive treatments with chloropicrin or methyl bromide in the hope of even greater profit. The even more expensive use of plastic tarps over treated soil to retain the gases was adopted later. The same law seems to govern the use of propagator-grown pathogen-free stock. At first, a grower may buy a small number of clean cuttings, grow them, and use cuttings from this stock for a few years. When it becomes clear that the second- or third-year crop has severe disease losses, he may decide to propagate from them for only one year. It finally becomes evident to the grower that it is better to have the propagator produce the cuttings, and for him to raise the crop. New growers, even today, tend to progress through such stages in adopting a control practice. A procedure that cannot be adopted in stages wins acceptance more slowly than one that can. A similar progression in application is evident in the adoption on a low-value crop of a control practice found to be profitable on a high-value crop.

The environmental tolerance of a plant is diminished by plant disease, making the crop more difficult to grow. Heather (*Erica* spp.) grown in soil infested with *Phytophthora cinnamomi* must be grown at as low soil moisture as possible. This is difficult to do, and plant growth is reduced because the moisture level is below that best for that crop. Foliage plants in soil infested with *Pythium* spp., *Rhizoctonia solani,* or *Xanthomonas* spp. sustain heavy losses when propagated or grown under overhead mist irrigation. Eliminating the pathogens, however, allows these plants to be grown under warm, moist, fertile conditions without disease development. If *Verticillium albo-atrum* is present in garden soil, only resistant varieties of chrysanthemum can be successfully grown, a severe restriction. When *Pythium ultimum, Rhizoctonia solani,* and excess soluble salts occur together in a soil, it is impossible, simply by controlling irrigation, to produce seedlings in it, since excessive moisture favors *Pythium,* deficient moisture augments salinity injury, and intermediate moisture favors *Rhizoctonia* (2).

Modern mechanized plant production requires close scheduling of the crop, an impossibility with capricious disease occurrence. Furthermore, it is impossible to properly evaluate the efficacy of a culture practice if the plants are diseased. It is as impossible to evaluate fertilizer or irrigation practices on plants that have root disease as it would be to conduct nutritional studies for normal humans on subjects with stomach ulcers. Growers cannot properly evaluate culture and learn from experience unless the plants are free of pathogens. The better growers always get greater benefits from disease control than do poor growers because the plants make maximal growth in response to freedom from disease.

Some chemical applications only temporarily inhibit a pathogen (e.g. Dexon on *Pythium* and *Phytophthora* spp., PCNB on *Rhizoctonia solani*), which resumes growth when the chemical breaks down. Such suppression of a soilborne pathogen by a nurseryman (even out of ignorance) is unethical and should be illegal, because the disease is simply checked in the nursery, only to cause greater damage in the uncontrolled environment of the garden when the plants are sold. Even worse, the garden soil may become infested, impairing plant growth there more or less permanently.

Mobility and Multiple Handling of the Living Plants

Living ornamental plants, and the propagules used to produce them, are shipped around far more than are other agricultural crops. With most other crops it is the produce (e.g. grain, tubers, fruit) that is shipped, and these products are inherently more durable than are living plants.

The well-known production of bulbs in the Netherlands for sale and use in the temperate areas of the world has required special handling and storage to avoid diseases. The Netherlands Bulb Research Center at Lisse was established to study these problems.

In recent years, there has been deep and growing concern in the floriculture industry about the volume of cut flowers (carnations, chrysanthemums, roses) grown in Colombia and marketed in the United States. In 1977, 34% of the carnations and 1.8% of the roses marketed in the United States came from Colombia. Israel is rapidly increasing its shipments of carnations and roses to the United States (50). This development has come about because of cheaper labor and lower heating costs than in the United States. In 1976, flowers and foliage worth $42 million were shipped into the United States from 33 countries, particularly Columbia, Guatemala, and Mexico (23). Orchids, anthuriums, and other tropical flowers and foliage are grown in Hawaii and airlifted to the mainland. Flowers grown in North Africa and Israel are marketed in Europe.

This movement of living plants and propagules provides a method, par excellence, for transport of plant pathogens and insects. For example, it was by this means that ascochyta ray blight (*Mycosphaerella ligulicola*) of chrysanthemum spread over the world. Confined to a small area in North Carolina, South Carolina, and Mississippi for over 40 years, the pathogen began to spread in the 1950s. This disease spread began when Florida started to ship chrysanthemum flowers in the late 1940s, and intensified with the development of a major propagation industry there after 1957 (6).

Australia has taken the sensible position that only six cuttings of a given plant variety may be imported; these are kept under quarantine until found to be free of pathogens, and then are used for propagation.

The shipment of symptomless carriers is particularly hazardous. Many viruses or viroids (e.g. chrysanthemum stunt) have long symptomless incubation periods, and fungal (*Cylindrocladium* on azalea) or bacterial (bacterial blight of geranium) pathogens may be carried externally, or may have infected the plant with symptoms not yet evident. On the other hand, some pathogens may destroy the plant during shipment. *Phytophthora cinnamomi* may kill choisya and daphne plants during marketing, and *Xanthomonas pelargonii* may kill a geranium before it can be planted, or soon thereafter.

During growing and shipping, plants pass through numerous hands (propagator, grower, wholesaler, shipper, retailer, and consumer) and are exposed to a variety of environmental conditions, many of them unfavorable. Plants may be over- or underwatered, exposed to insufficient light, excessively dry air, excessive or deficient nutrients, air pollutants (smog, ethylene), and temperatures that are too high or too low. These stresses weaken the plant and may favor attack by pathogens. For example, overwatering may favor *Pythium* root pathogens, and ethylene may hasten senescence and predispose a plant to *Botrytis cinerea*. Because of the number of agencies involved in moving a plant to the consumer, it often becomes legally impossible to determine the one responsible for the disease condition.

PROBLEMS OF THE URBAN ORNAMENTAL INDUSTRY

Many establishments producing ornamental plants, originally well out in the country, have become engulfed by urban sprawl. This has produced special problems unique to the floriculture and nursery industry. The establishment, surrounded by homes, may be taxed at the residence rate to force them to leave. The area may be rezoned for residences, also forcing a move. Homeowners in the area may resent the truck traffic necessary to the business, and may resort to one of these procedures to correct it. The surrounding homes often preclude expansion of the business, making it either remain static or move. If it remains, space becomes a limiting factor, precluding the possibility of storage of soil and other materials. A Sacramento, California grower has not paved the level area where his container-grown plants are placed, despite the obvious advantages of avoiding contamination by soilborne pathogens and of diminishing mud and dust, because to do so would place the establishment in a higher tax frame.

The use of soil fumigation in urban areas is so risky that insurance against suits by neighbors is essentially unobtainable. Some growers have been subjected to lawsuits from neighbors with fancied or real respiratory problems attributed to chemicals used by the establishment. Prior to the adoption of noncomposted light-weight soil mixes, it was customary to have compost piles containing manure. This led to objectionable odors and flies that aroused opposition from neighbors. This was often an important factor in causing the establishment to adopt noncomposted mixes. This change also made it unnecessary to stockpile large quantities of materials that monopolized a substantial part of the available space.

Ornamental plants frequently are so adversely affected by atmospheric pollutants from industry or from automobile traffic that the kinds of crops grown are sharply restricted. For example, *Cattleya* orchids are very sensitive to ethylene, a common pollutant, the flowers developing dry-sepal disease and becoming unmarketable. In one instance the concentration of this gas was related to the use of gas heaters in surrounding homes during cold weather.

Fortunately, if an establishment is forced to move, the increased value of the property when sold for residences usually makes possible starting anew in a modern well-equipped facility.

Trees and shrubs planted in city streets and parks are subjected to unusually severe unfavorable or injurious environmental stresses such as high temperatures, air and water pollution, soil compaction and high moisture content, wind stress, construction damage, salting of streets, and light effects. These present unique and difficult problems for the urban plant pathologist. Breeding and selection of trees and shrubs for these special stress sites is under way (26, 31a).

CONCLUSIONS AND PROSPECTS

Pathologists, faced with the great diversity and transitory nature of ornamental plant cultivars, and the necessity of producing an unblemished disease-free product, have emphasized the planting of pathogen-free propagules in treated soil, reinforced with careful sanitation and manipulation of the environment for disease control. Many formerly important diseases of major ornamentals have thus become exceedingly rare and reduced to unimportance.

The high valuation of most ornamentals and the carefully regulated conditions for their growth make possible a wide range of unique procedures for disease control. A number of techniques thus developed (e.g. culture-indexed cuttings, apical meristem cultures, improvements in tissue culture methods, virus indexing by grafting on indicator plants, soil steaming, and the development of aerated steam for treating soil and propagules) have also proved useful for controlling diseases of other crop plants.

The pathology of ornamental plants includes a very wide range of diseases and crops, with great opportunity for imaginative research and exploration of new methods of disease control. The pathologist working with ornamentals will be challenged by the energy crisis and increasing restrictions on the use of pesticides. However, a bright prospect especially available to him is the potential application of biological control methods made possible by the regulated environment that allows for ecological manipulation of micoorganisms (5).

With the increasing world population and consequent food scarcity there will be pressures, as during World War II, to lessen research on ornamentals and devote it to food crops. However, with disappearing beauty in much of the world today, the demand for ornamentals around the home probably will increase. Diseases of ornamental plants will thus continue to provide a fruitful and stimulating field of study for plant pathologists.

Literature Cited

1. Bailey, L. H., Bailey, E. Z. 1976. *Hortus Third.* New York: Macmillan. 1290 pp.
2. Baker, K. F., ed. 1957. The U. C. system for producing healthy container-grown plants. *Calif. Agric. Exp. Stn. Man.* 23:1–332
3. Baker, K. F. 1958. The development of floricultural pathology in North America. *Plant Dis. Reptr.* 42:997–1010
4. Baker, K. F. 1970. Selective killing of soil microorganisms by aerated steam. In *Root Diseases and Soil-Borne Pathogens,* ed. T. A. Toussoun, R. V. Bega, P. E. Nelson, pp. 234–39. Berkeley: Univ. Calif. Press. 252 pp.
5. Baker, K. F., Cook, R. J. 1974. *Biological Control of Plant Pathogens.* San Francisco: Freeman. 433 pp.
6. Baker, K. F., Dimock, A. W., Davis, L. H. 1961. Cause and prevention of the rapid spread of the Ascochyta disease of chrysanthemum. *Phytopathology* 51: 96–101
7. Baker, K. F., Olsen, C. M. 1960. Aerated steam for soil treatment. *Phytopathology* 50:82
8. Ball, G. J. 1948. The mum stunt situation. *Grower Talks* 12(7):6
9. Ball, V. 1947. A new mum disease. *Grower Talks* 11(7):20–21
10. Ball, V. 1950. Mum stunt—two years later. *Grower Talks* 13(9):14–19
11. Ball, V. 1976. Early American horticulture. *Grower Talks* 40(3):1–57
12. Brierley, P., Olson, C. J. 1956. Development and production of virus-free chrysanthemum propagative material. *Plant Dis. Reptr. Suppl.* 238:63–67
13. Brierley, P., Smith, F. F. 1940. Mosaic and streak diseases of rose. *J. Agric. Res.* 61:625–60
14. Broadbent, P., Baker, K. F., Franks, N., Holland, J. 1977. Effect of *Bacillus* spp. on increased growth of seedlings in steamed and in nontreated soil. *Phytopathology* 67:1027–34
15. Broadbent, P., Baker, K. F., Waterworth, Y. 1971. Bacteria and actinomycetes antagonistic to fungal root pathogens in Australian soils. *Aust. J. Biol. Sci.* 24:925–44
16. Calif. Agric. Exp. Stn. 1965. *Estimates of Crop Losses and Disease-Control Costs in California,* 1963. Calif. Agric. Exp. Stn. 102 pp.
17. Cook, M. T. 1916. The pathology of ornamental plants. *Bot. Gaz.* 61:67–69
18. Coyier, D. L., Picchi, J. 1976. Volatile action of a new fungicide for control of rose powdery mildew. *Proc. Am. Phytopathol. Soc.* 3:289
19. Crop Reporting Board. Econ., Stat., and Coop. Serv. 1978. Floriculture crops. Production area and sales, 1976 and 1977. Intentions for 1978. *US Dep. Agric. Crop Rep. Board.* SpCr 6–1(78): 1–27
20. Dimock, A. W. 1943. A method of establishing Verticillium-free clones of perennial plants. *Phytopathology* 33:3
21. Dimock, A. W. 1947. Chrysanthemum stunt. *NY State Flower Grow. Bull.* 26:2
22. Feder, W. A., Naegele, J. A., Cathey, H. M., Piringer, A. A. 1976. A guide to research in amenity horticulture. *Mass. Agric. Exp. Stn. Res. Bull.* 638:1–35
23. Florists' Rev. 1978. New ITC report lets private sector assess seriousness of imports' impact on U.S. horticulture. *Flor. Rev.* 162(4201):27–30, 66–73
24. Fossum, M. T. 1973. *Trends in Commercial Floriculture Crop Production and Distribution. A Statistical Compendium for the United States, 1945–1970.* Washington DC: Soc. Am. Flor. Endowment. 98 pp. Supplements to Summaries, Tables 76–99. 21 pp. 1975
25. Fossum, M. T. 1977. *Economic Trends and Projections for Commercial Floriculture. United States, 1951–1986.* Washington DC: Marketing Facts for Floriculture, Ltd. 9 pp.
26. Foster, R. S. 1977. Roots: Caring for city trees. *Techol. Rev.* 79(8):29–35
27. Gloeckner, F. C. 1974. *Gloeckner Chrysanthemum Manual.* New York: Fred C. Gloeckner Co. 175 pp.
28. Haagen-Smit, A. J., Darley, E. F., Zaitlin, M., Hull, H., Noble, W. 1952. Investigation on injury to plants from air pollution in the Los Angeles area. *Plant Physiol.* 27:18–34
29. Hedrick, U. P. 1950. *A History of Horticulture in American to 1860.* New York: Oxford Univ. Press. 551 pp.

30. Hepting, G. H., Cowling, E. B. 1977. Forest pathology: Unique features and prospects. *Ann. Rev. Phytopathol.* 15: 431–50

31. Hills, F. J., Hall, D. H., Kontaxis, D. G. 1975. Effect of powdery mildew on sugarbeet production. *Plant Dis. Reptr.* 59:513–15

31a. Himelick, E. B. 1976. Disease stresses of urban trees. *Better Trees for Metrop. Landscape Symp. Proc., US Dep. Agric. For. Serv. Gen. Tech. Rept.* NE-22:113–25.

32. Hoitink, H. A. J., Schmitthenner, A. F. 1974. Resistance of *Rhododendron* species and hybrids to *Phytophthora* root rot. *Plant Dis Reptr.* 58:650–53

33. Keller, J. R. 1953. Investigations on chrysanthemum stunt virus and chrysanthemum virus Q. *NY Agric. Exp. Stn. Ithaca Mem.* 324:1–40

34. Kunkel, L. O. 1941. Heat cure of aster yellows in periwinkles. *Am. J. Bot.* 28:761–69

35. Linderman, R. G., 1974. The role of abscised Cylindrocladium-infected azalea leaves in the epidemiology of Cylindrocladium wilt of azalea. *Phytopathology* 64:481–85

36. Linderman, R. G., Zeitoun, F. 1977. *Phytophthora cinnamomi* causing root rot and wilt of nursery-grown native western azalea and salal. *Plant Dis. Reptr.* 61:1045–48

37. Martin, N. E., Hendrix, J. W. 1967. Comparison of root systems produced by healthy and stripe rust-inoculated wheat in mist-, water-, and sand-culture. *Plant Dis. Reptr.* 51:1074–76

38. Milbrath, D. G. 1930. A discussion of the reported infectious chlorosis of the rose. *Calif. Dep. Agric. Mo. Bull.* 19: 535–44

39. Milbrath, D. G. 1930. Infectious chlorosis of the rose. *West. Flor. Nurs. Seedsman* 13(29):29–30

40. Moore, L. W. 1977. Prevention of crown gall on *Prunus* roots by bacterial antagonists. *Phytopathology* 67:139–44

41. Nelson, P. E., Pennypacker, B. W., Toussoun, T. A., Horst, R. K. 1975. Fusarium stub dieback of carnation. *Phytopathology* 65:575–81

42. Nelson, R. 1930. Infectious chlorosis of the rose. *Phytopathology* 20:130

43. Nursery and Seed Serv., Calif. Dep. Food and Agric. 1977. California nursery industry licenses and gross F. O. B. farm value. *Calif. Dep. Food Agric. Mimeo.* 10 pp.

44. Olson, C. J. 1949. Intensive program conducted to whip chrysanthemum stunt. *Flor. Rev.* 105:(2711):33–34

45. Post, K. 1949. *Florist Crop Production and Marketing.* New York: Orange Judd. 891 pp.

46. Quak, F. 1957. Meristeemcultuur, gecombineerd met warmtebehandeling, voor het verkrijgen van virusvrije anjerplanten. *Tijdschr. Plantenziekten* 63: 13–14

47. Rudd, W. N. 1893. Killing grubs in soil. *Am. Florist Chicago* 9(278):171

48. US Dep. Agric. 1977. *Agricultural Statistics 1977.* Washington DC: US Dep. Agric. 614 pp.

49. Voigt, A. O. 1976. Status of the industry. In *Bedding Plants,* ed. J. W. Mastalerz, pp. 1–3. University Park, Penn.: Penn. Flower Growers. 515 pp. 2nd ed.

50. Voigt, A. O. 1978. Facts and figures about imports and leisure activities. *Ill. State Flor. Assoc. Bull.* 379:8–9

51. Weeds Trees and Turf 1978. Sod producers plant fewer acres in 1978. *Weeds Trees Turf* 17(8):19–20

52. Weiss, F., McWhorter, F. P. 1930. Pacific Coast survey for rose mosaic. *Plant Dis. Reptr.* 14:203–5

53. White, R. P. 1930. A future outlet for pathological service. *Phytopathology* 20:112 [*Also as* Opportunities for plant pathology in the field of ornamental horticulture. *Flor. Exch.* 73(1):9, 34. 1930.]

54. White, R. P. 1930. Infectious chlorosis of the rose. *Flor. Exch.* 73(4):46

55. White, R. P. 1930. Quarantines and rose chlorosis. *Flor. Exch.* 73(11):50A, 54

56. Wildon, C. E. 1930. Michigan rose men discuss infectious chlorosis. *Flor. Exch.* 73(10):58

57. Yoder Brothers Inc. 1973. *Yoder Date Finders for Pot and Cut Mums.* Barberton, Ohio: Yoder Brothers Inc. 101 pp.

Ann. Rev. Phytopathol. 1979. 17:279–99
Copyright © 1979 by Annual Reviews Inc. All rights reserved

RELATIONSHIP OF PHYSICAL AND CHEMICAL FACTORS TO POPULATIONS OF PLANT-PARASITIC NEMATODES[1]

♦3711

Don C. Norton

Department of Botany and Plant Pathology, Iowa State University,
Ames, Iowa 50011

INTRODUCTION

Ecosystems consist of living communities of organisms and the nonliving environment (44, 88). Because these two components do not act independently, all experimental manipulations of single environmental parameters are unnatural. A diverse community of organisms exhibits characteristics that differ from those of its discrete constituents. Also, organismal behavior is modified by interactions within the environment. Thus, in most instances collections of facts concerning the effects of one environmental parameter on a given species should not be construed as definitive. Nevertheless, ecological work with a single or a few species relative to one or a few environmental parameters may help provide at least a partial, if never total, explanation of the presence and population density of a given nematode species. Distribution and abundance are different aspects of the same problem (2). Thus when we discuss factors that limit densities in a local area, we are also discussing factors that limit distribution.

The impact of a nematode community on a population of host plants is the result of all interactions, known and unknown, internal and external, of that community with the environment, including those with the host plants. The occurrence, size, and density of a nematode population need to be explained. They did not happen fortuitously. The physical-chemical aspects of the environments constitute one set of parameters governing

[1]Journal Paper No. J-9359 of the Iowa Agriculture and Home Economics Experiment Station, Ames, Iowa 50011. Project 2119.

0066-4286/79/0901-0279$01.00

nematode populations, but the nematode's behavior in the nematode community cannot be divorced from other biological parameters. Because of the parasitic nature of plant-parasitic nematodes, the ecology of the host is of major importance. Nematode populations are influenced by many edaphic factors, but cause and effect relationships are more difficult to delineate. One often is interested in the parameters that have the greatest influence on a given population but some correlated factors may be more apparent than real.

Exponential and logistic growth curves are largely textbook examples; few field populations fit these idealized mathematical models. Most of the well-defined curves, such as the exponential one of *Helicotylenchus pseudorobustus* (60) and the logistic curve of *Paratylenchus hamatus* (21), were derived from controlled experiments in the greenhouse or growth chamber. Variation in environmental factors in the field greatly alters the expression of the biotic potential so that the resulting ecological growth curve does not resemble a textbook curve for long.

Some edaphic factors, such as pH and texture, change very little unless they are altered by man through application of chemicals, cultivation, or other farm management practices or by the rhizosphere. In many agricultural areas temperature and moisture change drastically and are difficult to predict. Each affects the other and in turn affects aeration, osmotic pressure, and biological activity. The interrelationships of many parameters make interpretations difficult.

Populations of plant-parasitic nematodes can be studied in two major ways: (1) as potential pathogens that are to be controlled because they can be detrimental to food and fiber production; or (2) as animals in a complex biological system without regard to the alteration of the ecosystem to Man's benefit or detriment. In the latter situation the terms "good" and "bad" are not pertinent. But, in economic nematology the goal is to maintain or reduce the growth rates of pathogens to somewhere below an injury threshold.

In this review many specific growth rates were calculated from original data. The equation used was $\Delta N/N \Delta t$ where ΔN equals the change in number of nematodes from the beginning of the experiment (or at the low point in the cycle, usually at the beginning of the season) until the end of the experiment or some arbitrary time interval such as 90 days (Δt). Where data were taken from graphs, some small errors in exact numbers were unavoidable, but this was considered negligible. This paper is not intended to be a complete review of the literature; it is an attempt to assess the role of major environmental parameters in the growth and decline of nematode populations. The reader is referred to other papers (61, 109, 110, 115) for further discussions.

POLYSPECIFIC COMMUNITIES

In the past, the study of single populations of nematodes in sterilized soil provided much information on the effects of specific environmental parameters on growth. Data of this sort are not always relevant to the field. Also, a study of one nematode species in the field is only one facet of a complex polyspecific community. Currently, communities are being studied at one or more trophic levels—not so much as individual species but as communities so as to understand the relationship to species or groups of species to one another and to the environment. In order to understand the realized niche, polyspecific communities usually must be analyzed. As pointed out by Pielou (68), it is probably better to work on a given trophic level than to try to cover all organisms of a kind. The clustering of nematodes occurring in a given trophic level still relates all biotic and abiotic parameters. Although different types of analyses are made of nematode communities relative to classification and structure, it is important that the abiotic factors also be analyzed. It should not be implied however that monocultures are no longer of value. They become even more valuable because they may help explain changes at the community level.

Most work with plant-parasitic nematodes at the community level has been done in undisturbed habitats, with or without edaphic data or observations. As examples in natural areas, *Helicotylenchus platyurus* was prominent in hydromesophytic and swamp forest communities in Indiana but *Tylenchorhynchus silvaticus* was a prominent species in well-drained podzolic soils (32, 33). In these as well as other studies, the vegetation varies and the role of the host in distribution is yet to be determined. Similarly, in studying Iowa prairies, Schmitt & Norton (76) found that certain nematodes such as *Helicotylenchus hydrophilus, Hoplolaimus galeatus,* and *Xiphinema americanum* were restricted to potholes and their boundaries. Other species such as *Helicotylenchus pseudorobustus, H. leiocephalus, Tylenchorhynchus maximus,* and *Xiphinema americanum* occurred more on the low but well-drained slopes. The discovery of *X. americanum* in these areas (where moisture was adequate but not so excessive as to limit aeration) is possibly explained by the work of Van Gundy et al (101). They found that this nematode was sensitive to low oxygen concentration. Of 15 species of the Criconematinae examined in fields and woodlands in a 16.4 ha site in Michigan, *Macroposthonia rustica* had the highest prominence values in poorly drained loam and silt loam soils, while *Lobocriconema thornei* and *Criconema octangulare* had higher values in well drained sandy loam soils than in the other types of soils. *Crossonema menzeli* had its highest prominence value in sandy loam soils as compared with loamy sand, and loam and silt loam soils (40).

DEPTH

It is well known that soil properties change with depth. Although the top 20 cm of cultivated soil is more homogeneous than layers below the plow depth, there are differences in moisture and temperature. Studies of nematode distribution in soil are confounded by differences in root distribution (41, 75, 111). Interactions have been found between depth and many other soil properties.

Richter (71) recorded data from an experiment using a mixture of two species of *Tylenchorhynchus*. The mean specific growth rates at the 0–20 cm, 20–40 cm, and 40–60 cm depths from July 1964, which was the low point in the population, to December 1965 were 3.3, 4.1, and 5.4, respectively. Thus the specific growth rates of this combination of nematodes increased with depth even though the total numbers were greater in the upper layers of soil. During the same period, however, numbers of *Trichodorus viruliferus* and *T. pachydermus* were least at the 0–20 cm depth but had the highest specific growth rate, 42.7. Calculated specific growth rates at the 20–40 cm, 40–60 cm, and 60–100 cm depths were 35.5, 6.7, and 1.4, respectively. The largest population was at the 20–40 cm depth. Even though the author (71) states that these *Trichodorus* species are typically inhabitants of the deeper soil layers, it appears that the specific growth rates are less there. Differences in survival at different depths probably account for the paradox.

Griffin & Darling (23) worked with spruce in nurseries in Wisconson. I calculated that the specific growth rate of *Xiphinema americanum* was about the same at all depths down to 50 cm even though populations were larger in the upper than in the lower soil horizons. This evenness of the overall specific growth rates and the fluctuations involved suggest that the environment, or roots, in the upper layers were more conducive to greater and faster nematode buildup, but also to greater and faster declines than at the lower levels. During the ascending (spring increase) phase at the upper layers, the specific growth rates were greater than at the lower levels. During the ascending (fall increase) phase, the specific growth rates at lower depths were equal to or greater than those in the upper soil horizons. This assumes that there was no migration, perhaps not a good assumption.

Working with *X. americanum* and alfalfa in Iowa, Norton (59) also found that the largest populations occurred in the upper rather than the lower soil depths at two stations, one being a lowland station and the other about 150 m away on a knoll. The specific growth rates from mid-May into November were greater in samples centered at the 10 cm depth than those at the 25 cm depth at both stations. The average specific growth rate at the 10 cm depth at the lowland station was twice that at the 25 cm depth, being 1.46 and 0.78, respectively. The growth rates at the knoll station were 2.69

and 2.20 at the 10 cm and 25 cm depth, respectively. It is doubtful that any of the populations were density dependent because the alfalfa was only in the third year and populations several times those obtained were recorded elsewhere in the vicinity around alfalfa.

TILLAGE

Although tillage practices may not effect major changes in many soil properties such as pH and texture, they certainly influence temperature (114) and moisture (97). The importance of these changes on nematode densities relative to the importance of the effects of dilution or placement by disturbance during cultivation has not been determined.

Information on nematode populations and tillage is relatively meager. With some nematodes, populations generally decreased in plots with less tillage (83, 91). In other tests (1, 14), there was little difference in nematode numbers and tillage, or the results varied; sometimes fewer nematodes were present in nonplowed than in plowed land while the opposite was true in some instances.

The annual specific growth rate often does not change much from year to year. Otherwise, populations could become enormous in some fields. Thus, the management philosophy should be to use tillage or other management practices that will not increase (or preferably decrease) the specific growth rate, especially if the original population was large. Thomas (91) studied nematode populations under seven tillage regimes from May through October. The highest specific growth rate of a species did not always occur in the tillage regimes with the largest numbers of nematodes, but was governed partly by the population in May. For example, *Xiphinema americanum* had the highest specific growth rate (20.46) in the no-till-flat treatment and the lowest specific growth rate (0.21) in the fall-plow treatment in corn. On the other hand, *Helicotylenchus pseudorobustus* in which fluctuations of densities were much less than in *X. americanum,* had the highest specific growth rate (0.38) in the fall-plow treatment and the lowest (−0.01) in the till-plant treatment. Although some of the specific growth rates might reflect the initial population in May, it is doubtful that the densities ever became density dependent. Survival rates over the winter and other factors must be considered in the overview.

TEXTURE AND MINERAL COMPONENTS OF SOILS

Certain nematodes develop more frequently and more abundantly or cause greater damage in certain soil types or textures than in others. For example, the sting (*Belonolaimus* spp.), root-knot (*Meloidogyne* spp.), needle (*Longidorus* spp.), and stubby root (*Trichodorus christiei*) nematodes are found most frequently and in greatest densities in highly sandy or otherwise porous soils. Some root lesion nematodes such as *Pratylenchus zeae* are most common in sandy soils, but others of the same genus, such as *Pratylenchus hexincisus,* are abundant in the medium to heavy textured soils. Other nematodes are found in a variety of soils with no apparent relation to soil texture.

Probably more has been written about nematode populations relative to soil texture than any other edaphic parameter. Fortunately, it also is becoming more common in nematological work to report the percentages of each soil particle size, based upon the USDA or other standard soil analysis systems. This is important because the percentages of the various particle sizes can vary considerably within a textural classification. Calculated specific growth rates for some nematodes in different soils are presented in Table 1.

At least for some nematodes, densities or specific growth rates of other than the root-knot, sting, and needle nematodes are greater in sandy soils than in less sandy ones (12, 78, 100). For example, increases in population density of *Criconemoides xenoplax* on grape were significant only in sandy loam soil compared with silty loam and loam soils at a temperature of 22–26°C (Table 1). Sometimes the specific growth rate is less in a soil classified as sand than in sandy loams, as with *Pratylenchus brachyurus* (19). As pointed out by Ponchillia (69) and Wallace (106), this might be due in part to the size of the nematode especially in large soil pores in which leverage for movement is a problem.

Robbins & Barker (72) speculated that soil texture and temperature are the primary factors limiting the distribution of *Belonolaimus longicaudatus.* The nematode does best in soils with a minimum of 80% sand and a maximum of 10% clay (51, 72). Reproduction on strawberries was greater in sandy than in finer textured soils (72). With different grades of sand, however, reproduction on soybeans was greatest in the fine textured sand than in the medium and coarse textured sands (Table 1). When moisture was considered in tests with soybeans growing in 65-mesh silica sand, reproduction after 35 days was greater at 7% than at 30% or 2% moisture. Reproduction occurred at 30% moisture only when the nutrient solution was aerated.

Some nematodes are distributed widely and occur in many different soil textures but the density of a population will generally vary with texture. For example, *Xiphinema americanum* is common and widely distributed in the United States. The densest populations, however, occur in coarse soils containing less clay than in more heavily textured ones (59, 62, 69). But Schmitt (75) has shown with this nematode that if differences in soil texture between sites are small, then other factors may prevail over textural effects. Aeration is probably interrelated with texture in governing this nematode because survival decreased with decreasing amounts of oxygen (21, 7, 2, and 0%) applied to the soil (101).

Significant differences in numbers of five nematode species occurred in the greenhouse between a clay loam and a sandy loam or a fine sandy loam [McGlohon et al (49)]. These differences generally were not apparent during

Table 1   Calculated specific growth rates of nematodes in different soils

Soil	$\Delta N/N\Delta t$	Plant	N	t
Nematode: _Pratylenchus brachyuras_[a]				
Sand	−0.65			
Norfolk sandy loam	4.2	Strawberry		
Portsmouth loam	1.4			
Cecil clay loam	0.3		500	3 months
Sand	0.6			
Norfolk sandy loam	27.9	Cotton		
Portsmouth loam	8.4			
Cecil clay loam	0.65			
Nematode: _Hemicyliophora arenaria_[b]				
100% sand	69			
1:1 sand:topsoil	4	Tomato	10,000	5 months
100% topsoil	0			
Nematode: _Heterodera avenae_[c]				
Solonized brown soil	2.43			1st crop
Heavy gray soil	0.14			(7 months)
Red brown earth	0.43			
Solonized brown soil	4.29		3,500	2nd crop
Heavy gray soil	2.14	Wheat	(eggs)	(soil replanted)
Red brown earth	0.43			
Solonized brown soil	7.29			3rd crop
Heavy gray soil	3.00			(soil replanted)
Red brown earth	0.00			
Nematode: _Belonolainus longicaudatus_[d]				
Fine textured sand	18			
Medium textured sand	5	Soybean	1,000	90 days
Coarse textured sand	<0			
Nematode: _Criconemoides xenoplex_[e]				
Yolo sandy loam	72.7			
Yolo silty loam	29.6	Grape	300	3 months
Yolo loam	26.4			

[a] Endo 1959 (19).
[b] Van Gundy & Rackham 1961 (100).
[c] Meagher 1968 (50).
[d] Robbins & Barker 1974 (72).
[e] Seshadri 1964 (78).

one season in the field where soil was fumigated and the nematode species added. For example, the nine-month specific growth rate of *Helicotylenchus dihystera* on *Trifolium repens* in the greenhouse was 933 in a Cecil clay loam but only 232 in a Norfolk sandy loam. There was a slight, opposite trend in the field which evidently was not significant.

Four-month specific growth rates calculated from data of Chang & Raski (12) for *Hemicriconemoides chitwoodi* on Thompson seedless grapes were 39.42 in a fine sandy loam, 4.78 in a loamy sand, and 0.98 in a clay loam at 25°C. At 20°C the maximum growth rate of 2.94 occurred in the fine sandy loam.

It is now becoming generally recognized that pore sizes associated with different crumb sizes are as important or more important than sizes of the individual particles. Jones et al (36) and Jones & Thomasson (37) believe that nematodes generally only occupy pores greater than 30 μm in diameter and that nematodes are not found in pores less than 30 μm in diameter. *Xiphinema diversicaudatum,* a large species, was seen only in the macropores (greater than 30 μm in diameter) between the aggregates (36). It is well known that many nematodes such as *Xiphinema americanum, Longidorus,* and *Trichodorus* generally have their largest populations in coarser or more porous soils (13, 62).

These few examples demonstrate that certain nematodes are favored by some soil textures more than others. But most soils are heterogeneous and contain a diversity of taxa, no doubt a reflection of the heterogeneity of the soil texture, porosity, and other physical parameters. It is also true that if one mineral or crumb size dominates a soil, the qualitative and quantitative diversity of taxa may be less.

TEMPERATURE

Temperature is one of the most thoroughly studied edaphic factors that affect nematodes. This probably is because of the relatively accurate instrumentation available. No matter how accurate the instruments are, however, care must be taken in interpretation of results. Gradations in temperature may occur laterally in the field as well as vertically where there is a lag in diurnal fluctuation from the surface to the deeper layers. The degree of fluctuation and time lag at different depths are strongly influenced by texture and moisture. Usually there is little fluctuation in temperatures 1 m below the soil surface.

Griffin & Darling (23) found that the population of *Xiphinema americanum* was smaller and fluctuated less at deep than at shallow depths. Fewer roots at the deeper layers would not account for the lack of great fluctuations. Fluctuation among adults, gravid females, and juveniles at deeper layers were about equal. At deeper layers adults matured more slowly than those nearer the surface.

Most plant-parasitic nematodes are most active between 25 and 30°C. However, it frequently happens that smaller numbers of nematodes cause more damage at one temperature than larger numbers at another. As an example, Ferris (22) found that 100 *Pratylenchus penetrans* per gram of root could cause damage at 7–13°C, but over 400 nematodes per gram of root were required to cause similar damage at 16–25°C. Some nematodes do best at temperatures above 30°C. These include *Hemicycliophora arenaria* (100) and *Longidorus africanus* (42). *Ditylenchus dipsaci* is especially favored by cooler temperatures and abundant moisture (1, 24, 95). In alfalfa, for example damage is usually worse in the spring and fall than in the summer.

Although there are times when a single population peak occurs, and it is associated with temperature, there are frequently additional seasonal peaks as with *Pratylenchus scribneri* (92). Not all of these peaks are temperature dependent, however; reproduction is favored over a broad temperature range.

Accumulated heat units are frequently used to study nematode development (35, 85). These units are usually calculated as the number of degrees centigrade times the number of hours above a temperature below which nematode development does not occur or is negligible. The threshold temperature will vary with the nematode species, the host, and probably the environment. Using 10°C as the threshold temperature Tyler (96), found that 6500 to 8000 heat units were required for *Meloidogyne* sp. to develop to egg laying. Milne & DuPlessis (54) found that *M. javanica* completed development in 9300-C hr from the second stage juvenile to the egg in the field when 7.5°C was used as the threshold temperature. The beginning of egg production of *M. hapla* on lettuce in the field could be predicted within four days using heat units and a threshold temperature of 5 or 7°C (85).

In studying low temperature and nematode growth, Vrain et al (103) found that *M. incognita* reproduced at 20°C but not at 16°C or below. *M. hapla* reproduced at 20, 16, and 12°C but not at 8°C. They found that the threshold temperature for *M. incognita* on *Trifolium repens,* white "Dutch," was 10.1°C and 8.8°C for *M. hapla*. *M. hapla* required more heat units at 20, 16, or 12°C than did *M. incognita* to develop into the undifferentiated second stage juvenile. *M. incognita,* however, required more heat units than *M. hapla* to develop from the undifferentiated second stage juvenile to the second stage female. Vrain et al also found that reproduction of *M. incognita* and *M. hapla* was evident in March or April in North Carolina after inoculation and accumulation of 8500 to 11,250 degree hours in *T. repens* inoculated and transplanted to the field the previous October or November. Mugniery (57) found that *Globodera pallida* required fewer degree days for development than *G. rostochiensis* at 9.5°C, but there was a reverse trend at temperatures of 18, 19, or 24°C. The basal temperatures for *G. rostochiensis* and *G. pallida* were 6.2°C and 3.9°C, respectively.

MOISTURE

Moisture and temperature often interact. Consequently it is usually difficult to separate the effect of the two. Overall, however, moisture is the most important abiotic parameter governing nematode populations, directly or indirectly. Although there are many references to the use of Weather Bureau data in plant pathological and nematological literature, care must be exercised in relying on rainfall data unless it is recorded at the plots. This is especially true in areas where storms are erratic and local. In addition, the moisture available to a nematode varies greatly with the type of soil.

Recovery and maturation of nematodes are greatly influenced by the method. Rössner (73) and Simons (79) found that recovery of nematodes from dry soil was greater if the soil was moistened prior to sampling. Dryness also may promote maturation of nematodes. A higher percentage of brown cysts and more emergence of juveniles from brown cysts occurred when *Heterodera glycines* was subjected to moisture stress after a period of favorable growth than when not (25). Similar results were found by Wallace (107) who reported that maximum emergence of *H. schachtii* occurred when most soil pores were empty of water.

As a general rule, optimum plant growth occurs between 75 and 100% of field capacity. It might be expected that where nematodes reach large populations the growth requirements relative to moisture and other abiotic parameters are those similar to the host. But Townshend (94) found that more invasion of corn roots by *Pratylenchus penetrans* and *P. minyus* occurred at moisture concentrations above field capacity than below field capacity. This does not preclude that later nematode development is better within the root at greater soil moisture tensions. The fact that populations of *P. penetrans* are often greatest in sandy soils (46, 56, 65, 82, 84) would indicate this. Even though some nematodes can survive dry conditions for months or years, their establishment must come when moisture is favorable. Their distribution and thus their populations seem to be limited at least partly by lack of moisture during a critical stage in the nematode's life cycle. Thus, Seinhorst (77) reported that the disease caused by *Ditylenchus dipsaci* on onion persisted in heavy clay soils but not in the sandy ones in Holland. Wallace (108) found that the surface populations of *D. dipsaci* would decrease as rainfall declined, but that increased moisture would allow for greater lateral transport. Restriction of *Anguina tritici* in the United Stated to the southeastern states seems to be explained best by the abundance of rainfall in this area. Except for a report from Texas (89), the nematode has never been reported from the great wheat belt in the midwestern United States, an area considerably less humid than the southeastern states.

Soil moisture often regulates the amount of aeration. Populations of *Xiphinema americanum* were greatest in well-drained soils (76) and survival was the least at low oxygen concentrations (101). There was an inverse correlation of numbers of *Tylenchorhynchus* spp. with the amount of flooding and rainfall in Louisiana (28). Other scientists also have demonstrated that excessive moisture reduces nematode populations (7, 15, 99, 112). Although some of the reduction in nematode numbers might be due simply to a dilution of the number of nematodes, much of it is believed to be due to suffocation and buildup of toxic substances. Van Gundy et al (99) found that decrease in numbers of *Hemicycliophora arenaria* in irrigated citrus was linked to oxygen diffusion rates. Three days between irrigations were insufficient to allow enough oxygen diffusion for nematode reproduction but seven days between irrigations were.

Although some nematodes can withstand a wide range of moisture stresses, naturally there are limits. Root-knot juveniles and egg masses were susceptible to desiccation but egg masses were also intolerant of excessive moisture (67). Laboratory results were confirmed in the field when cultivation reduced soil moisture to the point where the nematode population declined. It is well known, however, that nematodes such as *Ditylenchus dipsaci, Anguina tritici,* and the bud and leaf nematodes of the genus *Aphelenchoides* can withstand desiccation for months to years.

Constant soil moisture is difficult to maintain and thus there are few direct observations on the effect of moisture on nematode populations. Interactions with other parameters occur. The greatest increase of *Pratylenchus penetrans* occurred at moderate soil moisture tension of pF 2 to 3, but with greater amounts of silt and clay, greater soil moisture tension was necessary for good plant growth (38).

AERATION

Aeration in the soil is regulated by pore space, moisture, depth, and temperature. Most soils are very heterogeneous allowing for a diversity of organisms, including nematodes. Where rainfall is erratic, the pore space fluctuates greatly. This, in turn, affects nematode development. Extremes of oxygen and carbon dioxide doubtless govern the occurrence of some soil and plant nematodes. Such influences can start early in the life of a nematode. The minimum for optimal hatching range for *Aphelenchoides compost-icola* and *Ditylenchus myceliophagus* eggs was 1.5 and 2.0% oxygen, respectively. Juvenile development was greatest above 5% oxygen (58). Bhatt & Rohde (6) found that respiration of five species generally was stimulated by the most dominating environmental factor of the natural habitat. Soil-inhabiting nematodes utilized oxygen most rapidly with 1 to 2% carbon dioxide, but *Aphelenchoides ritzemabosi* did so at 0.03% carbon dioxide. Klekowski et al (39) summarized respiration data on 68 nematode species and, although there was some overlapping, found that groups such as plant parasites, predators, and free-living marine nematodes generally clustered in separate groups along a regression line.

OSMOTIC PRESSURE

Moisture partly governs the osmotic pressure in the soil. If moisture is adequate, the osmotic pressure of the soil usually does not exceed 2 bars. Thus, for many purposes the osmotic pressure of the soil solution probably is not a major factor in infection processes. Plant cells commonly have osmotic pressures between 10 and 20 bars, however. High osmotic pressures may curtail nematode establishment or development of a population under high salt concentrations with adequate rainfall for the crop, or where inadequate rainfall results in an increase in salt concentrations. In the latter situation it is difficult to tell whether the inhibition of nematode development is due to lack of moisture per se or the increased osmotic pressure.

Because different nematode species occur and reproduce in different habitats, it is only natural to expect that some nematodes can withstand certain osmotic pressures better than others. *Belonolaimus maritimus* and *Meloidogyne spartinae,* sometimes found in brackish waters, apparently can tolerate higher osmotic pressures better than can other members of these genera (20, 70). Part of this might be due to the effect of hatching as there was a greater hatch of *M. spartinae* eggs in salt solutions of 0.2, 0.3, and 0.4 M than in solutions of 0.0, 0.1, and 1.0 M. Dropkin et al (16) found a marked inhibition in the hatch of *Meloidigyne arenaria* and *Heterodera rostochiensis* eggs in solutions of 0.2 M and larger molarities.

Viglierchio et al (102) found that mobility of several nematodes was not affected until osmotic pressures of about 10 bars were obtained. *Ditylenchus dipsaci* was the most tolerant and *Rhabditis* spp. were the least tolerant to increasing osmotic pressures. *Pratylenchus vulnus, Hemicycliophora arenaria, Tylenchulus semipenetrans,* and *Meloidogyne hapla* were intermediate in tolerance to osmotic pressure.

There seems to be some correlation between survival and tolerance to osmotic pressure. Bhatt & Rohde (6) found that *Anguina tritici, A. agrostis, Ditylenchus dipsaci,* and *Pratylenchus penetrans* respired well at osmotic pressures between 0.224 and 44.8 bars. *Anguina tritici* and *A. agrostis,* which can withstand desiccation the longest, respired more than the others at high osmotic pressures, and *P. penetrans,* not especially known to survive long periods, respired the least. *Ditylenchus dipsaci,* known for intermediate survival, was intermediate in respiration, but was not greatly different from *P. penetrans.*

ORGANIC MATTER

Residues

Organic matter serves many important functions in the soil. Among these, the soil acts as a substrate for microorganisms that are parasitic on nematodes or that produce toxins, directly or through breakdown products, that are detrimental to nematodes. Soil microorganisms also compete with nematodes for substrate, gases, and other life-sustaining elements.

Linford et al (45) reported that reduced galling occurred after chopped plants were incorporated into root-knot infested soil. The belief was that plant decomposition probably resulted in increased total nematode numbers which in turn supported life destructive to harmful nematodes. The parasitisms and predations that result from adding organic amendments are important in controlling nematode populations, but greater attention has been given in recent years to other explanations for nematode control after adding crop residues. Following on the work of Linford et al, many other amendments have been used in attempts to reduce numbers of plant-parasitic nematodes. Results have been variable but encouraging enough to warrant further study. Organic amendments have included ground tobacco stems (53); forages (31, 48); peat moss (64); plant products such as sawdust (80), paper, cottonseed meal (53), hardwood bark (47), cornmeal, soybean meal and oils (104), oil cakes (81), cotton waste, sugar beet pulp, castor pomace (43, 48), chitin, cellulose, and mycelial residues (52); animal manure or products (48); and other organic amendments (26).

As a result of the addition of some amendments, organisms that are antagonistic to their parasitic associates increase. Bergeson et al (5) found that an increase in actinomycetes in chitin-amended soil was associated with a decrease of secondary invasion of the root-knot galls in okra.

A period of "incubation" was sometimes necessary to obtain the best control from organic amendments. Incubation periods have varied from short for herbaceous materials to up to one year for maximum nematode reduction with highly cellulosic materials (29, 53, 105).

Mechanisms for Control

Besides an increasing interest in practical results in nematode control by chemicals in the soil, there also is increasing interest in the mechanisms of control.

Thorne (93) was one of the first to suggest that nematotoxic gases were produced during the decomposition of plants, sweet clover in his instance. Since then many other materials produced during decay of plant tissues have been demonstrated or suggested to have toxic effects on nematodes. These include tannins from raspberry against *Longidorus elongatus* (90), and oil cakes affecting hatching of *Meloidogyne javanica* eggs (81). Extracts of decomposing rye were selectively nematicidal in that higher concentrations were required to immobilize some saprobic nematodes than needed to immobilize *M. incognita* and *Pratylenchus penetrans* (66). Similar results were found by Sayre et al (74).

Organic acids have been implicated as toxic agents (3, 18, 28, 34, 69, 74, 86). Sometimes organic acids were present in greater quantities in decaying residue in the laboratory than in the field (30, 74). Sayre et al (74) reported that butyric acid was active only at pH 4.0–5.3. Hollis (27) found that N-butyric acid and smaller quantities of propionic acid increased rapidly in cornmeal-treated soil. At concentrations comparable to those found in cornmeal treated soil, butyric acid killed all nematodes, mainly *Tylenchorhynchus martini,* within a few hours. Nontoxic concentrations of propionic acid had an additive effect. The bacterium, *Clostridium butyricum,* which was present in the tests, is known to produce these two organic acids.

Stephenson (86) found that mineral acids had a greater killing effect on *Rhabditis terrestris* than organic acids. Of the 16 acids tested, nitric acid killed the nematodes in about 1.5 min while succinic acid took more than 16 min at comparable normalities. The writer attributed lethality partly to the lowering of the pH and partly to the dissimilar effects of undissociated ions. Formic, acetic, propionic, butyric, and valeric acids were toxic to *Dorylaimus* at concentrations of 1 to 0.001 M in distilled water. Formic acid was less toxic than the others. Fulvic acid or the fulvic acid fraction of soils has been shown to be deleterious to *Aphelenchoides goodeyi, Helicotylenchus pseudorobustus,* and *Xiphinema americanum* (18, 69).

Although control of nematodes by organic amendments is promising, implementation has not been achieved on a practical scale. The quantity of amendments necessary to bring about effective control often exceeds practical supply. Also, Johnson et al (31) found that nematicides were generally superior in reducing disease caused by *Meloidogyne incognita* than was the incorporation of crop residues. Sometimes it has been difficult to distinguish the yield increases due to nematode control from increase due to fertility added by the organic matter.

Control mechanisms are not understood in most instances. Biological control is occurring all of the time and the organic matter in the soil forms much of the substrate for these competing organisms. Man's task is to alter the environment so as to encourage competing organisms and yet not be detrimental to the environment for agricultural purposes. Spectacular success should not be expected, but modest achievements may complement more direct control measures.

pH

Numerous reports suggest that pH does not influence nematode populations in all situations (87, 116–118). This idea has led to the assumption that changes in pH affect the host more than it does the nematode. In view of the importance of pH in biological systems, it would be surprising if nematodes are not influenced by it. Many correlations of pH with nematode phenomena were developed in experiments that lacked adequate controls or were based on so few samples that cause and effect relationships could not be determined. Currently, however, the literature suggests that the pH of the soil may well be a significant factor in nematode behavior.

From 250 soil samples from around cabbage roots in Poland, Brzeski (8) found that *Tylenchorhynchus dubius* and *Pratylenchus crenatus* occurred more frequently ($P = 0.01$) in acid than in neutral or alkaline soils. *Heterodera schachtii* was associated with either neutral or acid soils. Trends were independent of soil type. This was confirmed for *T. dubius* in pot experiments in which the soil was adjusted to different pH levels with sulfuric acid (10). The three-month specific growth rates at pH 4.0, 5.5, 6.2, and 7.2 were 2.6, 2.2, 1.4, and 1.0 respectively. The highest hatching rate in deep well slides occurred at pH 7.0. In 1970 Brzeski (9) reported that frequencies (i.e. how often a species occurs among samples) among nematodes in 335 soil samples around carrot varied with pH. Frequencies of *T. dubius* and *P. crenatus* decreased as pH increased, these trends holding true in either sandy or silty soils.

Using initial pH values of 4.0, 6.0, and 8.0, Burns (11) found that the greatest colonization of soybean roots by *Pratylenchus alleni* was at pH 6.0 ($P = 0.01$). *Hoplolaimus galeatus* survived in greatest numbers at pH 6.0 ($P = 0.05$) as did *Xiphinema americanum* but the latter was not significant at $P = 0.05$. Morgan & MacLean (55) found that *Pratylenchus penetrans* grew best in vetch roots at pH 5.5–5.8 and declined rapidly at pH 6.6 and above. The root systems apparently were not affected by the pH treatments (ranging from pH 4.8 to 7.5) but there was some inhibition of top growth at pH 7.1 and above.

The 90-day specific growth rates for *Tylenchorhynchus vulgaris* on corn in soil adjusted to pH 5.5, 6.5, 7.7, 8.5, and 9.5 were 54.1, 52.5, 46.5, 38.3, and 30.9, respectively (98). Thus the specific growth rate decreased with increasing pH over the pH values studied. In greenhouse studies, 30-week specific growth rates of *Pratylenchus penetrans* in alfalfa were 64.4 and 47.1 at pH 5.2 and 6.4 respectively, but only 4.1 and 2.9 at pH 4.4 and 7.3, respectively (113). Although root growth was poor at pH 4.4, it was good at pH 7.3, 5.2, and 6.4. In composite samples collected over the season in 40 soybean fields over two years, Norton et al (62) found significant negative correlations of *Hoplolaimus galeatus* and *Tylenchorhynchus nudus* with pH, but a significant positive correlation with *Helicotylenchus pseudorobustus*. Interdependence with other parameters no doubt was involved. Duggan (17) collected 625 field soil samples with a pH range from 5.0 to 7.5 and found a positive correlation of *Heterodera avenae* with pH.

Studies of natural communities usually are more complicated than those in monocultural ecosystems. But sometimes other factors can be eliminated as major influences where one might think that pH is partly a controlling factor or at least correlated indirectly. For example, large numbers of *Xiphinema americanum* can be found around *Betula papyrifera* in Iowa soils with a pH averaging 6.1 (range 4.4–7.8), but the nematode was rare in woodlands in the northeastern United States where the pH averaged 4.0 (range 2.8–6.9) (63). Moisture and temperature probably were not limiting in either case, but the effect of soil texture was not determined. Even where mineral soils were sampled, the nematode was not found at low pH values in the northeastern woodlands.

SUMMARY

Populations of plant-parasitic nematodes that spend most of their life in the soil are influenced greatly by the soil environment. Nematodes entirely within the roots are controlled by the root environment, and indirectly by the soil environment because the latter affects plant growth.

Physiological requirements for different species of nematodes vary; for this reason nematodes function in different niches and occupy different habitats. Although some namatodes are known to occur and increase more rapidly in some habitats than in others, our knowledge is far from complete, especially where differences in soil characteristics are not obvious. Increasing our knowledge of nematode-soil relationships will increase our ability to predict nematode occurrence. The complex ecosystem in the soil makes a complete understanding of nematode-soil relationships difficult. Knowledge of the physical-chemical factors in the soil will give only partial answers. Biological relationships of the soil biota, including that of the host, must be evaluated along with the physical-chemical ones.

Some of the physical-chemical parameters studied most are temperature, moisture, percentage organic matter, pH, and texture. Our knowledge of these factors is still very incomplete. Other parameters also may be important. The plant nematologist must keep abreast with advancements in soil physics and the ecology of other soil biota to gain an adequate perspective of the ecology of nematodes. This is not an easy task. Specialization, with the danger of losing the overview, probably is inevitable. Integrated studies of ecology of the soil biota must be made.

Literature Cited

1. All, J. N., Kuhn, C. W., Gallaher, R. N., Jellum, M. D., Hussey, R. S. 1977. Influence of no-tillage-cropping, carbofuran, and hybrid resistance on dynamics of maize chlorotic dwarf and maize dwarf mosaic diseases of corn. *J. Econ. Entomol.* 70:221–25

2. Andrewartha, H. G., Birch, L. C. 1954. *The Distribution and Abundance of Animals.* Chicago: Univ. Chicago Press. 782 pp.

3. Banage, W. B., Visser, S. A. 1965. The effect of some fatty acids and pH on a soil nematode. *Nematologica* 11:255–62

4. Barker, K. R., Sasser, J. N. 1959. Biology and control of the stem nematode, *Ditylenchus dipsaci. Phytopathology* 49:664–70

5. Bergeson, G. B., van Gundy, S. D., Thomason, I. J. 1970. Effect of *Meloidogyne javanica* on rhizosphere microflora and fusarium wilt of tomato. *Phytopathology* 60:1245–49

6. Bhatt, B. D., Rohde, R. A. 1970. The influence of environmental factors on the respiration of plant-parasitic nematodes. *J. Nematol.* 2:277–85

7. Brown, L. N. 1933. Flooding to control root-knot nematodes. *J. Agric. Res.* 47:883–88

8. Brzeski, M. W. 1969. Nematodes associated with cabbage in Poland. II. The effect of soil factors on the frequency of nematode occurrence. *Ekol. Pol. Ser. A* 17:205–25

9. Brzeski, M. 1970. Plant parasitic nematodes associated with carrot in Poland. *Rocz. Nauk. Roln. Ser. E* 1:93–102

10. Brzeski, M. W., Dowe, A. 1969. Effect of pH on *Tylenchorhynchus dubius* (Nematoda, Tylenchidae). *Nematologica* 15:403–7

11. Burns, N. C. 1971. Soil pH effects on nematode populations associated with soybeans. *J. Nematol.* 3:238–45

12. Chang, H. Y., Raski, D. J. 1972. *Hemicriconemoides chitwoodi* on grapevines. *Plant Dis. Reptr.* 56:1028–30

13. Cohn, E. 1969. The occurrence and distribution of species of *Xiphinema* and *Longidorus* in Israel. *Nematologica* 15: 179–92

14. Corbett, D. C. M., Webb, R. M. 1970. Plant and soil nematode population changes in wheat grown continuously in ploughed and in unploughed soil. *Ann. Appl. Biol.* 65:327–35

15. Cralley, E. M. 1957. The effect of seeding methods on the severity of white tip of rice. *Phytopathology* 47:7 (Abstr.)

16. Dropkin, V. H., Martin, G. C., Johnson, R. W. 1958. Effect of osmotic concentration on hatching of some plant parasitic nematodes. *Nematologica* 3: 115–26

17. Duggan, J. J. 1963. Relationship between intensity of cereal root eelworm (*Heterodera avenae* Wollenweber 1924) infestation and pH value of soil. *Ir. J. Agric. Res.* 2:105–10

18. Elmiligy, I. A., Norton, D. C. 1973. Survival and reproduction of some nematodes as affected by muck and organic acids. *J. Nematol.* 5:50–54

19. Endo, B. Y. 1959. Responses of rootlesion nematodes, *Pratylenchus brachyurus* and *P. zeae,* to various plants and soil types. *Phytopathology* 49: 417–21

20. Fassuliotis, G., Rau, G. J. 1966. Observations on the embryogeny and histopathology of *Hypsoperine spartinae* on smooth cordgrass roots, *Spartina alterniflora. Nematologica* 12:90 (Abstr.)

21. Faulkner, L. R. 1964. Pathogenicity and population dynamics of *Paratylenchus hamatus* on *Mentha* spp. *Phytopathology* 54:344–48

22. Ferris, J. M. 1970. Soil temperature effects on onion seedling injury by *Pratylenchus penetrans. J. Nematol.* 2:248–51

23. Griffin, G. D., Darling, H. M. 1964. An ecological study of *Xiphinema americanum* Cobb in an ornamental spruce nursery. *Nematologica* 10:471–79

24. Grundbacher, F. J., Stanford, E. J. 1962. Effect of temperature on resistance of alfalfa to the stem nematode (*Ditylenchus dipsaci*). *Phytopathology* 52:791–94

25. Hamblen, M. L., Slack, D. A. 1959. Factors influencing the emergence of larvae from cysts of *Heterodera glycines* Ichinohe. Cyst development, condition, and variability. *Phytopathology* 49:317 (Abstr.)

26. Heald, C. M., Burton, G. W. 1968. Effect of organic and inorganic nitrogen on nematode populations on turf. *Plant Dis. Reptr.* 52:46–48

27. Hollis, J. P. 1958. Induced swarming of a nematode as a means of isolation. *Nature* 182:956–57

28. Hollis, J. P., Fielding, M. J. 1958. Population behavior of plant parasitic nematodes in soil fumigation experiments. *La. Agric. Exp. Stn. Bull.* 515:1–23

29. Johnson, L. F. 1959. Effect of the addition of organic amendments to soil on root knot of tomatoes. I. Preliminary report. *Plant Dis. Reptr.* 43:1059–62

30. Johnson, L. F. 1974. Extraction of oat straw, flax, and amended soil to detect substances toxic to the root-knot nematode. *Phytopathology* 64:1471–73

31. Johnson, L. F., Chambers, A. Y., Reed, H. E. 1967. Reduction of root knot of tomatoes with crop residue amendments in field experiments. *Plant Dis. Reptr.* 51:219–22

32. Johnson, S. R., Ferris, J. M., Ferris, V. R. 1973. Nematode community structure in forest woodlots. II. Ordination of nematode communities. *J. Nematol.* 5:95–107

33. Johnson, S. R., Ferris, J. M., Ferris, V. R. 1974. Nematode community structure of forest woodlots. III. Ordinations of taxonomic groups and biomass. *J. Nematol.* 6:118–26

34. Johnston, T. M. 1959. 'ʃfect of fatty acids mixtures on the rice stylet nematode (*Tylenchorhynchus martini* Fielding, 1956). *Nature* 183:1392

35. Jones, F. G. W. 1975. Accumulated temperature and rainfall as measures of nematode development and activity. *Nematologica* 21:62–70

36. Jones, F. G. W., Larbey, D. W., Parrott, D. M. 1969. The influence of soil structure and moisture on nematodes, especially *Xiphinema, Longidorus, Trichodorus* and *Heterodera* spp. *Soil Biol. Biochem.* 1:153–65

37. Jones, F. G. W., Thomasson, A. J. 1976. Bulk density as an indicator of pore space in soils usable by nematodes. *Nematologica* 22:133–37

38. Kable, P. F., Mai, W. F. 1968. Influence of soil moisture on *Pratylenchus penetrans. Nematologica* 14:101–22

39. Klekowski, R. Z., Wasilewska, L., Paplinska, E. 1972. Oxygen consumption by soil-inhabiting nematodes. *Nematologica* 18:391–403

40. Knobloch, N., Bird, G. W. 1978. Criconematinae habitats and *Lobocriconema thornei* n. sp. (Criconematidae: Nematoda). *J. Nematol.* 10: 61–70

41. Krebill, R. G., Barker, K. R., Patton, R. F. 1967. Plant-parasitic nematodes of jack and red pine stands in Wisconsin. *Nematologica* 13:33–42

42. Lamberti, F. 1969. Effect of temperature on the reproduction rate of *Longidorus africanus. Plant Dis. Reptr.* 53:559

43. Lear, B. 1959. Application of castor pomace and cropping of castor beans to soil to reduce nematode populations. *Plant Dis. Reptr.* 43:459–60

44. Lindeman, R. L. 1942. The trophic-dynamic aspect of ecology. *Ecology* 23:399–418

45. Linford, M. B., Yap, F., Oliveira, J. M. 1938. Reductions of soil populations of root-knot nematode during decomposition of organic matter. *Soil Sci.* 45: 127–41

46. Mai, W. F., Parker, K. G. 1956. Evidence that the nematode *Pratylenchus penetrans* causes losses in New York State cherry orchards. *Phytopathology* 46:19 (Abstr.)

47. Malek, R. B., Gartner, J. B. 1975. Hardwood bark as a soil amendment for suppression of plant parasitic nematodes on container-grown plants. *Hortic. Sci.* 10:33–35

48. Mankau, R., Minteer, R. J. 1962. Reduction of soil populations of the citrus nematode by the addition of organic materials. *Plant Dis. Reptr.* 46:375–78

49. McGlohon, N. E., Sasser, J. N., Sherwood, R. T. 1961. Investigations of

plant parasitic nematodes associated with forage crops in North Carolina. *NC Agric. Exp. Stn. Tech. Bull.* 148: 1–39

50. Meagher, J. W. 1968. The distribution of the cereal cyst nematode (*Heterodera avenae*) in Victoria and its relation to soil type. *Aust. J. Exp. Agric. Anim. Husb.* 8:637–40

51. Miller, L. I. 1972. The influence of soil texture on the survival of *Belonolaimus longicaudatus*. *Phytopathology* 62:670–71 (Abstr.)

52. Miller, P. M., Sands, D. C., Rich, S. 1973. Effect of industrial mycelial residues, wood waste fibers, and chitin on plant-parasitic nematodes and some soilborne diseases. *Plant Dis. Reptr.* 57:438–42

53. Miller, P. M., Taylor, G. S., Wihrheim, S. E. 1968. Effects of cellulosic soil amendments and fertilizers on *Heterodera tabacum*. *Plant Dis. Reptr.* 52: 441–45

54. Milne, D. L., DuPlessis, D. P. 1964. Development of *Meloidogyne javanica* (Treub.) Chit., on tobacco under fluctuating soil temperatures. *S. Afr. J. Agric. Sci.* 7:673–80

55. Morgan, G. T., MacLean, A. A. 1968. Influence of soil pH on an introduced population of *Pratylenchus penetrans*. *Nematologica* 14:311–12

56. Mountain, W. B., Boyce, H. R. 1958. The peach replant problem in Ontario. V. The relation of parasitic nematodes to regional differences in severity of peach replant failure. *Can. J. Bot.* 36:125–34

57. Mugniery, D. 1978. Vitesse de développement, en fonction de la température, de *Globodera rostochiensis* et *G. pallida* (Nematoda: Heteroderidae). *Rev. Nematol.* 1:3–12

58. Nikandrow, A., Blake, C. D. 1972. Oxygen and the hatch of eggs and development of larvae of *Aphelenchoides composticola* and *Ditylenchus myceliophagus*. *Nematologica* 18:309–19

59. Norton, D. C. 1963. Population fluctuations of *Xiphinema americanum* in Iowa. *Phytopathology* 53:66–68

60. Norton, D. C. 1977. *Helicotylenchus pseudorobustus* as a pathogen on corn, and its densities on corn and soybean. *Iowa State J. Res.* 51:279–85

61. Norton, D. C. 1978. *Ecology of Plant-Parasitic Nematodes.* New York: Wiley-Interscience. 268 pp.

62. Norton, D. C., Frederick, L. R., Ponchillia, P. E., Nyhan, J. W. 1971. Correlations of nematodes and soil properties in soybean fields. *J. Nematol.* 3:154–63

63. Norton, D. C., Hoffmann, J. K. 1974. Distribution of selected plant parasitic nematodes relative to vegetation and edaphic factors. *J. Nematol.* 6:81–86

64. O'Bannon, J. H. 1968. The influence of an organic soil amendment on infectivity and reproduction of *Tylenchulus semipenetrans* on two citrus rootstocks. *Phytopathology* 58:597–601

65. Parker, K. G., Mai, W. F. 1956. Damage to tree fruits in New York by root lesion nematodes. *Plant Dis. Reptr.* 40:694–99

66. Patrick, Z. A., Sayre, R. M., Thorpe, H. J. 1965. Nematocidal substances selective for plant-parasitic nematodes in extracts of decomposing rye. *Phytopathology* 55:702–4

67. Peacock, F. C. 1957. Studies on root knot nematodes of the genus *Meloidogyne* in the Gold Coast. II. The effect of soil moisture content on survival of the organism. *Nematologica* 2:114–22

68. Pielou, E. C. 1974. *Population and Community Ecology. Principles and Methods.* New York: Gordon & Breach. 424 pp.

69. Ponchillia, P. E. 1972. *Xiphinema americanum* as affected by soil organic matter and porosity. *J. Nematol.* 4: 189–93

70. Rau, G. J. 1963. Three new species of *Belonolaimus* (Nematoda: Tylenchida) with additional data on *B. longicaudatus* and *G. gracilis. Proc. Helminthol. Soc. Wash.* 30:119–28

71. Richter, C. 1969. Zur Vertikalen Verteilung von Nematoden in einem Sandboden. *Nematologica* 15:44–54

72. Robbins, R. T., Barker, K. R. 1974. The effects of soil type, particle size, temperature, and moisture on reproduction of *Belonolaimus longicaudatus. J. Nematol.* 6:1–6

73. Rössner, J. 1971. Einfluss der Austrocknung des Bodens auf Wandernde Wurzelnematoden. *Nematologica* 17: 127–44

74. Sayre, R. M., Patrick, Z. A., Thorpe, H. J. 1965. Identification of a selective nematicidal component in extracts of plant residues decomposing in soil. *Nematologica* 11:263–68

75. Schmitt, D. P. 1973. Soil property influences on *Xiphinema americanum* populations as related to maturity of loess-derived soils. *J. Nematol.* 5:234–40

76. Schmitt, D. P., Norton, D. C. 1972. Relationships of plant parasitic nematodes to sites in native Iowa prairies. *J. Nematol.* 4:200–6

77. Seinhorst, J. W. 1956. Population studies on stem eelworms (*Ditylenchus dipsaci*). *Nematologica* 1:159–64

78. Seshadri, A. R. 1964. Investigations on the biology and life cycle of *Criconemoides xenoplax* Raski, 1952 (Nematoda: Criconematidae). *Nematologica* 10:540–62

79. Simons, W. R. 1973. Nematode survival in relation to soil moisture. *Meded. Landbouwhogesch. Wageningen* 73-3:1–85

80. Singh, R. S., Singh, B., Beniwal, S. P. S. 1967. Observations on the effect of sawdust on incidence of root knot and on yield of okra and tomatoes in nematode-infested soil. *Plant Dis. Reptr.* 51:861–63

81. Singh, R. S., Sitaramaiah, K. 1966. Incidence of root knot of okra and tomatoes in oil-cake amended soil. *Plant Dis. Reptr.* 50:668–72

82. Slootweg, A. F. G. 1956. Rootrot of bulbs caused by *Pratylenchus* and *Hoplolaimus* spp. *Nematologica* 1:192–201

83. Southards, C. J. 1971. Effect of fall tillage and selected hosts on the population density of *Meloidogyne incognita* and *Pratylenchus zeae. Plant Dis. Reptr.* 55:41–44

84. Springer, J. K. 1964. Nematodes associated with plants in cultivated woody plant nurseries and uncultivated woodland areas of New Jersey. *NJ Dep. Agric. Cir.* 429:1–40

85. Starr, J. L., Mai, W. F. 1976. Predicting on-set of egg production by *Meloidogyne hapla* on lettuce from field soil temperatures. *J. Nematol.* 8:87–88

86. Stephenson, W. 1945. The effect of acids on a soil nematode. *Parasitology* 36:158–64

87. Stöckli, A. 1952. Studien über Bodennematoden mit besonderer Berücksichtigung des Nematodengehaltes von Wald-, Grünland- und ackerbaulich genutzten Boden. *Z. Pflanzenernaeh. Dueng. Bodenkd.* 59:97–139

88. Tansley, A. G. 1935. The use and abuse of vegetational concepts and terms. *Ecology* 16:284–307

89. Taubenhaus, J. J., Ezekiel, W. N. 1933. Checklist of diseases of plants in Texas. *Trans. Tex. Acad. Sci.* 16:5–118

90. Taylor, C. E., Murant, A. F. 1966. Nematicidal activity of aqueous extracts from raspberry canes and roots. *Nematologica* 12:488–94

91. Thomas, S. H. 1978. Population densities of nematodes under seven tillage regimes. *J. Nematol.* 10:24–27

92. Thomason, I. J., O'Melia, F. C. 1962. Pathogenicity of *Pratylenchus scribneri* of crop plants. *Phytopathology* 52:755 (Abstr.)

93. Thorne, G. 1926. Control of sugar-beet nematode by crop rotation *US Dep. Agric. Farmer's Bull. 1514.* 2 pp.

94. Townshend, J. L. 1972. Influence of edaphic factors on penetration of corn roots by *Pratylenchus penetrans* and *P. minyus* in three Ontario soils. *Nematologica* 18:201–12

95. Tseng, S. T., Allred, K. R., Griffin, G. D. 1968. A soil population study of *Ditylenchus dipsaci* (Kühn) Filipjev in an alfalfa field. *Proc. Helminthol. Soc. Wash.* 35:57–62

96. Tyler, J. 1933. Development of the root-knot nematode as affected by temperature. *Hilgardia* 7:391–415

97. Unger, P. W., Phillips, R. E. 1973. Soil water evaporation and storage. Conservation tillage. *Soil Conserv. Soc. Am.* 1973:42–54

98. Upadhyay, K. D., Swarup, G. 1972. Culturing, host range and factors affecting multiplication of *Tylenchorhynchus vulgaris* on maize. *Indian J. Nematol.* 2:139–45

99. Van Gundy, S. D., McElroy, F. D., Cooper, A. F., Stolzy, L. H. 1968. Influence of soil temperature, irrigation and aeration on *Hemicycliophora arenaria. Soil Sci.* 106:270–74

100. Van Gundy, S. D., Rackham, R. L. 1961. Studies on the biology and pathogenicity of *Hemicycliophora arenaria. Phytopathology* 51:393–97

101. Van Gundy, S. D., Stolzy, L. H., Szuszkiewicz, T. E., Rackham, R. L. 1962. Influence of oxygen supply on survival of plant parasitic nematodes in soil. *Phytopathology* 52:628–32

102. Viglierchio, D. R., Croll, N. A., Gortz, J. H. 1969. The physiological response of nematodes to osmotic stress and an osmotic treatment for separating nematodes. *Nematologica* 15:15–21

103. Vrain, T. C., Barker, K. R., Holtzman, G. I. 1978. Influence of low temperature on rate of development of *Meloidogyne incognita* and *M. hapla* larvae. *J. Nematol.* 10:166–71

104. Walker, J. T. 1969. *Pratylenchus penetrans* (Cobb) populations as influenced by microorganisms and soil amendments. *J. Nematol.* 1:260–64

105. Walker, J. T., Specht, C. H., Mavrodineau, S. 1967. Reduction of lesion nematodes in soybean meal and oil-amended soils. *Plant Dis. Reptr.* 51:1021–24

106. Wallace, H. R. 1958. Movement of eelworms. I. The influence of pore size and moisture content of the soil on the migration of larvae of the beet eelworm, *Heterodera schachtii* Schmidt. *Ann. App. Biol.* 46:74–85
107. Wallace, H. R. 1959. Further observations on some factors influencing the emergence of larvae from cysts of the beet eelworm, *Heterodera schachtii* Schmidt. *Nematologica* 4:245–52
108. Wallace, H. R. 1962. Observations on the behaviour of *Ditylenchus dipsaci* in soil. *Nematologica* 7:91–101
109. Wallace, H. R. 1963. *The Biology of Plant Parasitic Nematodes.* London: Arnold. 280 pp.
110. Wallace, H. R. 1973. *Nematode Ecology and Plant Disease.* London: Arnold. 228 pp.
111. Wallace, H. R., Greet, D. N. 1964. Observations on the taxonomy and biology of *Tylenchorhynchus macrurus* (Goodey, 1932) Filipjev, 1936 and *Tylenchorhynchus icarus* sp. nov. *Parasitology* 54:129–44
112. Watson, J. R. 1921. Control of rootknot. II. *Fla. Agric. Exp. Stn. Bull.* 159:30–44
113. Willis, C. B. 1972. Effects of soil pH on reproduction of *Pratylenchus penetrans* and forage yield of alfalfa. *J. Nematol.* 4:291–95
114. Willis, W. O., Amemiya, M. 1973. Tillage management principles: Soil temperature effects. Conservation tillage. *Soil Conserv. Soc. Am.* 1973:22–42
115. Winslow, R. D. 1960. Some aspects of the ecology of free-living and plant-parasitic nematodes. In *Nematology*, ed. J. N. Sasser, W. R. Jenkins, pp. 341–415. Chapel Hill: Univ. North Carolina Press
116. Yeates, G. W. 1968. An analysis of annual variation of the nematode fauna in sand dune at Himatangi Beach, New Zealand. *Pedobiologia* 8:173–207
117. Yeates, G. W. 1973. Abundance and distribution of soil nematodes in samples from the New Hebrides. *NZ J. Sci.* 16:727–36
118. Yeates, G. W. 1974. Studies on a climosequence of soils in tussock grasslands. 2. Nematodes. *NZ J. Zool.* 1:171–77

Ann. Rev. Phytopathol. 1979. 17:301–10

BIOLOGICAL WEED CONTROL   ♦3712
WITH MYCOHERBICIDES[1,2]

George E. Templeton and David O. TeBeest

Department of Plant Pathology, University of Arkansas,
Fayetteville, Arkansas 72701

Roy J. Smith, Jr.

US Department of Agriculture, Science and Education Administration—
Agricultural Research, Stuttgart, Arkansas 72160

INTRODUCTION

Chemical herbicides are the most effective immediate solution to most weed problems but they are not the only or necessarily the best solution. Certain fungi can be used to control weeds and in some cases are more efficaceous than chemicals. We accept universally that plant pathogens can kill plants. Previous reviews have illustrated the need for and theorized on the potential of using plant pathogens for weed control (35, 37). We now have empirically determined that pathogens can control weeds when properly employed either as classical biological control agents or as microbial herbicides. Previously, we have addressed some of the economic, regulatory, and technological constraints to their development for general use (32). We foresee removal, or at least reduction, of the deterrents and suggest it is now appropriate to focus on the biotic and abiotic constraints to diseases in nature that impact on safety and efficacy of biological agents for weed control. This should

[1] The US Government has the right to retain a nonexclusive, royalty-free license in and to any copyright covering this paper.

[2] Cooperative investigations of the University of Arkansas Agricultural Experiment Station and the Agriculture Research, Science and Education Administration, US Department of Agriculture. Published with the approval of the Director of the Arkansas Agricultural Experiment Station.

clarify our perspective of the research that is needed on the fundamental ecology of plant pathogens in nature to guide further empirical development in this new cooperative research arena of plant pathology and weed science.

CURRENT TACTICS IN BIOLOGICAL CONTROL

Current efforts to control weeds with plant pathogens fall into two categories—the classic tactic and the bioherbicide tactic (32).

The Classical Tactic

The classic biocontrol tactic, in essence, is the importation of a pathogen from the area of co-evolution with its host, and its release into a new geographic area where the host already exists and has become a weedy pest in the absence of its pathogen. Control of the target host is dependent upon self-perpetuation and natural dispersal of the pathogen.

Classic biological control with plant pathogenic fungi is being studied and has been successfully used to control at least two weeds of economic importance. Chondrilla rust, *Puccinia chondrillina* Bubak and Syd., which is indigenous in the Mediterranean area, has been introduced into Australia; there it has controlled rush skeletonweed, *Chondrilla juncea* L., which had invaded millions of acres of wheat (*Triticum aestivum* L.) and ranges (8). Similar efforts with chondrilla rust are under way in the western United States under the auspices of the US Department of Agriculture's Plant Protection Laboratory in Frederick, Maryland and the state Departments of Agriculture in Oregon, Washington, Idaho, and California (C. H. Kingsolver, unpublished).

The same classic strategy is being actively pursued in Florida by the University of Florida and the US Corps of Engineers to control water hyacinth [*Eichornia crassipes* (Mart.) Solms.] with a foreign rust (not yet released) and a native but geographically restricted species of *Cercospora* (4, 5). The level of water hyacinth control with the indigenous *Cercospora* has not equalled that obtained with chemical herbicides, but timely applications of this pathogen have reduced water hyacinth biomass (6).

In Chile the autoecious blackberry rust *Phragmidium violaceum* (Schultz) Winter introduced from Europe suppressed blackberries (*Rubus constrictus* Lef. et M. and *R. ulmifolius* Schott.) encroaching on ranges and pastures (20).

In Hawaii introduction of *Cercosporella ageratinae nomen nudem (C. adenophorum* Spreng.) from Jamaica controlled pamakani weed (*Agerarina riparia-Eupatorium riparia*) in ranges and pastures (E. E. Trujillo, unpublished information).

The classical biocontrol strategy will require further research to determine its potential for controlling weeds. Classic weed control with fungal pathogens is comparable to classic biocontrol with insects, and many of the principles apply to both. The best targets for the use of exotic pathogens are weeds introduced into the United States from other areas, and the geographical origin of the weed is the best source of natural enemies. Weed pathogens should be especially beneficial on introduced perennial weeds that grow in dense stands and infest large land areas, particularly when small residual populations of the host do not cause economic losses and other weed control practices are not economically or environmentally sound (17). Certainly, the use of pathogens in the classic approach has potential for controlling alien weeds that are geographically isolated from their parasites. Plants introduced into new regions or disturbed areas without their fungal parasites often spread and increase unimpeded to become weedy pests. Many of the aquatic weeds, such as water hyacinth, alligator-weed [*Alternanthera philoxeroides* (Mart.) Griseb.], hydrilla (*Hydrilla verticillata* Royle) and eurasian watermilfoil (*Myriophyllum spicatum* L.) are good candidate weeds for the classical strategy of biocontrol as are many range weeds, including mesquite (*Prosopis* spp.), rush skeletonweed, lantana (*Lantana camara* L.), thistles (*Crisium* spp. and *Sonchus* spp.) and Texas persimmon (*Diospyros texana* Scheele) and common persimmon (*D. virginiana* L.). The permanency, environmental safety, and economy of the classic strategy are strong inducements for further research by public agencies to find and evaluate pathogens for use in that strategy. Successful control of weeds with fungi used in the classic approach should stimulate additional research.

The Bioherbicide Tactic

Fortunately, contemporary biological control research is expanding. New strategies are being sought for alternative methods of weed control for integration into weed management systems. One new strategy is to employ microbes as herbicides by applying them to target weeds in a manner similar to chemical herbicides (2, 9, 10, 11, 14, 18, 21, 22, 24, 29, 30, 33). Fungal plant pathogens applied as sprays that uniformly kill or suppress weed growth are logically termed *mycoherbicides* (31). Exotic and indigenous fungi have potential for use as mycoherbicides, but up to now only indigenous fungi have been researched. Both alien and indigenous weeds may be controlled by mycoherbicides. Weed pathogens used as mycoherbicides have the greatest potential for use on hard-to-control weeds in annual crops where, specificity, immediacy, and completeness of control are paramount. Classical biocontrol operates too slowly and may not reduce pests below economic thresholds. Mycoherbicides are also being developed for use against perennial weeds in orchard crops (22); they may also have potential for control of range and aquatic weeds if chemical herbicides cannot be used.

The concept of using fungi as mycoherbicides is biologically feasible with several host-pathogen combinations. Northern jointvetch [*Aeschynomene virginica* (L.) B.S.P.] and winged waterprimrose [*Jussiaea decurrens* (Walt.) DC.] were killed in rice (*Oryza sativa* L.) with host-specific strains of *Colletotrichum gloeosporioides* (Penz.) Sacc. (2, 9, 10, 11). Strangler vine (*Morrenia odorata* Lindl.) has been controlled successfully in citrus groves with strains of *Phytophthora citrophthora* (R. et E. Smith) Leonian (22). The oak wilt fungus, *Ceratocystis fagacearum* (Bretz) Hunt, can be used as a selective silvicide (15). Common persimmons are killed on range land in southern Oklahoma by the persimmon wilt fungus, *Cephalosporium diospyri* Crandall (C. A. Griffith, unpublished). All of these fungi are apparently indigenous to the geographical area in which they are being researched and they incite endemic diseases of their respective hosts. Applying them in massive inoculations as mycoherbicides overcomes or compensates for natural constraints to their epidemic development in nature. Applied as mycoherbicides they commonly kill 95–100% of the weeds. Rarely do epidemics of this intensity occur naturally either in undisturbed ecosystems or in agroecosystems of cultivated crops. On rare occasions soilborne root diseases may achieve this level of epidemic intensity, but pathogens developing on aerial plant parts rarely destroy their host. Even in woody perennials, where individual plant kill is very noticeable and occurs over large geographical areas, control of all plants of a species by a pathogen is an uncommon phenomenon (7). Does this mean that all "aerial" pathogens are useful as mycoherbicides, provided technical difficulties of mass spore production can be overcome? Experience tells us to avoid such generalizations in biology. Each pathogen must be considered on an individual basis. A look at the natural constraints on epidemic development of individual pathogens should eventually provide better criteria for selecting those pathogens with greatest potential as mycoherbicides.

NATURAL CONSTRAINTS ON NORTHERN JOINTVETCH ANTHRACNOSE

Natural constraints that sustain the northern jointvetch anthracnose disease at endemic levels in nature, but permit it to be elevated to epidemic levels by manipulation as a mycoherbicide, have been investigated and provide insight to selection of pathogens as biological control agents.

The Pathogen

Control of northern jointvetch with a host-specific strain of the indigenous fungus *Colletotrichum gloeosporioides* has been intensively studied and extensively field tested since its serendipitous discovery in 1969 (2, 9, 11,

23–28, 30). This pathogen has co-evolved with its host in an exquisitely balanced relationship. The host-pathogen relationship provides the fungus, a facultative saprophyte, with an ecological niche in which it does not have to compete with saprobes for food and provides a mechanism for its survival during seasonal or longer periods of disappearance of its host. The disease, an endemic anthracnose, occurs naturally each year on the native population of its leguminous host, but it rarely kills this herbaceous, hard-seeded annual. Peak levels of the pathogen occur as the weed matures when seeds become infected as pods are shed. In addition to seed infection, the fungus may overwinter as mycelium in dead stems in natural weed colonies. Infected weed debris, a source of primary inoculum, is destroyed in agricultural lands by tillage practices that return the plant refuse to the soil where it is colonized and assimilated by saprobes (26).

Certain features of the fungus contribute to the balanced host-pathogen relationship or the endemicity of the disease. The pathogen reproduces exclusively by asexual conidia that persist only during the growing season. Wind dissemination of spores is reduced by their sticky nature and by the host epidermis that frequently covers the (nonsetose) acervuli where spores are produced. Although insects are attracted to the sticky spore masses, no vector relationships are believed to exist. Spore dissemination by insects, however, may contribute to spread of this disease to overcome the restriction of its natural spread, which is by splashing rain and low level of seedborne inoculum. Dissemination of the pathogen parallels the host plants' ability to spread long distances by floating seed pods or individual loments. Lack of a resistant spore stage for overwintering in the soil, reliance upon low levels of seed infection for primary inoculum in the spring (25), and poor aerial dissemination of spores restrict the fungus and contribute to its lag in the parasite-host growth cycle. The balanced growth relationship between weed and pathogen cause the disease to be endemic. These restrictive features of the pathogen are counterbalanced by its extreme virulence (host-susceptibility) and a steep inoculum response curve (rapid increases in disease severity occur with slight increases in inoculum concentrations). Interestingly, host susceptibility and high sensitivity to inoculum levels are not apparent in nature, but they can be demonstrated experimentally.

It is clear that poor inoculum production and dispersal mechanisms are key natural constraints on the northern jointvetch anthracnose pathogen that sustain this host-pathogen interaction at the endemic level.

The Host

The host plant appears to have few properties that serve as natural constraints to epidemic buildup of the disease. It grows in thick colonies in uncultivated environments and is uniformly susceptible to the pathogen. The presence of crop plants when the host is growing as a weed in rice or soybeans [*Glycine max* (L.) Merr.], however, impedes natural spread of the fungus. Consequently, single plants of a colony in untreated rice or soybean fields are often severely diseased at maturity, while other plants in the colony are free of disease or only have small lesions on tissue above the crop canopy.

Northern jointvetch is extremely susceptible to *C. gloeosporioides*. The pathogen is also highly specific and stable. Such extreme susceptibility masks variation in pathogenicity or genetic variations in resistance of the host. Indian jointvetch (*Aeschynomene indica* L.) is the only other plant infected after inoculation with *C. gloeosporioides*. In nature, infection of this weed has not been observed even though it occurs as a cohabitant with the susceptible species, is subjected to the same environment, and is surrounded by the same inoculum sources as the susceptible species (23).

The Environment

Conidial germination, appressoria formation and penetration, and production of secondary inoculum of *C. gloeosporioides* occur in a wide range of enviornments. Temperatures for spore germination and infection of northern jointvetch range from 16 to 36°C. The disease develops in this temperature range, but the optimum temperature for fungus growth is 28°C (27). Free moisture on the weeds is required for conidial germination and aids dissemination of spores. Appressoria form rapidly as moisture levels diminish and are probably resistant to periodic drying. Although infection, spore germination, symptom development, and production of secondary inoculum occur over a wide range of moisture levels, they do not seem to require particular sequential environmental regimes. Because northern jointvetch grows in low, damp areas and in thick colonies in nature, microclimates are usually favorable to the pathogen. Disease development is favored in the field microclimate because crop plants are interspersed with the host weed. Thus, interspersed crop plants restrict spread in nature, but they enhance control with mycoherbicides by improving microclimate for disease development.

IDENTIFYING KEY CONSTRAINTS

From the foregoing discussion, we conclude that low carryover and poor dissemination of inoculum are important constraints that hold the northern jointvetch anthracnose disease to endemic levels in nature. Host resistance, environment, and spatial isolation of the host undoubtedly are contributing constraints, but they are not as important for the endemic pathogen of northern jointvetch as for pathogens of economic crops that primarily cause epidemics (1, 2, 4, 12, 13, 19, 34, 36). Furthermore, disease cycles of all fungi used as mycoherbicides have common characteristics. All spread slowly naturally and do not build up to lethal levels early enough in the life cycle of their hosts to kill it before economic thresholds are reached. However, it cannot be assumed that disease cycles of all endemic pathogens operate in this manner; too few examples exist to make generalizations. Some endemic pathogens may be constrained by host resistance (low virulence), narrow environmental requirements, spore dormancy, or long incubation periods. We can probably find examples among the endemic weed diseases that have constraining elements to represent a continuum of every feature in disease cycles. Research is required to determine whether these constraints could be overcome by massive inoculations with spores. For example, can narrow environmental requirements for infection be overcome by timely applications of high levels of inoculum? Can timely application of the pathogen, at particularly susceptible growth stages of the weed, compensate for low pathogen virulence or high levels of host resistance? Would variation in host genotype be matched by virulent genotypes of the pathogen? Could the pathogen be manipulated by genetic engineering to provide the necessary virulence and specificity? Does genetic variations in host and pathogen reduce weed control below economic threshold levels?

Genetic variation is a universal phenomenon of living organisms. Host plants and pathogens are no exception. This fact must be addressed when considering biological control whether it be with exotic organisms used in the classic control approach or with endemic organisms used as mycoherbicides. Variation relates to both effectiveness and safety and is probably the most important component. Once released, an organism cannot be recovered, nor likely eradicated, and may be self-perpetuating to become a pest itself. We are concerned about genetic variations in introduced or exotic organisms whose responses in a new enviornment cannot be reliably predicted. We are also concerned about genetic variations in endemic organisms because they increase the probability of selecting an undesirable mutant or overloading the biotic or geographic constraints that have held the organism in check.

The potential impact of genetic variations on effectiveness of biological control is particularly obvious to plant pathologists. Plant pathologists have long studied and relied on host resistance as a major means of disease control in crop plants and have repeatedly faced the specter of new races of pathogens overcoming host resistance with ultimate epidemics in agricultural crops (7, 34). Information about specificity, variation, and roles of host resistance in plant pathogens is derived almost exclusively from epidemic diseases, principally with obligate parasites, in agroecosystems. In these systems, man has deployed host-resistant genes to protect the crop that grows in massive monocultures. Instability of the host-pathogen interaction is greatly magnified in such systems and the role of host-resistance in controlling diseases has been highlighted. We must be careful not to transpose concepts of disease resistance in crops to those of endemic diseases of weeds in natural ecosystems from which pathogens, principally facultative saprophytes, are obtained for use as mycoherbicides. Disease development in natural ecosystems is probably influenced more by biotic and abiotic factors other than host resistance. Consequently, genetic uniformity and stability of host and pathogen will probably be more common with endemic diseases of weeds than with epidemic disease of crops in agroecosystems.

CONCLUSIONS

The future development of mycoherbicides for use in integrated pest management systems is dependent on research directed to (a) finding endemic pathogens of major weeds—a task already begun, (b) developing methods for mass production of stable spores, and (c) determining disease cycles in an effort to understand the principal constraints responsible for endemicity of individual diseases. When the limited resources that are devoted to these problems and microbial ecology of plant pathogens are considered, it is unlikely that a complete continuum of constraining elements will be identified in the foreseeable future. However, much can be gained from our knowledge of epidemic diseases of crops in agroecosystems. We must, however, keep clearly in mind that the microbial ecology of epidemicity and endemicity are distinctly different, just as microbial ecology in natural ecosystems is distinct from that in agroecosystems. Epidemics in agroecosystems are characterized as a struggle for survival between two organisms (13), but endemicity in natural ecosystems is better characterized as one of harmony and balance (16). Conceivably, endemic diseases of plants may be beneficial to host survival, by thinning plant stands and reducing competition for nutrients and light. Endemic diseases also may increase species diversity thus reducing the threat of disease epidemics or of total plant destruction by sedentary insects. The few mycoherbicides researched, thus far, support these beneficial concepts, but the information that has been generated is insufficient to support generalizations.

Understanding the level of parasitism of a pathogen, or where it ranks on the scale of parasitism between obligate parasite and saprophyte, can aid in predicting the potential of a plant pathogen as a mycoherbicide. At one end of the parasitism scale are the highly host-specific pathogens that harm their obligate hosts relatively little, if any. Although this type of pathogen is widespread, it is held well below epidemic levels in nature—it is not lethal because it is restrained by environment, host resistance, and spatial isolation of host plants in the natural vegetation. At the other end of the scale are the facultative parasites—a group of generalist pathogens that have escaped from the channel of specialization. They have a wide host range, are often lethal to young plants, are capable of living on dead organic matter in the soil, and are most damaging when their hosts are under stresses caused by the environment or other factors (16). These are two extremes and many pathogens occupy an intermediate position of parasitism. The development of pathogens as mycoherbicides depends on finding pathogens of major weeds that are high enough on the scale of parasitism to be specific, but low enough on the scale to be lethal when applied in inundative inoculations. Additional pathogens must be found and evaluated before we can understand the cardinal levels of specificity and virulence on the parasitism scale. Although research will greatly enhance the selection of effective pathogens for weed control, it should, also, contribute substantially to understanding diseases of economic crops and improve the manipulative use of pathogens in integrated weed and disease control programs.

Literature Cited

1. Adams, M. W., Ellingboe, A. H., Rossman, E. C. 1971. Biological uniformity and disease epidemics. *Bioscience* 21:1067–70
2. Boyette, C. D. 1978. *Biological control of winged waterprimrose in rice with an endemic fungal pathogen.* MS thesis. Univ. Arkansas, Fayetteville. 27 pp.
3. Browning, J. A. 1974. Relevance of knowledge about natural ecosystems to development of pest management programs for agro-ecosystems. *Proc. Am. Phytopathol. Soc.* 1:191–99
4. Charudattan, R., McKinney, D. E., Cordo, H. A., Silveira-Guido, A. 1976. *Uredo eichhorniae,* a potential biocontrol agent for waterhyacinth. In *Proc. IV Int. Symp. Biol. Control of Weeds.* ed. T. E. Freeman. pp. 210–13. Gainesville: Univ. Florida. 299 pp.
5. Conway, K. E. 1976. Evaluation of *Cercospora rodmanii* as a biological control of waterhyacinths. *Phytopathology* 66: 914–17
6. Conway, K. E., Freeman, T. E. 1976. The potential of *Cercospora rodmanii* as a biological control for waterhyacinths. See Ref. 4, pp. 207–9
7. Cowling, E. B. 1978. Agricultural and forest practices that favor epidemics. In *Plant Disease: An Advanced Treatise,* Vol. 2, ed. J. G. Horsfall, E. B. Cowling, 361–82. New York: Academic. 436 pp.

8. Cullen, J. M. 1976. Evaluating the success of the programme for the biological control of *Chondrilla juncea* L. See Ref. 4, pp. 117–21
9. Daniel, J. T. 1972. *Biological control of northern jointvetch in rice with a newly discovered Gloeosporium.* MS thesis. Univ. Arkansas, Fayetteville. 27 pp.
10. Daniel, J. T., Templeton, G. E., Smith, R. J. Jr. 1974. Control of *Aeschynomene* sp. with *Colletotrichum gloeosporioides* Penz. f. sp. *aeschynomene.* US Patent No. 3,849,104
11. Daniel, J. T., Templeton, G. E., Smith, R. J. Jr., Fox, W. T. 1973. Biological control of northern jointvetch in rice with an endemic fungal disease. *Weed Sci.* 21:303–7
12. Day, P. R. 1973. Genetic variability of crops. *Ann. Rev. Phytopathol.* 11:293–312
13. Day, P. R. 1974. *Genetics of Host-Parasite Interaction.* San Francisco: Freeman. 238 pp.
14. Freeman, T. E., Charudattan, R., Conway, K. E. 1976. Status of the use of plant pathogens in the biological control of weeds. See Ref. 4, pp. 201–6
15. French, D. W., Schroeder, D. B. 1969. Oak wilt fungus, *Ceratocystis fagacearum,* as a selective silvicide. *For. Sci.* 15:198–203
16. Harper, J. L. 1977. *Population Biol-*

ogy of Plants. New York: Academic. 892 pp.

17. Huffaker, C. B., Messenger, P. S., ed. 1976. *Theory and Practice of Biological Control.* New York: Academic. 788 pp.

18. Kirkpatrick, T. L. 1978. *Bioherbicidal potential of Colletotrichum malvarum for prickly sida control.* MS thesis. Univ. Arkansas, Fayetteville. 47 pp.

19. Nelson, R. R. 1978. Genetics of horizontal resistance to plant diseases. *Ann. Rev. Phytopathol.* 16:359–78

20. Oehrens, E. 1977. Biological control of the blackberry through the introduction of rust, *Phragmidium violaceum,* in Chile. *FAO Plant Prot. Bull.* 25(1): 26–28

21. Ohr, H. D. 1974. Plant disease impact on weeds in the natural ecosystem. *Proc. Am. Phytopathol. Soc.* 1:181–84

22. Ridings, W. H., Mitchell, D. J., Shoulties, C. L., El-Ghell, N. E. 1976. Biological control of milkweed vine in Florida citrus groves with a pathotype of *Phytophthora citrophthora.* See Ref. 4, pp. 224–40

23. Smith, R. J. Jr., Daniel, J. T., Fox, W. T., Templeton, G. E. 1973. Distribution in Arkansas of a fungus disease used for biocontrol of northern jointvetch in rice. *Plant Dis. Reptr.* 57:695–97

24. Smith, R. J. Jr., Fox, W. T., Daniel, J. T., Templeton, G. E. 1973. Can plant diseases be used to control weeds? *Ark. Farm Res.* 22(4)12

25. TeBeest, D. O., Brumley, J. M. 1978. *Colletotrichum gloeosporioides* borne within the seed of *Aeschynomene virginica. Plant Dis. Reptr.* 62:675–78

26. TeBeest, D. O., Templeton, G. E., Smith, R. J. Jr. 1978. Decline of a biocontrol fungus in field soil during winter. *Ark. Farm Res.* 27(1):12

27. TeBeest, D. O., Templeton, G. E., Smith, R. J. Jr. 1978. Temperature and moisture requirements for development of anthracnose on northern jointvetch. *Phytopathology* 68:389–93

28. TeBeest, D. O., Templeton, G. E., Smith, R. J. Jr. 1978. Histopathology of *Colletotrichum gloeosporioides* f. sp. *aeschynomene* in northern jointvetch. *Phytopathology* 68:1271–75

29. Templeton, G. E. 1974. Endemic fungus disease for control of prickly sida in cotton and soybeans. *Ark. Farm Res.* 23(1):12

30. Templeton, G. E. 1976. *Colletotrichum malvarum* spore concentration and agricultural process. US Patent 3,999,973

31. Templeton, G. E., TeBeest, D. O., Smith, R. J. Jr. 1976. Development of an endemic fungal pathogen as a mycoherbicide for biocontrol of northern jointvetch in rice. See Ref. 4, pp. 214–20

32. Templeton, G. E., Smith, R. J. Jr. 1977. Managing weeds with pathogens. In *Plant Disease: An Advanced Treatise,* Vol. 1, ed. J. G. Horsfall, E. B. Cowling, pp. 167–76. New York: Academic. 465 pp.

33. Trujillo, E. E., Obrero, F. P. 1976. *Cephalosporium* wilt of *Cassia surattensis* in Hawaii. See Ref. 4, pp. 217–20

34. Van der Plank, J. E. 1975. *Principles of Plant Infection.* New York: Academic. 216 pp.

35. Wilson, C. L. 1969. Use of plant pathogens in weed control. *Ann. Rev. Phytopathol.* 7:411–34

36. Wolfe, M. S., Barrett, J. A. 1977. Population genetics of powdery mildew epidemics. In *The Genetic Basis of Epidemics in Agriculture,* ed. P. R. Day. *Ann. NY Acad. Sci.* 287:151–63

37. Zettler, F. W., Freeman, T. E. 1972. Plant pathogens as biocontrols of aquatic weeds. *Ann. Rev. Phytopathol.* 10:455–70

Ann. Rev. Phytopathol. 1979. 17:311–24
Copyright © 1979 by Annual Reviews Inc. All rights reserved

VEGETABLE CROP PROTECTION IN THE PEOPLE'S REPUBLIC OF CHINA

❖3713

Paul H. Williams

Department of Plant Pathology, University of Wisconsin-Madison, Madison, Wisconsin 53706

INTRODUCTION

Vegetables constitute a particularly important part of the daily food of the Chinese. Perhaps nowhere else is such a diversity of vegetables utilized in the diet. As part of the plan to improve the nutritional status of her people, the People's Republic of China (PRC) has established vegetable production goals based on an average daily per capita consumption of 0.5 kg per day. For a nation with a population estimated at over 900 million this constitutes a prodigious production effort.

In June and July 1977 the author traveled in the PRC as a member of a United States delegation on vegetable farming systems. Our visit was sponsored by the Committee on Scholarly Communications with the PRC —a joint effort of three organizations in the US: the National Academy of Sciences, the Social Sciences Research Council, and the Council of Learned Societies.

The host organization for our visit within the PRC was the Chinese Association of Agriculture and Forestry; we visited the Provinces of Hopei, Shansi, Shensi, Shantung, Kiangsu, Chekiang, Kwangtung, and Kwangsi and the municipalities of Peking and Shanghai. In an earlier article in the *Annual Review of Phytopathology,* Kelman & Cook (21) provide an excellent overview of agriculture and plant pathology in the PRC. I, therefore, restrict myself largely to protection of vegetable crops against diseases, insects, and nematodes. A more detailed review of this subject may be obtained in the full report of the Vegetable Farming Systems Delegation (22).

311

Various practices designed primarily to protect the crops from a wide range of diseases, insects, and nematodes are integral parts of Chinese vegetable farming systems. Concepts of plant protection are well understood by vegetable farmers. The general vigor of their crops attests to the effective deployment of plant protection schemes. Only through a continuing effort have the effects of numerous pathogens and insect pests found under intense vegetable culture been eliminated or minimized. The same basic quality criteria for vegetables exist in China as throughout the world and, thus, only disease- and blemish-free products are considered to be marketable. As a consequence, great emphasis is placed on the production of high-quality vegetables.

Plant protection in China embodies a fascinating mix of ancient practices derived empirically over centuries and modern methodology of integrated control requiring precise timing in the introduction of biological agents and chemical pesticides.

IDENTIFICATION OF DISEASES AND PESTS

Accurate identification of disease agents and insect pests is essential to effective plant protection. Within the past six years, a number of publications containing excellent colored illustrations of the main diseases and pests of all major crops have been made available to plant protection specialists on production brigades and teams (1, 3, 6–8, 15–18). Copies of many of these fine publications were obtained during the trip and are available for study in the National Agricultural Library, Washington DC, and the library of the Department of Plant Pathology, the University of Wisconsin-Madison. Manuals have detailed diagnostic information on the life history of the disease agents, disease cycles, and various methods used in control including cropping practices, use of resistant cultivars, and chemical and biological methods. Other manuals also include methods of surveying, recording, reporting, and forecasting disease and insect outbreaks (5, 14).

Most agricultural colleges and academies have specimen collections of economic insects and plant diseases. These collections are used in extension, teaching, and research. An important activity of the Kiangsu Provincial Institute of Agricultural Sciences is the production of teaching specimens of many pests of vegetables and other crops. Mounts contain preserved specimens of the adults, immature stages, and eggs. Mounted along with the pests are specimens of their predators and parasites. Over 200,000 of these specimen mounts have been produced by the Institute for distribution to communes throughout the province.

Monitoring and forecasting of disease and insect outbreaks are an important part of plant protection in China (5, 14). Each production team has five to six individuals who work as a group in plant protection. A group may have specialists in diseases, insects, and weeds. Each specialist is responsible for observations on the incidence of the pests and both the selection and execution of the appropriate control measures. General observations of disease incidence on the crops are made every two to three days, and reports are passed on to the plant protection specialists in the research group of the commune. Once a month, communes report to a municipal or county protection station where data are tabulated for the county. Regional data are collected by provincial institutes. At monthly intervals, plant protectionists of the communes meet with county or municipal specialists to assess general problems at the county level. Notification of outbreaks of serious insects or spreading diseases, such as potato late blight (*Phytophthora infestans*), is given directly to the commune and county institutes who announce the outbreaks by telephone and over the radio and loudspeakers to the production teams.

Vegetable communes have disease forecasting stations which have been set up to monitor insect and pathogen occurrence and to record climatic conditions that favor their buildup. Pest monitoring devices are often used, both for detection and control. Black light traps, bait stations, and greased yellow plastic sheets for aphids are commonly used. Collected insects are examined for their sexual maturity in order to precisely time insecticide applications. Spore collection devices are used on most communes and are primarily for the detection of aerial inocula of rice and wheat pathogens. In vegetable production areas, inoculum buildup is normally monitored by continual inspection of the crops themselves. Farmers and plant protectionists know from experience the kinds of conditions that favor outbreaks of various diseases and pests, and employ preventative cultural or protective measures to avoid damaging outbreaks. Although experimental teams on each commune keep weather and climatic data, forecasting of disease and insect outbreaks in vegetables through predictive modeling is not widely practiced.

DISEASE AND PEST MANAGEMENT

Prevention and Sanitation Methods

As in most other parts of the world, prevention is considered the most important precept in plant protection. Quarantines and exclusionary methods are less important in vegetable production than other preventative controls primarily because of the high degree of self-sufficiency practiced at the commune level. The fact that virtually all vegetable seed and vegetatively propagated plant parts for a commune are produced and remain on that commune greatly reduces the need for inspectors of plant materials.

The limited localized transport of vegetables to neighboring towns or villages minimizes the long-distance spread of insect and pathogens on plant parts. Whenever vegetable seed is imported from abroad, phytosanitary certificates are required and inspections are made by county and provincial personnel. Seed potatoes moving from one province or county to another are the only vegetables subject to quarantine. Potatoes are inspected primarily for the presence of ring rot (*Corynebacterium sepedonicum*).

Sanitation measures play an important role in vegetable-disease prevention. The general practice of removal of all leafy plant parts immediately upon harvest of the crop greatly reduces the potential for inoculum and insect increase. When crops such as cabbage are harvested, excess leaves are removed from the field and used fresh or are dried for cattle, pig, or fish feed. Leaves and vines of legumes, cucurbits, and solanaceous crops are also removed and either used as feed or composted. Rigorously practiced weed control is important in minimizing the buildup of pests (22).

Cultural Methods

Many of the practices used in vegetable cultivation are carried out primarily for the purpose of preventive control of disease or insect pests.

Rotations are the most important cultural control on the intensively cropped vegetable communes of the near suburbs. The wide range of vegetables grown on any commune provides the production teams with the option of rotating with unrelated crops. Many of the complex rotational patterns have taken into account the need to avoid consecutive cropping of vegetables susceptible to the same pathogen. In North China, three- to five-year rotations of the same crop on the same ground are practiced. However, in Kiangsu, buildup of root rot, *Fusarium solani* f. sp. *cucurbitae,* on cucumbers was attributed to the use of winter melon and other cucurbits in the rotation. Similarly, the increase of *Phytophthora parasitica* and *Pythium aphanidermatum* root rots on cucurbits, tomatoes, and eggplant was attributed to inadequate rotations.

I was amazed to find that clubroot, induced by *Plasmodiophora brassicae,* was rarely observed as a problem although crucifers constitute more than half of the total vegetable production in China. Considering the importance of clubroot in various other countries of Asia there must be some fundamental aspect in the cropping of crucifers which relates to clubroot control.

In contrast to the reports of the earlier American Plant Studies (21, 25) and Wheat Studies Delegations who found relatively little damage from root diseases on cereal crops, numerous root diseases were observed particularly on cucumbers and eggplant. In Kiangsu, root diseases were more serious than foliar problems on vegetables and they were reported as increasing. Root diseases are less of a problem on vegetable communes in the far suburbs where rotations of vegetables and cereals are possible.

In the Lower Yangtze and South China, where water vegetables are an integral part of the cropping, flooding plays an important part in the rotation. Every two to three years, the beds upon which vegetables are grown are broken down and the land is converted to one or two crops of rice or planted to water spinach, water chestnut, water bamboo, or lotus. In regions where frost limits vegetable production during midwinter, fields are deep-plowed to expose insects to freezing. Where bacterial rot is a serious problem on Chinese cabbage, crop residues are plowed under and the soil surface is dried and exposed to the sun for three to five days before the replanting commences.

One of the most significant deterrents to the spread of bacterial diseases is the exclusive use of ditch and furrow irrigation rather than overhead sprinkling. The almost complete lack of seedborne bacterial diseases of legumes, crucifers, and solanaceous crops in the drier regions of North China attests to the effectiveness of furrow irrigation. In the Lower Yangtze and South China, considerably more disease was seen on seed crops of onion, crucifers, and cucurbits. Deep furrows and high ridges are used to manage bacterial soft rot in Chinese cabbage by rapidly draining excess soil moisture after heavy rains.

Intercropping of various vegetables maximizes land use and localizes development and spread of some foliar and soilborne pathogens and insects. Turnip mosaic virus, downy mildew (*Peronospora parasitica*), and bacterial soft rot (*Erwinia carotovora*) are difficult to control in South China when Chinese cabbage, leaf mustard, cabbage, and cauliflower are intercropped or grown in sequential overlapping rotations. More commonly however, intercropping involves widely divergent crops such as cabbage, eggplant, cucumber, yard-long beans, and celery.

The Chinese emphasize the importance of proper soil fertility and timely applications of compost, manure, and chemical fertilizers as important factors in keeping plants growing vigorously, particularly to minimize the effects of virus diseases of tomatoes, peppers, and crucifers.

Avoidance of disease by growing certain vegetables earlier or later in the season than is usual is a common practice. In North China and the Lower Yangtze regions, eggplant and peppers are planted early under plastic, permitting the crops to come into full productivity in July, a month earlier than normal. In this way the most destructive effects of phytophthora, pythium, and phomopsis fruit rots are avoided. In Kwangtung, Chinese cabbages are normally sown about August 5th; however, by delaying seeding until late August or early September, the most serious effects of mosaic virus and bacterial soft rot can be avoided.

Chemical Methods

Although emphasis in vegetable growing is placed on various biological and cultural methods of disease and insect management, production of high-quality vegetables is heavily dependent upon the application of chemical pesticides (2, 9, 11). The Chinese are well aware of the problems of chemical residues on fresh vegetables and are striving to minimize their use. Organochlorines are not used in vegetables, although there still is limited use of DDT on cotton and for mosquito control (20). Benzene hexachloride and parathion are not used because of their toxicity to humans. Virtually all organic pesticides used in vegetables are manufactured in China and supplied to the communes by state companies.

Pesticides are applied with various types of 10–20 liter hand-operated tank sprayers usually by teams of three to eight women. Spray pressure is generated either by charging the tank with a built-in hand pump or by a continuously operated hand lever pump. Even larger areas of potatoes and cotton were sprayed by teams using hand-operated pumps. Several versions of small power-drive sprayers and some tractor-drawn boom-type sprayers are in limited use.

Although pesticide applicators' handbooks stress the need for adequate protection while handling and applying chemicals (11), we observed that pesticide applicators wore relatively little additional protective clothing and few protective devices. Rubber gloves were used when dispensing some of the concentrated organophosphate insecticides. Despite the continual exposure to applications of pesticides, blood samples are not taken from the applicators and examined for the buildup of toxicants.

The most widely applied insecticides in vegetables are the organophosphates, trichlorfon, dichlorvos, and dimethoate. All are popular for vegetables because of their nonpersistence, environmental degradability, and low cost (20). They are widely used to control flies in markets and in poison baits and traps, and as sprays for chewing insects and aphids. Massive numbers of adult *Pieris rapae* and severe cabbage worm damage to cabbage and kohlrabi in Shensi and Shantung were attributed to buildup of populations resistant to both dichlorvos and trichlorfon. In these regions, farmers were turning to the use of *Bacillus thuringiensis*. Dimethoate is the preferred chemical for aphid control on vegetables. To prevent the transmission of turnip mosaic virus by turnip aphid, *Rhopalosiphum pseudobrassicae,* and cabbage aphid, *Brevicoryne brassicae,* Chinese cabbage seedbeds are sprayed prior to seeding and then at seven nine-day intervals until the crop has headed.

Nematodes are not a serious problem in vegetables though occasionally in North China root knot nematodes (*Meloidogyne* sp.), can affect tomato production. In these cases, soil is fumigated with dichloropropane-dichloro-propene.

The most widely used fungicide on vegetables throughout China is Bordeaux mixture. Bordeaux is mixed (1:1:100) in the field in large glazed earthenware vessels. The copper sulfate is procured from the state, and quick lime is obtained locally from kilns on most communes. Bordeaux is used on potatoes for control of both early blight (*Alternaria solani*), and late blight; it is applied whenever the first signs of disease appear in the spring. In Chekiang and Kwangtung, weekly applications of Bordeaux are required to protect tomatoes from early blight and the leaf spot reduced by *Septoria lycopersici*. In Shensi and Shantung, cucumbers are sprayed with Bordeaux to control both downy mildew and powdery mildew (*Sphaerotheca fuliginea*), although organic fungicides are now more commonly used in mildew control.

The addition of sulfur to the roots of poorly growing eggplant dates back 1400 years in Shantung agriculture. Whether these applications of sulfur corrected minor element deficiencies by increasing soil acidity or whether it acted fungicidally is not known. In Chekiang, sulfur and lime are incorporated into the soil where verticillium wilt, (*Verticillium albo-atrum*) infected eggplants were removed. Potassium permanganate, $KMnO_4$, is used as a 1% soak to rid tomato seeds of tobacco mosaic virus (TMV) and in Kwangtung, 0.05% $KMnO_4$ is applied to the foliage to cucurbit vegetable as a downy mildew preventative. The only use of mercury in vegetables culture was for seed treatment of cucumbers for *Fusarium solani* f. sp. *cucurbitae*. Seeds were soaked 4 hr in 0.1% mercury bichloride.

The bis-dithiocarbamates maneb and zineb are widely used on vegetables, particularly on cucurbits for downy and powdery mildew and on tomatoes for early blight and leaf mold induced by *Cladosporium fulvum*, and for Taro leaf blight induced by *Phytophthora calocasia*. Daconyl and thiophanate-methyl are also used for cucumber downy mildew control. Chlorothalanyl, tuzet, bavistin, and benomyl are all being used experimentally.

The Chinese have developed a number of antibiotics for the control of plant pathogens on rice, fruits, and other major crops; however, none of these are presently used on vegetables. The recent development of qingfengmycin ("Qingfeng" means to celebrate the bumper harvest) by the Agricultural Antibiotics Laboratory of the Shanghai Institute of Plant Physiology serves as an example of how discoveries made in research institutes can be rapidly developed for use at the commune level. In 1973, a compound produced by a new species *Streptomyces qingfengmyciticus* n. sp. was effective in controlling rice blast, *Pyricularia oryzae*. The basic research on production, chemical characterization, biological activity, and toxicity was published in 1974 and 1975 (10, 13). At the same time, researchers at the

Institute, together with members from four communes in the Shanghai area, developed and published (12) simplified methods for the production and application of qingfengmycin by the production teams.

Biological Methods

In the narrow sense of directly applying parasites or predators, biological control is used almost exclusively for insects in China. Nowhere did we learn of efforts to control pathogens, nematodes, or weeds by directly using biological agents. Biological control of insects in China is widely practiced among pests of many crops, including forest trees. In vegetables, extensive use of pesticides makes it difficult to use most predators and parasites. This is probably why on vegetables *Bacillus thuringiensis* is the major insect parasite used in China. Water suspensions of *B. thuringiensis* are sprayed on crucifers to control the larvae of common cabbage worm (*Pieris rapae*). These suspensions are commonly produced by production brigades using semisolid media and cultures of bacteria obtained from provincial plant protection institutes. Newly hatched silkworm larvae are used in assays to standardize the potency of *B. thuringiensis* (19).

Plant protection units on communes utilize various other means of biocontrol depending in part on the availability of the agents and partially on their ability to raise parasites and predators. On the Evergreen Commune in the suburbs near Peking, lace wing flies, (*Chrysopa sinica*), were being reared for release for aphid control in vegetables. In Shensi, wheat fields are swept with nets to collect lady beetles, *Coccinellidae,* which are released into vegetable crops. At the time of release, insecticides are withheld so as to permit effective predation by the beetles. The egg parasites, *Trichogramma* sp., are reared on silkworm eggs and used widely on rice leaf roller, (*Cnaphalocrocis medinalis*), sugarcane borer, (*Argyroploce schistaceana*), and corn borer, (*Ostrinia nubilalis*), in northeastern China. As with *Bacillus thuringiensis*, effective methods have been developed so that brigades may rear massive numbers of *Trichogramma* sp. (20). Priority is placed on the development of new and improved forms of biological control that will reduce chemical pesticide application. At several agricultural universities and institutes, we observed that various other alternatives to chemical control of insects were being explored (22), including nuclear polyhedrosis viruses and cytoplasmic polyhedrosis viruses (22).

Throughout China, large numbers of chickens run free, particularly on vegetable communes, in small villages, and in private plots; it is likely that they play an important role in keeping populations of many insect pests to a minimum.

Ducks are also widely observed throughout China, particularly in the countryside. Flocks of ducks ranging from 50 to 1200 are tended by young boys who herd them through rice paddies where they consume large numbers of insects. Ducklings may be purchased in the markets and many families have their own ducks as well as chickens, which run loose.

Miscellaneous Forms of Control

Besides the traditional forms of plant protection, the Chinese use a number of other methods, some of which are intensively deployed at certain seasons. Insect trapping is widely practiced in vegetable farming. Black light traps, similar to those used in monitoring for insect populations, are widely used. Lights are often suspended above large vessels filled with water from the cooking of rice; insects falling into the vessels of rice water serve as a high protein feed for fish or pigs or as organic nitrogen fertilizer.

During March and April, when adults of the tomato fruit worm, *Heliothus armigera,* are flying, farmers cut young willow or poplar twigs about 0.5–1.0 cm in diameter, let them wilt for two days in the sun, then tie them in bundles of 10 and place them among the vegetables at densities of 150 bundles per hectare. During the night, adult moths are attracted to the wilted twigs and crawl into the bundles. Each morning, bundles are examined and the adult moths destroyed. Yellow-colored aphid traps are widely deployed in vegetable fields. Half-meter squares of polyethylene are painted yellow, and smeared with grease to trap attracted aphids. The sheets are staked vertically and placed throughout the crops at a density of 15–30/ha; their use can reduce the overall application of chemicals for aphid control by up to one half.

Insect baiting is widely practiced on vegetable communes during periods when adults are known to be flying. Mixtures of sugar, wine, vinegar, water, and small amounts of trichlorfon are placed in shallow bowls or pails on tripods among the vegetables.

An effective cutworm control in Chekiang is the use of bundles of rice straw (8 cm X 30 cm) suspended vertically on sticks about 1 m above the vegetables. The straws serve as an attractive site for adults to rest during the day and the bundles are baited with wine, sugar, and trichlorfon solution to kill the visiting adults. Members of the plant protection group inspect the bundles periodically to record the numbers of trapped insects and recharge the traps by spraying the straw with fresh solutions of poisoned bait. As many as 300–450 straw bait traps may be used per hectare. Attractive features of poisoned bait traps are their low cost and the fact that insecticides used are not applied to the crop directly. The effectiveness of baiting is dependent on the proper timing and placement of the bait in the field. Thus, baiting is coordinated closely with information on insect hatches and flights gathered from black light traps.

Hot-water treatment of seed for eradication of seedborne pathogens is used in a number of provinces. Cucumber seed is treated at 50–55°C for 15 min to kill *Fusarium solani* and *Colletotrichum lagenarium*. Pepper seed is commonly soaked for 15 min at 55°C to eliminate damping-off fungi and bacterial spot induced by *Xanthomonas vesicatoria*. Nowhere did we hear of crucifer seed being hot-water treated for black rot, (*Xanthomonas campestris,*) or black leg (*Phoma lignam*) control, nor was celery seed treated for late blight control, (*Septoria apii*).

Integrated Pest Management

Chinese vegetable farmers are well grounded in the concepts of integrated pest management, and peasant farmers have undoubtedly used a variety of disease and insect management practices for many centuries. Although many of the ancient methods of baiting and trapping are still being used today, it is impressive to see how they are used to reduce the application of modern chemicals. All of the following attest to the degree to which integrated pest management is part of Chinese vegetable production (22): 1) the use of disease and insect monitoring as a guide to timing chemical applications; 2) the coordination of chemical applications so as not to diminish the effects of biocontrol agents released into crops; 3) the employment of cultural practices which minimize pest buildup; and 4) the use of genetic resistance.

Resistance to Diseases and Insects

The development of disease resistant varieties has high priority in the programs of all vegetable production brigades, research institutes, and academies throughout China. Mao Tse Tung's admonitions to "develop improved seed" is clearly understood to mean develop genetic resistance in crops. Virtually every brigade, institute, and university was involved in "improving the seed." A major effort is under way, primarily via mass selection, to develop more uniformity and productivity within existing cultivars. Inherent in these seed improvement programs is selection toward types carrying resistance to various diseases and insects. Collections of local cultivars of all vegetable crops in approximately 50,000 communes provides a rich source of genetic diversity.

The relative lack of serious damage to most crops indicates that functional genetic resistance is available in many vegetable crops (24). The relative absence of cucumber mosaic virus (CMV) indicates that high levels of CMV resistance have long existed in cucumbers in China; this idea is confirmed by the use of Chinese cucumbers as a source of CMV resistance for US cucumber breeders in the 1920s (23).

Different specific breeding approaches are being used at the various provincial and municipal institutes. Plant pathologists work directly with breeders in providing incoculum, in helping to develop incubation environments, and in assisting with the scoring of disease reactions (22).

Active programs on the development of mosaic resistant tomatoes, downy mildew resistant cucumbers, and mosaic virus and downy mildew resistant Chinese cabbage exist in virtually every province. In the North, TMV is frequently a limiting factor in tomato production and numerous crosses have been made between local varieties and foreign accessions. In an effort to locate resistance to the tomato streak disease, breeders in Shantung and Kiangsu are turning to *Lycopersicon hirsutum* and *L. pimpinellifolium*. Breeders in Szechwan also had sought resistance to the bacterial wilt reduced by *Pseudomonas solanacearum* in *L. pimpinellifolium*. The Institute of Vegetable Research for Shantung Province has developed a number of F_1 hybrids of Chinese cabbages that are resistant to mosaic virus; a few lines also have some resistance to soft rot and downy mildew. Soft rot resistant types are primarily those with more elongated heads and are classified as *Brassica campestris* ssp. *pekinensis.*

An important part of potato improvement in various provinces is breeding for virus and blight resistance. With the help of research personnel at provincial and municipal institutes, potato improvement groups in the communes are encouraged to make crosses between superior potato varieties and to select for locally adapted types. Tubers of well adapted southern types may be sent north to the Potato Research Institute in Heilungkiang, where crosses are made and the seeds returned to various provinces for selection under prevailing local conditions. Often potato breeders will take their stocks to the Potato Research Institute and make their crosses under supervision of Institute personnel.

Little progress has been made in developing resistance to the severe soilborne root or fruit rot organisms, *Phytophthora, Pythium,* and *Rhizoctonia;* these fungi limit production of eggplant and cucurbits during the wet, hot season in late July and August. Though high moisture and temperatures in these months favor crop growth, the destructive effects of these fungi severely reduce numbers of marketable fruits. White cultivars of eggplant, *Solanum xanthocarpum,* are reported to be more resistant to *Phomopsis vexans* than purple forms, *S. melongena.*

Prevention of Potato Degeneration

Considerable effort is spent on minimizing the effects of virus degeneration of white potato (4). Tuber production and selection schemes have resulted in sustained yields over seven generations of potato production. In Kiangsu, two crops of potatoes are grown each year. A spring planting is made in February which serves both as a source of tubers for consumption and, primarily, as a source of seed tubers for the main fall crop. Robust tubers are selected from the previous fall's crop and are planted in hills. As the warm season progresses in April and May, multiplication of virus is favored and only those plants with low virus titer or resistance will produce sustained yields during the season of increasing temperatures. The spring crop is harvested in early June and tubers from each hill are weighed and seed tubers saved only from those hills showing high production. Seed potatoes are stored over the summer, sprouted in August and September with the aid of gibberellic acid, and then rooted sprouts are grown on the shaded north sides of tall ridges to keep plants cool and thus reducing virus multiplication. The fall potato crop is harvested in November and December, and shows less degeneration than the spring crop. The best fall tubers are saved for the spring seed planting. A similar two-crop system for potato production is being used successfully in Shantung and other provinces.

In addition to maintenance programs for seed potatoes at the commune level, at the Laboratory of Cytology and Tissue Culture at the Institute of Botany in Peking a significant effort is under way to produce virus-free seed stocks. Meristem culture techniques and tuber indexing are aided by serologic procedures. Young mericlone-plants are sent to the cool region of Inner Mongolia where the plants are grown in the field and individually indexed serologically for the presence of Potato Virus X and Potato Virus Y.

CONCLUSIONS

The general supply and quality of vegetables to the markets in China has never been higher. Also, an excellent job is being done in protecting crops from diseases and pests. Nevertheless, the maintenance of a stable future supply of vegetables remains uncertain. It is clear that the intensity of cropping has accelerated the buildup of various soilborne diseases. Likewise, the use of cultivars of tomatoes and vine crops requiring intensive hand labor will favor spread of mechanically transmitted viruses. The need to intensify vegetable production around expanding industrialized centers may also produce significant problems with respect to atmospheric and water pollution.

At present, the practice of self-sufficiency in vegetable and vegetable seed production by the People's communes has both strong positive and negative aspects in relation to plant protection. Although self-sufficiency has served to restrict the long-distance spread of diseases and pests on vegetables and seeds, this same principle has limited the full exploitation of drier regions in North China such Shensi and Shantung; they appear to be ideal for the production of disease-free vegetable seed. At the same time, self-sufficiency in seed production has produced a diversity of germ plasm in each vegetable species that is likely unparalleled anywhere in the world; this in itself provides important long-term stability derived from genetic heterogeneity. The stability provided by genetic diversity could be maintained even with F_1 production if China adheres strictly to the policy of self-sufficiency at the commune level.

Perhaps more than any other nation, the Chinese are in a better position to provide alternatives to chemical control of vegetable diseases and pests. Although there is heavy reliance on pesticides at present, applied research on the integration of biological and other alternative control practices not only should minimize the use of pesticides but also reduce the selection pressures toward chemical resistant pests. Within the diversity of traditional vegetable culture in China lies a great resource of knowledge which needs to be understood more fully in terms of modern plant protection.

ACKNOWLEDGMENTS

I would like to express my appreciation to Mr. Halsey Beemer of the committee on Scholarly Communications with the People's Republic of China for his guidance and assistance provided prior to and during the trip. I would also like to thank Professor Arthur Kelman for continuing interest, encouragement, and assistance during the preparation of this manuscript, and to Dr. S. H. Ou for assistance with the manuscript and literature citations. I am particularly indebted to many colleagues in the Chinese Association of Agriculture and Forestry who through their hospitality over the course of my month in the PRC gave me the opportunity of becoming their friend.

Literature Cited

1. Anonymous 1959. *Plant Pathology and Mycology Methods.* Peking: People's Publ. Soc. 1133 pp. (In Chinese)
2. Anonymous 1971. *Agricultural Chemicals.* Shanghai: Shanghai People's Press. 137 pp. (In Chinese)
3. Anonymous 1973. *Control of Soil Insects.* Peking: Agric. Publ. Soc. 67 pp. (In Chinese)
4. Anonymous 1973. *How to Prevent Potato Degeneration.* Kwantung: Chin. Acad. Sci., Inst. Genet. 104 pp. (In Chinese, English transl. available)
5. Anonymous 1973. *Mass Participation in Survey and Forecasting Agricultural Pests and Diseases.* Shanghai: Shanghai People's Press. 380 pp. (In Chinese)
6. Anonymous 1973. *Plant Protection Handbook.* Peking. 110 pp. (In Chinese)
7. Anonymous 1974. *Colored Illustrations of Pests and Diseases of Agricultural Crops.* Shanghai: People's Publ. Soc. 160 pp. (In Chinese)

8. Anonymous 1974. *Handbook for Plant Protection Workers.* Shanghai: Shanghai People's Press. 518 pp. (In Chinese)
9. Anonymous 1974. *New Agricultural Chemicals.* Peking: Agric. Publ. Soc. 142 pp. (In Chinese)
10. Anonymous 1974. Studies on Qingfengmycin. I. Identification of the producing strain. *Acta Microbiol. Sin.* 14:42–46 (In Chinese)
11. Anonymous 1975. *Agricultural Chemicals.* Shanghai: Shanghai People's Press. 275 pp. 2nd ed. (In Chinese)
12. Anonymous 1975. *Qingfengmycin— Methods of Production and Use.* Shanghai: Shanghai People's Press. 47 pp. (In Chinese)
13. Anonymous 1975. Studies on Qingfengmycin. II. Isolation, purification and characterization. *Acta Microbiol. Sin.* 15:101–9 (In Chinese, English summ.)
14. Anonymous 1976. *Forecasting Pests and Diseases of Agricultural Crops.* Peking:

People's Educ. Publ. Soc. 166 pp. (In Chinese)

15. Anonymous 1976. *Knowledge of Vegetable Disease Control.* Shanghai: Shanghai People's Press. 53 pp. (In Chinese)

16. Anonymous 1976. *Plant Protection.* Shanghai: Shanghai People's Press. 264 pp. (In Chinese)

17. Anonymous 1976. *Vegetable Pest and Disease Control.* Chekiang People's Press. 103 pp. (In Chinese)

18. Anonymous 1976. *Vegetable Diseases and Insects and Their Control.* Peking: Agric. Publ. Soc. 150 pp. (In Chinese)

19. Anonymous. 1977. A study on the standardization of *Bacillus thuringiensis* products by using newly-hatched silkworm larvae as test insects. *Acta Entomol. Sin.* 20:5–13 (In Chinese, English summ.)

20. Guyer, G. E., Chairman. 1977. *Insect Control in the People's Republic of China: A Trip Report of the American Insect Control Delegation. CSCPRC Rept. No. 2.* Washington DC: NAS. 218 pp.

21. Kelman, A., Cook, R. J. 1977. Plant pathology in the People's Republic of China. *Ann. Rev. Phytopathol.* 17: 409–29

22. Plucknett, D. 1979. *Vegetable Farming Systems in the People's Republic of China: A Trip Report of the American Vegetable Farming Systems Delegation.* Washington DC: NAS. In press

23. Porter, R. H. 1929. Reaction of Chinese cucumbers to mosaic. *Phytopathology* 19:85–87 (Abstr.)

24. Provvidenti, R. 1977. Evaluation of the vegetable introductions from the People's Republic of China for resistance to viral diseases. *Plant Dis. Reptr.* 61: 851–55

25. Wortman, S., Chairman. 1975. *Plant Studies in the People's Republic of China: A Trip Report of the American Plant Studies Delegation.* Washington DC: NAS. 205 pp.

Ann. Rev. Phytopathol. 1979. 17:325–41
Copyright © 1979 by Annual Reviews Inc. All rights reserved

LICHENS AS BIOLOGICAL INDICATORS OF AIR POLLUTION

♦3714

Erik Skye

Institute of Ecological Botany, University of Uppsala, Box 559, S–751 22
Uppsala, Sweden

INTRODUCTION

The effect of air pollution on plants has attracted the interest of many research workers and is the subject of a number of handbooks published in Europe and the United States (4, 6, 13, 16, 18, 31, 36).

Because of their sensitivity lichens are particularly significant biological indicators of air pollution. The first general survey on this subject was published in England in 1973: *Air Pollution and Lichens* (13). Since then a number of other publications on this subject have appeared (33).

Nylander was the first to point out that lichens react to air pollution. He noticed the poor, sparse lichen vegetation in the peripheral parts of the Luxembourg garden in Paris (26). In 1866, he wrote: "Les lichens donnent à leur manière la mesure de salubrité de l'air et constitutent une sorte d'hygiomètre très sensible." This observation evidently aroused some attention but it was not until the turn of the century that changes in the composition and appearance of the lichen flora in a number of cities and towns in Europa were studied in detail.

Lichens occupy a special place in the plant world. Each lichen is a double organism comprising a fungus and one or sometimes two algae. The fungus and the alga are generally assumed to live symbiotically although there is some debate about possible parasitism on the part of the fungus.

The appearance of the lichen thallus is determined by the fungus in the large majority of cases. An outer cortex of tightly packed fungal hyphae usually form the upper surface of the lichen thallus and sometimes the lower surface as well. In between there is a medulla of more loosely woven hyphae. The algal cells occur in a protected position within the medulla. Since the fungi have no chlorophyll, the algae are alone responsible for photosynthesis.

325

0066-4286/79/0901-0325$01.00

The fungi have their own reproductive organs, called apothecia, in which spores are produced. If a germinating fungal spore is to grow into a lichen thallus, the right kind of alga must be present, together with suitable conditions for the development of symbiosis.

The lichen as such spreads vegetatively, by fragmentation and by means of special reproductive organs, called isidia and soredia. These structures contain both fungus and alga and thus assumes some uniformity in the genetic make up of the lichens—an advantage in the use of lichens as indicators of air pollution.

Lichens have no roots and generally no other special organs for nutrient uptake. All nutrients are taken up directly through the thallus.

Nevertheless, they are very effective in active uptake of substances from weak aqueous solutions, and can concentrate and store many different substances in the thallus. The thallus may then serve as a storage place for longer or shorter periods. Metabolic turnover and growth are slow. In temperate regions, spring and autumn are the growing seasons for lichens. Growth is probably limited by drought in summer and by low temperatures in winter.

Lichens occur in all climatic regions of the earth, from the tropics to the arctic tundra and ice-free areas of Antarctica, from sea level to the highest mountain tops. Many species are widespread while others are restricted in range. Lichens are generally classified according to the substrate on which they grow, e.g. on soil, on rocks and stones, on wood, or on bark. Some lichens are confined to a particular substrate, whereas others can grow on almost any substrate and are therefore included in all the groups.

This review deals mainly with the species growing on bark, the epiphytes. Most epiphytic lichens also grow on other substrates. In an investigation from the Stockholm region of Sweden, it was shown that 40% of the bark species grew almost exclusively on bark. Almost 15% occurred on bark or on wood, about 25% on bark and on stone, while less than 20% occurred on bark, wood, and stone.

THE GENERAL ECOLOGY OF LICHENS

Not all kinds of bark are suitable for all species of lichens. The surface of different barks differs considerably in both water-holding capacity and chemical nature. Different parts of a tree also provide completely different environments for lichens in terms of temperature, radiation, moisture, protection from snow, etc.

Du Rietz (10) tried to introduce some order into this variety by dividing the epiphyte vegetation into poor-bark and rich-bark vegetation, in a parallel manner as for fen vegetation. As for fens, this concept would be based on species richness or poorness. He also observed that poor-bark vegetation grew on bark that ranged from very acid to acid, while rich-bark vegetation grew on bark that ranged from moderately acid to almost neutral. Dust from roads and arable fields affects the substrate by increasing the pH, in some cases by several units. In a corresponding way, the pH may decrease in areas polluted by acid gases or aerosols.

Du Rietz's classification of the epiphytic vegetation has been criticized, for example by Almborn (1, 2). Almborn considers that the correlation between pH value and number of species obtains only under certain special conditions and then only for epiphyte communities with a high light requirement. In other cases, the so-called rich-bark vegetation may include only a few species. The terms *rich bark* and *poor bark* as based on the number of species in disturbed plant communities are therefore misleading and unsuitable. Almborn considers that the question is mainly one of dust impregnation. Dust-free plant localities, for example within forests, support the species Du Rietz included among poor-bark vegetation while trees along roadsides and in other dusty places support rich-bark vegetation.

Almborn's point is of course valid. It is not possible to make comparisons between different localities without regard to environmental factors [see also (7, 33)]. For example in investigating the lichen flora of an area in relation to air pollution, it is important to take both solar radiation and the dust effect into account.

It is possible that Almborn underestimates the significance of the normal acidity and buffering capacity of bark. Even with no effect of dust, broad-leaved trees such as elm and ash support a different epiphyte flora from the demanding broad-leaved species, such as birch and aspen or conifers. A large dust effect is necessary before conifers, for example, support a so-called rich-bark vegetation. Du Rietz's formulation of species richness or species poorness (10) has undoubtably led to misinterpretations; the fact that his division into poor-bark and rich-bark communities was based on indicator species has been ignored. The term *rich-bark* has more and more come to mean the bark of the more demanding broad-leaved trees such as elm and ash whereas *poor-bark* is applied to bark of conifers and less demanding broad-leaved trees. This is so, whether a particular tree of the poor-bark type supports a more species-rich lichen vegetation than a particular tree of a rich-bark type. In these circumstances, rich-bark species are those that normally occur on rich bark; while poor-bark species normally occur on poor bark. In dusty places, for example near a dirt road or arable land, some rich-bark species may occur on poor bark. The opposite change of substrate also occurs, as Skye (38) has shown.

USE OF LICHENS IN AIR POLLUTION SURVEYS

Investigations of the lichen flora and lichen vegetation in polluted areas have been made in a number of European countries and in the United States. Natho (23) summarized the then published literature on the lichen flora in towns. Subsequent literature was reviewed in Skye's work (38), which also deals with the literature about air-polluted areas outside towns, in Barkmann (3, 3a, 4) and in Domrös (8). Other additions to the literature are discussed by De Sloover & LeBlanc (6) and De Wit (7).

Rydzak (e.g. 29–31) reacted strongly against the idea of the damaging effect of air pollutants, as did Natho (23–25), Klement (15–17), and Steiner & Schulze-Horn (40). The authors denied that there was any effect of air pollutants. Rydzak, for example, wrote (29) "Daraus ersehen wir, dass die Annahme einer Einwirkung von SO_2 auf die Dislokation von Flechten in den Städten zu einem Unsinn führt und ein Beispiel für eine kollektive, wissenschaftliche Suggestion ist." In 1969 (31), he summarized his opinions as follows:

> The toxic hypothesis does not suffice to explain the condition of the lichen flora in towns; it also makes further studies of the lichen ecology difficult. On the other hand, the ecological hypothesis, also called drought hypothesis, gives a uniform view of the problem and stimulates investigations of the ecology of lichens not only in towns, but also in their natural habitats. The conflict between these two hypotheses has been so far of profit for the advance of lichenology.

Klement (17) goes so far as to explain that several ground-living species of lichens are dying out in nonurban areas of northern Germany because of a long-term change in microclimate, which has not been measured or otherwise demonstrated. In large parts of western Europe the natural lichen vegetation has disappeared from tree trunks, for example, and has been replaced by other types of vegetation (3a). The new vegetation is considerably more tolerant of substrate acidity than the lichens which have disappeared (see 7, 38). These effects occur far outside the area influenced by the urban microclimate, for example at Närkes Kvarntorp, (34), in a completely rural area subject to pollution by atmospheric SO_2. These observations do not support the theory of a drought effect.

In limestone areas, and other areas with lime-rich soils, there is rich-bark vegetation, because of the effect of lime-rich dust on substrates which otherwise would be acid. Thus the reaction of lichens to airborne substances need not result in wholly or almost wholly lichen-free areas. A special lichen flora favored by the high pH of the substrate or able in some other way to utilize the water-soluble components of the dust may develop in the vicinity of dust-producing industrial activities. Investigations in Sweden (11, 21, 36,

37) provide evidence of this. A number of nitrophilous plants also increase markedly around certain industrial areas. In such conditions, and without laboratory tests, it may be difficult to determine whether a lichen species, such as *Xanthoria parietina*, has increased because the pH of the substrate has increased or because of the availability of suitable nitrogenous compounds. *Xanthoria parietina* is one of the species that are found around manure heaps and on stones covered by bird droppings (see 9).

Lichens grow very slowly and are difficult to cultivate. They are therefore rather poor materials for laboratory investigations. For this reason, there have been very few laboratory experiments to explain the effects of air pollutants on lichens. A further difficulty is the very low concentrations in which air pollutants occur. Pearson & Skye (27) carried out a series of experiments with the lichen *Parmelia sulcata*. Rao & LeBlanc (28) demonstrated that *X. parietina* exposed to 5 ppm SO_2 for 1 hr in air of high humidity showed clear morphological and physiological changes. These were particularly marked in the algal layers. Similar damage occurred in examples of *X. parietina* which were inplanted in heavily air-polluted areas (18; see also 19). De Wit (7) gave an account of an interesting series of experiments with gases. Four series of fumigations with a duration of 3 to 6 months were performed. These gases were HF (two experiments), SO_2, O_3, and C_2H_4 (one experiment) and SO_2, O_3, and $SO_2 O_3$ (one experiment). A control group of lichens was also placed in an untreated glasshouse.

De Wit could confirm that lichens are more sensitive to pollutants under moist than under dry conditions. Otherwise the results from these experiments were somewhat difficult to interpret and supported the above statement that laboratory experiments with lichens are difficult and rather unrewarding.

No changes at all brought about by drying out could be demonstrated in these laboratory experiments, whereas SO_2 had a definite effect.

An experiment with gaseous SO_2 applied to lichens in the field (38) showed that *Parmelia sulcata* had a significantly lower chlorophyll content after the experiment than before. Knowledge gained from laboratory experiments in conjunction with results of field investigations suggests that the absence of lichens, or changes in the lichen flora around industries or in towns is mainly caused by air pollution.

Unfortunately, lichens are used very little in practical clean-air work. De Wit (7) points out some of the difficulties that arise in the correlation of the content of certain pollutants in the air with the occurrence of lichens. As always, in working with biological materials individual differences must be taken into account. Some individuals of a given species of lichens are more sensitive than others. There is an advantage here in that lichens reproduce almost entirely asexually. In temperate regions the content of pollutants in the air would be expected to be most important for lichen damage in autumn and spring.

An investigation of the lichen flora around the shale oil installation in Kvarntorp took up the question of indicator species for the first time in Sweden (34). Skye specified that an indicator species should be identifiable even by a nonspecialist and should occur generally over a large area — except of course close to the source of pollution. Also, the species must not be too sensitive, nor too insensitive, to the pollutants in question. The following five species were suggested: *Anaptychia ciliaris, Evernia prunastri, Parmelia acetabulum, Ramalina fraxinea,* and *X. parietina.* All these normally occur on rich bark. *Evernia prunastri* has a somewhat wider ecological range than the other species and also occurs on poor bark. A new investigation was made in the Kvarntorp area in 1967 (39). It is of interest to examine the changes in distribution pattern of the different species [Figures 3–7 in (39)]. In all cases there was a continued decrease. This seemed to be most marked in *A. ciliaris* and *R. fraxinea,* less marked in *P. acetabulum* and *X. parietina,* and least in *E. prunastri.*

The relative change for these species, i.e. a decrease in comparison with all the other species, is illustrated by Tables 1 and 2. It is apparent that on rich bark *E. prunastri* maintained the same position as previously, *X. parietina* and *P. acetabulum* decreased less than other species, while *A. ciliaris* decreased most. However, on poor bark *E. prunastri* decreased considerably.

If poor-bark species had been chosen as indicators the picture would have been different.

We see from Table 2 that new species have come in since 1951– 1953. These are lichens which in normal circumstances are absent or rare on rich bark in dusty areas (which is the situation in this instance). Because of the acidification of the substrate which has taken place in the Kvarntorp area since the extraction of shale oil began, the pH of the bark in some areas is now such as to allow colonization by poor-bark species. Shale oil extraction ceased at Kvarntorp on October 1, 1966. It will be interesting to follow the continued development there.

In other areas where there is acidification of the substrate, further interesting observations have been made. Skye (35, 38) and Lundström (20) showed that *Lecanora conizaeoides* occurred on trees otherwise free of lichens in the Stockholm area. This was true of both rich-bark and poor-bark trees. This species seems not to be damaged either by substrate acidification or by any direct poisonous effect of sulfur compounds. *Lecanora conizaeoides* has a widespread distribution in the most air-polluted areas in northwest Europe. The species can therefore be regarded as a positive indicator of acid air pollutants. At least one further tolerant species is apparent from the Stockholm data, *Lecidea scalaris* (particularly on poor bark). In the same way, certain other poor-bark species can be positive indicators of acidification of rich bark. Moberg (22) stated that some species especially tolerant of air pollution do not occur on poor bark in Köpman-holmen, in central Sweden. On the other hand, *Alectoria implexa, Cetraria*

Table 1 Relative shifts between different species on slightly acid to almost neutral bark at Kvarntorp, southern Sweden

1953	1967
Parmelia sulcata	Hypogymnia physodes
Hypogymnia physodes	Physcia tenella
Physcia tenella	Parmelia sulcata
Evernia prunastri	Evernia prunastri
	Cetraria chlorophylla
	Parmeliopsis ambigua [a]
	Cladonia spp. [a]
	Alectoria jubata [a]
	Cetraria pinastri [a]
Ramalina	Xanthoria parietina
Anaptychia ciliaris	Ramalina
Physcia "grisea"	Physcia enteroxantha
Physcia aipolia	Physcia orbicularis
Cetraria chlorophylla	Physcia adscendens
Physcia adscendens	Physcia pulverulenta
Xanthoria parietina	Parmelia acetabulum
Physcia pulverulenta	Parmelia exasperatula
Parmelia exasperatula	Anaptychia ciliaris
Parmelia glabratula v.	Parmelia glabratula v.
Parmelia fuliginosa	Parmelia fuliginosa
Xanthoria polycarpa	Physcia aipolia
Parmelia acetabulum	Usnea spp.
Physcia orbicularis	Xanthoria polycarpa
Usnea spp.	Physcia farrea
Xanthoria fallax	Xanthoria fallax
Pseudevernia furfuracea	Pseudevernia furfuracea
	Physcia nigricans
	Parmelia subargentifera

[a] After some names = new colonizers.

Table 2 Relative shifts between different species on acid bark at Kvarntorp, southern Sweden

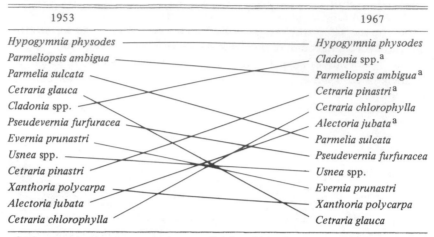

1953	1967
Hypogymnia physodes	Hypogymnia physodes
Parmeliopsis ambigua	Cladonia spp.[a]
Parmelia sulcata	Parmeliopsis ambigua[a]
Cetraria glauca	Cetraria pinastri[a]
Cladonia spp.	Cetraria chlorophylla
Pseudevernia furfuracea	Alectoria jubata[a]
Evernia prunastri	Parmelia sulcata
Usnea spp.	Pseudevernia furfuracea
Cetraria pinastri	Usnea spp.
Xanthoria polycarpa	Evernia prunastri
Alectoria jubata	Xanthoria polycarpa
Cetraria chlorophylla	Cetraria glauca

[a] After some names = new colonizers on slightly acid to almost neutral bark.

glauca, Parmeliopsis aleurites, and *P. hyperopta* seemed to be very sensitive. In the Kvarntorp area, *Cetraria glauca* suffered the greatest relative decrease on poor bark, followed by *Parmelia sulcata, Evernia prunastri, Pseudevernia furfuracea,* and *Parmeliopsis ambigua,* among others. Moberg's opinion of the sensitivity of these species is also supported by the observations in the Stockholm area.

Near Sundsvall there are flourine-containing air pollutants in addition to sulfur dioxide. Eriksson (12) found a statistically significant relationship between the fluorine content in pine needles and the presence of *P. furfuracea* and *A. implexa.* Both species were absent where the fluorine content of old pine needles was 41 ppm or greater. In Moberg's investigation *P. furfuracea* was not especially sensitive to air pollutants from the pulp mill at Köpmanholmen, but *A. implexa* was. *Pseudevernia furfuracea* thus seems to be more sensitive to air pollutants containing fluorine than to sulfur dioxide–containing pollutants.

Pollutants that raise the pH of the substrate affect the lichen flora to a large extent, particularly on poor bark (37). Ericson (11) has investigated the lichen flora in the district near Köping, southern Sweden; he found that *Hypogymnia physodes* and *P. furfuracea,* which are both abundant there, are absent in the industrial area itself and down wind from it, where the effect of fertilizers is too great. If this distribution is compared with that of *X. parietina, Physcia tenella,* or *P. ascendens,* it is apparent that these three rich-bark species occur on poor bark, in the places where the *H. physodes* and *P. furfuracea* are excluded. Similar observations have been made by

Magnusson (21) concerning species near the cement works at Slite on Gotland.

An attempt has been made to use lichens to study the regional distribution of air pollutants in the extreme south of Sweden. The following 14 epiphytic lichen species were chosen: *Anaptychia ciliaris, Cetraria chlorophylla, Evernia prunastri, Hypogymnia physodes, Lecanora conizaeoides, Lecidea scalaris, Parmelia acetabulum, P. exasperatula, P. subargentifera, P. sulcata, Physcia pulverulenta, Pseudevernia furfuracea, Ramalina fraxinea,* and *Xanthoria parietina.*

These were chosen to include some that are sensitive to air pollutants, e.g. *A. ciliaris, P. subargentifera, P. pulverulenta,* and *R. fraxinea,* and others that are insensitive, e.g. *H. physodes* and *P. sulcata.* The latter two, and particularly *H. physodes,* are poor-bark species that can colonize rich bark when it is acidified. *Lecanora conizaeoides* and *Lecidea scalaris* can best be regarded as poor-bark species with a high resistance to air pollutants. They have the capacity to exploit the absence of other species and spread considerably in areas of air pollution. Within this area there is reason to expect that the western parts are the most polluted and that southwest winds disperse the pollutants toward the northeast.

A network of 110 sampling sites was used. At each site, which was represented by the tree most rich in lichens, records were made of the tree species, their approximate age, appearance of crown and trunk, habitat conditions, etc.

The abundance, community relationships, and reproductive potential of each lichen species also were recorded. The percentage of the trunk surface, from the ground up to 3 m, covered by epiphytic lichens and mosses was estimated. The results of this investigation will be published during the winter of 1978/1979. An account of some of the results that illustrate the usefulness of the method are given below.

Not unexpectedly, *P. sulcata* has the widest range; it is present at 46 stations. In contrast, *P. subargentifera* is present at only one of the 110 stations.

The distribution of the individual lichen species does not give a complete picture of the conditions in the various localities. The presence, and sometimes also the absence, of certain species certainly gives some information. Compare, for example, the distribution of *L. conizaeoides* and *L. scalaris* (Figures 1*a* and 2*a*) with that of *P. subargentifera* and *R. fraxinea* (Figures 3*a* and 4*a*). The picture becomes more complete if the abundance of the species is also given (cf maps 1*b* and 5*b*).

The distribution of *H. physodes* is interesting. This species is typical of acid bark with a pH of about 4.5. It is usually absent from rich bark or occurs only very sparsely. Figures 5*a* and *b* show that it becomes more abundant to the east and north within the investigated area. This suggests that the normal lichen flora is becoming impoverished, and that the substrates are becoming acidified, over the whole area.

The same is true of *P. sulcata,* although it is not so restricted to poor bark as *H. physodes.*

The reproductive potential of the different species, as shown by the fertility conditions and development of vegetative reproductive organs (soredia and isidia) is also of importance in estimating the effect of the environment on the species (Figures 1*c, d* and 6*c, d*).

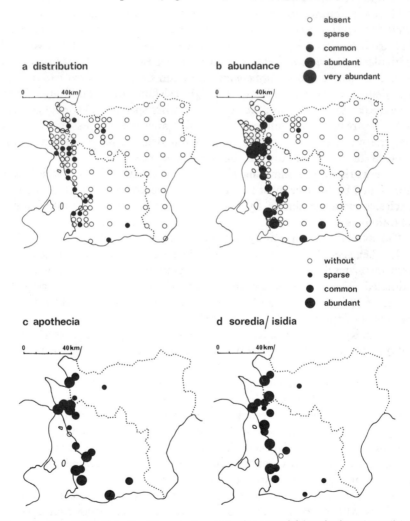

Figure 1   The distribution, abundance, and reproductive potential (apothecia and soredia) of *Lecanora conizaeoides* Cromb. in southern Sweden.

LIMITATIONS ON THE USE OF LICHENS
AS INDICATORS OF AIR POLLUTION

Some criticism has been directed at the use of individual species as indicators of air pollutants. De Sloover & LeBlanc (6), for example, wrote (p. 46): "We believe that these methods although interesting, open the door to an impoverishment of the information desired; furthermore, because of our limited knowledge of the autecology of epiphytes, these methods are sub-

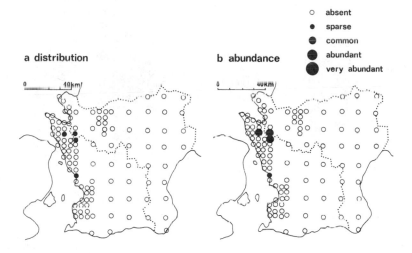

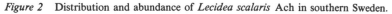

Figure 2 Distribution and abundance of *Lecidea scalaris* Ach in southern Sweden.

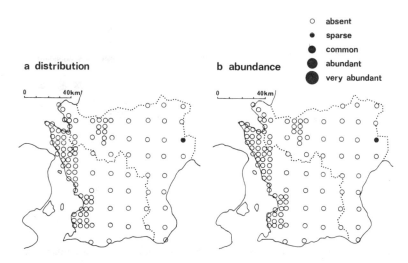

Figure 3 Distribution and abundance of *Parmelia subargentifera* Nyl in southern Sweden.

jected to so many imponderables that it is rather difficult to justify the interpretation based on data gathered from a few species." Such criticism is partly justified. However, the idea behind the use of indicator species, as implied by the above, is that nonbotanists such as planners, health inspectors, and countryside rangers should be able to estimate the type, intensity, and distribution of air pollutants over a limited area. Thus indicator species can be used for practical air pollution surveys even though many questions

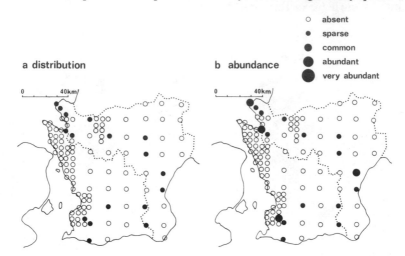

Figure 4   Distribution and abundance of *Ramalina fraxinea* L. Ach in southern Sweden.

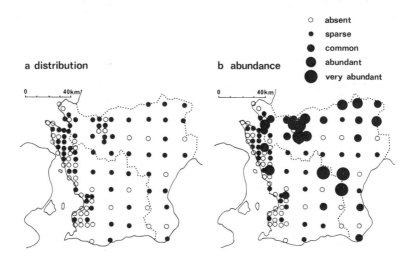

Figure 5   Distribution and abundance of *Hypogymnia physodes* L. Nyl in southern Sweden.

remain about the specific physiological mechanisms of the pollutant induced stress on the lichens (see also 7).

The total number of species at each site gives some idea of the zonation
around a harmful source. However, counting these is a very laborious
procedure which requires a good knowledge of lichens and is therefore not
practical [see, for example, (4, 37)]. The method can be used in certain
special circumstances, for example if the pollutants make the substrate more
acid than usual and poor-bark species are being worked with. However, if

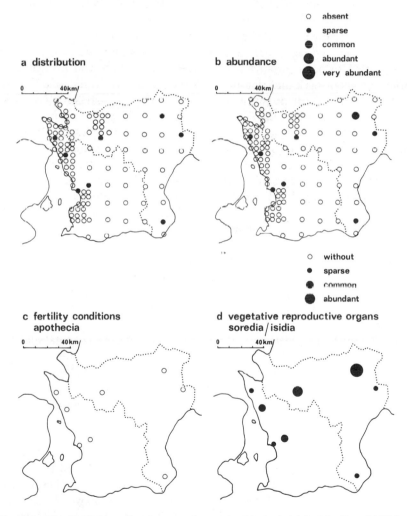

Figure 6   The distribution, abundance, and reproductive potential (apothecia and isidia) of
Anaptychia ciliaris L. Körb. in southern Sweden.

there is new colonization of rich-bark trees by poor-bark species of lichens, or the reverse, the number of species at a station, taken alone, will be meaningless.

Barkmann (4) considered that the distribution and appearance of the epiphytic plant communities gives a clear indication of the degree of air pollution and in this respect is better than the distribution of individual species. Against the background of the above, it need only be said that Barkmann is correct. However, this is so only if the composition and ecological requirements of the epiphyte vegetation are known and understood. For this reason, combinations of species, and not distinct plant communities, were used as the basis for the survey maps (p. 24) in Skye (38). The results should be similar, since pure plant communities would hardly be expected to occur around a harmful source. There will always be a mixture of poor-bark and rich-bark communities.

De Sloover & LeBlanc (6) noticed the difficulty of correlating the presence and distribution of the epiphyte vegetation of an area with the exact zones of pollution. They consider that three basic principles should guide the ecologist in this type of work: (a) to take a large number of quantitative criteria affecting the epiphyte vegetation into consideration; (b) to find habitats as ecologically similar as possible for the sites to be compared; (c) to present a simplified readily accessible, cartographic account. Maps have been published by many authors, but the majority are so generalized that only very superficial conclusions can be drawn from them, according to De Sloover & LeBlanc.

For various reasons, I have come to the conclusion that bark is a better substrate for lichens than stone or soil, and that records should be restricted to a single tree species, which must then be common and widespread over the area to be investigated. The sampling sites must be chosen with care and be as homologous as possible.

The following parameters are thought to be significant: (a) total number of species, (b) degree of cover of each species, (c) frequency of each species in the particular unit (e.g. a quadrat, a single tree trunk), (d) vitality, and (e) fertility [(see also (14) and Barkmann (3a)].

The degree of cover for each species is difficult to determine, so De Sloover & LeBlanc have preferred to combine frequency and abundance in a single scale.

Through studying the sequence of disappearance of lichens and mosses from a polluted area, a picture of their sensitivity to pollution (degree of toxiphoby) is built up. This must also be based on comprehensive studies, and the local climate must be taken into account. De Sloover & LeBlanc have listed the epiphytes in order of sensitivity and assigned an "index of toxiphoby" to each species according to its place in the list. The data obtained on the number of species, index of toxiphoby, and frequency-abundance are then used to derive a numerical value for the index of atmospheric purity (IAP).

Another attempt to derive a numerical expression of the lichens' reaction to air pollution was made by Trass (41). Like Barkmann, he pointed out that the results were more reliable for work with groups of species, i.e. plant communities, rather than single species. In addition, the degree of cover of the different species and their tolerance to an urban environment should also be studied. Using such analyses, he calculated an index, P, for each plant community by a simple mathematical formula. Trass also attempted to correlate his P-values with the SO_2 contents of the air.

Hawksworth (in 13) gave a detailed account of different methods of mapping lichen distribution. De Wit (7) made a critical assessment of the different methods and found that the relatively complicated mathematical formulae used by some authors (e.g. De Sloover, Trass) rarely yielded more useful information than did simple observations of numbers and composition of species at individual sites.

USE OF TRANSPLANTED LICHENS IN AIR POLLUTION SURVEYS

In 1961, Brodo published results of transplanting experiments with lichens by a new method that he had developed (5). In brief, this involves cutting out circular discs of bark on which lichens are growing from healthy trees in an area unaffected by air pollution and transferring them to trees in polluted areas. The development of the transplanted lichens can then be followed closely, e.g. by photographing the bark discs from time to time. By moving lichens within the unpolluted area, the effect of the treatment itself on the experimental plants can be checked. The method has been used to a limited extent in Sudbury, Ontario, Canada by LeBlanc & Rao (18) and later on a larger scale in the same area.

In Europe, Schönbeck has been the major user of transplants as test organisms for air pollution (32). He cuts out lichens from healthy trees in areas unaffected by air pollution, fixes 10 such samples on a board, and places these boards at a large number of sampling sites. He works mainly with *H. physodes*. By having sampling sites in unpolluted areas over the whole of Germany, with very different types of climate, Schönbeck has been able to demonstrate that the climate had absolutely no negative effect on the survival of the lichens. Nor was there any effect of how or where the lichens were placed, or where they were collected, as long as the air was unpolluted. However, the lichens were damaged to some extent at the sampling sites in areas of polluted air. The extent of the damage could be recorded by infrared photography and planimetry of healthy and damaged surface areas. Statistical treatment was possible because Schönbeck had 10 lichen samples at each sampling site.

Experiments with transplanted lichens in glasshouses have shown that damage became apparent after four days at a concentration of 2 μg SO_2 per m^3. The lichens were not killed, although damage could be seen. As expected, dry lichens were less affected by air pollution than were damp lichens.

CONCLUSIONS

Thus with the help of lichens — whether growing naturally on tree trunks or whether transplanted in some way — an assessment can readily be made of the extent of air pollution. In this context, because of their sensitivity, and because they normally occur almost everywhere, they are far better than any measuring instrument. However, the difficulty lies in translating the picture of lichen distribution into exact air pollution values, a difficulty that is certainly not insuperable. In the initial phase of investigation of damage around a source of pollution, lichens can assist in the assessment of the type of air pollution involved. If change of substrate of the lichens, as well as their general appearance, is interpreted correctly, a good picture of the air pollution situation is already available.

Literature Cited

1. Almborn, O. 1948. Distribution and ecology of some south Scandinavian lichens. *Bot. Not. Suppl.* 1:2
2. Almborn, O. 1953. Some aspects of the sociology of epiphytic lichen communities. *Proc. 7th Int. Bot. Congr.*, Stockholm, 1950. Stockholm: Almquist & Wiksell
3. Barkmann, J. J. 1961. De verarming van de cryptogamenflora in ons kand gedurende de laatste honderd jaar. *Natura* (10):141–51, 58
3a. Barkmann, J. J. 1958. *Phytosociology and Ecology of Cryptogamic Epiphytes.* Van Gorcum & Comp. Assen
4. Barkmann, J. J. 1963. De epifyten-flora en-vegetatic van Midden-Limburg (België). *Verh. K. Ned. Akad. Wet., Afd Natuurkd. 2 Reeks* 54:1–46
5. Brodo, I. M. 1961. Transplant experiments with corticolous lichens using a new technique. *Ecology* 42(4):838–41
6. De Sloover, J., LeBlanc, F. 1968. Mapping of atmospheric pollution on the basis of lichen sensitivity. *Proc. Symp. Recent Adv. Trop. Ecol.*, Faridabad
7. De Wit, T. 1976. Epiphytic lichens and air pollution in the Netherlands. *Verhandeling. 8. Rijksinstituut voor Natuurbeheer Research Institute for Nature Management.* Leersum
8. Domrös, M. 1966. Luftverunreinigung und Stadtklima ein Rheinisch-Westfälischen Industriegebiet und ihr Auswirkung auf den Flechtenbewuchs der Bäume. *Arb. z. rheinischen Landeskunde* 23. Bonn: Ferd. Dümmlers
9. Du Rietz, G. E. 1932. Zur Vegetationsökologie der ostschwedischen Küstenfelsen. *Beih. Bot. Centralbl.* 49:62–112
10. Du Rietz, G. E. 1945. Om fattigbark- och rikbarksamhällen *Sven. Bot. Tidskr.* 39:147
11. Ericson, K. 1969. *Luftföroreningars inverkan på epifytiska lavar i trakten av Köping.* Uppsala: Växtbiol. Inst.
12. Eriksson, O. 1966. *Lavar och luftföroreningar i Sundsvallstrakten.* Uppsala: Växtbiol. Inst.
13. Ferry, B. W., Baddeley, M. S., Hawksworth, D. L. 1973. *Air Pollution and Lichens.* Bristol: Athlone Press Univ. London

14. Jones, E. W. 1952. Some observations on the lichen flora of tree boles, with special reference to the effect of smoke. *Rev. Bryol. Lichenol.* 21:96–115

15. Klement, O. 1956. Zur Flechtenflora des Kölner Domes. *Descheniana* 109: 87–90

16. Klement, O. 1958. Die Flechtenvegetation der Stadt Hannover. *Beitr. Naturk. Niedersachsens* 11H3:56–60

17. Klement, O. 1966. Vom Flechtensterben im nördlichen Deutschland. *Ber. Naturhist. Ges.* 110:55–66

18. Le Blanc, F., Rao, D. N. 1966. Réaction de quelques lichens et mousses épiphytiques a l'anhydride sulfreux dans la région de Sudbury Ontario *Bryologist* 69(3):338–46

19. Le Blanc, F., Rao, D. N. 1966. Influence of an ironsintering plant on the epiphytic vegetation in Wawa, Ontario. *Bryologist* 70:141–57

20. Lundström, H. 1968. Luftföroreningars inverkan på epifytfloran hos barrträd i Stockholmsområdet. *Stud. For. Suec.* 56:5–55

21. Magnusson, B. 1969. *Luftföroreningars inverkan på epifytiska lavar i Slite.* Uppsala: Växtbiol. Inst.

22. Moberg, R. 1968. Luftföroreningars inverkan på epifytiska lavar i Köpmanholmen. *Sven. Bot. Tidskr.* 52(1): 169–96

23. Natho, G. 1964. Die Verbreitung der epixylen Flechten und Algen im demokratischen Berlin. *Wiss. Z. Humboldt Univ. Berlin, Math. Naturwiss. Reihe* 13:53–75

24. Natho, G. 1964. Zur Verbreitung rindenbewohnender Flechten in Kleinstädten-Ostseebad Kühlungsborn. *Wiss. Z. Humboldt Univ. Berlin, Math. Naturwiss. Reihe* 13:639–43

25. Natho, G. 1964. Flechtenentwicklung in Städten. (Ein überblick) *Drudea* 4(1):33–44

26. Nylander, W. 1866. Les lichens du jardin du Luxembourg. *Bull. Soc. Bot. Fr.* 13:364–72

27. Pearson, L., Skye, E. 1965. Air pollution affects pattern of photosynthesis in *Parmelia sulcata*, a corticolous lichen. *Science* 148(3677):1600–2

28. Rao, D. N., LeBlanc, F. 1966. Effects of sulfur dioxide on the lichen algae, with special reference to chlorophyll. *Bryologist* 69:69–75

29. Rydzak, J. 1954. Rozmieszczenie i ekologia porostów miasta Lublin. (Dislokation und Ökologie von Flechten der Stadt Lublin.) *Ann. Univ. Mariae Curie-Sklodowska Sect. C* 1953:233–336

30. Rydzak, J. 1959. Influence of small towns on the lichen vegetation. Part VII. Discussion and general conclusions. *Ann. Univ. Mariae Curie-Sklodowska Sect. C* 1958:275–323

31. Rydzak, J. 1969. Lichens as indicators of the ecological conditions of the habitat. *Ann. Univ. Mariae Curie-Sklodowska Sect. C* 1968:131–64

32. Schönbeck, J. 1969. Eine Methode zur Erfassung der biologischen Wirkung von Luftverunreinigungen durch transplantierte Flechten. *Staub Reinhalt. Luft* 29:1

33. Seaward, M. R. D. 1977. *Lichen Ecology.*, pp. 31–68. London: Academic

34. Skye, E. 1958. Luftföroreningars inverkan på busk- och bladlavfloran kring skifferoljeverket i Närkes Kvarntorp. *Sven. Bot. Tidskr.* 52:133–90

35. Skye, E. 1964. Epifytfloran och luftföroreningarna. *Svensk Naturvetensk.*, pp. 327–32

36. Skye, E. 1965. Botanical indications of air pollution. The plant cover of Sweden. *Acta Phytogeogr. Suec.* 50. 285–87. Uppsala

37. Skye, E. 1967. Lavar mäter lortluft. *Forskning och framsteg.* H 1. 3–6. Stockholm

38. Skye, E. 1968. Lichens and Air Pollution. A study of cryptogamic epiphytes and environment in the Stockholm region. *Acta Phytogeogr. Suec.* 52: 8–123

39. Skye, E., Hallberg, I. 1969. Changes in the lichen flora following air pollution. *Oikos* 20:547–52

40. Steiner, M., Schulze-Horn, D. 1955. Über die Verbreitung und Expositionsabhängigkeit der Rindenepiphyten im Stadtgebiet von Bonn. *Decheniana* 108:1–16

41. Trass, H. 1968. Indeks samlikurühmituste kasutamiseks öhu saastatuse määramisel *Eesti Loodustead. Arh. Ser.* 11:628

Ann. Rev. Phytopathol. 1979. 17:343–66
Copyright © 1979 by Annual Reviews Inc. All rights reserved

CONTROL OF STORAGE DISEASES OF GRAIN[1]

♦3715

John Tuite

Department of Plant Pathology, Purdue University, West Lafayette, Indiana 47907

George H. Foster

Department of Agricultural Engineering, Purdue University, West Lafayette, Indiana 47907

The study of the control of storage diseases of grain is multidisciplinary because a variety of factors influence the microbial deterioration of grain. In fact it may be possible to demonstrate that storage can be affected adversely by events occurring from the time the grain is planted through its harvest. These subtle and sometimes profound effects are mitigated by good storage practices, but they can be important in commercial practices. In the course of this discussion some of these effects on storability are noted.

Any substantial improvement in the practical control of storage diseases of grain will depend on (a) the determination of losses induced by fungal growth, (b) application of new and old technology, (c) extensive educational programs, and (d) better rewards for grain quality. Present knowledge of the extent of mold-induced mycotoxin losses with the exception of aflatoxin is insufficient to motivate substantial improvement in control. While considerable work has been done on mycotoxins, there are probably many more toxic metabolites to be discovered and of those known, only a few have been adequately surveyed for and tested on livestock to properly assess their significance. The aflatoxin problem has demonstrated the seriousness of a potent mycotoxin but may have led to the overemphasis of the acute effects of mycotoxins to the detriment of interest in chronic effects and less potent

[1]Purdue Agricultural Experiment Station Journal Paper No. 7553.

0066-4286/79/0901-0343$01.00

mycotoxins. As recent research indicates, these latter subclinical problems, although not as obvious, may be significant (104). An interesting development in appraisal in the marketplace is the assay of suspicious looking kidneys of pigs in Denmark for ochratoxin (79, 80). If the kidney is positive for ochratoxin, the animal carcass is destroyed coupling a survey with quality control (80). Such studies, particularly when assay methods are increased in their sensitivity so that toxins may be detected in animal tissue, need to be emphasized.

Losses caused by fungal activity are obscured by the grain grading system used in the United States. For example, No. 2 corn may contain up to 5% damaged kernels but no more is paid for a lot of No. 2 corn with damage as low as 3.1%. Therefore mold or other damaged kernels are usually blended with lots of sound corn to the upper limit of allowable damage. As a result, mold losses may be hidden by dilution. In this way penalties are not, or not severely, assessed, for spoiled grain and may even be a source of profit, as when highly damaged corn is purchased at a reduced price and subsequently blended into a lot that will sell at a higher price. It cannot be assumed, however, that blending of every type of mold-damaged grain is objectionable and will result in measurable negative effects on animals. These assessments of feeding losses incurred by the blending of different kinds of molded grain should receive priority.

Proceeding to a consideration of control of storage diseases we emphasize the experiences of midwestern United States and corn, with an indication of some of the new concepts or technologies. Technological changes of harvesting and handling of corn in the United States has solved some problems and created others. The shift to the picker sheller and the combine in the Midwest, where 90% of the corn in the United States and over 40% of the world's corn is produced, has relieved farmers of many inefficient harvesting, handling, and storage problems inherent in ear corn. For example, shelled corn requires only 50% of the volume needed by ear corn, allowing for the reduction of storage and transportation needs (25). Shelled corn is a process-reliable and predictable material, superior to ear corn (25). Field losses (ear rots and downed ears) that are aggravated by the late harvest, the latter usually necessary to ear corn storage, are lessened by the early combine harvest. Ear corn stored in cribs is also more subject to rodents and insects than shelled corn stored in metal bins. Attack by molds in cribs was not uncommon and depended on harvest moisture, weather, building maintenance, and design. The advantage of cribbed corn with respect to mold growth was that the natural aeration usually prevented grain heating, which is both the cause and result of accelerated fungal growth. Harvesting shelled corn, however, produces large amounts of physically damaged and mold-susceptible moist corn over a relatively short period that is then stored in a confined area. This rapid early harvest was apparently necessary to keep up with the doubling of corn yields that occurred in the United States after World War II. Thus, the problems of corn storage have increased considerably.

Prevention of inhibition of mold growth in storage is usually accomplished by modification of the interseed environment (moisture, temperature, and atmosphere). Chemical preservatives are used to a small degree, and their use in conjunction with environmental methods is now being suggested (100). Ideally, moisture or temperature, alone or in combination, is modified to inhibit all fungus growth. Hermetic storage is an example of environmental modification of the atmosphere that prevents only undesirable fungal growth but not harmless yeast and bacterial growth. There are problems with the general application of this concept of favoring harmless over toxic organisms.

DIAGNOSTIC TESTS

A potentially important tool for controlling grain storage diseases is the rapid assessment of grain condition so as to predict storability. Storability is viewed in terms of the grain's ability to resist microbial attack but may be gauged by the ability to maintain germination or to maintain grade. Thus, measurement of the physical condition of the grain (broken seed coat, stress cracks, cracked grain, and foreign material), insect infestation, amount of microbial infection, and seed germination, and assessment of the storage conditions (temperature, moisture) can help decide how the grain should be handled, e.g. dried, cleaned, fumigated, aerated. The extent of physical damage to the seed coat, indicating potential vulnerability to mold attack, can be measured visually and/or aided by a fast green staining technique (78). The method has been quantitated by a rating system (32) and by use of a colorimetric procedure based on differential absorption of fast green by damaged and nondamaged kernels (33, 35). Examining grain for stress cracks, induced mainly by high temperature drying (140, 156) but also by combine harvesting at high moisture (34), gives an indication of susceptibility to breakage. A sample breakage test has been devised to measure susceptibility to breakage (133), and improvements in breakage tests are being attempted (65). Insects enhance mold development because they increase moisture and temperature, open areas of the grain for attack, and supply inoculum (38). Visual inspection, staining, and X-ray techniques reveal the kind and amount of infestation.

Measurement of the extent of fungal growth and therefore of the extent of deterioration is promising and experimentally useful (144). A fungal product such as ergosterol (122) or chitin, the later a cell wall component of most storage fungi, has been related to the amount of fungal growth (46, 56, 124). Unfortunately, these methods are not entirely applicable, because ergosterol is not readily produced by the important storage organism, *Aspergillus glaucus*, and an increase in chitin is not detected in early stages of growth (124). Neither method distinguishes between preharvest or postharvest fungi. Both methods are also unsuitable for routine determination because of their complexity, although the ergosterol assay requires only one hour. Another test such as fat acidity, a result of fungal activity, is not sensitive enough in corn and small grains (89) but may be useful for rapeseed (93).

Glutamic acid decarboxylase activity has predicted storability of corn seed as determined by germination (58) and appears to be a sensitive index of viability and deterioration (11). Optical measurements of grain quality are attractive as they are rapid and nondestructive. There appears to be little progress in this area since the work of Birth & Johnson (15), who distinguished mold-damaged corn from sound corn by a difference in fluorescence. Molds quenched the natural fluorescence of sound corn. Detection of volatile products associated with fungal growth, although specific to certain fungi (74), is of uncertain use (45, 81).

Cultural techniques, e.g. plating and dilution counts of fungi, reveal the numbers and kinds of fungi that have grown in and on the seed. As fungi have specific environmental requirements one can often determine the previous storage conditions (36–38) and what mycotoxins may be present by the numbers and kinds of fungi isolated. It is often necessary to use proper media lest important components of the mycoflora are missed (36–38, 92). Selective media for important storage fungi such as the *Aspergillus glaucus* group and *A. flavus* are available (13, 18, 38). The percent infection by mycotoxogenic and other fungi as revealed by plating has been used to assess grain condition (38, 48, 82, 130), but since the plating technique does not quantitate mold growth and only anticipates mycotoxins, it is not a precise measure of grain deterioration. Prior heavy invasion by field fungi may not indicate susceptibility to fungus invasion in storage (123). Culture techniques are not entirely applicable to commercial operations as the fungi may have died or been killed by high temperature drying and the technique takes at least 4–5 days. Visual inspection for molds, with magnification, is useful but is only a partial substitute for culture techniques. Despite the inadequacies of measuring fungi in grain it is important that new procedures in drying, storing, and handling of grain be evaluated with microbiological tests. More floristic and ecological studies of grain fungi are needed to improve the value of microbiological tests.

Measurement of carbon dioxide production, to appraise loss in quality and predict storability of corn, has been widely accepted. Steele et al (131) and Saul (117) have determined that 7.4 g CO_2/kg of dry matter of corn will result in about 0.5% loss in dry matter. Saul used this amount as the limit of acceptable quality. More than this amount of CO_2 or 0.5% dry matter loss, he believes, will result in a loss in grade (117). From monitoring of CO_2 production of stored corn, Saul constructed a table using grain moisture and temperature to predict storability (117, 152). The data are also presented in graph form and are well known as *Saul's curves*. Saul (117) cautioned that these data do not account for mycotoxin production and there is little leeway in the allowable storage times. He emphasizes that in using the curves in predicting storability one should be aware that any prior storage of the corn will have used some part of storage life. This may account for some of the difficulty encountered with the application of the data to commercial practice. For example, it is not uncommon for freshly harvested corn to remain in a truck overnight prior to its transport to the elevator, and elevator operators would have no knowledge of this history. Also, as emphasized by Steele & Saul, physical kernel damage is important in increasing the rate of deterioration; however, this factor is rarely taken into account (117, 119, 131). Saul & Steele's data are based on about 30% mechanical damage, as determined by the method of Schmidt et al (20), which is expected to occur when 28% moisture corn is combined. They employ a multiplier which permits adjustment of storability prediction according to damage (118, 131). Saul's curves have been used extensively in almost all storage simulations and storage modeling. The evaluation of these data, as well as their correlation with all the quality criteria that can be marshaled is long overdue. Measurement of microbial activities, especially mycotoxigenic fungi and mycotoxin production, is especially important.

MYCOTOXIN ANALYSES

An important development in the improvement of grain quality and control of storage diseases is the chemical determination of mycotoxins. The presence of toxic metabolites may reflect the deterioration of grain and indicate possible feed and food hazards. The regulatory aspects of aflatoxin contamination (110) have resulted in a greater awareness of the problems of mold deterioration and have had a positive effect on storage practices. Mycotoxin surveys (84, 127) have focused on mold-induced problems. Three kinds of mycotoxin methods have been of use to the industry: 1. Rapid semiquantitative methods, usually a minicolumn method (68, 149), are suitable for the elevator or feed mill but the toxic and flammable solvents used require adequate safeguards. 2. Multitoxin methods (158), which quantitatively detect several mycotoxins, require several hours, a chemical laboratory, and experienced personnel. 3. Analytical methods, which employ expensive high pressure liquid or gas chromatography equipment, identify and quantitate mycotoxins precisely. There are several experienced commercial laboratories that will do all three types of analysis. In addition, the black light test pioneered by Shotwell and her associates (126) has been a valuable rapid presumptive test for aflatoxin in corn, particularly in the Midwest where incidence is low. A vibrating table with a viewer (10), which permits rapid inspection of corn kernels BGYF (bright green yellow fluorescence) associated with growth of *A. flavus*, may eliminate the need to grind the sample before viewing for BGYF. Sensing of specific wavelengths of BGYF automatically has been suggested (109) but probably is not economic although it may be for high cash crops such as pistachio nuts (88). Misuse of the black light technique has created problems in the industry (145), but they can be overcome by proper training and following procedures. As chemical assay methods become further refined and adapted they are likely to be an integral part of assessing and maintaining grain quality. It is unlikely that mycotoxins will be a factor in grain grading, but their absence or presence at amounts below a certain level may be a required notation on grain certificates by certain buyers.

MOISTURE AND TEMPERATURE DETERMINATION

An important concept in the control of grain storage diseases is one emphasized by Christensen (37, 38); that is, variation in grain moisture is the rule in bulk storage, and fungi grow where moisture is suitable and not according to the average moisture content (MC). The latter is emphasized in grading and merchandising. Therefore it is important to know the variation in grain moisture, something more readily said than done. However, automatic grain samplers will conveniently obtain good samples of moving grain, and pneumatic samplers, although not official and possibly biased, are helpful in obtaining samples of wet and dry grain in deep bulks. A substantial problem is accurate determination of grain moistures (37, 39, 75), particularly in excess of 23–25%; at these moistures electronic meters may differ from oven determinations by as much as 2–3 percentage points. Also, there appear to be many meters in elevators that are inaccurate and not regularly calibrated (22). Near-infrared techniques are reputed as accurate in moisture determinations (135), including higher moistures, but the cost of the equipment is prohibitive if purchased only for moisture determinations. There is evidence that the official oven method for corn (103°C for 72 hr) does not agree with the Motomco®, the official USDA approved meter, or the Karl Fisher method (37). The latter is considered as a standard because of its accuracy (99). Small differences (0.2%) in moisture may determine fungal growth at critical levels. The equilibrium moisture content

(EMC) of grain is influenced by genotype (69), previous drying temperatures (146), storage temperature (107), hysteresis (69), and whether ground or unground (146). As a result of these variables, even if accurate moistures are determined, the prediction of microbial growth is imprecise.

It is the interseed relative humidity (water activity, a_w) that determines fungal activity (36–38). Storage fungi do not readily grow below 72% RH regardless of substrate, and species of storage fungi have specific minimum humidity requirements (36, 38). Monitoring of interseed relative humidity has been suggested to appraise storage conditions (57, 71), but the technology has not been applied to commercial practice.

Monitoring of grain temperature, usually by thermocouples, has been important to the management of grain (26). Knowledge of temperature and moisture is essential to predicting storability. Localized temperature increases reflect insect and microbial activity. Heat production by microorganisms means that the grain is already deteriorated. Thus, sensing of temperature increase is important to management of the entire grain bulk so that steps may be taken to prevent deterioration beyond the hot spots.

HARVESTING DAMAGE

Combined corn is frequently heavily damaged (broken grain or breaks in the seed coat) (7, 29). Some of the damage may be invisible to the unaided eye or without the use of staining technique (78). As Chowdhury & Buchele (32) point out, unless this damage results in broken corn or material that passes through a 12/64 in (4.76mm) round hole sieve it is not recognized in US Official Grading Standards (2). What is significant is, that according to Saul, the deterioration rate of damaged corn may be 2 to 3.5 times the rate of handshelled corn (117, 119). Mechanical damage is generally correlated with increased harvest moisture although very dry grain is highly vulnerable (29). Associated with combine harvest are stress cracks in the endosperm that make the grain vulnerable to breakage on handling (34). The extent and significance of stress cracks developed in harvesting, as compared to those originating from high temperature drying, is much less. It is suggested that harvest damage may be decreased by selecting varieties with kernels that detach from the cob more readily and have greater kernel strength (155). Improvements in combine design are being attempted and include rubber belts that squeeze the kernels off the cob (117) and rotary combines that attempt to lessen the frequency and magnitude of kernel impacts (155). It appears that commercial combines adjusted and operated according to manufacturers' directions do much less damage than that found in ordinary practice (24). Therefore farmer harvesting practices are significant in influencing the storability of the crop.

DRYING

High Temperature–High Speed

High temperature drying in the Midwest is the method of choice in conditioning corn for storage and marketing. It is used because of its high capacity and rapidity, making the grain immediately suitable for movement in commercial channels. Grain dried at high temperatures is decreased in hygroscopicity and as a result is in equilibrium with a higher relative humidity than low temperature dried grain. It is therefore recommended that dent corn that reached temperatures of 83°C and above be stored at moistures of 0.5 to 1.0% lower than corn dried at lower temperatures (146). Dryers, particularly the cross flow and batch, operated at too high a temperature and a high drying rate give brittle kernels caused by the formation of stress cracks in the endosperm (102, 140). As a result BCFM (broken corn foreign material) is increased and constitutes a storage hazard. This fine material tends to accumulate in spout lines and its localization prevents uniform aeration, as well as generally decreasing air flow (62, 134). Fines are also hygroscopic and vulnerable to mold growth. Screenings may contain 6–7 times as many spores as their companion kernels (143), and spores are common to grain dust (87). The increased interest in dust and its relation to grain elevator explosions has added impetus to making grain less brittle, which should have storage benefits. The collection of dust has not always helped storage problems as frequently the dust is added back to the grain; many operators consider it wasteful to discard the dust, and dust is difficult to handle and store by itself.

New developments in dryers, such as the concurrent and/or the incorporation of a tempering cycle in the dryer (23), have improved the physical properties of the grain and reduced energy requirements (9). In a concurrent dryer, grain temperatures may be lower and are more uniform than in a conventional cross flow dryer (9, 23). There is a premium on performance and capacity of dryers. Higher drying temperatures are more efficient but should not be used at the sacrifice of grain quality (23, 97). Efforts are being made to standardize the testing and evaluation of dryers so that performance may be evaluated objectively (8, 76).

Removal of moisture from grain is more efficient at higher moistures (97). High temperature drying of grain to moistures below 19% is not only less efficient but enhances development of stress cracks (140). Both rate of drying and heat time are important in causing stress cracks (59, 140). In an effort to increase capacity and efficiency of dryers and to decrease stress cracks, Foster developed a procedure known as dryeration (50). The wet corn is usually dried to 16% MC, or up to 20% if the original moisture is very high, and transferred without cooling to a tempering bin for 4–10 hr.

After tempering, it is aerated until cooled, which usually reduces moisture an additional 1 to 2%. The stored heat in the grain is utilized in absorbing moisture from the corn when it is aerated. Because one third of the normal dryer operation involves grain cooling, dryer capacity is considerably increased. After cooling, the grain is transferred to another bin, as surface grain is often very wet because of moisture condensation on the bin roof and resulting runoff. Drying efficiency by dryeration may be increased 25% and dryer capacity doubled (50). Dryeration is suited for both elevator and on-farm operations.

Another procedure which is finding many adherents is combination drying (25, 59). This consists of two drying methods: higher temperature followed by natural or low heat drying (slow drying). It is energy efficient and has most of the virtues of dryeration. It is not well suited to commercial elevators because grain moisture is not reduced quickly enough. However, it eliminates one extra handling, and seed germination is higher at the end of the high temperature drying cycle than dryeration. Typically the corn is dried to 20–22% and transferred to a bin for cooling and slow drying. The corn can be high temperature dried to lower moistures if slow drying is of some risk (59). The slow drying process is discussed in a subsequent section.

SLOW DRYING

Natural or Ambient Air Drying

Aeration is equated with forcing unheated (natural) air through a grain bulk (28) generally at airflow rates of 0.1 to 0.05 CFM/BU. Initially it was used to cool grain so as to slow fungal and insect growth and to prevent moisture migration. Moisture migration occurs when temperature gradients develop in grain bulks and is an important cause of moisture accumulation and resultant molding (24, 99). Aeration was developed as an alternative to "turning" the grain (moving the grain from bin to bin) because it saves money and time and prevents damage to the grain when it is moved. With the pressures of high moisture harvests it is used to hold corn prior to high temperature drying; however, slow drying may be accomplished with higher air flow rates. Usually with these high flow rates (0.5 to 4 CFM/Bu) with unheated air the process is called *natural air drying*. Molding develops when the original moisture of the corn is too high and not matched with a sufficiently high air flow. There are practical and economic limits to air-flow rates but air flow is of prime importance (96, 105). Delay of harvest can be helpful as storage temperatures are usually decreased (48) but may incur increased field losses. Problems occur in periods of warm or wet weather and when fines accumulate in the grain and interfere with aeration. Predictions of safe storage based on weather records, relative humidity, and temperature can be made (103, 105), but because weather varies there is an inherent risk. Upward air movement through warm grain in cold weather causes condensation on bin roofs and moisture accumulation on sides and top of the grain mass, and as a result encourages molding. As a consequence some operators prefer to draw the air downward through the bulk (26). Opening the hatches of bins lessens condensation with an upward flow.

A number of computer simulations (40, 103, 141, 142) and field studies (4, 59, 128, 147) have been done to determine the limits of natural air drying of corn at various locations. Natural air drying in central Indiana appears to be reliable with corn of no higher than 20–21% MC (147). Humidistat or clock control of the fan can reduce energy requirements (103). The latter may also increase dry matter loss (103). Continuous aeration interrupted only by below freezing weather has proven workable (96, 142, 147). No heat drying saves important gas drying fuels and spread electrical demands over many months. It also preserves the physical quality of grain as stress cracks are not increased (128). Its initial moisture requirements (20–21%) for a full bin are too low to be used in most harvest seasons in the Midwest as a sole drying method unless modified by partial filling of the bin (96) or by stirring (96) or used in combination with high temperature drying (59).

LOW HEAT DRYING (LOW TEMPERATURE DRYING)

A development of no heat drying is low heat drying. The objective is, usually by a 1.6–5.5°C increase in air temperature, to dry grain in the bin more quickly and to be more independent of weather than natural air drying. One simulation study did not indicate any particular advantage to low heat drying in locations where relative humidities are not high (105). The bin may be filled initially or filled in layers, the latter permiting higher moisture grain. As in natural air drying, depth of grain and fan capacity (both determining air flow rates) are the main factors that determine which moistures may be dried before deterioration takes place (105). Geographical location and the time of harvest are also important as they determine, in part, storage temperatures (48). Recommendations for Indiana and Illinois are available (4, 128). Low heat drying appears to work in central Indiana (5), and simulation studies confirm field experiences (105). It appears that moistures of about 23–24% can be dried by this method, and perhaps slightly higher moistures are feasible with layer drying. An early harvest, warmer climates than central Indiana, or unusually warm years may permit mold deterioration at the upper layers before they are reached by the drying front. Therefore, the method is unpredictable in full bins with moistures of 24–25% because of the unpredictability of weather. A stirrer increases capacity (assists air flow and avoids overdrying if air temperature is increased) but requires more energy and is yet another mechanical device that requires maintenance. Combination drying makes low heat drying more reliable. The fungi that have developed in laboratory and field studies are organisms that are favored by low temperatures and moderate-to-high moistures e.g. *Alternaria, Penicillium* spp., and *Gibberella zeae* (48) although *Fusarium moniliforme* was found in one study (59). Zearalenone and ochratoxin have been found in laboratory studies (48).

SOLAR DRYING

Solar drying has been confined to low heat drying where air temperature is increased as much as 16–17°C. It is based on an intermittent and a not always reliable source of energy. The collector area required for high temperature drying is unrealistic. Solar drying appears more rapid than natural air drying (41), particularly in areas with high relative humidity (105). It may be possible to safely dry corn with moisture contents of 24%, close to ordinary harvest moistures, but more tests are needed. Use of electricity is increased by more fan operation, but solar drying requires less total energy when compared to high temperature drying (42). Solar drying appears more efficient than no heat drying in cold weather (42). Solar drying, however, is not economically competitive with today's fuel costs but approaches feasibility compared with costs of electricity (77). Tax incentives and the use of collectors for other tasks (159), such as in drying winter wheat in midsummer (77) or in heating buildings, may make solar drying more competitive (77). Mold growth is possible because of increased grain temperatures and includes *Aspergillus flavus* (41, 42). Certain warmer locations in the country (Virginia, southern Pennsylvania, Illinois, much of Oklahoma, and the Texas panhandle) are considered areas of potential aflatoxin contamination with solar dryers (111). Overdrying of grain is also a problem (41) but can be corrected by stirring devices (106).

HERMETIC STORAGE

This is a proven method of preserving high moisture corn (HMC) for feed (66). It was developed initially in France (153) and has been used in the Midwest with corn primarily fed to beef cattle (12, 73). HMC may be stored in glass-lined bins (52), or other kinds of airtight bins that have a breathing system to prevent structural failure because of differential pressures and to limit the exchange of in-storage gas with outdoor air (72). The grain undergoes fermentation, O_2 is depleted, and CO_2 is increased by the respiration of the grain, yeast, and bacteria (27, 72). HMC is also stored in concrete silos and in trenches (bunkers) covered with plastic. The corn in bunkers is usually cracked and packed tightly so that anaerobic conditions may develop more rapidly and more completely. The problem with these systems is one of maintaining a high CO_2–low O_2 condition, especially as the grain is fed out. A number of fungi grow at anaerobic or semianaerobic conditions including *F. moniliforme* and *Penicillium roqueforti;* they ordinarily do not proliferate in properly sealed storage, possibly because of competition with yeasts and bacteria and temperature effects (148). Once the ensiled corn is removed from storage it is highly vulnerable to mold growth with only a few days of allowable storage time (144a, 157). Exposed corn in bunkers and silos must be fed out rapidly to keep ahead of the mold growth. The method is an energy efficient system but limits management flexibility, because the grain must be used for feed at or near the storage site.

Recently, barley at low moisture (ca 13–14%) has been successfully stored under nitrogen to maintain high viability and insect-free conditions (125).

CHEMICAL CONTROL

This area has attracted the unscrupulous and the naive. Compounds have been promoted without adequate testing. Research in chemical control should proceed with an awareness of the stringent economic, regulatory, and microbial inhibitory requirements of a chemical preservative. Methods used for evaluating treated samples are critical in determining the efficacy of chemicals. While no one method of evaluation is completely adequate, plating kernels is essential. Evaluations that rely solely on visual inspection, dilution technique, or mycotoxin analysis are inadequate and can result in erroneous conclusions. In vitro tests may be misleading because some chemicals are considerably more toxic on a culture medium than in or on the grain. If this occurs the chemical should be removed or inactivated before the kernel is plated.

A major problem of chemically treated grain is that it cannot move in ordinary commercial channels. Therefore, treated corn usually remains on the farm or its identity is preserved (stored in a separate bin) in an elevator as an added expense. The industrial users of corn resist the concept of chemical control as the treated grain may be comingled with untreated grain and may affect the quality and yield of their milled products (54).

The search for chemical preservatives is encouraged by the potential flexibility, low capital outlay, and high capacity they lend to preserving grain. The energy crises with high costs and uncertain supplies of propane and natural gas for drying has given impetus to chemical treatment. A successful chemical must have very low mammalian toxicity but wide and usually long-lasting microbial-inhibiting properties. The latter property may be partly overcome by either periodic addition or combination of chemical treatment with slow drying systems. A significant finding in chemcal control of storage fungi was the demonstrated efficacy of certain organic acids. Propionic acid (PA) has been one of the most effective and the most widely used (70, 73, 114). Its value was demonstrated in England in the early 1960s on barley and wheat (70, 73) and it has since been used commercially on corn in the United States. Acetic and formic acid are much less inhibitory but are used in combination with propionic acid (61). Methylene bis-propinate, which breaks down into formaldehyde and propionic acid, is equal or superior to PA (17, 116) but may have some negative effects on animal weight gain (17). Isobutyrate and ammonium butyrate are promising volatile fatty acids (16). The calcium and sodium salts of propionic acid are inhibitory in feeds but of less primary value in shelled corn (114). Sorbic acid is effective in an ethanolic solution but not as a powder when applied to grain (114). Its fungicidal value is lessened greatly at a pH above 5 (14).

There have been various proprietary mixtures of PA with butylated hydroytoluene (BHT), benzoic acid, formaldehyde, or formic acid. These additives do not appear to increase the efficacy of PA (150). The claim that there is considerable synergism between a variety of volatile fatty acids (64) has not been confirmed (114). It is an attractive concept as there are fungi and bacteria that have varying degrees of tolerance to PA and other organic acids. Therefore, a combination of compounds would more likely inhibit all the organisms present. This is an area where more work is needed.

Fungi that are often found in failing acid treatments include *Monascus, Paeciliomyces varioti, Penicillium* spp., *Aspergillus glaucus, A. fumigatus,* and occasionally *A. flavus* (19, 116, 150). The presence of the first two fungi, by their rarity in moldy grain and prevalence in acid-treated grain, indicates some tolerance to organic acids. *Paeciliomyces varioti* was less sensitive to PA than other fungi tested in in vitro tests (94) and there is some fear it may become a problem where grain is repeatedly treated at low dosages. The occurrence of the four other fungi may simply reflect favoring environments of temperature, moisture, and atmosphere. Mycotoxins may or may not be found in treated grain that has molded (19). Mycotoxins seem unaffected by common chemical fumigants (21) or preservatives (144a) except for ammonia which is used to destroy aflatoxin (20, 151).

Acid preservation of grain has three main problems: cost, metal corrosion, and loss of commercial grade. The latter is due to odor and the fact that moisture cannot be determined with a meter. These problems usually are all overcome by storage of treated grain in exposed piles for cattle feed in the drier regions of the United States such as Colorado, Kansas, and Nebraska. Rainfall in the Midwest, however, leaches the acid and increases kernel moisture in uncovered piles to allow mold growth. Covering the grain with plastic has been unsuccessful because of condensed moisture and difficulty in maintaining cover. Therefore in the Midwest, corn treated with acid appears to be less attractive as a preservative but is occasionally used on the farm, particularly in beef cattle feeding operations where high moisture corn is feed efficient (12). Initial promotion of organic acid treatment of grain indicated a carefree operation with little management. As a result some early users were disappointed. They were unaware that certain requirements had to be met, e.g. grain moisture accurately known so as to apply the correct dose, uniform chemical treatment on clean grain, and aeration with large bulk storage.

A recent development and promising use of PA involves treatment of high moisture corn in a low heat drying system (129). The corn (23–28%) is treated with 0.3% wt/wt of PA, an amount that leaves no detectable odor. This procedure increases allowable storage time and as a result the capability of low heat drying systems. The grain may also be marketed in commercial channels.

Another feature of PA-treated corn is the increased storability of the resulting mixed feed—feed made with treated corn can be safely stored for several months (144a).

Either aqueous or anhydrous ammonia has been used experimentally to preserve grain (16,100). Both forms at 1.5–2% inactivate aflatoxin (20, 52); destruction of aflatoxin is enhanced by roasting the NH_3-treated grain (43). Ammonia is not as effective in long-term storage as is PA (16), but it is considerably cheaper and may increase the nutritive value of the treated grain. Ammonia appears to be less effective against bacteria than PA (16, 100), and the fungus *Scopulariopsis brevicaulis* is often found in unsuccessful treatments (16). Anhydrous ammonia can complement low heat or natural air drying systems (100). Small amounts are trickled into the system to hold the grain while it is being dried. The lower amount of NH_3 reduces the amount of discolored grain ordinarily associated with NH_3 treatment (100).

Some compounds inhibit the production of a mycotoxin or mycotoxins; for example, the organic phosphorus insecticide dichlorvos inhibits aflatoxin production (121). Aside from contributing to a knowledge of the metabolism of a mycotoxin this type of compound has limited application because of its specificity. A statisfactory chemical must prevent the growth of most storage organisms and not just the production of a toxic metabolite.

Gentian violet was advocated and used in the poultry industry without being very inhibitory to storage molds in feed. The dye is highly fungistatic in vitro (31, 136) but not in feed (137, 144a); however, when poultry are fed the treated feed their performance appears to be better in some field reports (137). The efficacy may be a result of the antimycotic and antibacterial effects of the dye in the animal or lessening of mycotoxin effects (138). Recently, FDA has requested that the material be withdrawn from the market because of alleged carcinogenicity.

BLENDING

Warehousemen and farmers usually blend corn of different moistures and qualities. The warehouse operator does this more knowingly and more skillfully. This practice is to blend grain to obtain a grade, to condition high moisture corn without artificial drying, and to utilize dried corn. It is a common practice but is recognized as one of risk (26, 113). The mixing of high and low moisture grain does not result in grain of uniform moisture. Because of hysteresis the original lots remain somewhat different. The initial higher moisture corn, although now lower, is at a higher MC than the original lower sample. The magnitude of difference is related to the magnitude of the original difference in moisture (69). If the average moisture content of the mixture has a equilibrium relative humidity below that allowing mold growth, molding will not occur (115), although there is a report to the contrary (83). However, if the original moisture spread is large or the mixing inadequate there appears to be a chance of molding. Unfortunately there is not sufficient research to indicate specific recommendations on moisture spread. A difference of 3% in moisture between 2 lots of corn to be blended would be the most tolerated by one operator (26). The previous condition of the grain, time of year, BCFM, etc are important factors to consider in blending.

TEMPORARY STORAGE

Bumper crops result in corn piling up at elevators and farms in the Midwest. A shortage of rail cars aggravates the problem. It is unlikely that elevators and farmers will build sufficient storage space to meet the storage needs of such years. As a consequence, in some years harvested corn is stored in outdoor piles or in emergency type structures. It appears that, particularly in cool and wet postharvest years, considerable molding occurs in outdoor piles. Empirical observations have been made which are embodied in recommendations to lessen or avoid mold-induced losses (53, 90). Information is given on how to build and maintain a grain pile (should be dried corn) so that maximum shedding of rain occurs and when, how, and if to aerate (53).

HANDLING AND TRANSPORTATION-ASSOCIATED STORAGE PROBLEMS

Substantial amounts of corn are shipped from the Midwest to southern states and to export ports, the latter located largely in the South. Physical damage and BCFM increases each time the grain is handled (51), and case histories of overseas shipments revealed 1–7% increase in BCFM during the transfer of corn from elevator to ship and ship to elevator (67, 101). The basic problem is the brittleness of the grain, which is caused primarily by high temperature drying (140). The situation is aggravated by long drops and impact on hard surfaces particularly at low temperatures (49). Flow retarding devices can reduce handling damage somewhat (44, 49). Operating equipment at full capacity also is helpful (60). Containerization has been suggested as a possible solution (132). As the BCFM is not uniformly distributed, its localization in a grain bulk creates marketing problems and increases the potential of microbial deterioration. Chilled corn arriving in the South from the Midwest may increase in moisture at least from 0.1–0.2% when exposed to warm humid air of the gulf ports (1, 154). A preliminary survey of export corn indicates that most samples of corn were infected with *Aspergillus glaucus* and that one fifth of the samples had some insect infestation (139). The effect of these factors, combined with the lack of effective aeration in shipholds and other units of transportation, increases the possibility of molding, particularly if the moisture of some of the corn is above 15.5–16%. Corn to be shipped to the tropics or subtropical climates would be particularly vulnerable (39). Barge and rail shipments in the United States may also become molded, especially if delayed in warm weather.

GENETIC IMPROVEMENT

The control of storage diseases can be aided indirectly and directly through plant breeding. Indirect means include the development of varieties with kernel resistance to insect attack both in the field and in storage (47, 112), rapid drying rate (108), and resistance to harvest and handling damage (155). Direct means involve resistance to fungal attack (field and storage) (30, 95) or inhibition of mycotoxin production (159). This latter feature ordinarily is of limited value, but not if preharvest contamination of afla-toxin is involved (160), because of the severe regulatory aspects and potency of the toxin (110). Usually there are too many fungal but nonmycotoxin-associated losses (heating, must odor, discolored grain) in storage (38) to select only for inhibition of a specific mycotoxin.

In considering ways to prevent storage losses the implications and ramifications must be examined because a desirable feature may have liabilities. For example, if the structural characters (63, 108) that result in rapid dry-down are related to a thin and possibly weak pericarp the seed may be easily mechanically damaged (86) or penetrated by fungi. It is important to investigate the properties that permit rapid drying and to assess their relation to handling and harvest damage and mold invasion. Genotypes that lose moisture rapidly might absorb it similarly, and therefore would be less desirable when a harvest is interrupted by periodic rains. Another example of different manifestations of a desirable character concerns resistance to insects conferred by tight and long husks (47, 55). The feature results in slow field drying. In southeastern United States the field drying rate is not usually a problem, but insect attack associated with *Aspergillus flavus* is (91). Tight husk varieties are suitable and are widely used there, but in the Midwest the reverse is true. Resistance to insects may be best sought in resistance internal to the pericarp or to a pericarp with no undesirable drying or industrial use characteristics. If incorporation of these characters into commercial hybrids is to be done, breeders and hybrid seed companies must be supplied with suitable screening methods and be convinced of the merits of these characters in relation to clearly desirable features such as high yields and standability.

CULTURAL PRACTICES

This is perhaps the most speculative area of this discussion. Some evidence leads to the conclusion that the cultural conditions of the crop may influence its storability. The obvious example is found with *Aspergillus flavus* in corn where the fungus can be both a field and storage organism, particularly in the southeastern United States (85). Heavy field infections are expected to result in more rapid storage growth than when infection occurs after harvest. Therefore, cultural and environmental conditions that predispose or are associated with field infections of storage organisms are important considerations (84, 85) and are being intensively studied. These include weed competition, planting date, site, moisture, temperature, nutrients, host population density, and insect attack. Insect attack appears to be a significant factor in field infection and aflatoxin contamination (91). Conditions that stress the plant are associated with more field infection (84), and it is suspected that environmental conditions during the critical period of seed development are important (85). Heavy infection by field fungi such as *Alternaria* prior to harvest, however, does not necessarily make the grain more vulnerable to attack by storage molds (123). The study of preharvest factors is worthy of attention and may reveal why crops of some years seem to store better than others and may suggest ways of improving the storability of grains.

CONCLUSIONS

Prevention of microbial deterioration of stored grains is varied and can be complex, and methods employed are strongly influenced by economic incentives. The recognition of mycotoxins has stimulated control efforts by emphasizing the potential losses caused by fungal attack. The energy crisis (price and availability of fossil fuels) has prompted agricultural engineers to intensify their interests in the design of more efficient high speed dryers as well as to seek alternative low energy methods of conditioning grain such as those involving solar energy, unheated air, or partially heated air. Indications are that these alternative drying methods result in a physically better grain; however, because drying is slower and dependent on weather, some of the methods involve more risk of molding. Computer simulation studies are extensively used to evaluate and design these new strategies but lack adequate information on mold activities. The challenge to the plant pathologist is to keep pace, supplying the important microbiological-mycotoxin evaluations of these new methods and procedures. Rapid methods of estimating the extent of deterioration to help predict storability are needed. In additon, improved varieties that dry more rapidly, damage less when handled, and resist insect and mold attack in storage and the field are needed. It remains to be seen if these desirable characters are mutually exclusive.

Finally, more attention needs to be given to the education of the grower as his role is increasingly important in determining the storability and quality of grain. Economic incentives appear to be a necessary part of any viable program.

ACKNOWLEDGMENTS

Thanks and appreciation are given to B. Ashman, L. Bauman, C. M. Christensen, R. W. Curtis, P. Pecknold, D. B. Sauer, D. Scott, R. Wyatt, and D. Wilson for help in preparation of this review.

Literature Cited

1. Aldis, D. F., Foster, G. H. 1977. *Moisture Changes in Grain from Exposure to Ambient Air*. ASAE Pap. No. 77–3524 St. Joseph, Mich.: ASAE
2. Anonymous 1970. *Official Grain Standards of the United States*. US Dep. Agric. Consum. Mark. Serv., Grain Div.
3. Arbelaez, G. 1971. *Moisture and temperature requirements of certain field fungi of corn*. MS thesis. Purdue Univ. West Lafayette, Ind.
4. Arnholt, D. J., Rupp, J. A., Hufferd, W. T, 1974. *Farmers Guide for Electric Grain Drying*. Plainfield, Ind.: Public Serv. Indiana
5. Arnholt, D. J., Tuite, J. 1976. *Mold study results with electric drying—a progress report*. Presented at 1976 Farm Electrification Conf., Purdue Univ., West Lafayette, Ind.
6. Ayerst, G. 1969. The effects of moisture and temperature on growth and spore germination in some fungi. *J. Stored Prod. Res.* 5:127–41
7. Ayres, G. A., Babcock, C. E., Hull, D. O. 1972. Corn combine field performance in Iowa. *In Grain Damage Symp.* Ohio State Univ. Columbus
8. Bakker-Arkema, F. W., Brook, R. C., Brooker, D. B. 1978. *Energy and Capacity Performance Evaluation of Grain Dryers. ASAE Pap. No. 78–3523.* St. Joseph, Mich.: ASAE
9. Bakker-Arkema, F. W., Brook, R. C., Walker, L. P., Kalchik, S. J., Ahmadnia-Sokhansanj, A. 1978. Concurrent-flow grain drying-grain quality aspects. *Proc. 1977. Corn Quality Conf., Univ. Ill. AE–4454*
10. Barabolak, R., Colburn, C. R., Just, D. E., Kurtz, F. A., Schleichert, E. A. 1978. Apparatus for rapid inspection of corn for aflatoxin contamination. *Cereal Chem.* 55:1065–67
11. Bautista, G. M., Lugay, J. C., Cruz, L. J., Juliano, B. O. 1964. Glutamic acid decarboxylase activity as a viability index of artificially dried and stored rice. *Cereal Chem.* 41:188–91
12. Beeson, W. M., Perry, T. W. 1958. The comparative feeding value of high moisture corn and low moisture corn with different feed additives for fattening beef cattle. *J. Anim. Sci.* 17:368–73
13. Bell, D. K., Crawford, J. L. 1967. A botran-amended medium for isolating *Aspergillus flavus* from peanuts and soil. *Phytopathology* 57:939–41
14. Bell, T. A., Etchells, J. L., Borg, A. F. 1959. Influence of sorbic acid on the growth of certain species of bacteria, yeast, and filamentous fungi. *J. Bacteriol.* 77:573–80
15. Birth, G. S., Johnson, R. M. 1970. Detection of mold contamination in corn by optical measurements. *J. Assoc. Anal. Chem.* 53:931–36
16. Bothast, R. J., Adams, G. H., Hatfield, E. E., Lancaster, E. B. 1975. Preservation of high-moisture corn: A microbiological evaluation. *J. Dairy Sci.* 58:386–91
17. Bothast, R. J., Black, L. T., Wilson, L. L., Hatfield, E. E. 1978. Methylene bis propionate preservation of high-moisture corn. *J. Anim. Sci.* 46:484–89
18. Bothast, R. J., Fennell, D. I. 1974. A medium for rapid identification and enumeration of *Aspergillus flavus* and related organisms. *Mycologia* 66: 365–69
19. Bothast, R. J., Goulden, M. L., Shotwell, O. L., Hesseltine, C. W. 1976. *Aspergillus flavus* and aflatoxin in acid-treated maize. *J. Stored Prod. Res.* 12:177–83
20. Brekke, O. L., Peplinski, A. J., Lancaster, E. B. 1975. *Aflatoxin Inactivation in Corn by Aqua Ammonia. ASAE Pap. No. 75–3507.* St. Joseph, Mich.: ASAE
21. Brekke, O. L., Stringfellow, A. C. 1978 Aflatoxin in corn: A note on ineffectiveness of several fumigants as inactivating agents. *Cereal Chem.* 55:518–20
22. Brickenkamp, C. S. 1978. NBS and state cooperation to improve the accuracy of moisture meters. *Proc. 1977 Corn Quality Conf. Univ. Ill., AE–4454*
23. Brook, R. C., Bakker-Arkema, F. W. 1978. Simulation for design of commercial concurrent flow grain dryers. *Trans. ASAE* 21:978–81
24. Brooker, D. B., Bakker-Arkema, F. W., Hall, C. W. 1974. *Drying Cereal Grains*. Westport, Conn.: AVI Publ. Co., 265 pp.
25. Brooker, D. B., McKenzie, B. A., Johnson, H. K. 1978. *The Present Status of On-Farm Grain Drying. ASAE Tech. Pap. 78–3007.* St. Joseph, Mich.: ASAE
26. Brucker, D. G. 1978. Management of wet corn and storage. *Inn Alternatives for Grain Conditioning and Storage*, ed. G. C. Shove, pp. 40–42. Urbana-Champaign, Ill.: Univ. Illinois
27. Burmeister, H. R., Hartman, P. A., Saul, R. A. 1966. Ensiled high-moisture corn. *Appl. Microbiol.* 14:31–34
28. Burrell, N. J. 1974. Aeration. *In Storage of Cereal Grains and Their Products*, ed. Clyde M. Christensen, pp. 454–80. Am.

Assoc. Cereal Chem. St. Paul, Minn.: 549 pp.

29. Byg, D. M., Hall, G. E. 1968. Corn losses and kernel damage in field shelling of corn. *Trans. ASAE* 11:164–66

30. Calvert, O. H., Lillehoj, E. B., Kwolek, W. F., Zuber, M. S. 1978. Aflatoxin B_1 and G_1 production in developing *Zea mays* kernels from mixed inocula of *Aspergillus flavus* and *A. parasiticus. Phytopathology* 68:501–6

31. Chen, T. C., Day, E. J. 1974. Gentian violet as a possible fungal inhibitor in poultry feed: Plate assays on its antifungal activity. *Poultry Sci.* 53:1791–95

32. Chowdhury, M. H., Buchele, W. F. 1976. Development of a numerical index for critical evaluation of mechanical damage of corn. *Trans. ASAE* 19: 428–32

33. Chowdhury, M. H., Buchele, W. F. 1976. Colorimetric determination of grain damage. *Trans. ASAE* 19:807–8, 811

34. Chowdhury, M. H., Kline, G. L. 1978. *Stress Cracks in Corn Kernels From Compression Loading. ASAE Pap. No. 78–3541.* St. Joseph, Mich.: ASAE

35. Chowdhury, M. H., Marley, S. J., Buchele, W. F. 1976. Effects of different bio-parameters for colorimetric evaluation of grain damage. *Trans. ASAE* 19:1019–21

36. Christensen, C. M. 1974. *Storage of Cereal Grains and Their Products.* St. Paul, Minn.: Am. Assoc. Cereal Chem. 549 pp.

37. Christensen, C. M. 1978. Moisture and seed decay. In *Water Deficits and Plant Growth,* Vol. V, *Water and Plant Disease,* ed. T. T. Kozlowski, pp. 199–219 New York: Academic. 323 pp.

38. Christensen, C. M., Kaufmann, H. H. 1969. *Grain Storage: The Role of Fungi in Quality Loss.* Minneapolis: Univ. Minn. Press. 153 pp.

39. Christensen, C. M., Kaufmann, H. H. 1977. *Spoilage of corn in ship transport.* Unpublished report

40. Colliver, D. G., Brook, R. C., Peart, R. M. 1978. *Optimal Management Procedures for Solar Grain Drying. ASAE Pap. No. 78–3512.* St. Joseph, Mich.: ASAE

41. Converse, H. H., Foster, G. H., Sauer, D. B. 1978. Low temperature grain drying with solar heat. *Trans. ASAE* 21:170–75

42. Converse, H. H., Lai, F. F., Sauer, D. B. 1978. *In-bin Grain Drying Energy Savings From Solar Heat. ASAE Pap. No. 78–35 29.* St. Joseph, Mich: ASAE

43. Conway, H. F., Anderson, R. A., Bagley, E. B. 1978. Detoxification of aflatoxin-contaminated corn by roasting. *Cereal Chem.* 55:115–17

44. Ditzenberger, D. 1972. Methods of reducing grain damage in handling. In *Grain Damage Symp.,* pp. 120–33, Ohio State Univ., Columbus, Ohio

45. Donald, W. W. 1974. *Organic volatiles as a measure of fungal growth in stored soybean and corn seed.* MS thesis. Univ. Minn., Minneapolis. 138 pp.

46. Donald, W. W., Mirocha, C. J. 1977. Chitin as a measure of fungal growth in stored corn and soybean seed. *Cereal Chem.* 54:466–74

47. Eden, W. G. 1952. Effects of kernel characteristics and components of husk cover on rice weevil damage to corn. *J. Econ. Entomol.* 45:1084–85

48. Felkel, C., Tuite, J., Brook, R. 1978. *Storability of High Moisture Corn When Stored at Simulated Post-Harvest Temperatures of Three Locations in Indiana. ASAE Pap. no 78–3515.* St. Joseph, Mich.: ASAE

49. Fiscus, D. E., Foster, G. H., Kaufmann, H. H. 1971. *Physical Damage of Grain Caused by Various Handling Techniques. Trans. ASAE,* 14:480–85, 491

50. Foster, G. H. 1973. Heated-air grain drying. In *Grain Storage: Part of a system,* ed. R. M. Sinha, W. E. Muir, pp. 189–208. Westport, Conn.: AVI Publ. Co. 481 pp.

51. Foster, G. H., Holman, L. E. 1973. Grain breakage caused by commercial handling methods. *Mark. Res. Rept. No. 968,* ARS, USDA

52. Foster, G. H., Kaler, H. A., Whistler, R. L. 1955. Effects on corn of storage in airtight bins. *J. Agric. Food Chem.* 3:682–86

53. Foster, G. H., McKenzie, B. A., Brook, R. C., Tuite, J. 1978. *Temporary Corn Storage in Outdoor Piles. AE–91.* West Lafayette, Ind.: Purdue Univ.

54. Freeman, J. E. 1973. Quality factors affecting value of corn for wet milling. *Trans. ASAE* 16:671–78, 682

55. Giles, P. H., Ashman, F. 1971. A study of pre-harvest infestation of maize by *Sitophilus zeamais* Motch. (Coleoptera, Curculionidae) in the Kenya highlands. *J. Stored Prod. Res.* 7:69–83

56. Golubchuk, M., Cuendet, L. S., Geddes, W. F. 1960. Grain storage studies XXX. Chitin content of wheat as an index of mold contamination and

wheat deterioration. *Cereal Chem.* 37: 405–11

57. Gough, M. C. 1976. Remote measurement of moisture content in bulk grain using an air extraction technique. *J. Agric. Eng. Res.* 21:217–19

58. Grabe, D. F. 1965. Prediction of relative storability of corn seed lots. *Proc. Assoc. Off. Seed Anal.* 54:100–9

59. Gustafson, R. J., Morey, R. V., Christensen, C. M., Meronuck, R. A. 1976. Quality changes during high-low drying. *Trans. ASAE* 21:162–69

60. Hall, G. E. 1974. Damage during handling of shelled corn and soybeans. *Trans. ASAE* 17:335–38

61. Hall, G. E., Hill, L. D., Hatfield, E. E. Jensen, A. H. 1974. Propionic-acetic acid for high-moisture corn preservation. *Trans. ASAE* 17:379–82, 387

62. Haque, E., Foster, G. H., Chung, D. S., Lai, F. S. 1978. Static pressure drop across a bed of corn mixed with fines. *Trans. ASAE* 21:997–1000

63. Helm, J. L., Zuber, M. S. 1972. Inheritance of pericarp thickness in corn belt maize. *Crop Sci.* 12:428–30

64. Herting, D. C., Drury, E. E. 1974. Antifungal activity of volatile fatty acids on grains. *Cereal Chem.* 51:74–83

65. Herum, F. 1978. Equipment for measuring corn susceptibility to mechnical damage. *Proc. 1977 Corn Quality Conf., Ill. Univ. AE–4454*

66. Hill, L. D. 1978. High moisture grain storage in alternatives for grain conditioning and storage. *Conf. Proc.*, ed. G. C. Shove, Univ. Illinois Dep. Agric. Eng., Urbana, Ill.

67. Hill, L. D. 1978. Changes in grain quality during transport. *Proc. 1977 Corn Quality Conf.*, Univ. Ill. AE–4454

68. Holaday, C. E., Landsden, J. 1975. Rapid screening methods for aflatoxin in a number of products. *J. Agric. Food Chem.* 23:1134–36

69. Hubbard, J. E., Earle, F. R., Senti, F. R. 1957. Moisture relations in wheat and corn. *Cereal Chem.* 34:422–33

70. Huitson, J. J. 1968. Cereals preservation with propionic acid. *Proc. Biochem.* 3:31–32

71. Hunt, W. H., Pixton, S. W. 1974. Moisture—its significance, behavior, and measurement. See Ref. 36, pp. 1–45

72. Isaacs, G. W., Ross, I. J., Tuite, J. 1959. *A Zero-Pressure venting System for Air Tight Storages, ASAE Pap. No. 59–810.* St. Joseph Mich.: ASAE

73. Jones, G. M., Mowat, D. N., Elliot, J. I., Moran, E. J. Jr. 1974. Organic acid preservation of high moisture corn and other grains and the nutritional value: A Review. *Can. J. Anim. Sci.* 54:499–517

74. Kaminski, E., Stawicki, S., Wasowicz, E. 1974. Volatile flavor compounds produced by molds of *Aspergillus, Penicillium,* and *Fungi imperfecti. Appl. Microbiol.* 27:1001–4

75. Kaufmann, H. H., Christensen, C. M. 1970. *Storage Environment and Mold Growth. ASAE Pap. No. 70–301.* St., Mich.: ASAE

76. Keener, H. M., Glenn, T. L. 1978. *Measuring Performance of Grain Drying Systems. ASAE Pap. No. 78–3521.* St. Joseph, Mich.: ASAE

77. Kline, G. L., Oderkirk, W. L. 1978. *Solar Collector Costs for Low-Temperature Grain Drying. ASAE Pap. No. 78–3508.* St. Joseph, Mich.: ASAE

78. Koehler, B. 1957. Pericarp injuries in seed corn. Prevalence in dent corn and relation to seedling blights. *Ill. Agric. Exp. St. Bull. 617*

79. Krogh, P. 1977. Ochratoxin A residues in tissues of slaughter pigs with nephropathy. *Nord. Veterinaermed.* 29: 402–5

80. Krogh, P. 1978. Ochratoxin A in slaughter animals: Field observations and residue kinetics. *Int. Conf. "Mycotoxins"* August 1978, Munich (Abstr.)

81. Lee, L. S., Cucullu, A. F., Pons, W. A., Ir 1973 Gas-liquid chromatographic detection of actively metabolized *Aspergillus parasiticus* in peanut stocks. *J. Agric. Food Chem.* 21:470–73

82. Lichtwardt, R. W., Barron, G. L. 1959. A quantitative deterioration rating scale for shelled corn. *Iowa State J. Sci.* 34:139–46

83. Lillehoj, E. B., Fennell, D. I., Hesseltine, C. W. 1976. *Aspergillus flavus* infection and aflatoxin production in mixtures of high-moisture and dry maize. *J. Stored Prod. Res.* 12:11–18

84. Lillehoj, E. B., Fennell, D. I., Kwolek, W. F. 1977. Aflatoxin and *Aspergillus flavus* occurrence in 1975. Corn at harvest from a limited region of Iowa. *Cereal Chem.* 54:366–72

85. Lillehoj, E. B., Fennell, D. I., Kwolek, W. F., Adams, G. L., Zuber, M. S., Horner, E. S., Widstrom, N. W., Warren, H., Guthrie, W. D., Sauer, D. B., Findley, W. R., Manwiller, A., Josephson, L. M., Bockholt, A. J. 1978. Aflatoxin contamination of corn before harvest: *Aspergillus flavus association with insects collected from developing ears. Crop Sci.* 18:921–24

86. Mahmoud, A. R., Kline, G. L. 1972. Effect of pericarp thickness on corn kernel damage. See Ref., pp. 187–97

87. Martin, C. R., Sauer, D. B. 1975. *Physical and Biological Characteristics of Grain Dust. ASAE Pap. No. 75–4060.* St. Joseph, Mich.: ASAE

88. McClure, W. F., Farsaie, A. 1977. *Dual-wave Length Fluorescence Fiber-Optic Photometer for Sorting Aflatoxin Contaminated Pistachio Nuts. ASAE Pap. No. 77–5503,* St. Joseph, Mich.: ASAE

89. McGee. D. C., Christensen, C. M. 1970. Storage fungi and fatty acids in seeds held thirty days at moisture contents of fourteen and sixteen percent. *Phytopathology* 60:1775–77

90. McKenzie, B. A. Emergency grain storage in existing buildings. AE–92. Purdue Univ., West Lafayette, Ind.

91. McMillian, W. W., Wilson, D. M., Widstrom, N. W. 1978. Insect damage, *Aspergillus flavus* ear mold, and aflatoxin contamination in south Georgia corn fields in 1977. *J. Environ. Qual.* 7:564–66

92. Mills, J. T., Sinha, R. N., Wallace, H. A. H. 1978. Multivariate evaluation of isolation techniques for fungi associated with stored rapeseed. *Phytopathology* 68:1520–25

93. Mills, J. T., Sinha, R. N., Wallace, H. A. H. 1978. Assessment of quality criteria of stored rapeseed—a multivariate study. *J. Stored Prod. Res.* 14:121–33

94. Milward, Z. 1976. Further experiments to determine the toxicity of propionic acid to fungi infesting stored grain. *Trans. Br. Mycol. Soc.* 66:319–24

95. Moreno-Martinez, E., Christensen, C. M. 1971. Differences among lines and varieties of maize in susceptibility to damage by storage fungi. *Phytopathology* 61:1498–1500

96. Morey, R. V., Cloud, H. A., Gustafson, R. J., Petersen, D. N. 1978. *Fan Management for Ambient Air Drying Systems. ASAE Pap. No. 78–3005.* St. Joseph, Mich.: ASAE

97. Morey, R. V., Cloud, H. A., Lueschen, W. E. 1976. Practices for the efficient utilization of energy for drying corn. *Trans. ASAE* 19:151–55

98. Morey, R. V., Keener, H. M., Thompson, T. L., White, G. M., Bakker-Arkema, F. W. 1978. *The Present Status of Grain Drying Simulation. ASAE Pap. No. 78–3009.* St. Joseph, Mich.: ASAE

99. Muir, W. E. 1973. *Temperature and Moisture in Grain Storages. In Grain Storage: Part of a System,* ed. R. N.

Sinha, W. E. Muir. Westport, Conn.: AVI Publ.

100. Nofsinger, G. W., Bothast, R. J., Lancaster, E. B., Bagley, E. B. 1977. Amonia-supplemented ambient temperature drying of high moisture corn. *Trans. ASAE* 20:1151–54, 1159

101. Paulsen, M. R., Hill, L. D. 1977. Corn breakage in overseas shipments—two case studies. *Trans. ASAE* 20:550–57

102. Peplinski, A. J., Brekke, O. L., Griffin, E. E., Hall, G., Hill, L. D. 1975. Corn quality as influenced by harvest and drying conditions. *Cereal Foods World* 20:145–49, 154

103. Pfost, H. B., Maurer, S. G., Grosh, L. E., Chung, D. S., Foster, G. 1977. *Fan Management Systems for Natural Air Dryers. ASAE Pap. No. 77–3526.* St. Joseph, Mich.: ASAE

104. Pier, A. C., Cysewski, S. J., Richard, J. L., Thurston, J. R. 1977. Mycotoxins as a veterinary problem. In *Mycotoxins in Human and Animal Health,* ed. J. V. Rodricks, C. W. Hesseltine, M. A. Mehlman, pp. 745–50. Park Forest South, Ill.: Pathotox Publ. 807 pp.

105. Pierce, R. O., Thompson, T. L. 1978. *Management of Solar and Low-Temperature Grain Drying Systems. I. Minimum Airflow Rates Supplemental Heat and Fan Operation Strategies with Full Bin. ASAE Pap. No. 78–3513,* St. Joseph, Mich.: ASAE

106. Pierce, R. O., Thompson, T. L. 1978. *Management of Solar and Low-Temperature Grain Drying Systems. II. Layer Drying and Solution to the Overdrying Problem. ASAE Pap. No. 78–3514,* St. Joseph, Mich.: ASAE

107. Pixton, S. W., Warburton, S. 1971. Moisture content/relative humidity equilibrium of some cereal grains at different temperatures. *J. Stored Prod. Res.* 6:283–93

108. Purdy, J. L., Crane, P. L. 1967. Inheritance of drying rate in "mature" corn (*Zea mays* L.). *Crop Sci.* 7:294–97

109. Rambo, G. W., Zachariah, G., Parente, A, Tuite, J. 1976. Spectral analyses of fluorescences in dent maize infected with *Aspergillus flavus* and *Aspergillus parasiticus. J. Stored Prod. Res.* 12: 229–34

110. Rodricks, J. V. 1978. Regulatory aspects of the mycotoxin problem in the United States. In *Mycotoxic Fungi, Mycotoxins, Mycotoxicoses,* Vol. 3, ed. T. D. Wyllie, L. G. Morehouse. pp. 159–71. New York: Dekker. 202 pp.

111. Ross, I. J., Loewer, O. J., White, G. M. 1978. *Potential for Aflatoxin Develop-*

ment in Low Temperature Drying Systems. ASAE Paper No. 78–3516. St. Joseph, Mich.: ASAE

112. Santos, J. P. 1977. A Brazilian corn germ plasm collection screened for resistance to Sitophilus zeamais Motschulsky (Coleoptera: Curculionidae) and Sitotroga cerealella (Oliver) (Lepidoptera: Gelechiidae). MS thesis. Purdue Univ., West Lafayette, Ind.

113. Sauer, D. B. 1978. Contamination by mycotoxins: when it occurs and how to prevent it. See Ref. 110, pp. 145–58

114. Sauer, D. B., Burroughs, R. 1974. Efficacy of various chemicals as grain mold inhibitors. Trans. ASAE 17.557 59

115. Sauer, D. B., Burroughs, R. 1978. Fungal growth, aflatoxin production, and moisture equilibriation in mixtures of wet and dry corn. Unpublished

116. Sauer, D. B., Hodges, T. O., Burroughs, R., Converse, H. H. 1975. Comparison of propionic acid and methylene bis propionate as grain preservatives. Trans. ASAE 18:1162–64

117. Saul, R. A. 1968. Effects of harvest and handling on corn storage. In Proc. 23rd Ann. Corn and Sorghum Res. Conf. pp 33–36. Washington DC: Am. Seed Trade Assoc. 192 pp.

118. Saul, R. A. 1970. Deterioration Rate of Moist Shelled Corn at Low Temperatures. ASAE Pap. No. 70–302, St. Joseph, Mich.: ASAE

119. Saul, R. A., Steele, J. L. 1966. Why damaged shelled corn costs more to dry. Agric. Eng. 47:326–29, 337

120. Schmidt, J. L., Saul, R. A., Steele, J. L. 1968. Precision of estimating mechanical damage in shelled corn. US Dep. Agric. ARS 42–142

121. Schroeder, H. W., Cole, R. J., Grigsby, R. D., Hein, H. Jr. 1974. Inhibition of aflatoxin production and tentative identification of an aflatoxin intermediate "versiconal acetate" from treatment with dichlorvos. Appl. Microbiol. 27: 394–99

122. Seitz, L. M., Mohr, H. E., Burroughs, R., Sauer, D. B. 1977. Ergosterol as an indicator of fungal invasion in grains. Cereal Chem. 54:1207–17

123. Seitz, L. M., Sauer, D. B., Mohr, H. E. Burroughs, R. 1975. Weathered grain sorghum: Natural occurrence of alternariols and storability of the grain. Phytopathology 65:1259–63

124. Seitz, L. M., Sauer, D. B., Mohr, H. E., Burroughs, R., Hubbard, J. D. 1978. Aspergillus growth on grain measured by ergosterol, chitin and aflatoxin assays. Int. Cereal and Bread Congr., Winnipeg, Canada (Abstr.)

125. Shejbal, L. 1978. Preservation of cereal grains at various moisture contents in nitrogen. 3rd Int. Congr. Plant Pathol., Munchen, August 1978 (Abstr.)

126. Shotwell, O. L., Goulden, M. L., Hesseltine, C. W. 1972. Aflatoxin contamination: Association with foreign material and characteristic fluorescence in damaged corn kernels. Cereal Chem. 49:458–65

127. Shotwell, O. L., Hesseltine, C. W., Goulden, M. L. 1973. Incidence of aflatoxin in southern corn, 1969–1970. Cereal Sci. Today 18:192–95

128. Shove, G. C. 1973. Low temperature drying. Proc. Grain Conditioning Conf., Univ. Illinois, Urbana, Ill.

129. Shove, G. C., Walter, M. F. 1974. Grain Preservative Extends Allowable Drying Time. ASAE Pap. No. 74–3533. St. Joseph, Mich.: ASAE

130. Sorger-Domenigg, H., Cuendet, L. S., Christensen, C. M., Geddes, W. F. 1955. Grain storage studies; XVII. Effect of mold growth during temporary exposure of wheat to high moisture contents upon the development of germ damage and other indices of deterioration during subsequent storage. Cereal Chem. 32:270–85

131. Steele, J. L., Saul, R. A., Hukill, W. V. 1969. Deterioration of shelled corn as measured by carbon dioxide production. Trans. ASAE 12:685–89

132. Stephens, L. E., Foster, G. H. 1977. Reducing Damage to Corn Handled Through Gravity Spouts. Trans. ASAE 20:367–71

133. Stephens, L. E., Foster, G. H. 1976. Breakage tester predicts handling damage in corn. US Dep. Agric. ARS-NS-49

134. Stephens, L. E., Foster, G. H. 1976. Grain bulk properties as affected by mechanical grain spreaders. Trans. ASAE 19:354–58, 363

135. Stermer, R. A., Pomeranz, Y., McGinty, R. J., 1977. Infrared refectance spectroscopy for estimation of moisture of whole grain. Cereal Chem. 54:345–51

136. Steward, R. G., Wyatt, R. D., Ashmore, M. D. 1977. The effect of various antifungal agents on aflatoxin production and growth characteristics of Aspergillus flavus and Aspergillus parasiticus in liquid medium. Poultry Sci. 56:1630–35

137. Stewart, R. G., Wyatt, R. D., Kiker, J. 1977. Effect of commercial antifungal compounds on the performance of

broiler chickens. *Poultry Sci.* 56: 1664–66

138. Stewart, R. G., Wyatt, R. D., Lanza, G. M., Edwards, H. M., Ruff, M. D. 1979. Physiological effects of gentian violet in broiler chickens. Unpublished ms

139. Storey, C. L., Sauer, D. B. 1978. *Survey of Insect and Microbial Activity in the United States Wheat and Corn Exports.* US Grain Mark. Res. Lab., Manhattan, Kans. Unpublished rept.

140. Thompson, R. A., Foster, G. H. 1963. Stress cracks and breakage in artificially dried corn. *USDA Mark. Res. Rept. No. 631*

141. Thompson, T. L. 1972. Temporary storage of high-moisture shelled corn using continuous aeration. *Trans. ASAE* 15:333–37

142. Thompson, T. L., Villa, L. G., Cross, O. E. 1969. *Simulated and Experimental Performance of Temperature Control Systems for Chilled High-Moisture Grain Storage. ASAE Paper No. 69–856.* St. Joseph, Mich.: ASAE

143. Tuite, J. 1975. Mold profiles of commercial samples of whole corn kernels and screenings. *Anderson Grain Quality Prog. Rept.* Purdue, West Lafayette, Ind.

144. Tuite, J. 1978. Use of fungi and their metabolites as criteria of grain quality and storage systems. *Proc. 1977 Corn Quality Conf. Univ. Ill. AE 4454*

144a. Tuite, J. 1979. Unpublished results

145. Tuite, J., Brook, R., Scott, D., Park, E. 1978. Report of a survey of the grain industry in Indiana in 1977 for aflatoxin control practices used. *Purdue Univ. Stn. Bull. 201*

146. Tuite, J., Foster, G. H. 1963. Effect of artificial drying on the hygroscopic properties of corn. *Cereal Chem.* 40:630–37

147. Tuite, J., Foster, G. H., Thompson, R. A. 1970. *Moisture Limits for Storage of Corn Aerated with Natural and Refrigerated Air. ASAE Pap. NO. 70–306.* St. Joseph, Mich.: ASAE

148. Tuite, J., Haugh, C. G., Isaacs, G. W., Huxsoll, C. C. 1967. Growth and effects of molds in the storage of high moisture corn. *Trans. ASAE* 10:730–32, 737

149. Tuite, J., Scott, D. H. 1978. *Rapid Screening Methods for Aflatoxin in Corn. Purdue Univ. BP 5–22.* West Lafayette, Ind.

150. Deleted in proof

151. US Dep. Agric. 1977. *Proposed process for treatment of corn with ammonia to reduce aflatoxin content.* North. Reg. Res. Cent., Peoria, Ill.

152. US Dept. Agric. 1968. Guidelines for mold control in high-moisture corn. *Farmers Bull. No. 2283.* 16 pp.

153. Vayssiere, P. 1948. Hermetic storage, the process of the future for the conservation of foodstuffs. *UN Food and Agric. Organ. Agric. Stud.* 2:115–22

154. Wade, F. J., Christensen, C. M. 1977. *Condensation of Cold Corn Shipped to Reserve, Louisiana. ASAE Pap. No. 77–3523.* St. Joseph, Mich.: ASAE

155. Waelti, M., Buchele, W. F. 1969. Factors affecting kernal damage in combine cylinders. *Trans ASAE* 12:55–59

156. White, G. M., Ross, I. J. 1972. Discoloration and stress cracking in white corn as affected by drying temperature and cooling rate. *Trans. ASAE* 15: 504–7

157. Wilson, D. M., Huang, L. H., Jay, E. 1975. Survival of *Aspergillus flavus* and *Fusarium moniliforme* in high-moisture corn stored under modified atmospheres. *Appl. Microbiol.* 30:592–95

158. Wilson, D. M., Tabor, W. H., Trucksess, M. W. 1976. Screening method for the detection of aflatoxin, ochratoxin, zearalenone, penicillic acid, and citrinin. *J. Assoc. Off. Anal. Chem.* 59: 125–27

159. Zink, H., Brook, R. C., Peart, R. M. 1978. *Engineering Analysis of Energy Sources for Low Temperature Drying. ASAE Pap. No. 78–3517*

160. Zuber, M. S. 1977. Influence of plant genetics in toxin production in corn. In *Mycotoxins in Human and Animal Health*, ed. J. V. Rodricks, C. W. Hessletine, N. A. Mehlman, pp. 173–79. Park Forest South, Ill.: Pathotox Publ. 807 pp.

Ann. Rev. Phytopathol. 1979. 17:367–403
Copyright © 1979 by Annual Reviews Inc. All rights reserved

COEVOLUTION OF THE RUST FUNGI ON GRAMINEAE AND LILIACEAE AND THEIR HOSTS

◆3716

Y. Anikster and I. Wahl

Division of Mycology and Plant Pathology, Tel Aviv University, Tel Aviv, Israel

Israel is one of the centers of the genetic diversification in Gramineae and Liliaceae families (70, 158, 162), and the indigenous species of both families are important components of the native vegetation countrywide. Very prevalent are the grass genera *Triticum, Aegilops, Hordeum,* and *Avena,* which comprise some of the progenitors of cultivated small grain crops.

Gramineae plants throughout the world harbor rust populations representing a wide range of morphologic and biologic heterogeneity. Calculations based on Cummins' (28) data show that members of the Gramineae are hosts of over 380 *Puccinia* and *Uromyces* species, 83 of them heteroecious. About 20% of both *Puccinia* and *Uromyces* spp. on grasses have the alterate stage on Liliaceae plants. Studies in Israel on the coevolution of rust fungi on Gramineae and Liliaceae and their hosts are expected to contribute to a better understanding of various aspects of the concept of host-parasite coevolution.

Our study has been carried out for a period of 28 years. In the initial stages we dealt mainly with screening selections resistant to rust diseases from the populations of native wild barley, wheat, and oats, and concomitantly explored the physiologic specialization of the fungi. Gradually the scope of our interest and research has expanded and embraced the evolution of morphology, taxonomy, life cycles, and nuclear history of the rust organisms.

This review is confined to host-parasite coevolution in natural ecosystems undisturbed by human interference, and does not include "man-guided"

367

0066-4286/79/0901-0367$01.00

(77) evolution in agrosystems. Rusts of Liliaceae are treated mainly in connection with the ontogeny and phylogeny of rusts on Gramineae.

The taxonomic identity of plants native to Israel was determined with the aid of the analytical key by Eig et al (49).

HOST-RUST FUNGI COEVOLUTION: CONCEPTS AND EVIDENCE

The rust fungi belong to the order Uredinales in the Basidiomycetes. They comprise over 1000 species distributed among 100 genera (73). In nature, rusts are obligate parasites on nearly 200 plant families and occur worldwide except in Antarctica. They represent a very ancient group of organisms whose progenitors were well established as parasites on ferns of the Carboniferous Age (12). During 250–300 million years of phylogenetic history (57, 115), the Uredinales have undergone profound changes in morphologic and biologic traits. "The forms persisting to the present day might well be regarded as living fossils" (57). By being obligate parasites the rusts have evolved hand in hand with their hosts (9, 32, 34, 44, 45, 54, 57, 61, 65, 74, 83, 89, 96, 110, 119, 120, 122, 144, 159). This concept, first expressed by Dietel in 1904 (32), has gained general acceptance. It implies that either component in the host-parasite system has decisively influenced its counterpart (11, 32, 34, 44, 46, 69, 83, 86, 96, 120, 122, 144, 149, 159). According to Savile (122), "hosts and parasites substantially reflect each other's ages of origin."

The idea of host-parasite coevolution is supported by ample evidence from different fields. Here reference is made to three groups of evidence: (a) occurrence of primitive rusts on ancient plant forms, (b) restriction of older rust genera to a narrow host range, while younger rusts inhabit a broader range of hosts that have originated in more recent geologic ages, and (c) evolution of "correlated species." Certain rusts of similar or different life cycles occurring on related hosts display close resemblance in morphologic and biologic attributes. Accumulation of species that exhibit correlation have sufficient points of resemblance to indicate descent from a common ancestor (14).

The following examples illustrate each of the three groups of evidence. The old fern family Osmundaceae extending back into the Triassic and even the late Paleozoic periods (12) hosts primitive rusts of the genus *Uredinopsis*. All spore forms in the genus are colorless, the teliospores are scattered in the mesophyll, and telia are not organized. *Uredinopsis* species are heteroecious and develop pycnia and aecia on trees of the *Abies* spp. Also the other primitive rust genera *Milesia* and *Hyalopsora* produce uredia and telia on ferns and aecia on trees of *Abies,* which is an old genus of coniferous plants. Both important characters of the described rusts, pleomorphism and heteroecism (host alternation), are universal among the more primitive genera of Uredinales harbored on ancient hosts. Obviously, they evolved at the early stages of the phylogenetic history of the fungi.

The older rusts have a narrow host range, as shown by the following examples. The genera *Hamaspora* and *Gymnoconia* and virtually all species of *Phragmidium* and *Gymnosporangium* undergo all or a part of their life cycle on members of the Rosaceae, suggesting that they evolved concurrently with the evolution of Rosaceae from Polycarpicae (57). Aecia of the genera *Uredinopsis, Hyalopsora,* and *Milesia* are formed on *Abies,* while *Pucciniastrum* produces aecia on *Abies* and the related genus *Pica.* Likewise, aecia of the rust genus *Chrysomyxa* develop on *Picea,* whereas the aecia of *Melampsoridium* develop on *Larix,* and those of *Coleosporium* as well as *Cronartium* are produced on *Pinus.* All heteroecious species in the genus *Melampsora* produce telia on *Populus* or *Salix,* while the autoecious species have a wider host range (27).

In contrast, the two genera of more recent origin, *Puccinia* and *Uromyces* (artificially separated from each other), which include nearly one half of all known species of Uredinales (86), are of worldwide distribution on plants of numerous families of dicots and monocots, but absent on the family Abietineae. However, they too show some degree of preference for particular host groups. For example, many rusts of Leguminosae (Papilionatae) belong to the genus *Uromyces* (31, 83). The same is true of rusts on *Euphorbia* spp. (138). Conspicuously, *Uromyces* rusts are absent on sympetalous plants (83).

The expansion of the host range of rust fungi is most important to the understanding of their phylogenetic history and sometimes sheds light on the evolution of the hosts. According to Dietel (32), the transition to new hosts proceeds in the direction of the progressive evolution of the vascular plant, or to plants that originated at the same time as the source host, but not to plants that had evolved in earlier periods. The transference of the rust from the primary to secondary hosts, termed by Leppik (88) *biogenic radiation,* proceeds in heteroecious organisms by moving to the new host either the haploid or the dicaryotic generation, but never by simultaneous transfer of both generations. This sequence facilitates tracing the coevolution of rusts and their hosts. Klebahn (81) described two pathways of fungus expansion:

1. Examples of expansion from the primary aecial host to a wider range of telial hosts, are the radiation of *P. coronata* from *Rhamnus* to grasses, of *P. graminis* from *Berberis* to Gramineae, and of *P. recondita* from *Thalictrum* to grasses.

2. Expansion of the primary telial stage to a wider range of aecial hosts, is exemplified by *P. isiacae* Wint. with the dicaryotic stage on *Phragmites* spp. and aecia on plants distributed among some 22 families of Angiospermae (28), or *P. aristidae* Tracy (*P. subnitens* Diet.) with uredia and telia on several grasses and aecia on about 100 species of 24 families (14).

Correlations between rust organisms are manifested in various ways (2, 13, 14, 36, 51, 65, 74, 101, 102, 137, 138). Particularly relevant to our subject are (*a*) correlations between heteroecious *Puccinia* and *Uromyces* species, discussed by Orton (101); (*b*) correlations between heteroecious and autoecious rusts, or between *hemi*-forms and *eu*-autoecious species, as illustrated by *P. holcina* Guyot and *P. blasdalei* Diet. & Holw. (64), and other organisms (74); (*c*) correlations between heteroecious rusts and species with reduced life cycles inhabiting the alternate host of the macrocyclic ancestors (74).

Telia of the short-cycle rusts frequently simulate the habit of aecia of the parental macrocyclic species (27). If the aecial stage develops systemic mycelium, telia of the microcyclic species also arise from mycelium of the same type (27, 35, 138), and may cause malformations similar to those produced on the aecial host (35, 138).

Teliospores in the correlated species are morphologically much alike. Uredinologists agree that the microcyclic rusts descended from the corresponding macrocyclic species. The presence of peridial cells, rudimentary aeciospores, and urediospores in the telia of the microforms, and the morphologic resemblance of these inclusions with the respective structures of the macrocyclic rusts involved not only emphasize the relationship of the two rust species so correlated but also attest to the origin of the microcyclic rust from the macrocyclic one. The regressive evolution sometimes results in more than one microcyclic progeny from the same parental species. For example, *Rhamnus* plants harbor two microcyclic *Puccinia* species with teliospores characterized by epical, digitate projections exactly like those on teliospores of *P. coronata*, i.e. *P. mesnieriana* Thüm in Europe, California, Israel, and Syria, and *P. schweinfurthii* (P. Henn.) Magn. in Ethiopia and tropical East Africa.

In 1887, Dietel (30, 36) was the first to call attention to the similarity of teliospores of macrocyclic and microcyclic rusts [loc. cit. (14)] and to recognize such species as "parallel organisms" (36). He described this phenomenon in *P. coronata* and *P. mesnieriana* and other species (30, 36). Fischer (51) furnished additional examples of "parallelism" between macrocyclic and microcyclic species and suggested that similarity of their teliospores reveals natural relationships. Tranzschel (137) formulated the hypothesis, known as Tranzschel's law (27, 85) postulating that the aecia of *hemi*-rusts, suspected of being heteroecious, should be looked for on plants harboring a microcyclic rust with teliospores similar to those of the *hemi*-form involved. Likewise, a given aecium is genetically linked with the *hemi*-rust with teliospores morphologically resembling those formed by microcyclic rusts borne on the host of the aecium or on closely related plants.

Tranzchel's law has greatly facilitated identification of unknown alternate hosts of many heteroecious rusts and contributed to the understanding of the evolution of the rust organisms.

The hypothesis advanced by Tranzschel (137, 138), Jackson (74), and others that telia of microcyclic rusts have descended from aecia of the heteroecious macrocyclic precursor gained support from the studies of Johnson & Newton (80). They produced cultures of *P. graminis tritici* by inbreeding that were incapable of developing aecia on *Berberis* but formed uredia and telia on the alternate host; they assumed that these abnormalities "may have some significance in explaining the origin of short-cycle rusts from heteroecious long-cycle rusts."

In Tranzschel's opinion (139), the phenomenon described by Johnson and Newton duplicated what already had taken place in nature in ancient times.

Savile (119, 122) proposed an alternate route of evolution of microcyclic rusts. He envisioned that a macrocyclic autoecious rust may have developed on the aecial host of the parental heteroecious species. They often gave rise to lineages with considerable diversity, leading to a highly advanced microcyclic species. According to his hypothesis, the *P. recondita* heteroecious group of grass rusts gave rise to a large group of autoecious rusts on *Allium* (aecial host of some of the grass rusts), which ultimately developed microcyclic rusts on *Scilla* and allied genera.

Adverse environmental conditions exert preferential selection pressure favoring the evolution of microcyclic rusts (5, 52, 72, 93, 102, 113, 121–123). Fischer (52) pointed out that the microcyclic rusts constitute 39.7% at the higher elevations—above the forest region, but only 20.3% of the total Uredinales flora of Switzerland. They also predominate in artic and alpine regions (121, 122).

Savulescu (123) reported that in Rumania fungi with simplified life cycles are more common in mountainous regions, in arid steppes, and in areas with increased soil salinity than in other parts of the country.

The microcyclic rusts inhabit the alternate hosts, which frequently possess a higher survivability than the main host. This is especially true of Gramineae rusts alternating with woody or geophytic plants (23, 146). Most of the microcyclic rusts lack pycnia. They often produce dicaryotic mycelia as a result of infection with a single basidiospore that contains two sister nuclei (74, 75, 91). The tendency of microcyclic rusts to develop homothallism (4, 8, 20, 75, 121, 122) enhances their adaptability to unfavorable environments.

STEM RUST

Life Cycle

Stem rust *Puccinia graminis* Persoon is the most extensively studied rust because of its great economic importance and worldwide distribution. This heteroecious and complex "mammoth" species (58) forms uredia and telia on Gramineae and its alternate stages on a wide range of *Berberis* species or, less commonly, on *Mahonia.* According to Gäumann (58), over 70 *Berberis* species are hosts of the fungus, which develops the dicaryotic stage in 356 Gramineae species of 53 genera. The fungus represents a very broad spectrum of morphologic and biologic variability. The heterogenic taxon has been divided into smaller units on the basis of taxonomic traits (68, 142), while Eriksson subdivided it into varieties—formae speciales—which reflect their parasitic adaptation to certain genera (59, 72, 131). They are dealt with later in this review. The evolution of *P. graminis* with the main and alternate hosts has been discussed by Leppik (87, 88). He and other researchers (32, 81) consider *Berberis vulgaris* L. to be the primary host of the fungus. This species is of Asiatic origin and has been subsequently introduced through the Mediterranean countries to Europe, Africa, and North America. Stem rust has followed its aecial host and radiated to the secondary hosts in the Gramineae family. Worldwide distribution has become possible because of its establishment on grasses. About 90% of the congenial gramineous plants belong to the subfamily Festucoideae which is concentrated mainly in the northern hemisphere. "The gene centers of our grain crops, which are the main hosts of the present-day races, are all included in the area of Festucoideae. This area could also be the ancestral homeland of the present day stem rust" (87).

In Australia, New Zealand, South Africa, and South America, the evolution of stem rust has become disassociated from the aecial host, and its survival and development are dependent on the uredial stage. Sometimes the organisms lose spore forms other than urediospores; the loss at first is functional but may become morphological with further evolution (65). Similar processes also take place in other parts of the globe when the interconnection with the aecial host is severed. Thus, in Eastern Siberia where *Berberis sibirica* Pall. occurs only in the mountains, races of *P. graminis* have evolved that rarely produce telia. They hibernate in the uredial stage on rye (83).

Role of Berberis in Fungus Evolution

EVOLUTION OF VARIETIES (FORMAE SPECIALES) AND PHYSIOLOGIC RACES   Stakman and co-workers (132) demonstrated that at least five varieties of stem rust, namely, *tritici, secalis, avenae, agrostitidis,* and *poae* alternate with *Berberis.* Green (62) speculated that "the *formae speciales* of *P. graminis* have evolved from a rust that attacked barberry and certain gramineous hosts." They probably originated through gene recombination that increased virulence in certain of these hosts at the expense of virulence in others. Thus, a hybrid between the varieties could be expected to resemble the ancestral type more closely than the specialized forms of today. Luig & Watson (92) arrived at a similar conclusion. They reported that in Australia some hybrids between *P. graminis tritici* and *P. graminis secalis* are more compatible with wild grasses than with cereals. Assuming that cereal rusts are progenies of wild grass rusts (89), the preferred adaptation of the artificial hybrids to wild grasses "would constitute a step towards the more 'primitive' state" (92). On the other hand, some of the crosses of *P. graminis tritici* with *P. graminis secalis* produced on *Berberis* represent hitherto unrecognized varieties of *P. graminis,* narrowly specialized on cereals such as barley and noninfectious on plants of other genera (90). Similar results were obtained by crossing *P. graminis tritici* with *P. graminis agrostitidis* (22).

The role of *Berberis* in the origin of new physiologic races and in the unlocking of new genes of virulence in *P. graminis* was amply documented (130, 131).

According to A. P. Roelfs (unpublished information) over 80 physiologic races of *P. graminis tritici* and about 10 races of *P. graminis avenae* very likely originated from *Berberis* in North America. They include race 15B of wheat stem rust and race 7 of oat stem rust. The prevalence of varieties and races of stem rust in any barberry region is conditioned to a considerable extent by cereals and wild grasses grown there (132).

ABNORMALITIES IN THE LIFE CYCLE   Jackson (74) contended that deviations from the normal life cycle of a rust organism are likely to reveal evolutionary tendencies of the fungus. The following types of aberration are noteworthy.

1. Johnson & Green (79) related that infection experiments with a culture of wheat stem rust race 59, involving the infection of barberry and the subsequent inoculation of wheat seedlings with aeciospores, resulted in formation of uredia in some tests and of abortive pycnia in others. Gäumann (58) interpreted this phenomenon as a "fragmentary" break in the normal cycle of aecial host–telial host alternation.

2. As already mentioned, Johnson & Newton (80) obtained by inbreeding of physiologic races of *P. graminis tritici* by means of selfing of certain selected isolates for several successive generations, *brachy*-forms that developed pycnia, uredia, and telia on *Berberis*. This finding supports the contentions of those who like Jackson (74) hold that in the course of the more recent evolution of the Uredinales, there has been a derivation of autoecious species from heteroecious species and of microcyclic species from macrocyclic species (80).

3. Critopoulos (23) reported the existence of an autoecious macrocyclic form of *P. graminis* on *Berberis cretica* in Greece under natural conditions in the mountains. He considered the production of teliospores and urediospores on the perennial alternate host as an adjustment of the fungus to produce all spore stages on a host having its habitat at high altitudes, where conditions are more suitable for the survival and germinability of teliospores.

SIMPLIFIED FORMS ON BERBERIS   As indicated above, the development of simplified correlated species is presumably an outcome of retrogressive evolution. Johnson & Newton (80) pointed out that *P. graminis* might in theory have a microcyclic correlated species in *Berberis vulgaris*. Grove (65) suggested that *P. berberidis* Mont. on *Berberis glauca* Kunth. in Chile and *P. meyerialberti* Magn. on *P. buxifolia* Lam. in Chile and Patagonia are presumably correlated with *Puccinia graminis*. The same may be true of *P. berberidis trifoliae* Diet. & Holw. on *Mahonis swasey* (Buckl.) Fedde. Leppik (87) surmised that these species with simplified life cycles are ancient rusts that evolved on the *Berberis-Mahonia* complex in the northern hemisphere.

Evolution of Puccinia graminis in the Absence of Alternate Host

The alternate host of *P. graminis* is uncommon and of no significance in some parts of the world where stem rust is widespread. Consequently, the survival and evolution of the fungus relies mainly on the urediospores and dicaryotic mycelium while other spore forms become functionless, or even disappear (65). Under such conditions the evolution of *P. graminis* depends on somatic hybridization and mutation (155, 156), but also parasexualism should be considered (155, 156). Somatic hybridization involves exchange of nuclei between fusing hyphae in the dicaryotic stage. This process yields new varieties of *P. graminis* (92) as well as new physiologic races (15). The prerequisite for the occurrence of somatic hybridization is the availability of a congenial host for the hybridized rust organisms. In Australia (92),

Agropyron scabrum Beauv. is a source of somatic hybrids between *P. graminis tritici* and *P. graminis secalis* that are compatible with *A. scabrum* and noninfectious on commercial wheat and rye. Also, other grasses can serve as media for somatic hybridization. Watson & Luig (156) provided a diagrammatic scheme of the gradual incorporation of new genes into the Australian wheat stem rust populations by somatic hybridization and mutation. They concluded that in North America and Europe where the alternate host plays an important role many standard races occur. In Australia where barberries are practically nonexistent, fewer standard races occur, but the number of components within these races appears to be higher than any recorded elsewhere. The multiplicity of such variants or strains has presumably resulted from mutation and somatic recombinations.

Significantly, in North America also, wheat stem rust races like 15B became independent of the alternate host (78), though 15B was originally isolated from barberry (71, 131). The evolutionary tendency of dropping out the aecial stage is noteworthy. Conceivably, races incompatible with *Berberis* develop in regions free of the alternate host. Several variants of race 15B have been identified (71, 130). They may have arisen not only from sexual processes but also by other mechanisms (130).

Gramineae–Puccinia graminis Coevolution in Israel

The problem has been investigated by Gerechter-Amitai (59) in extensive studies of stem rust. Israel's native Gramineae flora is very diversified, being represented mainly by the subfamily Festucoideae. It includes the wild progenitors and relatives of wheat, barley, and oats (162). Notably, Israel is located in the extensive area of Festucoideae distribution (87). The very broad host spectrum of *P. graminis* among native grasses is presumably an outcome of a prolonged host-parasite coevolution involving the dicaryotic stage of the fungus—in the absence of the alternate host. Plants belonging to about 108 species of 45 genera out of the total number of 157 species of 68 genera tested were compatible in varying degrees with *P. graminis* in natural and artificial inoculation tests, allowing urediospore formation. In nature, stem rust was isolated from 89 species of 40 genera out of 150 investigated species that embraced 65 genera. Practically all congenial grasses belonged to Festucoideae. Obviously, Gerechter-Amitai's results show good agreement with Leppik's findings.

P. graminis on native grasses comprises four varieties: *avenae, tritici, secalis,* and *lolii.* The first-named variety predominated throughout the country and inhabited mainly wild *Avena* species. *P. graminis tritici,* ranking second in order of prevalence, inhabited primarily wild *Hordeum* species, less frequently wild *Triticum, Aegilops,* and *Bromus* spp., and rarely grasses of other genera. The commonest hosts of *P. graminis* among native grasses represented genera to which the cultivated small grain crops belong (59).

Numerous native grasses serve as common hosts to more than one variety of stem rust. Seedlings of the genera *Ammophilia, Boissiera,* and *Eremopyrum* were receptive to any one of the four stem rust varieties. In some species, like *Poa sinaica* Steud., all four varieties sporulated on the same host. Common hosts presumably facilitate somatic hybridization between varieties. Consequently, a number of the rust isolates showed affinity to more than one variety.

Nevertheless, a definite distribution pattern of the main stem rust varieties among the grass tribes could be recognized. Three tribes, Agrostideae, Phalarideae, and Aveneae, were receptive particularly to *P. graminis avenae.* Plants of the tribe Festuceae have harbored mostly the variety *avenae* but also the variety *tritici,* whereas grasses of the tribe Hordeae were principally compatible with the stem rust variety *tritici.* Gerechter-Amitai (59) pointed out that affinity between host ranges of *P. graminis* varieties is sometimes associated with resemblance in morphologic attributes of the fungus and may indicate phylogenetic kinship. Practically all wild grasses supporting sporulation of *P. graminis tritici* behave likewise with respect to *P. graminis secalis.* Similarity in the host range of the varieties *tritici* and *secalis* in the United States (133) and Europe (68) has been reported. On the basis of morphologic resemblance Guyot et al (68) classified both varieties in the same subspecies, *Puccinia graminis major.*

Practically all hosts of *P. graminis lolii* in Israel were congenial with the variety *avenae.* In Australia also (153), about half of the hosts of the variety *lolii* are receptive in nature to the variety *avenae.* Similarity in the dimensions of teliospores of both varieties has been indicated (153). Gerechter-Amitai concluded that "as indicated by our work, the parasitic specialization of obligate parasites may be another useful criterion in the study of the evolution of Gramineae."

BROWN LEAF RUST COMPLEX, *PUCCINIA RECONDITA*

Complexity of Species

The brown leaf rust fungus *Puccinia recondita sensu lato* Rob. ex Desm. [*P. rubigo-vera* (DC) Wint.] of grasses including small grains is a variable pluriverus species (29, 47, 48, 58, 72, 142). Cummins (26) listed over 60 synonymous species included in *P. recondita* which comprises a multiplicity of telial-aecial host relationships. The dicaryotic stage inhabits mainly hor-

deaceous plants and alternates with species of Ranunculaceae, Boraginaceae, Balsaminaceae, Compositeae, Crassulaceae, Hydrophyllaceae, and
Monocotyledones (48, 72). Dupias [Fig. 15 in ref. (48)] suggested a scheme
of phylogenetic relationships among the mentioned groups of brown leaf
rust. The fungus exhibits parasitic specialization on both the main and
alternate hosts.

Our discussion deals primarily with the constituents of the *P. recondita*
complex parasitic on small grains.

Brown Leaf Rust of Wheat

P. recondita var. *tritici* Erikss. develops the dicaryotic generation primarily
on *Triticum,* but occasionally also attacks *Hordeum* spp., rye, and grasses
of the genera *Aegilops, Agropyrum,* and *Bromus* (72, 100, 142). In some
countries the grass hosts serve as important reservoirs of inoculum and
prompt the origination of new pathogenic forms (142).

P. recondita tritici consists of at least three physiologic groups: (*a*) the
common group completing the life cycle on *Thalictrum* native to Europe,
Asia, and North America (47, 76, 111); (*b*) the other group alternating in
Portugal with *Anchusa, Lycopus,* and *Echinospermum* of the Boraginaceae
family (47); and (*c*) the Siberian group alternating with *Isopyrum fumarioides* L. (47, 72, 100, 142). Sibilia (126) reported that *Clematis vitalba* L.
also serves as the aecial host of the fungus harbored by *Triticum.*

The two groups of rust cultures completing the life cycle on *Triticum*—
Anchusa and *Triticum*—*Thalictrum,* respectively, are cross-incompatible,
and were not infectious on *I. fumarioides.* Plants of the latter species were
resistant also to cultures isolated from wheat in Canada. Results of a large
number of inoculations (47) revealed that only species of *Thalictrum* native
in the centers of origin of the genus *Triticum* or in areas where wheat was
cultivated since prehistoric times were congenial with *P. recondita tritici.*
Similarly only those species of Boraginaceae that have the same centers of
origin as the respective gramineous hosts were receptive to *P. recondita
tritici* from *Aegilops, Secale,* and *Triticum.* "Congeniality was found to
exist only when the gametophytic and the sporophytic hosts belonged to the
same center of origin" (43). Leaf rust has apparently followed two pathways
of evolution, one involving alternate hosts of the Ranunculaceae family
(*Thalictrum, Isopyrum, Clematis*) and the other including hosts of the
Boraginaceae family. This may explain the existence of sexual barriers
between the two groups, namely, *P. recondita tritici* linked with *Thalictrum*
and *P. recondita tritici* associated with *Anchusa.* A similar explanation can
be offered for the lack of such barriers between *P. recondita Thalictrum*—
Triticum and *P. recondita Thalictrum–Agropyrum,* or between *P. recondita
Anchusa-Triticum* and *P. recondita Anchusa-Secale* (47).

Studies in the United States (111) corroborated some of the conclusions of investigations in Portugal. Saari et al (111) reported that "relatively few infections occurred on North American thalictra, compared with the abundant infections on the European species of the genus. This might be expected, however, as *P. recondita* f. sp. *tritici* was introduced to the New World only within the last century or two." The most susceptible alternate hosts of the tested *Thalictrum* species native in the United States are *T. alpinum* L. and *T. sparsiflorum* Turcz., "since plants of these species are also found in the Old World" (111). Consequently, unlike in Italy (142) and Portugal (42, 44) where *Thalictrum* functions as an alternate host in nature, its significance in the United States in such a capacity seems to be slight (111, 157); however, the genetic recombinates are noteworthy from the phylogenetic viewpoint. One of the aeciospore cultures derived by mass mating pycniospores of *P. recondita tritici* was avirulent on wheat and compatible only with *Bromus commutatus* Schrad., while another culture was avirulent to "universally susceptible wheats, revealing thus the existence of resistance genes unknown heretofore" (111).

On the whole, sexual processes on *Thalictrum* seem to be less important in the evolution of new fungus forms than somatic hybridization or mutation (72).

The importance of *I. fumarioides* as an alternate host of *P. recondita tritici* is confined to eastern Siberia, where it plays a significant part in the perpetuation and propagation of the fungus (100).

Brown Rust of Rye

Puccinia recondita Rob. ex Desm.—*P. dispersa* Erikss. & Henn. (*P. rubigovera* Erikss. & Henn.)—forms the dicaryotic stage on rye worldwide and to a limited extent on *Bromus, Hordeum,* and *Elymus.* D'Oliveira (45) distinguished in this variety two biologic groups: the common group *secalis* with aecia on species of *Anchusa* and *Lycopsis* and another group designated *P. rubigo-vera clematis secalis* with aecia on *Clematis vitalba.*

Some species included in *P. recondita sensu lato* with pycnia and aecia on Ranunculaceae or Boraginaceae were described by Gäumann (58) and Dupias (48). They presumably constitute biologic groups that have followed different pathways of evolution under diverse ecologic conditions (47). Several rust species of the *Thalictrum* group, like *P. triticina* and *P. alternans* Arth. are cross-compatible, thus revealing their interrelationship (72). Probably they derive from the primary *Thalictrum* host (81) and are products of biogenic radiation.

The group with aecia on Boraginaceae comprises numerous aecia-telia lineages with dicaryotic state on *Aegilops* (21, 47, 58), *Agropyrum, Bromus, Elymus,* and *Secale* (47).

Correlated Species

Several correlated species have evolved from *P. recondita* in the process of evolution. They consist of microcylic *Puccinia* species (27, 30) and macrocylic *Uromyces* species, such as *U. dactylidis* Otth. (27) and *U. alopecuri* Seym. (48), both alternating with Ranunculaceae. The correlated *Puccinia* and *Uromyces* taxa are similar in morphology of teliospores and biology of the fungi (48). Savile (122) postulated that the *P. recondita* group of grass rusts gave rise to autoecious rust on *Allium,* "with progressive loss of paraphyses, shortening of life cycles, and evolution of deciduous teliospore pedicels."

CROWN RUST COMPLEX

The Fungus: Its Life History and Phylogeny

The fungus *Puccinia coronata* Corda' is a complex heteroecious species of worldwide distribution. The dicaryotic stage inhabits 290 species of Gramineae that belong to 72 genera representing 17 tribes in four subfamilies (127). As in the case of stem rust, most hosts fall in the subfamily Festucoideae. *P. coronata sensu lato* alternates with 34 species of *Rhamnus* (127) and a few other genera of Rhamnaccac and Eleagnaceae (14, 72, 127). Cummins (28) combined 21 synonyms in *P. coronata.* Our discussion is limited to *P. coronata* with the haploid stage on *Rhamnus.* Dietel (32), Klebahn (81), and Leppik (88) surmised that *Rhamnus* plants were the primary host of the fungus that radiated to secondary grass hosts [Figures 1 and 2 in ref. (88)].

P. coronata* alternating with buckthorn comprises a number of varieties, viz. *agropyri, agrostis, alopecuri, arrhenatheri, avenae, calamagrostidis, festucae, glyceriae, holci, lolii, phalaridis, secalis,* and a few others (72, 100, 103, 127). Distinction between varieties is rather arbitrary and indicates mainly their preferential parasitism on plants of genera of practical importance (72). Notably, the variety *secalis* is infectious on rye and barley but not on oats (72, 103).

ROLE OF ALTERNATE HOST   Extensive studies have demonstrated the great importance of the sexual stage of the fungus on *Rhamnus* in the evolution of physiologic races in *P. coronata avenae.* Dietz (38) showed that *Rhamnus* plants have influenced the populations of physiologic "forms" (races) of *P. coronata.* Murphy (98) postulated that continuous races population shifting in oat crown rust is largely due to hybridization and segregation of the fungus on the alternate host. Fleischmann's (53) studies brought out that in Canada the race composition of isolates procured from aecia on *R. cathartica* L. was correlated with the races on cultivated oats. Santiago

(112) attributed the evolution of variation in pathogenicity of *P. coronata avenae* in Portugal to the alternate buckthorn hosts which occur throughout the country. They appear to be a very important source of inoculum in cultivated oats.

CYTOLOGY McGinnis (95) found that germinating teliospores of *P. coronata* on *Agropyron repens* L. Beauv. have $n = 3$ chromosomes, while *P. graminis* has $n = 6$ chromosomes. He postulated that *P. coronata* is possibly a constituent species of *P. graminis,* "in which case the two species would have at least one genome in common."

Studies in Israel

Israel is located in the center of origin and diversification of the hexaploid wild species *Avena sterilis* L. (84, 162), the putative progenitor of cultivated oats. This species and *A. barbata* Pott. (84, 162) are ubiquitous and represent the predominant component of the annual herbaceous cover in winter and spring. The stands are highly variable and polymorphic. The high percentage of open pollination in *A. sterilis* presumably contributed to the genetic variability of this species (161, 162). The wild *Avena* populations are annually attacked by *P. coronata avenae* that alternates in nature with *Rhamnus alaternus* L. (106) and *R. palaestina* Boiss. (104). The latter is common and endemic in Israel and constitutes an important element of east Mediterranean vegetation.

The prevalence of the main and alternate hosts and their long-lasting association have stimulated extensive research on the coevolution of the *Avena–P. coronata–Rhamnus* system. The study was initiated by the junior author in 1950 and is being continued (16, 39, 147, 149, 150–152). It involves primarily the evolution of various types of host protection and virulence of the fungus.

VARIETAL HYBRIDIZATION Eshed (50) demonstrated that some plants serve as common hosts to 2–7 varieties of *P. coronata,* viz *agrostis, alopecuri, arrhenateri, avenae, festucae, holci, lolii,* and *phalaridis.* Such hosts are, conceivably, suitable sites for somatic hybridization between varieties. They also facilitate the survival and dissemination of hybrids produced on *R. palaestina* which are sometimes incompatible with one or both of the hosts of their parents. The common hosts belong mainly to the species *Pholiurus incurvatus* (L.) Schinz & Thell. and *Vulpia membranacea* (L.) Lk. Selfing of intervarietal hybrids unlocked pathogenic potentialities of *P. coronata* and produced cultures of broader pathogenic spectra than the parental cultures. The performance of varieties and their hybrids seems to be "a reflection of the evolution of host-parasite relationships started way back in the past and still going on at present in a natural ecosystem" (50).

PHYSIOLOGICAL SPECIALIZATION OF *PUCCINIA CORONATA AVENAE*
About 100 physiologic races of the fungus were identified in samples collected countrywide (151). Some races that had never been described before were discovered first in Israel (147, 148, 152). They include the dangerous races 264, 270, and races virulent on Santa Fe but avirulent on Landhafer (148). Studies by Wahl et al (152) have revealed distinct similarity in the composition of race populations in oat species and *Rhamnus palaestina*. For example, the "Landhafer races," group 263–264–276–277, the race group 202–203, race 286, and race 270 have appeared in the same order of prevalence on the main and alternate hosts (151, 152). At the same time the "Victoria races," group 216–217, that are rare in oats, were conspicuously absent from the aecial material. Several races like 237 were more common on buckthorn than on *Avena* plants, presumably because of lower competitiveness with other races on oats.

The alternate host appears to have a pronounced effect on diversification of oat crown rust races. Wahl et al (152) obtained one race in three collections from buckthorn and only one in eight from oats, and more than one race from a single aecial cup. Oat crown rust races selfed by Dinoor (39) were heterozygous making for more variability of the fungus (18). Eshed (50) concluded that heterozygosity of pathogenicity is widespread in the varieties of *P. coronata*. It was suggested (19) that in the dicaryotic rust fungi heterozygosity furnishes survival advantage to the organism.

The prevalence of heterozygosity in cultures of *P. coronata* seems to reflect the hybrid nature of their origin.

Significantly, despite the continuous production of new races, the composition of race populations has remained rather stable over more than two decades of race surveys. For example, the versatile race 276 has predominated annually in all parts of the country (16, 147–149, 151). This stability is attributable to the permanence in the composition of wild oats and other compatible native grasses, undisturbed by human interference.

COMMONNESS OF *PUCCINIA MESNIERIANA*   The concept of the evolution of *P. coronata avenae* in the Mediterranean region as reflected in the similarity of race composition in the main and alternate hosts gets additional verification from the continuous prevalence of *P. mesnieriana* in Israel (104, 152). As mentioned, this species is correlated with *P. coronata* and has descended in the process of retrogressive evolution. *P. mesnieriana* often causes hypertrophic deformation of twigs and young fruits, like those associated with aecia of *P. coronata*. Furthermore, this resemblance in host reactions is evidence of phylogenetic relationship between the heteroecious and microcyclic rust species inhabiting buckthorn (35).

Germinating teliospores of *P. mesnieriana* emit three-celled basidia, with each of the upper two cells producing a single basidiospore (Y Anikster, unpublished information). The basidial development of *P. mesnieriana* resembles that of most *Uromyces* species on barley in Israel and their correlated microforms on Liliaceae. Studies on the nuclear history of *P. mesnieriana* are in progress. The complete lack of pycnia suggests that *U. mesnieriana* is homothallic, since, according to Buller (20), throughout the Uredinales absence or imperfect development of pycnia is correlated with homothallism. The presumed homothallism of *P. mesnieriana* fits into the lineage that passes from the heterothallic *P. coronata avenae* with normal pycnia to the homothallic *P. coronata eleagni* with vestigial, functionless pycnia on *Eleagnus commutata* Bernh. The intermediate stages between the two extremities are *P. coronata calamagrostidis* with small, functional pycnia and *P. coronata bromi* with few, small, and possibly fertile pycnia (20).

PUCCINIA BROWN LEAF RUST OF BARLEY

Taxonomy and Life History

Puccinia brown leaf rust of barley is of worldwide occurrence and has become increasingly important in recent years (61, 72). The causal agent, *Puccinia hordei* Otth *sensu stricto,* is a heteroecious fungus with the dicaryotic stage limited in nature to *Hordeum* spp. and alternates primarily with *Ornithogalum* species. The organism produces a high rate of mesospores in the telium. Naumov (100) reported the existence of the dicaryon on several genera of Gramineae, and D'Oliveira (46) also obtained aecia by aritficial inoculation on *Dipcadi serotinum* (L.) Medic. In Kenya and England, aecia develop on the latter species in nature (72). A considerable pathogenic variability of the fungus in *Hordeum* species was reported (58, 72, 142).

The existence of the main host in the following *Hordeum* species has been experimentally ascertained (58): *H. distichum* L., *H. hexastichum* L., *H. maritimum* With., *H. spontaneum* C. Koch, and *H. tetrastichum* Kcke. D'Oliveira (40a) considered leaf rust on *H. bulbosum* L., *H. jubatum* L., *H. marinum* Huds., and *H. murinum* L. in Portugal as biologic forms of *P. hordei.* However, their linkage with *Ornithogalum* has not been confirmed experimentally. Tranzschel (139) indicated the importance of *H. bulbosum* in the hibernation of urediospores in the USSR. The brown leaf rust organism on *H. murinum* has been considered as an autonomous species *P. hordei murini* Buchw. (3), distinct mainly by a lower percentage of mesospores in a telium (less than 40%) than in *P. hordei.*

The role of *Ornithogalum* in the life cycle of *P. hordei* varies with the geographic region. It seems to be of no importance in Central Europe because of the lack of synchronization between teliospore germination and the appearance of *Ornithogalum* plants (58). The aecial stage is uncommon in North America despite the general distribution of *O. umbellulatum* L. In contrast, Critopoulos (24) found that in Greece (Attica) the life cycle of the fungus is completed in nature by basidiospore infection of the neighboring *O. umbellulatum*. This process is essential for the perpetuation of the fungus. Tranzschel (139, 140) reported that in the Crimea *P. hordei* develops aecia on *O. pyrenaicum* L. (= *O. narbonense* auct.) and sometimes reaches epidemic proportions. The fungus produces telia abundantly where *Ornithogalum* plants are present, while telial formation is scarce in Central USSR, which is devoid of the alternate host (139). D'Oliveira has conducted fundamental studies on the host-parasite coevolution of the double host-parasite system, *Hordeum–P. hordei–Ornithogalum*. He demonstrated (46) that 32 species of *Ornithogalum* were compatible with *P. hordei*. The center of origin and diversification of these species coincides at least partly with that of *H. spontaneum*—the putative progenitor of cultivated barley—or overlaps the regions where barley has been cultivated since remote antiquity. In contrast, the incompatible species of *Ornithogalum* originated in the regions where no native species of *Hordeum* are known to occur (46).

Studies in Israel

TAXONOMY AND LIFE HISTORY   Israel is located in the center of origin and genetic diversification of *H. spontaneum* (70, 162). Populations of this species are distributed throughout the country and represent a broad spectrum of morphologic and physiologic variation (70). In addition, the following species belong to the native wild *Hordeum* flora, *H. bulbosum* (tetraploid type, $2n = 28$), *H. marinum*, and *H. murinum–H. leporinum*. The *Hordeum* center of diversification overlaps the center of genetic diversification of the genus *Ornithogalum* comprising the species *O. narbonense* L., *O. brachystachys* C. Koch, *O. divergens* Bor., *O. eigii* Feinbr., *O. lanceolatum* Lab., *O. montanum* Cyr., and *O. trichophylum* Boiss. & Heldr. The latter species is limited to the arid Negev region. The *Ornithogalum* flora coexists in many areas with *Hordeum* plants, and particularly with *H. spontaneum* and *H. bulbosum*.

The aim of our study was to explore whether or not *P. hordei* completes its life cycle on *Ornithogalum* species and to estimate the impact of coevolution of the *Hordeum–P. hordei–Ornithogalum* system on the taxonomy and biology of the rust organism. The morphologic characteristics of teliospores and urediospores of *Puccinia* brown leaf rust isolates from *H. spontaneum*, *H. bulbosum*, and *H. murinum* are much alike, and the percentage of mesospores per telium has a similar range of variation (3). Furthermore, Y. Anikster (unpublished information) has experimentally proved compatibility of cultures isolated from *H. murinum* with at least five *Ornithogalum* species. These findings suggest that *P. hordei murini* is not an autonomous species and should be incorporated in the species *P. hordei* Otth. The same approach was adopted by Cummins (28). It is noteworthy that teliospores of *P. hordei,* regardless of their source host, often show the type of striation and angular shape [Fig. 2 in ref. (3)] as found by Guyot (67) in *P. hordei* (particularly in the Mediterranean region) and other *Puccinia-Uromyces* species alternating with Liliaceae.

All tested cultures of *P. hordei* have proven to be compatible with *Ornithogalum* plants in artificial inoculation tests. They have exhibited distinct parasitic specialization respective to the main host species and to a lesser extent also to the alternate host species. Only reciprocal inoculations with cultures from *H. spontaneum* and *H. vulgare* L. were successful, while infectivity of cultures derived, respectively, from *H. bulbosum* and *H. murinum* was confined to the source host species. Anikster's (2) method of teliospore germination enabled inoculation of *Ornithogalum* species and other Liliaceae plants with cultures isolated from the mentioned *Hordeum* species. The results (Y. Anikster, J. G. Moseman, and I. Wahl, unpublished information) demonstrated parasitic specialization of *P. hordei* on *Ornithogalum* species. Congeniality of *Dipcadi erythreaum* Webb. & Bert. and *Leopoldia eburnea* Eig & Feinbr. with *P. hordei* revealed in the tests had not been known before.

Ornithogalum species are alternate hosts of *P. hordei* in nature, and aeciospores isolated from various species of the genus have displayed parasitic specialization with respect to *Hordeum* species. For example, all 3472 aecial clusters from *O. montanum* and 2875 clusters from *O. lanceolatum* were infectious only on *H. bulbosum,* while about 66% of 2141 aecial clusters secured from *O. narbonense* were compatible with plants of the *H. spontaneum–H. vulgare* group. *O. eigii* and *O. trichophyllum* have served as alternate hosts of *P. hordei* cultures from *H. spontaneum* group, *H. bulbosum,* and *H. murinum.*

PERPETUATION OF THE FUNGUS   Barley plants dry up in nature at the onset of the rainless season at the end of May and beginning of June. Dormant teliospores on barley stubble preserve the viability of the fungus. They start to germinate with the beginning of the ensuing rainy season in November, releasing basidiospores which infect the foliage of *Ornithogalum* plants that emerge at the same time. Aeciospores thus formed complete the life cycle of the fungus by infecting seedlings of wild *Hordeum* species and *H. vulgare. H. spontaneum, H. bulbosum,* and *H. murinum* are ubiquitous and play a decisive role in the increase and country-wide dissemination of inoculum. Rusted plants of *H. spontaneum* influence epidemics of *P. hordei* on cultivated barley. Conceivably, the coordinated sequence of development of the various stages of the life cycle, their compatibility with a wide range of main or alternate hosts, and their adaptibility to specific environments are an outcome of long-lasting host-parasite coevolution.

EVOLUTION OF VIRULENCE   D'Oliveira (41) reported evolution of physiologic races of *P. hordei* on the *Ornithogalum* alternate hosts. The importance of *Ornithogalum* in Israel in the life history and perpetuation of *P. hordei* prompted studies on the role of *Ornithogalum* in the physiologic specialization of the fungus.

Studies by Golan et al (61) and others (Y. Anikster, J. G. Moseman, and I. Wahl, unpublished information) involved cultures isolated from *H. spontaneum* and *H. vulgare* as well as cultures from *O. brachystachys, O. eigii,* and *O. narbonense.* They led to the following conclusions:

1. All pathogenic forms found on barley were present among the aeciospore cultures.
2. New pathogenic entities never reported before were identified in the tested fungus material.
3. Some isolates have rendered ineffective all known genes of resistance to *P. hordei* (7), including gene *Pa7,* incorporated in the cultivar Cebada Capa—CI 6193 (61).
4. Isolates virulent on barley accessions with gene *Pa7* were found first on the alternate host, and thereafter on *H. spontaneum.*

HOST-PARASITE COEVOLUTION OF *UROMYCES* RUSTS ON BARLEY AND LILIACEAE

Taxonomy and Life History

Knowledge of the taxonomy and biology of barley *Uromyces* rusts is very scanty. Some information on the subject was summarized briefly (2, 9, 10, 14, 145). Viennot-Bourgin (146) analyzed the morphologic and phylogenic problems of barley *Uromyces* in the Near East.

All *Uromyces* species on *Hordeum* belong to the section *Angulati* (66) by having angular teliospores with uniformly thick and ridged walls devoid of visible germ pores (10). Germ tubes emerge by rupturing the cell wall. Prior to our studies, only the alternate host of *U. hordeinus* Arth. was recognized and identified as *Nothoscordium bivalve* L. Britt. (14, 145). Researchers (66, 67, 145) speculated that *Uromyces* species of barley alternate with Liliaceae species. They based the hypothesis on Tranzschel's law, taking into account the morphologic resemblance between teliospores of *Uromyces* on *Hordeum* and in microcyclic species on Liliaceae, such as *U. scillarum sensu lato* (Grev.) Wint. The latter species was subdivided by

Schneider (116, 124) into a number of subspecies, elevated by Gäumann (58) to the rank of "small species."

Since some *Hordeum* species and Liliaceae plants bearing *U. scillarum* are endemic in Israel (104, 105) and grow in close association, we based our study on the working hypothesis that host-parasite coevolution has yielded heteroecious *Uromyces* species on barley, alternating with Liliaceae plants and correlated with microcyclic Uromyces species that inhabit alternate hosts of the respective macrocyclic organisms. Four heteroecious species and several varieties (formae speciales) were identified on wild barley (2, 5, 8, 9).

1. *U. viennot-bourginii* Wahl & Anikst. (*a*) var. *bellevaliae eigii* Anikst. (stages 0,I on *Bellevalia eigii* Feinbr.; stages II,III on *H. spontaneum*), correlated with *hemi*-rust *U. oliveirae* Anikst. & Wahl on *B. eigii.* (*b*) var. *bellevaliae flexuosae* Anikst. (0,I on *B. flexuosae* Boiss.; II,III on *H. spontaneum*), correlated with *U. scillarum* (2, 8). Pycnia are rare in both varieties.

2. *U. reichertii* Anikst. & Wahl (0,I on *Scilla hyacinthoides* L.; II,III on *H. bulbosum*), correlated with microcyclic *U. rayssii* Anikst. & Wahl on *S. hyacinthoides.*

3. *U. christensenii* Anikst. & Wahl (I on *Muscari parviflorum* Desf.; II,III on *H. bulbosum*), correlated with *U. scillarum* on *M. parviflorum.*

4. *U. hordeastri* Guyot (*a*) var. *bulbosi-bellevaliae flexuosae* Anikst. (0,I on *B. flexuosa;* II,III on *H. bulbosum*), correlated with *U. scillarum* on *B. flexuosa,* pycnia rare. (*b*) var. *bulbosi-scillae autumnalidis* Anikst. (0,I on *Scilla autumnalis* L.; II,III on *H. bulbosum*), correlated with *U. scillarum* on *S. autumnalis* L. (*c*) third variety with stages II and III on *H. marinum* Huds. and the alternate host unknown.

Descriptions of the listed rust species and varieties were given elsewhere (2, 8–10). Conceivably, the aecial stage is responsible for their origin and differentiation. Viennot-Bourgin (146) labeled them "small units" that are distinct in biologic traits rather than in morphologic characters. Their narrow parasitic specialization is more pronounced on the alternate host than on the main host.

Adaptation to Adverse Environment

MAIN HOST–ALTERNATE HOST RELATIONS Completion of life cycles of heteroecious *Uromyces* fungi on barley in arid and semi-arid regions in Israel is facilitated by coordinated seasonal and spatial development of the main and alternate hosts. Both hosts grow in close proximity [Figure 1 in (9)] which facilitates reciprocal infection. Savile (121, 122) reported similar host disposition in alpine and arctic regions. In Israel dormant teliospores preserve viability of the fungi in the absence of congenial hosts during the hot and rainless summer season. They germinate readily at the onset of the rainy winter season and cause foliar infection of the geophytic alternate host that emerges at the same time. Teliospore germination is spread over a period of time enhancing thus the survivability of the rust organism. The described factors enabling alteration of generations in the fungus are presumably an outcome of selection pressure inherent in the host-parasite coevolution in adverse environments.

PREVALENCE OF SHORT-CYCLED RUSTS   As emphasized above, evolution of short-cycled rust forms is generally considered to be an adaptation to various types of adverse environment. In Israel short-cycled *Uromyces* rusts on Liliaceae are of country-wide distribution (Figure 1) and inhabit *Bellevalia desertorum* Eig. & Feinbr., *B. eigii, B. flexuosa, Leopoldia maritima* (Desf.) Parl., *Muscari commutatum* Guss., *M. parviflorum, M. racemosum* (L.) Mill., *Ornithogalum eigii, O. trichophyllum, Scilla autumnalis, S. hyacinthoides, Urginea maritima* (L.) Bak., and *Pancratium parviflorum* Dec. of the Amaryllidaceae family.

Some of the listed species parasitize the alternate hosts of the heteroecious *Uromyces* species on barley and are correlated with the listed microforms. The affinity of the correlated heteroecious species and microforms is reflected in the morphology of the teliospores, their germinability, and nuclear history of the basidia and basidiospores (2, 5, 6). Teliospore germination is difficult in *U. reichertii* and its short-cycle counterpart *U. rayssii* but relatively easy in other heteroecious *Uromyces* species of barley and their microcyclic descendants. Four mononucleate basidiospores per basidium are formed on *U. reichertii* and *U. rayssii,* while two binucleate basidiospores per basidium develop, for example, in *U. christensenii* and the correlated microcyclic *U. scillarum.*

HOMOTHALLISM   Evolution of homothallic derivatives from heterothallic ancestors appears to be an effective adaptation of rust fungi to unfavorable environments. Savile (121, 122) reported that most rusts in alpine and arctic regions are homothallic and produce dicaryotic mycelium in the host as a result of a single basidiospore infection. This process telescopes the life cycle facilitating its completion under adverse conditions. New genetic forms originate by exchange of nuclei of genetically different mycelia through hyphal fusion (121, 122). The microcyclic *Puccinia cruciferarum* Rud., devoid of pycnia, embodies such a gene flow (118).

In Israel the majority of microcyclic *Uromyces* species in Liliaceae are homothallic (Figure 1). The organisms produce two binucleate basidiospores per basidium and each basidiospore gives rise to dicaryotic mycelium which in turn yields teliospores (2, 4). The basidiospore contains sister nuclei and the fungus is homothallic. The existence of short-cycled rusts falling in the outlined ontogenic pattern is not uncommon, and exemplified by *Puccinia arenariae* (Schum.) Wint. or *U. scillarum* (74, 75, 91). According to Viennot-Bourgin (146), rusts with reduced life cycles are perpetuated

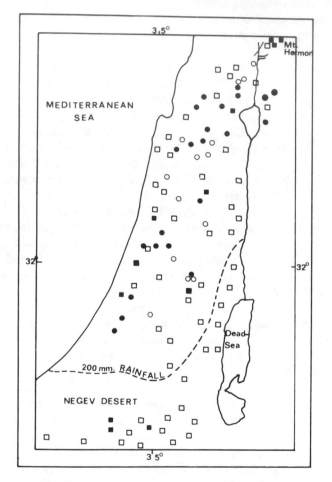

Figure 1   Geographic distribution of *Uromyces* species on wild barley and Liliaceae in Israel. Note the prevalence of homothallic rusts with two-binucleate basidiospores in the Negev desert with annual rainfall less than 200 mm, and at high elevations in the north (near Mount Hermon). ● = heteroecious, heterothallic *Uromyces* species with four-uninucleate basidiospores; ○ = microcyclic, heterothallic *Uromyces* species with four-uninucleate basidiospores; ■ = heteroecious, homothallic *Uromyces* species with two-dicaryotic basidiospores; and □ = microcyclic, homothallic *Uromyces* species with two-dicaryotic basidiospores. Telia of heteroecious organisms formed on barley; of microcyclic species developed on Liliaceae.

in Liliaceae by teliospores and dicaryotic mycelium in the parental tissue of the host. On the other hand, homothallic heteroecious rusts with binucleate basidiospores were rarely encountered abroad (8). In Israel, in contrast, heteroecious, homothallic *Uromyces* species on barley that produce binucleate basidiospores are widespread and prevail in regions with an annual rainfall of less than 200 mm or elevations of 1300 m and more above sea level (4). Their geographic distribution is shown in Figure 1.

Phylogeny

The process of host-parasite coevolution can be traced by examining the series of congruences manifesting the regressive development of some important characters of the fungus.

The pycnium regression lineage extends from normal pycnia in *U. reichertii* through their occasional appearance in *U. viennot-bourginii* to complete disappearance in *U. christensenii*. This pathway is related to the pattern of basidia and basidiospore development in the mentioned rust species (2, 5, 6, 8). *U. reichertii* has four-celled mononucleate basidia, each cell bearing a single uninucleate basidiospore. The organism is heterothallic. *U. christensenii* develops basidia with two binucleate cells, each cell producing one binucleate basidiospore. Monobasidiospore infection of the alternate host results in the formation of aecia without pycnia. Obviously the organism is homothallic. *U. viennot-bourginii* is a transitional stage between *U. reichertii* and *U. christensenii* (8). Most isolates of *U. viennot-bourginii* produce basidia and basidiospores as in *U. christensenii* and are homothallic. Several isolates gave rise to basidia comprising one binucleate cell bearing one binucleate basidiospore, and two mononucleate cells—each cell bearing a single mononucleate basidiospore. Isolates of the latter type produce pycnia on the alternate host. The uninucleate basidiospores are presumably responsible for pycnial formation.

Another lineage encompasses intergrading phases between heteroecious *Uromyces* species with a complete life cycle and the microcyclic species consisting of telia only. For example, *U. scillarum* on *Ornithogalum eigii, Leopoldia maritima,* and other species (2) contain vestigial urediospores in telial pustules, and *U. oliveirae* forms both uredial and telial pustules as separate structures (10). The urediospores are germinable but noninfectious on *O. eigii* or the main hosts of the corresponding heteroecious rust *U. viennot-bourginii* (2).

Savile (119, 122) postulated that a group of heteroecious rusts of grasses, including *P. recondita,* with angular ridged teliospores and lacking germ pores gave rise to autoecious rusts on *Allium* and eventually to microcyclic rusts on *Bellevalia, Muscari,* and *Scilla.* Our data suggest that the hypothetical group of grass rusts may have contained barley *Uromyces* rusts alternating with species of *Bellevalia, Muscari,* and *Scilla* and other Liliaceae genera. They probably gave rise to microcyclic species on Liliaceae hosts without the intermediary of *Allium.*

Guyot (67) speculated that barley *Uromyces* rusts are relicts of an ancient flora associated with climates of the Mediterranean type. In Israel, however, these rusts are not rudiments on the verge of extinction "but dynamic and plastic organisms" (5), correlated with numerous short-cycle organisms. The deciduous, moderately dispersive teliospores with sculptured walls in the heteroecious species and in the microcyclic descendants are phylogenetically advanced structures (83, 122). Despite their narrow host specialization, particularly in the Liliaceae hosts, the species involved have a common host in *Leopoldia eburnea* (Y. Anikster, unpublished information). This species, confined to dry southern regions, never rusts in nature, but in greenhouse inoculation tests it has shown congeniality with *U. viennot-bourginii* var. *bellevaliae eigii, U. reichertii, U. christensenii,* and *U. hordeastri* var. *bulbosi-scillae autumnalidis.* Aeciospores produced by any one of these species infected the *Hordeum* source hosts. Also, cultures of *P. hordei* isolated, respectively, from *H. spontaneum, H. bulbosum,* and *H. murinum* formed pycnia and aecia on *L. eburnea.* Aeciospores of each isolate were compatible with the corresponding *Hordeum* source hosts. Remarkably, also teliospores of *U. oliveirae* and *U. rayssii* and cultures of *U. scillarum* derived, respectively, from *Bellevalia desertorum, Muscari parviflorum, Leopoldia maritima, Ornithogalum trichophyllum, Urginea maritima,* and *Pancratium parviflorum* were compatible with *L. eburnea.* Teliospores formed on the latter species infected the source hosts and developed telia. The data may indicate phylogenetic relationships among the listed *Uromyces* and *Puccinia* rusts and open up new possibilities for the study of their genetics.

Gäumann (58) referred to *U. hordeastri* and *P. hordei* as parallel forms. Arthur & Cummins (14) indicated that *U. hordeinus* and *P. hordei* (*P. anomala*) are structurally correlated but physiologically distinct. We suggest that all recognized barley *Uromyces* species and also their short-cycle derivatives are correlated indirectly with *P. hordei.* All the organisms fit Orton's (101, 102) criteria of correlation in that they show morphologic resemblance in all existing spore states, parasitize the same or closely related plant species, and have similar geographic distribution. Notably, teliospores of *P. hordei* are angular and have ridged walls without visible germ pores [Figure 2 in (3)].

Ordinarily, *Puccinia* components in the correlated system are of wider geographic distribution than the *Uromyces* counterparts, presumably due to greater adaptibility (33, 102). This is not true in the case of *Uromyces* spp.–*P. hordei* system in Israel or Iran (145, 146).

Numerous short-cycle derivatives attest to the more ancient origin of heteroecious *Uromyces* species on barley. *P. hordei* has not yet reached the stage of simplification, and is presumably younger than the *Uromyces* counterpart.

TAXONOMIC ASPECTS OF HOST-PARASITE COEVOLUTION

Host Taxonomy

Fungi have aided in various ways in giving a better insight into the taxonomy and phylogeny of flowering plants. Savile (114) listed two categories of assistance rendered by mycology to the elucidation of this problem: (*a*) the contribution of small groups of fungi to exploration of the relationships and phylogenetic sequence within a family or even within a single genus and (*b*) the role of fungi in arranging major groups of plants. Gerechter-Amitai's (59) studies on *P. graminis* in Israel provided interesting illustrations of how parasitic specialization of the stem rust organism serves as a useful criterion in studies of evolution in Gramineae. One of the pertinent examples is the high degree of compatibility of *P. graminis avenae* with grasses of the tribes Agrostideae and Phalarideae, thus justifying the revised classification of Gramineae by Stebbins & Crampton (135) that included these two tribes in the tribe Aveneae.

Savile (114) found that more advanced rust types occur more frequently on plants of Liliflorae than of Glumniflorae and postulated that the first-named order is younger than the latter. The results of our studies on *Uromyces* species on *H. bulbosum* corroborate this hypothesis. These rusts comprise species with alternate hosts belonging to the genera *Bellevalia, Muscari,* and *Scilla* of the Liliaceae family. Conceivably, the fungi radiated from *H. bulbosum* to their alternate hosts. Dietel's (32) hypothesis concerning the host-rust coevolution implies that *Bellevalia, Muscari,* and *Scilla* are of more recent origin than *Hordeum* or that they originated at the same time. On the other hand, the results of our studies do not corroborate Savile's (117) arguments against Hutchinson's transferring the genus *Allium* to the Amaryllidaceae family. According to Savile, the closer proximity of *Allium* to Liliaceae than to Amaryllidaceae is manifested in the similarity between teliospores of rusts attacking plants of the genera *Allium, Bellevalia, Muscari,* and *Scilla.* The teliospores lack germ pores and are adorned with anastomosing ridges. "No rust is known on any amaryllidaceous host which shows these characters," stated Savile (117). However, Y. Anikster (unpublished information) has found in Israel on several occasions a *Uromyces* microcyclic rust on leaves of *Pancratium parviflorum* of the Amaryllidaceae family, producing teliospores with anastomosing ridges on the surface and lacking germ pores. Isolates of this rust are compatible in artificial inoculation tests with plants of *Leopoldia eburnea* of the Liliaceae family. Cultures reisolated from the latter host have proved to be congenial with the source host. The data neither support nor disprove the idea of associating *Allium* with Amaryllidaceae. They only demonstrate that, in the center of the Liliflorae and rusts coevolution, the broad spectrum of fungus variability embraces components with morphologic characters unknown elsewhere, and with infectivity transgressing the barrier separating Liliaceae and Amaryllidaceae. They may thus indicate affinity of the two families (25), though "jumps" of rusts to remote taxa are known (34, 73, 120, 122).

Taxonomy of Rust Fungi

The importance of integration of evolutionary aspects of host-parasite coexistence into taxonomic research of the implicated fungi was stressed by a number of investigators (2, 3, 13, 14, 27, 57, 58, 83).

Some of the results obtained in our own studies are discussed here to illustrate the problem.

On the basis of Tranzschel's law we assumed that native Liliaceae plants harboring microcyclic *Uromyces* rusts serve as alternate hosts of *Uromyces* species inhabiting the indigenous wild barleys with teliospores morphologically resembling those of the microforms. Experiments justified this surmise, and contributed to a better understanding of the taxonomy, biology, and phylogeny of the rusts involved.

Studies of *Puccinia* brown leaf rust organisms parasitic on the native *Hordeum* species resulted in revision of their taxonomy and relegation of the species *P. hordei murini* Buchw. to a component of the species *P. hordei* Otth. (3). By proving that native *Ornithogalum* plants are alternate hosts of *Puccinia* brown leaf rust cultures isolated from *H. murinum* (Y. Anikster, unpublished information), the "experimentum crucis" (58) was furnished to identify these cultures as *P. hordei.*

Studies on the nature of compound aecia-telia clusters on *Scilla hyacinthoides* and *Muscari parviflorum* (Figure 2) have ascertained that the aecial and telial constitutes of the cluster, despite their immediate proximity and even coalescence, belong to two separate though phylogenetically related species. Only inoculation tests can prove that aecia and telia of a putatively autoecious rust organism belong to the same life cycle. Thus it was experimentally demonstrated that the reportedly (58) autoecious demicyclic *Uromyces scillinus* (Durrieu & Montague) Herriot is actually a mixture of telia of *U. rayssii* and aecia of *U. reichertii* (2, 10).

These results raise the question whether teliospores formed on *Berberis cretica* in Greece belong to an ancient autoecious *eu*-form, as claimed by

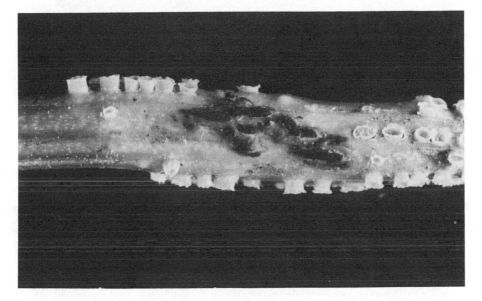

Figure 2 Compound cluster on *Muscari parviflorum*, consisting of aecia of *Uromyces christensenii* (light pustules) and telia of *Uromyces scillarum* (dark pustules) (X 16).

Critopoulos (23) or rather represent a microcyclic descendant of *P. graminis*. Critopoulos stated, "The telia occur in such proximity to aecia that there can be no doubt about the relationship of these two stages." Experience gained in Israel suggested that the "relationship" may involve not two stages in a life cycle of the same organism but rather two correlated species, one macrocyclic and the other microcyclic, representing two stages in a phylogenic history. Also Critopoulos realized that the appearance of telia on *B. cretica* might be interpreted as a step toward the evolution of a microcyclic organism.

Similar questions may be raised with respect to the existence of autoecious rusts on Gramineae. Rust species on grasses are heteroecious, with a few exceptions such as *P. graminella* Diet. & Holw. that are autoecious and parasitize *Stipa* species. A photograph of a leaf section (13, 36) shows teliospores of *P. graminella* arising from a "gametophytic mycelium" (13) at the sides of an aecium. However, to our knowledge the nature of the aecial-telial relationship has not been proven by inoculation tests. *P. graminella* is regarded as an autoecious, demicyclic organism (14, 63, 85) correlated with heteroecious demicyclic, microcyclic, and endocyclic species (85). Dietel (37) visualized that *"Puccinia graminella* is a *heteroecious species that, in addition to aecia on Malvaceae, also develops secondary aecia in the telial hosts"* (Dietel's italics).

Close proximity of aecia and telia or even formation of compound aecia-telia clusters does not prove that the implicated two stages belong to the same fungus. However, study of the compound structures may elucidate the mode of origin of microforms, and is therefore of phylogenetic significance. In some locations compound aecia-telia clusters have appeared for over 10 years in succession on young leaves of *S. hyacinthoides* that emerged annually from the perennial bulb cluster of single plant origin, after desiccation of the foliage from the preceding year. A parallel process has taken place in *M. parviflorum*.

COEVOLUTION OF HOST DEFENSE AND RUST VIRULENCE

Rust fungi as obligate parasites have evolved in interdependence with their hosts in the centers of origin and genetic diversification of the latter. Their coexistence lasting from antiquity was associated with reciprocal selection pressure exerted by the components interlocked in the host-parasite system. This interaction has resulted in "balanced polymorphism" (69, 96) in the populations of progenitors or relatives of cultivated crops and their respective parasites. As an outcome of correlated evolution, the host populations contain a multitude of distinct types and levels of protection matched by broad spectra of virulence in the pathogens (100a, 144). Parasitic variation in rust organisms becomes particularly pronounced in the regions where the distribution of functional alternate hosts coincides with that of the main host (150, 151). The wide range of host protection embraces specific resistance, general resistance, tolerance, avoidance (escape), etc (110, 144, 159, 160).

According to Zhukovsky (159), "wild relatives of cultivated crops in the centers of their origin as a rule do not possess absolute immunity. They are protected against pathogens by "field resistance" or tolerance. Ordinarily fungi attack only some plant portions, inducing necrosis and producing lower spore yields. This enables survival of the hosts and their parasites for millennia." The pathogens form there new virulent and agressive races. Similar conclusions regarding the evolution of strains with enhanced virulence were reached by Moseman (97). Mode (96) postulated that in balanced host-parasite systems moderate resistance prevails which is more stable and has substituted for populations of high resistance and shorter duration. Van der Plank (143) labeled diseases in natural ecosystems "endemic" and contended that they favor the evolution of a high degree of general resistance in the host and relatively low virulence in the pathogen. Similar evolutionary implications stemming from host-parasite coexistence were elaborated by Harlan (69) who contended that "genetic diversity and coevolution over time have generated an endemic balance," which can be viewed as the "evolutionary endpoint." In Flor's (55) opinion, "the primary gene center [of host-parasite coevolution] has been and probably will continue to be the plant breeder's principal source of both vertical and horizontal phylogenic resistance." Shattock (125) maintained that the greatest stability between hosts and their parasites is likely to occur in the centers of origin of plants where host and parasite have coexisted for the longest period of time. Nelson (100a) envisioned that "the long process of coevolution resulted in the ultimate accumulation of many resistance and virulence genes."

Consequently, there seems to be a general consensus that disease resistance should be sought mainly in the homeland of the hosts and their parasites (55, 89, 107, 110, 144, 154, 159, 160). Delineation of such centers is of paramount importance (88, 89, 160) in view of the ever-growing realization of the fact that germplasm reservoirs of the wild relatives of wild plants are and will be valuable sources of plant defense against diseases (55, 82).

Many of the conclusions drawn by researchers in other countries were corroborated by studies in Israel, which is a part of the centers of origin and genetic diversification of *Avena sterilis*—the putative ancestor of cultivated oats, *Hordeum spontaneum*—the assumed progenitor of cultivated barley, and wild emmer, *Triticum dicoccoides* Körn., which is one of the ancestors of cultivated wheat (70, 162). These wild species "represent a vast living gene pool, extremely variable, self-reproducing, and continually changing with the pressures of the environment" (161).

Sources of crown rust resistance were selected from *A. sterilis* (16, 17, 99, 149, 150, 151), which are very effective in North America (40, 56, 99, 108). A list of genes from *A. sterilis* conferring crown rust resistance was published by Simons et al (129). Also slow-crown rusting (16, 94) and tolerance (128, 147) were found in this wild oat species. Patterns of integration of all these types of plant defense in protection structures in natural oat stands sampled country-wide were investigated (16, 94). The relevance of knowledge of such structures for gene management in agro-ecosystems was emphasized by Browning (18). Resistance to stem rust in *A. sterilis* is mainly of the slow-rusting type (16, 136), though conventional race-specific resistance associated with hypersensitive reactions also has been selected from this oat species (40, 109).

Hordeum spontaneum has furnished effective resistance to *P. hordei* in Israel (7), the United States, (J. G. Moseman, unpublished information), and Germany (G. Fischbeck, unpublished information). The alternate *Ornithogalum* host plays an important role in the evolution and geographic distribution of this resistance (7). Notably, reactions elicited by *P. hordei* on the alternate host are distinctly less specialized than those on the main host (Y. Anikster, unpublished information). This finding is in agreement with Green's (62) observations on the performance of *P. graminis tritici* on barberry, as compared with reactions on wheat plants. The considerable uniformity in *Ornithogalum* reaction can be attributed not only to a lesser parasitic specialization in the monocaryon than in the dicaryon, as suggested by Green, but also to the presumably high degree of genetic homogeneity in asexually propagating plants.

Aaronson (1) discovered rust resistance in *T. dicoccoides* as early as 1913. This wild emmer provided resistance to many races of wheat stem rust and leaf rust (141), as well as very promising protection against stripe rust (60).

Significantly, disease resistance in the three wild species mentioned is scarce in arid regions, presumably due to inadequate selection pressure, since none of the rust diseases reaches epidemic proportions under dry conditions.

Studies on the physiologic specialization of *P. coronata avenae, P. graminis avenae,* and *P. hordei* reveal a broad spectrum of pathogenic variation in each of the rust species. The consistent, country-wide predominance of versatile races of crown rust like race 276, oat stem rust races 72 and 8 (16, 136), and *P. hordei* races (7, 61) cannot be ascribed to preferential selection pressure of the host, and does not support the concept that races with a wide range of "unneccessary" genes of virulence are inferior in their fitness to survive and compete (136, 149).

It is postulated that the gene pools of protection against rusts in wild barley, wheat, and oat species native to Israel, and the broad spectra of physiologic specialization of the rust organisms involved, are an outcome of a prolonged host-obligate parasite coevolution in the center of origin and genetic diversification of the hosts concerned. The protection structures in natural habitats incorporate conventional race-specific resistance, slow-rusting, and tolerance in proportions varying with the environmental conditions.

CONCLUDING REMARKS

Dietel (32), in formulating in 1904 the theory of host-rust coevolution was concerned with the role of obligate parasitism of the fungus in its phylogenetic history and the process of its expansion to new main and alternate hosts. A similar attitude was adopted by his eminent contemporaries, Fischer, Klebahn, Magnus, and Tranzschel. In the ensuing decades the theory has broadened in scope and exerted a great impact on other important problems like genetics of host-parasite coexistence (54, 55), cytology, morphology, and ecologic adaptability of the organism. The theory has become a pivotal base in studies on the origin and nature of plant defense against diseases and fungus virulence.

According to Nelson (100a), "coepicenters, geographic areas in which both host and parasite have evolved, most accurately depict the story of the evolution of genes for virulence and resistance." This statement is germane to the evolution of numerous important characters in the hosts and their rust parasites.

Studies on the nature and routes of such evolution processes are particularly interesting in the regions where the centers of genetic diversification of the main and alternate host overlap. There the aecial stage is most important (43, 45), and the balanced polymorphism is achieved thanks to the coordinated development of all constituents of the system in adaptation to specific ecologic conditions. Teliospores germinate readily in some of the rust components.

However, rust evolution has progressed along different pathways. Some processes have resulted in the origination of autoecious rusts with complete life cycles, but often autoecious forms with simplified life history have emerged. Organisms with reduced cycles include *hemi*-rusts and microforms. In the *hemi*-forms urediospores have assumed responsibility for the geographic distribution and partly also for the perpetuation of the fungus. In arid regions urediospore survival is enhanced by their larger size, thicker walls, and deeper pigmentation (104, 122).

Short-cycle rusts, which are often homothallic, are adjusted to unsuitable environments in the arctic as well as arid regions. This may explain their prominence in Israel.

The major trends in the evolution of grasses have apparently influenced the evolution of their rusts. According to Stebbins (134), "Most of the common species of grasses . . . contain in varying proportions gene combinations derived from two, three, four, or more separate and sometimes widely divergent ancestors."

Conceivably, owing to their genetic interrelationship, wild grasses serve as congenial hosts for rust fungi derived from a wide array of hosts, expedite their somatic hybridization, and thus enhance variability.

Plants receptive to a wide range of parasitic entities were found in Israel in arid regions, where rust development is weak. Apparently, lack of strong preferential selection pressure is responsible for the absence of differential reactions in the host to the tested rust cultures.

Uromyces rust populations on indigenous barleys in Israel alternating with native Liliaceae plants furnish numerous genetic lineages with intergrading stages. These series of congruences are suitable for tracing the pathways of evolution of morpholgic, cytologic, taxonomic, and physiologic attributes.

Studies on the types of plant defense against diseases in coepicenters of the host and parasite not only provide useful material for crop improvement, but may also help to clarify the controversial problems related to specific and general resistance. Equally important are the elucidation of defense structures against diseases in natural ecosystems, the integration and cohesion of the components, and their coevolution. Corollary research on fungus virulence may provide a deeper insight into the problems of stabilizing selection of parasitic entities, and a preview of what can be expected in other regions (147, 148).

Rust fungi of our time are not only "living fossils" but very dynamic, plastic organisms, progenitors of new morphologic and biologic forms, and of new shifty enemies.

ACKNOWLEDGMENTS

This review is respectfully dedicated to the memory of our teachers M. N. Levine, G. Minz, T. Rayss, and I. Reichert, founders of uredinology in Israel.

We gratefully acknowledge G. Viennot-Bourgin's most valuable participation in all phases of our studies, and are very thankful to J. A. Browning, the late E. E. Leppik, J. G. Moseman, and D. Zohary for inspiring discussions and suggestions over the years.

Literature Cited

1. Aaronson, A. 1913. The discovery of wild wheat. *City Club Bull*. 6(9):167–75
2. Anikster, Y. 1975. *Studies on the taxonomy, biology and evolution of the genus Uromyces on barley in Israel*. PhD thesis. Tel Aviv Univ., Tel Aviv. 134 pp.
3. Anikster, Y., Abraham, C., Greenberger, Y., Wahl, I. 1971. A contribution to the taxonomy of *Puccinia* brown leaf rust of barley in Israel. *Isr. J. Bot.* 20:1–12
4. Anikster, Y., Eshed, N., Wahl, I. 1978. Evolutionary aspects of epidemiology. In *Abstr. Pap., 3rd Int. Congr. Plant Pathol.*, p. 315. München, 16–23 August 1978
5. Anikster, Y., Moseman, J. G., Wahl, I. 1972. Life cycles and evolutionary trends of some *Uromyces* species inhabiting wild barleys and Liliaceae in Israel. *Proc. Eur. Mediterr. Cereal Rusts Conf.* Praha. 11:81–86
6. Anikster, Y., Moseman, J. G., Wahl, I. 1973. Nuclear development in *Uromyces sp.* on wild barleys and Liliaceae in Israel. *Abstr. Pap., 2nd Int. Congr. Plant Pathol. Minneapolis, Minn.* 5–12 Sept. 1973. Abstr. No. 0068
7. Anikster, Y., Moseman, J. G., Wahl, I. 1976. Parasitic specialization of *Puccinia hordei* Otth. and sources of resistance in *Hordeum spontaneum* C. Koch. *Barley Genet.* III:468–69. *Proc. 3rd Int. Barley Genet. Symp.*, Garching, 1975
8. Anikster, Y., Moseman, J. G., Wahl, I., 1977. *Uromyces viennot-bourginii*, its life cycle, pathogenicity and cytology. *Soc. France Phytopathol.*, pp. 9–17
9. Anikster, Y., Wahl, I. 1966. *Uromyces* rusts on barley in Israel. *Isr. J. Bot.* 15:91–105
10. Anikster, Y., Wahl, I. 1966. Quatre espéces nouvelles d'"*uromyces*" récoltées en Israël sur orges sauvages et sur Liliacées. *Bull. Soc. Mycol. Fr.* 82(4): 546–60
11. Arthur, J. C. 1906. Reasons for desiring a better classification of the Uredinales. *J. Mycol.* 12:149–54
12. Arthur, J. C. 1924. Fern rusts and their aecia. *Mycologia* 16:245–51
13. Arthur, J. C. 1929. *The Plant Rusts (Uredinales)*. New York: Wiley. 446 pp.
14. Arthur, J. C., Cummins, G. B. 1962. *Manual of the Rusts in United States and Canada*. New York: Hafner. pp. 438

15. Bridgmon, G. H. 1957. The production of new races of *Puccinia graminis* var. *tritici* by hyphal fusion on wheat. *Phytopathology* 47:517 (Abstr.)
16. Briggle, L. W., Wahl, I. 1978. Slow rusting of oats *Avena sterilis*, infected with crown rust and stem rust. *Final Res. Rep., Presented to US-Israel Binat. Sci. Found.* 71 pp.
17. Brodny, U., Briggle, L. W., Wahl, I. 1976. Reaction of elite US crown rust resistant oat selections and Israeli *Avena sterilis* selections to *Puccinia coronata* var. *avenae*. *Plant Dis. Reptr.* 60:902–6
18. Browning, J. A. 1974. Relevance of knowledge about natural ecosystems to development of pest management programs for agro-ecosystems. *Ann. Phytopathol. Soc. Proc.* 1:191–99
19. Browning, J. A., Frey, K. J. 1969. Multiline cultivars as a means of disease control. *Ann. Rev. Phytopathol.* 7: 355–82
20. Buller, A. H. R. 1950. *Researches on Fungi*, Vol. 7. Toronto Publ.: R. Soc. Can., Univ. Toronto Press. 458 pp.
21. Chabelska, H. 1938. Life-cycle of the rust on *Anchusa strigosa* Lab. *Palest. J. Bot. Jerusalem Ser.* 1:101–3
22. Cotter, R. U., Levine, M. N. 1938. Experiments in crossing varieties of *Puccinia graminis*. *Phytopathology* 28:6 (Abstr.)
23. Critopoulos, P. D. 1947. Production of teliospores and uredospores of *Puccinia graminis* on *Berberis cretica* in nature. *Mycologia* 39:145–51
24. Critopoulos, P. 1956. Perpetuation of the brown rust of barley in Attica. *Mycologia* 48:596–600
25. Cronquist, A. 1968. *The Evolution and Classification of Flowering Plants*. Boston: Houghton-Mifflin. 396 pp.
26. Cummins, G. B. 1956. Host index and morphological characterization of the grass rusts of the world. *Plant Dis. Reptr. Suppl.* 237:52 pp.
27. Cummins, G. B. 1959. *Illustrated Genera of Rust Fungi.* Minneapolis, Minn.: Burgess, 131 pp.
28. Cummins, G. B. 1971. *The Rust Fungi of Cereals, Grasses and Bamboos.* Berlin, Heidelberg & New York: Springer. 570 pp.
29. Cummins, G. B., Caldwell, R. M. 1956. The validity of binomials in the leaf rust fungus complex of cereals and grasses. *Phytopathology* 46:81–82
30. Dietel, P. 1899. Waren die Rostpilze in früheren Zeiten plurivor? *Bot. Centralbl.* 79:113–17
31. Dietel, P. 1903. Über die auf Leguminosen lebenden Rostpilze und die Verwandschaftsverhältnisse. *Ann. Mycol.* 1:3–14
32. Dietel, P. 1904. Betrachtungen über die Verteilung der Uredineen auf ihren Nährpflanzen. *Centralbl. Bakteriol. Parasitenkd. Infektionskr.* Abt. 2 *Allg. landwirtsch. Technol. Bakteriol. Gärungsphysiol. Pflanzenpathol.* 12: 218–34
33. Dietel, P. 1911. Einige Bemerkungen zur geographischen Verbreitung der Arten aus den Gattungen *Uromyces* and *Puccinia*. *Ann. Mycol.* 9:160–65
34. Dietel, P. 1914. Betrachtungen zur Systematik der Uredineen. I. *Mycol. Zentralbl.* 5(2):65–73
35. Dietel, P. 1918. Über die wirtswechselnden Rostpilze. *Centralbl. Bakteriol. Parasitenkd. Infektionskr. Abt. 2 Allg. Landwirtsch. Technol. Bakteriol. Gärungsphysiol. Pflanzenpathol. Pflanzenschutz* 48:470–50
36. Dietel, P. 1928. Uredinales. In *Natürliche Pflanzenfamilien*, ed. A. Engler, K. Prantl, 6:24–98. Leipzig: Englemann. 290 pp. 2nd ed.
37. Dietel, P. 1939. Betrachtungen über *Puccinia graminella* (Speg.) Diet. et Holw. *Uredineana* I(1938):27–32
38. Dietz, S. M., 1926. The alternate hosts of crown rust, *Puccinia coronata* Corda. *J. Agric. Res.* 33:953–70
39. Dinoor, A. 1967. *The role of cultivated and wild plants in the life cycle of Puccinia coronata Cda. var. avenae F & L and in the disease cycle of oat crown rust in Israel.* PhD thesis. Hebrew Univ., Jerusalem. 373 pp. (In Hebrew, with English summary)
40. Dinoor, A., Wahl, I. 1963. Reaction of non-cultivated oats from Israel to Canadian races of crown rust and stem rust. *Can. J. Plant Sci.* 43:468–70
40a. D'Oliveira, B. 1937. Brown rust of wild species of *Hordeum*. *Rev. Agron.* 25(3)-230–34
41. D'Oliveira, B. 1939. Studies on *Puccinia anomala* Rost. 1. Physiologic races on cultivated barleys. *Ann. Appl. Biol.* 26:56–82
42. D'Oliveira, B. 1940. Notas sobre a produção da fase aecidica de algumas ferrugens dos cereais em Portugal. *Rev. Agron.* 28:201–8
43. D'Oliveira, B. 1940. Sobre a necessidade da cooperação internacional dos paises mediterranicos no estudo das uredineas do cereais. *XIII Congr. Luso-Espanhol. Prog. Cienc. Natl. Lisboa.* 5:117–20

44. D'Oliveira, B. 1951. The centers of origin of cereals and the study of their rusts. *Agron. Lusit.* 13(3):221–26
45. D'Oliveira, B. 1960. Ideas concerning the evolution and distribution of cereal rusts. *Port. Acta Biol. Ser. A* 6:111–24
46. D'Oliveira, B. 1960. Host range of the aecial stage of *Puccinia hordei* Otth. *Melhoramento* 13:161–88
47. D'Oliveira, B., Samborski, D. J. 1966. Aecial stage of *Puccinia recondita* on Ranunculaceae and Boraginaceae in Portugal. In *Cereal Rust Conf. 1964*, ed. R. C. F. Macer, M. S. Wolfe, pp. 133–50. Cambridge: Plant Breed. Stn.
48. Dupias, G. 1971. Essai sur la biogéographie des Urédinées. Son apport a la systématique. *Bull. Soc. Mycol. Fr.* 87(2): 245–412
49. Eig, A., Zohary, M., Feinbrun, N. 1960. *Analytical Flora of Palestine*. Palestine: J. Bot. Jerusalem. 515 pp. 4th print. (In Hebrew)
50. Eshed, N. 1978. *The genetical basis of the pathogenicity of Puccinia coronata CDA in Israel.* PhD thesis. Hebrew Univ., Jerusalem. 190 pp.
51. Fischer, E. 1898. Entwicklungsgeschichtliche Untersuchungen über Rostpilze. 1. Über Beziehungen zwischen Uredineen, welche alle Sporenformen besizen und solchen von reduziertem Entwicklungsgang, pp. 109–120. *Beitr Kryptogamenflora Schweiz* 1(1):121 pp.
52. Fischer, E. 1904. Die Uredineen der Schweiz. *Beitr. Kryptogamenflora Schweiz* 2(2):590 + XCIV pp.
53. Fleischmann, G. 1967. Virulence of uredial and aecial isolates of *Puccinia coronata* Corda f. sp. *avenae* identified in Canada from 1952–1966. *Can. J. Bot.* 45:1693–1701
54. Flor, H. H. 1955. Host-parasite interaction in flax rust—its genetics and other implications. *Phytopathology* 45:680-85
55. Flor, H. H. 1971. Current status of the gene-for-gene concept. *Ann. Rev. Phytopathol* 9:275–96
56. Frey, K. J. 1976. Plant breeding in the seventies: Useful genes from wild plant species. *Egypt. J. Genet. Cytol.* 5(2): 460–82
57. Gäumann, E. A. 1952. *The Fungi.* New York & London: Hafner. 420 pp.
58. Gäumann, E. 1959. Die Rostpilze Mitteleuropas. *Beitr. Kryptogamenflora Schweiz* 12:7–1407
59. Gerechter-Amitai, Z. K., 1973. *Stem Rust, Puccinia graminis Pers. on cultivated and wild grasses in Israel.* PhD thesis. Hebrew Univ., Jerusalem. 420 pp. (In Hebrew, Engl. summ.)

60. Gerechter-Amitai, Z. K., Stubbs, R. W. 1970. A valuable source of yellow rust resistance in Israeli populations of wild emmer, *Triticum dicoccoides* Koern. *Euphytica* 19:12-21
61. Golan, T., Anikster, Y., Wahl, I. 1978. A new virulent strain of *Puccinia hordei. Euphytica* 27:185–89
62. Green, G. J. 1971. Hybridization between *Puccinia graminis tritici* and *Puccinia graminis secalis* and its evolutionary implications. *Can. J. Bot.* 49: 2089–95
63. Greene, H. C., Cummins, G. B. 1958. A synopsis of the Uredinales which parasitize grasses of the genera *Stipa* and *Nasella. Mycologia* 50:6–36
64. Greene, H. C., Cummins, G. B. 1967. *Puccinia holcina* and *P. poarum* redefined. *Mycologia* 59:47–57
65. Grove, W. B. 1913. The evolution of the higher Uredineae. *New Phytol.* 12:89–106
66. Guyot, A. L. 1938. Les Urédinées. I. Genre *Uromyces. Encyclopedie Mycologique*, Vol. 8. Paris: Lechevalier. 438 pp.
67. Guyot, A. L. 1939. De quelques Urédinées nouvelles. *Uredineana* 1:59–90
68. Guyot, A. L., Massenot, M., Saccas, A. 1945–46. Considérations morphologique et biologique sur l'espèce *Puccinia graminis* Pers. *sensu lato. Ann. Ec. Natl. Agric. Grignon* 5:82–146
69. Harlan, J. R. 1976. Diseases as a factor in plant evolution. *Ann. Rev. Phytopathol.* 14:31–51
70. Harlan, J. R., Zohary, D. 1966. Distribution of wild wheats and barley. *Science* 153:1074–80
71. Hart, H., 1955. Complexities of the wheat stem rust situation. *Trans. Am. Assoc. Cereal Chem.* 13:1–14
72. Hassebrauk, K. 1962. Uredinales (Rostpilze) In *Handbuch der Pflanzenkrankheiten. Basidiomycetes*, ed. H. Richter, Founder, P. Sorauer. 3:2–275. Berlin & Hamburg: Parey: 6th ed. 747 pp.
73. Hennen, J. F. 1975. Taxonomy and evolution of rust fungi. *J. C. Arthur Mem. Lect., 23 June 1975 Before North Cent. Sect., Am. Phytopathol. Soc.*
74. Jackson, H. S. 1931. Present evolutionary tendencies and the origin of life cycles in the Uredinales. *Mem. Bull. Torrey Bot. Club* 18:5–108
75. Jackson, H. S. 1935. The nuclear cycle in *Herpobasidium filicinum* with a discussion of the significance of homothallism in Basidiomycetes. *Mycologia* 27: 553–72

76. Jackson, H. S., Mains, E. B. 1921. Aecial stage of the orange leaf rust of wheat, *Puccinia triticina* Eriks. *J. Agric Res.* 22:151–71

77. Johnson, T. 1961. Man-guided evolution in plant rusts. *Science* 133(3450): 357–62

78. Johnson, T., Green, G. J. 1954. Resistance of common barberry (*Berberis vulgaris* L.) to race 15 B of wheat stem rust. *Can. J. Bot.* 32:378–79

79. Johnson, T., Green, G. J. 1954. The production by *Puccinia graminis* of abortive pycnia on wheat. *Can. J. Agric. Sci.* 34:313–15

80. Johnson, T., Newton, M. 1938. The origin of abnormal rust characteristics through the inbreeding of physiologic races of *Puccinia graminis tritici. Can. J. Res. Sect. C* 16:38–52

81. Klebahn, H. 1904. *Die Wirtswechselnden Rostpilze.* Berlin: Gebrüder Borntraeger. 447 pp.

82. Knott, D. R., Dvorák, J. 1976. Alien germ plasm as a source of resistance to disease. *Ann. Rev. Phytopathol.* 14: 211–35

83. Kuprevich, V. F., Tranzschel, V. G. 1957. *Rust Fungi (Melampsoraceae).* Moscow & Leningrad: Publ. Acad. Sci. USSR (In Russian, with Engl. transl.) 1970. Jerusalem, Isr. Program for Sci. Transl. 518 pp.

84. Ladizinsky, G. 1971. Biological flora of Israel. 2. *Avena* L. *Isr. J. Bot.* 20: 133–51

85. Laundon, G. F., 1973. Uredinales. In *The Fungi,* ed. G. C. Ainsworth, F. K. Sparrow, A. S. Sussman. IVB:247–79. New York & London: Academic. 504 pp.

86. Leppik, E. E. 1959. Some viewpoints on the phylogeny of rust fungi. III. Origin of grass rusts. *Mycologia* 51:512–28

87. Leppik, E. E. 1961. Some viewpoints on the phylogeny of rust fungi. IV. Stem rust genealogy. *Mycologia* 53:378–405

88. Leppik, E. E. 1967. Some viewpoints on the phylogeny of rust fungi. VI. Biogenic radiation. *Mycologia* 59:568–79

89. Leppik, E. E. 1970. Gene centers of plants as sources of disease resistance. *Ann. Rev. Phytopathol.* 8:323–44

90. Levine, M. N., Cotter, R. U., Stakman, E. C. 1934. Production of an apparently new variety of *Puccinia graminis* by hybridization on barberry. *Phytopathology* 24:13–14 (Abstr.)

91. Lindfors, T. 1924. Studien über den Entwicklungsverlauf bei einigen Rostpilzen aus zytologischen und anato-mischen Gesichtspunkten. *Sven. Bot. Tidsk.* 18:1–84

92. Luig, N. H., Watson, I. A. 1972. The role of wild and cultivated grasses in the hybridization of formae speciales of *Puccinia graminis. Aust. J. Biol. Sci.* 25:335–42

93. Magnus, P. 1893. Über die auf Compositen auftrenden Puccinien mit Teleosporen von Typus der *Puccinia hieracii*, nebst einigen Andeutungen über den Zusammenhang ihrer specifischen Entwicklung mit ihrer vertikalen Verbreitung. *Ber. Dtsch. Bot. Ges.* 11: 453–64

94. Manisterski, J., Segal, A., Wahl, I. 1978. Rust resistance in natural ecosystems of *Avena sterilis.* In *Abstr. Pap., 3rd Int. Congr. Plant Pathol,* p. 301. München, 16–23 August, 1978

95. McGinnis, R. C. 1954. Cytological studies of chromosomes of rust fungi. II. The mitotic chromosomes of *Puccinia coronata. Can. J. Bot.* 32:213–14

96. Mode, C. J. 1958. A mathematical model for the co-evolution of obligate parasites and their hosts. *Evolution* 12:158–65

97. Moseman, J. G. 1970. Co-evolution of host resistance and pathogen virulence. *Barley Genet.* II. *Proc. Int. Barley Genet. Symp.,* ed. R. A. Nilan, pp. 450–56. Pullman: Washington State Univ. Press

98. Murphy, H. C. 1935. Physiologic specialization in *Puccinia coronata avenae. US Dep. Agric. Tech. Bull.* 433:1–48

99. Murphy, H. C., Wahl, I., Dinoor, A., Miller, J. D., Morey, D. D., Luke, H. H., Sechler, D., Reyes, L. 1967. Resistance to crown rust and soilborne mosaic virus in *Avena sterilis. Plant Dis. Reptr.* 51:120–24

100. Naumov, N. A. 1939. *The Rusts of Cereals in the USSR.* Moscow & Leningrad: Selkhosgiz. 403 pp. (In Russian)

100a. Nelson, R. R. 1978. Genetics of horizontal resistance to plant diseases. *Ann. Rev. Phytopathol.* 16:359–78

101. Orton, C. R. 1912. Correlation between certain species of *Puccinia* and *Uromyces. Mycologia* 4:194–204

102. Orton, C. R. 1927. A working hypothesis on the origin of rusts with special reference to the phenomenon of heteroecism. *Bot. Gaz.* 84:113–38

103. Peturson, B. 1954. The relative prevalence of specialized forms of *Puccinia coronata* that occur on *Rhamnus cathartica* in Canada. *Can. J. Bot.* 32:40–47

104. Rayss, T. 1951. Nouvelle contribution á

la connaissance des Urédinées de Palestine. *Uredineana* 3:154–221

105. Rayss, T., Chabelska, C. 1966. Données
additionnelles à l'étude des Urédinées
de Palestine. *Uredineana* 6:289–98

106. Rayss, T., Habelska, H. 1942. *Rhamnus
palaestina* Boiss.—a new host of crown
rust. *Palest. J. Bot. Jerusalem Ser.*
2:250

107. Reichert, I. 1958. Fungi and plant diseases in relation to biogeography.
Trans. NY Acad. Sci. Ser. II. 20(4):
333–39

108. Reitz, L. P., Craddock, J. C. 1969.
Diversity of germ plasm in small grain
cereals. *Econ. Bot.* 23:315–23

109. Rothman, P. G. 1976. Registration of
oat germplasm. *Crop Sci.* 16:315

110. Rudorf, W. 1959. Problems of collection, maintainance and evaluation of
wild species of cultivated plants. *FAO
Plant Introd. Newsl.* 5:1–4

111. Saari, E. E., Young, H. C. Jr., Kern-
kamp, M. F. 1968. Infection of North
American *Thalictrum* spp. with *Puccinia recondita* f. sp. *tritici. Phytopathology* 58:939–43

112. Santiago, J. C. 1968. Physiologic specialization of the oat crown rust fungus
in Portugal. *Proc. Cereal Rusts Conf.,
Oeiras-Portugal,* pp. 89–91

113. Savile, D. B. O. 1953. Short-season ad-
aptations in the rust fungi. *Mycologia*
45:75–87

114. Savile, D. B. O. 1954. The fungi as aids
in the taxonomy of the flowering plants.
Science 120:583–85

115. Savile, D. B. O. 1955. A phylogeny of
the Basidiomycetes. *Can. J. Bot.* 33:60–
104

116. Savile, D. B. O. 1961. Some fungal parasites of Liliaceae. *Mycologia* 53:31–52

117. Savile, D. B. O. 1962. Taxonomic disposition of *Allium. Nature* 196(4856):792

118. Savile, D. B. O. 1964. Geographic variation and gene flow in *Puccinia cruciferarum. Mycologia* 56:240–48

119. Savile, D. B. O. 1971. Co-ordinated
studies of parasitic fungi and flowering
plants. *Nat. Can.* 98:535–52

120. Savile, D. B. O. 1971. Coevolution of
the rust fungi and their hosts. *Q. Rev.
Biol.* 46:211–18

121. Savile, D. B. O. 1972. *Arctic Adapta-
tions in Plants.* Monogr. 6. Ottawa: Res.
Branch, Can. Dep. Agric. 81 pp.

122. Savile, D. B. O. 1976. Evolution of the
rust fungi (Uredinales) as reflected by
their ecological problems. *Evol. Biol.*
9:137–207

123. Savulescu, T. 1954. Biologia si dis-
tributia Uredinalelor din Republica
Populara Romina. *An. Inst. Cercet.*
Agron. 21:3–41 (In Rumanian, with
French and Russian summaries)

124. Schneider, W. 1927. Zur Biologie
einiger Liliaceen bewohnenden Uredineen. *Zentralbl. Bakteriol. Parasitenkd,
Abt. II* 72:246–65

125. Shattock, R. C. 1977. The dynamics of
plant diseases. In *Origins of Pest, Parasite, Disease and Weed Problems. 18th
Symp. Br. Ecol. Soc.* ed. J. M. Cherrett,
G. R. Sagar, pp. 83–107. Oxford, London, Edinburgh & Melbourne: Blackwell Sci. Publ. 413 pp.

126. Sibilia, C. 1960. La forma ecidica della
ruggina bruna delle foglie di grano *Puccinia recondita* Rob. ex Desm. in Italia.
Boll. Stn. Patol. Veg. Roma, Ser. 3,
18:1–8

127. Simons, M. D. 1970. *Crown Rust of
Oats and Grasses.* Monogr. 5. Am.
Phytopathol. Soc. 47 pp.

128. Simons, M. D. 1972. Crown rust tolerance of *Avena sativa*—type oats derived
from wild *Avena sterilis. Phytopathology*
62:1444–46

129. Simons, M. D., Martens, J. W.,
McKenzie, R. I. H., Nishiyama, I.,
Sadanaga, K., Sebesta, J., Thomas, H.
1978. *Oats: A Standardized System of
Nomenclature for Genes and Chromo-
somes and Catalog of Genes Governing
Characters.* US Dep. Agric., Agric.
Hand. No. 509. 40 pp.

130. Stakman, E. C. 1954. Recent studies of
wheat stem rust in relation to breeding
resistant varieties. *Phytopathology* 44:
346–51

131. Stakman, E. C., Harrar, J. G. 1957.
Principles of Plant Pathology. New
York: Ronald. 581 pp.

132. Stakman, E. C., Levine, M. N., Cotter,
R. U., Hines, L. 1934. Relation of barberry to the origin and persistence of
physiologic forms of *Puccinia graminis.
J. Agric. Res.* 48:953–69

133. Stakman, E. C., Piemeisel, F. J. 1917.
Biologic forms of *Puccinia graminis* on
cereals and grasses. *J. Agric. Res.*
10:429–96

134. Stebbins, G. L. 1956. Cytogenetics and
evolution of the grass family. *Am. J.
Bot.* 43:890–905

135. Stebbins, G. L., Crampton, B. 1961. A
suggested revision of the grass genera of
temperate North America. *Recent Adv.
Bot.* (from *IX Int. Bot. Congr.* Mon-
treal, 1959) 1:135–45. Toronto: Univ.
Toronto Press

136. Sztejnberg, A., Wahl, I. 1976. Mechanisms and stability of slow stem rusting
resistance in *Avena sterilis. Phytopathology* 66:74–80

137. Tranzschel, W. 1904. Über die Möglichkeit, die Biologie wirtswechselnder Rostpilze auf Grund morphologischer Merkmale vorauszusehen (Vorlauf. Mitteil.) *Arb. St-Petersburger Naturf. Ges.* 35:311–12

138. Tranzschel, W. 1910. Die auf der Gattung *Euphorbia* auftretenden autoecischen *Uromyces-Arten. Ann. Mycol.* 8:1–35

139. Tranzschel, W. A. 1939. Sovremiennoe sostaianie znanii po biologii rshavchin khlebnykh zlakov (Present state of knowledge about the biology of cereal rusts). In *Rshavchina zernovykh kultur* (Rusts of cereal crops), ed. N. A. Naumov, A. K. Zubareva, pp. 29–41. Moscow: Selkhosgiz. 286 pp. (In Russian)

140. Tranzschel, W. 1939. *Conspectus Uredinalium USSR.* Moscow: Acad. Sci. URSS. 426 pp. (In Russian)

141. United States Dep. Agric. Agric. Handb. No. 165. Index of Plant Diseases in the United States. Washington DC: Crops Res. Div., ARS. 531 pp.

142. Urban, Z. 1969. Die Grasrostpilze Mitteleuropas mit besonderer Berücksichtigung der Tschechoslowakei. *Rozpr. Cesko. Akad. Ved. Praha: Acad. Naklad. Cesk. Akad. Ved. 79(6):3–104*

143. Van der Plank, J. E. 1975. *Principles of Plant Infection* New York, San Francisco & London: Academic. 216 pp.

144. Vavilov, N. I. 1938. Seleksiia ustoichovykh sortov kak osnovnoi metod borby s rhavchinoi (Selection of resistant varieties as a basic method for rust control). See Ref. 139

145. Viennot-Bourgin, G. 1958. Contribution a la connaissance des champignons parasites de l'Iran. *Ann. Epiphyties* 2:97–210

146. Viennot-Bourgin, G. 1969. Mission phytopathologique en Iran en 1968. *Ann. Phytopathol.* 1(1):5–36

147. Wahl, I. 1958. Studies on crown rust and stem rust on oats in Israel. *Bull. Res. Counc. Isr. Sect. D* 6:145–66

148. Wahl, I. 1959. Physiologic races of oat crown rust identified in Israel in 1956–59. *Bull. Res. Counc. Isr. Sect D* 8:25–30

149. Wahl, I. 1970. Prevalence and geographic distribution of resistance to crown rust in *Avena sterilis. Phytopathology* 60:746–49

150. Wahl, I. 1972. Evolution of host resistance to rusts. *Proc. Eur. Mediterr. Cereal Rusts Conf.* Praha 11:267–70

151. Wahl, I., Dinoor, A. 1967. The screening of collections of wild oats for resistance and tolerance to oat crown rust

and stem rust fungi. *Final Res. Rep. Presented to the US Dep. Agric. ARS PL 480 Res. Grant FG-Is-138, AIO-CR-20.* 154 pp.

152. Wahl, I., Dinoor, A., Halperin, J., Schreiter, S. 1960. The effect of *Rhamnus palaestina* on the origin and persistence of oat crown rust races. *Phytopathology* 50:562–67

153. Waterhouse, W. L. 1951. Australian rust studies. VIII. *Puccinia graminis lolii,* an undescribed rust of *Lolium* spp. and other grasses in Australia. *Proc. Linn. Soc. NSW* 76:57–64

154. Watson, I. A. 1970. The utilization of wild species in the breeding of cultivated crops resistant to plant diseases. In *IBP Handb.* No. 11, *Genetic resources in Plants—their Exploration and Conservation,* ed. O. H. Frenkel, E. Bennet, pp. 441–57. Oxford & Cambridge: Blackwell. 554 pp.

155. Watson, I. A. 1970. Changes in virulence and population shifts in plant pathogens. *Ann. Rev. Phytopathol.* 8:209–30

156. Watson, I. A., Luig, N. H. 1966. SR15 —a new gene for use in the classification of *Puccinia graminis* var. *tritici. Euphytica* 15:239–50

157. Young, H. C. Jr., 1970. Variation in virulence and its relation to the use of specific resistance for the control of wheat leaf rust. In *Plant Disease Problems. Proc. 1st Int. Symp. Plant Pathol. Indian Phytopathol. Soc.,* ed. S. P. Raychaudhuri et al, pp. 3–8. New Delhi. 915 pp.

158. Zeven, A. C., Zhukovsky, P. M. 1975. *Dictionary of Cultivated Plants and Their Centers of Diversity.* Wageningen: Cent. Agric. Publ. Doc. 219 pp.

159. Zhukovsky, P. M. 1959. Vzaimootnoshennia mezhdu khozainom i gribnym parazitom na ikh rodine i vne ee (Interrelations between host and fungus parasites in their origin and beyond it). *Vestni. Skh. Nauki Moscow* 4(6):25–34 (In Russian, with English summary)

160. Zhukovsky, P. M. 1961. Grundlagen der Introduktion der Pflanzen auf Resistenz gegen Krankheiten. *Züchter* 31:248–53

161. Zillinsky, F. J., Murphy, H. C. 1967. Wild oat species as sources of disease resistance for the improvement of cultivated oats. *Plant Dis. Reptr.* 51:391–95

162. Zohary, D. 1971. Origin of South-west Asiatic cereals: Wheats, barley, oats and rye. In *Plant Life of South-West Asia,* ed. P. H. Davis et al, pp. 235–63. Edinburgh: Bot. Soc.

Ann. Rev. Phytopathol. 1979. 17:405–29
Copyright © 1979 by Annual Reviews Inc. All rights reserved

MOVEMENT OF FUMIGANTS IN SOIL, DOSAGE RESPONSES, AND DIFFERENTIAL EFFECTS

♦3717

Donald E. Munnecke[1]

Department of Plant Pathology, University of California, Riverside, California 92521

Seymour D. Van Gundy[2]

Department of Nematology, University of California, Riverside, California 92521

INTRODUCTION

In this review we discuss some of the recent research on usage of fumigants and the application and expansion of the various principles involved, with emphasis on the quantitative aspects. Compounds that act systemically in controlling disease in plants, or those that do not act as gases in the soil are not included. This review is not comprehensive, since a number of reviews provide auxilliary material: Goring has (25–27) reviewed the physical and biological factors involved in soil fumigation, and Wilhelm et al (78) recently prepared an interesting and readable review on mixtures of fumigants. The books on herbicide research edited by Audus (6, 7) are outstanding, particularly the chapters by Hartley in both editions (30, 31) and by Kaufman & Kearney (34), Crosby (18), and Grossbard (29) in the 1976 edition. Rodriguez-Kabana, Backman & Curl (69) emphasized material not found in other reviews. Finally, reviews that reflect our views and interest are also included (57, 75).

[1]I wish to acknowledge the contributions of my co-workers, J. L. Bricker, W. D. Wilbur, and particularly, M. J. Kolbezen.

[2]I wish to acknowledge the contributions of my colleagues, C. E. Castro, M. V. McKenry, and I. J. Thomason.

405

ACTIVITY OF FUMIGANTS IN SOIL

The important factors influencing the movement of soil fumigants are the chemical and adsorptive characteristic of the toxicant, temperature, moisture, organic matter, soil texture, and soil profile variability. Assuming the same soil profile and soil conditions at equivalent application rates, the fastest and farthest reaching fumigants (Table 1) are, in order of activity, MB, CP, 1,3-D, EDB, MIT, and DBCP, respectively. The factors contributing to fumigant activity are chemical, physical, or biological, or combinations of the three.

Chemical Factors

Some important properties of the common soil fumigants are given in Table 1. Soil fumigants, with the exception of MIT and DBCP, are applied 10 to 90 cm below the soil surface as a liquid or gas either at a point source or as a line source. The first process that takes place in warm soils is the initial vaporization of the chemical from the liquid phase to the gaseous phase. The individual molecules now have an affinity for, and may be removed from, the pore spaces by the water, mineral, and organic phases of the soil. The soil is a mass infinite in two horizontal directions with a finite soil surface–air boundary in the vertical direction. Since the diffusion of gas molecules is greater in air above the soil surface than within the soil system, upward mass flow and diffusion is usually greater than downward movement. Also, diffusion is unaffected by gravity. Therefore, seals of some sort are necessary at the soil surface to retain diffusing molecules long enough

Table 1 Properties of fumigants

Property	Common abbreviation[a]					
	MB	CP	EDB	DBCP	1,3-D	MIT
Molecular weight	95	154	188	236	112	—
Specific gravity	1.73	1.65	2.17	2.08	1.2	—
Vapor pressure 20° C (mm Hg)	1,380	20.0	7.69	0.58	21.0	21.0
Boiling point (C)	4.6	112	132	196	104–112	—
Solubility in water (Percentage at 20° C)	1.6	0.195	0.337	0.123	0.27	0.76
Water/air distribution[b] (20° C)	4.1	10.8	42.7	163.8	20.2	88.0

[a] Chemical name: MB, methyl bromide, bromo methane; CP, chloropicrin, trichlornitro methane; EDB, 1,2-dibromo ethane; DBCP, 1,2-dibromo-3-chloropropane; 1,3-D, 1,3-dichloropropene (1:1 mixture *cis* and *trans*); MIT, methyl isothiocyanate.
[b] Henry's law constant.

to give maximum effectiveness against organisms located near the soil surface. Deep injection extend the vertical and horizontal overlap of diffusing patterns and also help to retain the fumigant in the soil longer.

Thus, the relative diffusion pattern of any fumigant, given a constant set of physical factors, is predictable based on their inherent physical and chemical characteristics.

VAPOR PHASE   The vapor pressure, or the rate at which the chemical leaves the liquid phase, is important in determining the movement of fumigants in soil. If the vapor pressure of the liquid is high, as in the case of MB (Table 1), the gaseous concentration of the chemical in the soil pore spaces increases so rapidly that movement of the chemical occurs first by mass flow and then by diffusion. In comparison, a fumigant such as EDB, has a low vapor pressure and moves slowly in the vapor phase and little if any mass flow occurs except in extremely warm soil or unless water is applied. DBCP has such a low vapor pressure that solubility and movement in water films is more important than movement in the vapor phase. Obviously then, soil temperature is important and may affect fumigant movement in several ways. In the case of the halogenated hydrocarbons, a rise in temperature increases vapor pressure and decreases solubility. This alters the phase distribution and results in an increase in the rate of diffusion of the fumigant through the soil (24, 25, 42, 43, 51). Generally, there is a threefold increase in the value of the ratio of solubility of concentration in water to the concentration in air for 1,3-D as the soil temperature decreases from 25°C to 5°C. On the other hand, increases in temperature also increase the rate of other chemical reactions in soil that tend to degrade, hydrolyze, or reduce the effectiveness of the chemical and its biological activity (15, 16, 44, 73). The one exception to the above statement appears to be DBCP where solubility is more important than gaseous diffusion, and where there is slow degradation in soil (33); however, DBCP does degrade faster at high temperatures.

SOLUBILITY AND EQUILIBRIUM DISTRIBUTION RATIO   The vaporized molecules of soil fumigants tend to be dissolved in the soil water films. There is a continuous reestablishment of a dynamic equilibrium between air and water. This equilibrium is governed by Henry's Law, which states that the ratio of the concentration of the fumigant in the soil water to the concentration of the fumigant in the soil air is a constant at a given temperature. The vapor pressure and solubility of MB in water is high while the reverse is true for DBCP (Table 1), yet the potential concentration of DBCP over time in the water is 40 times greater than for MB. DBCP is the most persistent in soil of all the fumigants. It also follows that if the soil water content for any fumigant is increased, both the rate of movement is slowed and the distance traveled in the soil pores are reduced.

The characteristics of MIT are lower adsorption, relatively greater partition into water from air, slower diffusion, and higher decomposition rate at high temperatures, as compared with other soil fumigants. These factors probably are responsible for its relatively poor control of organisms in large soil masses (44, 73).

Solubility is important because most soil microorganisms are bathed in water films protecting them from direct contact with the fumigants in vapor phase. The toxic molecules must be present in the soil solution in which the target pest is bathed. Marks et al (47) found that the dynamic equilibrium of EDB between five different nematodes and the bathing solution was reached after about 30 min exposure, and that the internal concentration was 2–20 times greater than the external bathing solution.

Moisture is important in the adsorption of fumigant on soil particles. Soils at or dryer than the wilting point (–15 bars) have few or no water molecules surrounding the soil particles, and adsorption of the fumigant molecules directly to the particle surface is increased. In moist soils, adsorption is reduced. In general, acceptable toxicant movement is best in moist soils with a water potential of –0.6 to –15 bar. Soils wetter than –0.6 bars have a number of air passageways blocked by water, and movement of the fumigant is decreased and diluted in the soil solution (51).

DIFFUSION INTO THE ATMOSPHERE   The loss of soil fumigants to the atmosphere should be considered in any material balance profile. Upward diffusion is greater than downward diffusion because the gas molecules diffuse more rapidly into the air above the soil surface than within the soil atmosphere. Polyethylene tarps of 25–100 μm thicknesses are used for MB and chloropicrin fumigations to slow the loss of fumigants to the atmosphere. Compaction of the soil surface or the application of water seals are used with the less volatile fumigants such as 1,3-D and EDB. McKenry & Thomason (51) have estimated the loss of cis-1,3-D to the atmosphere after a commercial application at a depth of 0.3 m in a warm, moist, sandy loam soil to be from 5–10%.

Little published information is available on the atmospheric dispersal, atmospheric chemistry, and ultimate dry or wet deposition of fumigants from the atmosphere into ecosystems. There is little doubt, however, that the fumigant molecules, or some degradation product of them, ultimately return to the soil, vegetation, or surface-water systems of the earth. For this reason it is important that more be learned through research about these environmental aspects of the use of fumigants (E. B. Cowling, personal communication).

Physical Factors

The various physical factors that influence the success or failure of fumigants in soil are so interrelated that it is difficult to measure their separate inputs. Even so, we can consider the factors separately, realizing that they are not acting singly in the absence of the other factors.

TEMPERATURE   Usually, fumigant effectiveness increases with increase in temperature. Fumigants such as MB are not very effective in soils at temperatures below 10°C, but the relationships were implied in the fungicide literature and only a few quantitative reports have been published. Gandy & Chanter (19) found that CT (concentration X time) for lethal doses for four fungi decreased with increased temperature in a fumigation chamber. *Agaricus bisporus*, the slowest growing fungus of those tested, was most sensitive to MB, whereas *Trichoderma koningi*, the fastest growing fungus, was least sensitive to MB. Munnecke & Bricker (59) used a system designed by Kolbezen & Abu-El-Haj (38) whereby a mixture containing 25,000 μl MB per liter of air was continuously circulated over discs of mycelial colonies of *Pythium ultimum* held for varying times at 5 temperatures between 5 and 30°C, inclusive. The hours required to kill 90% of the propagules at each temperature were linearly correlated with the temperature. It required approximately 3.8 times more MB to kill 90% of the fungus propagules at 5°C than at 30°C. This relationship is similar to one reported by Kenaga (36), who found that it took approximately 3.5 times more MB to kill the confused flour beetle at 5°C than at 30°C.

McKenry & Thomason (51) showed that when the temperature was increased from 5°C to 25°C, there was an 8-fold increase in toxicity of EDB, 4.4-fold increase for *cis* 1,3-D, and 4.7-fold increase for *trans* 1,3-D to *Meloidogyne javanica*. Since Marks, Thomason & Castro (47) showed that permeation of EDB was unaffected by temperature, it appears that the increased toxicity of EDB with rising temperatures is related to increased metabolic activity of the nematode. In the case of metham-sodium, Ashley, Leigh & Lloyd (5) reported that 25–50% greater concentration of MIT was released from soil incubated at 15°C than at 10°C.

The fact that soil fumigants are so much more toxic at higher temperatures is perhaps evidence to explain the success of field fumigations in soils of the warmer agricultural areas in the world. This principle applies particularly to organisms found in the upper 0.3 m of soil.

Soil fumigations are not always enhanced in warmer soil as compared to cold soils. Altman & Fitzgerald (4) found that they could increase yields of sugar beets in fields infested with sugar beet nematodes 2- to 3-fold by fumigating cold soils in the fall months with D-D. No direct comparisons were made with warm soil fumigations, however. Good & Rankin (23) used

SMDC (vapam), DMTT (mylone), 1,3-D, and MIT in winter fumigations in Georgia and got similar results, concluding that best results were obtained when soil temperatures were below 16°C, but above 0°C. The reasons for these successes were not known. It is possible that it could be due to seasonal differences in the population dynamics or life stages of beneficial or injurious organisms rather than just to the physics of fumigant dispersal. However, the exposure time in each case was several months and the target pests were nematodes in the upper 50 cm of soil, which are generally more susceptible to fumigants than fungi and bacteria (10). The half-life of 1,3-D, *cis* and *trans* in buffer solution was 90–100 days at 2°C and 11–13 days at 15°C (74). The toxic degradation product of 1,3-D, chloroallyl alcohol, had a corresponding extension at low soil temperatures, 142 days at 2°C, and 43 days at 15°C. Hydrolysis of 1,3-D may be slowed in cool soils if they are dry.

MOISTURE   Soil moisture is a very important factor governing success or failure of soil fumigation practices. We described how it controls the fate of fumigants in their distribution through the soil mass in the field; gas passes readily through a dry soil, but not through a water-saturated soil. The effect of moisture on the organisms is not so obvious. It is generally accepted that gaseous compounds such as MB are most effective in killing organisms when they are moist, but not saturated with water. Munnecke, Ludwig & Sampson (63) showed that there was a relationship between the relative humidity of the mixture of MB and air and the susceptibility of *Alternaria solani* spores to the toxicant. At low relative humidity (RH), the spores were virtually unaffected by the treatment; at 75–90% RH, however, optimum kill was obtained; and at near saturation, slightly less kill was obtained. No conclusive evidence was obtained but there were indications that kill was related to increased uptake, and presumably penetration, of MB into the spores at the high but not at the very low humidity conditions.

Similar responses have been shown by D. Freckman et al (unpublished) with nematodes surviving under dry conditions in the state of anhydrobiosis, wherein the worms were ten times more resistant to MB than when in active moist conditions. Presumably a water film surrounding the nematode is essential for the toxicant to be dissolved in and transported into the metabolically active nematode.

The effect of soil moisture on fumigation of soils with MB was studied by Munnecke et al (64), and a portion of the work has been adapted herein as Table 2. Most effective control of *Pythium ultimum* was obtained in moderately moist soil (12% water), next in very wet soil (37% water), and least in very dry soil (2% water). With *Rhizoctonia solani,* best control of damping-off was obtained in moderately moist soil (10% water). In contrast

Table 2 Effect of soil moisture content, concentration of methyl bromide, and duration of exposure to the gas on damping-off of peas[a]

Methyl bromide μl/liter	Days	Pythium ultimum, 21° C Moisture percentage			Rhizoctonia solani, 24° C Moisture percentage		
		2	12	37	2	10	37
600	1	85	95	70	95	98	93
	8	80	40	85	50	55	90
1,200	1	85	73	80	94	80	93
	8	75	0	0	15	12	45
2,300	1	70	12	82	80	83	80
	8	45	0	0	8	0	8

[a] Adapted from Figure 2 of Munnecke et al (64).

to results with *P. ultimum,* control of damping-off due to *R. solani* in very wet soil and in very dry soil was similar and only slightly poorer than that obtained in moderately moist soil.

McKenry & Thomason (51) clearly demonstrated the effect of soil moisture content of a silty clay loam on the diffusion pattern of 1,3-D. At a moisture content of 23% (dry-weight basis), 1,3-D did not reach a depth of 0.46 m in 20 days, while the 1,3-D went beyond 1.2 m at a moisture content of 7.7%. Unfortunately, in field situations, soil moisture usually increases with soil depth, impeding movement, and increasing the difficulty of control of soil pests on deep rooted perennials.

Soils, particularly below 0.5 m in depth, may be nearly saturated with water, especially if they are fine-textured; in such cases they are impenetrable. This was illustrated by Kolbezen et al (40). In the "wet" plots of Moreno sandy clay loam, the soil was kept free of all plants during the dry summer months, whereas in the "dry" plots, a planting of sudan grass was used to withdraw moisture from the soil. Some of the physical characteristics of the soils are reproduced from their paper in Table 3. Methyl bromide (195 g per m²) was applied into the air over the soil, covered, sealed carefully, and shaded from the sun. Samples were taken from the air at the surface and from soil atmospheres at 0.9, 1.8, and 2.7 m deep from each of four stations for periods up to 50 days after application of the gas, using techniques developed by Kolbezen & Abu-El-Haj (39).

Plots of the log concentration attained versus days after application showed that much better penetration of MB occurred in the dry soil compared to the wet soil. Most striking was the cumulative dosage (CT: μl per ml × days) at the various depths attained in 50 days. In the dry plots, the CT were 370 K (K = 1000), 240 K, and 125 K at depths of 0.9, 1.8, and

Table 3 Effect of moisture on some characteristics of a sandy clay loam soil[a]

Character	Condition of soil	Depth in soil (m)					
		0.3	0.6	0.9	1.5	1.8	2.1
Percentage of particle size	dry	52	46	70	60	54	47
smaller than 100μ	wet	52	46	70	60	54	47
Percentage of open pore space	dry	25	24	35	32	28	—
	wet	28	28	23	27	27	—
Volume percentage of water	dry	19	17	16	15	16	16
	wet	16	27	28	21	19	23

[a] From Kolbezen et al (40).

2.7 m, while in the wet plot they were 350 K, 135 K, and 45 K, respectively. They related the CT values to the physical factors shown in Table 3.

In another experiment with this soil, they demonstrated that when soil, initially so wet that it allowed only poor diffusion of MB, is dried there is a transition range in moisture wherein diffusion increases rapidly with small decreases in moisture. The CT increased from 80 K to 240 K when pore volume increased from 19.8 to 30% of the soil volumes. CTs for the other two plots with pore volumes intermediate between these extremes fell along a smooth curve that rose sharply as the pore volume approached 30%. Abdalla et al (1) studied distribution of MB in fields in which MB was applied with commercial rigs using chisels set 0.76–0.81 m deep, and 1.68 m apart. They found that sudan grass planting effectively dried a sandy clay loam soil down to depths of 2.44 m (fallow, 13.6%; sudan planting, 6.6% moisture). Much greater concentrations of gas were attained at the 1.83 and 2.44 m depths in the dried soil as compared to the wet fallow soil. The beneficial effect of drying has been equally apparent in sandy loam, sandy clay loam, and silt loam soils. In general, for any given soil texture, if the soil is dry, there is usually sufficient open pore space for the small molecule of the fumigant to pass through without excessive impediment of diffusion.

SOIL TEXTURE AND COMPOSITION   Soil texture is often directly related to the success or failure of a field fumigation. It is difficult to study the effect of soil texture on fumigant behavior in the field because other conditions, such as moisture content, temperature, variations in soil profile, and organic matter content all vary so much that it is virtually impossible to get two fields for comparison that differ only in their soil texture. Generally, coarse-textured soils are easy to fumigate successfully, whereas fine-textured soils, such as the clay loams and clays, are more difficult to fumigate successfully. In laboratory studies, it has been shown that the mineral content of the soils has no important effect on the movement and diffusion of fumigants (26).

The clay fraction does have some sorptive properties especially when dry, but the majority of the sorption that occurs in fumigated soils is attributed to the organic matter in the soils (26, 51). Of course, much, if not the majority, of the fumigant taken up by soils is by the soil water fraction.

The soil texture of many agricultural soils varies with depth. Agricultural soils may have clay lenses at various depths, or they may be compacted below the surface. Either of these factors seriously affects downward diffusion of soil fumigants. McKenry & Thomason (52, 53) demonstrated that very little MB passed through a compacted layer at 1.8 m below the surface of a loam soil, and a silt layer at 1.5 m in sandy loam soil. McKenry et al (54) also showed lack of penetration of 1,3-D into a silty clay layer at 0.9 m below the surface. These studies stimulated interest in new techniques for application of fumigants by modifying the soil profile. These techniques include slip plough tillage and back-hoe site preparation.

McKenry (48) has attempted to summarize the relationship of soil texture, percentage of moisture, and fumigant dosage to the killing of nematodes and other pathogens. He has studied many soils varying from sand to peat that were considered "too dry" or "too wet" to fumigate. The moisture levels for "too dry" ranged from approximately 2% for sand to 30% for peat. "Too wet" conditions vary from 6% for sand to over 40% for peat. In the ranges between "too dry" and "too wet," concentration of fumigant is estimated depending upon the moisture content of soil. The largest variation occurs with clay soils wherein safe moisture contents vary from 15–35%. Soil fumigation is generally not recommended in clay soils.

Biological Factors

Microbial transformation and degradation of pesticides commonly occurs in soil (34).

BIODEHALOGENATION   Biodehalogenation by soil organisms has been demonstrated for 1,3-D, EDB, and DBCP (11, 14, 16). The fumigant 1,3-D appeared to be chemically hydrolyzed to 3-chloroallyl alcohol and then converted to 3-chloroacrylic acid. The chlorine is removed and the intermediate products are converted to carbon dioxide and water. The rate of disappearance of 1,3-D at 15–20°C in sandy soil is 2–3.5% per day, and in clay soils up to 25% per day. The chloroallyl alcohol disappeared at rates of 20–60% per day at 15°C (74). In the laboratory, DBCP is converted to n-propanol, chloride, and bromide, and EDB is converted to ethylene and bromide. In contrast to 1,3-D, EDB and DBCP do not readily disappear from the soil (16, 32).

Altman (2) found that at least four common soil bacteria were able to use 1,3-D at concentrations up to 100 μl/liter as energy sources with a resulting increase in amino acid production.

In general, when fumigants are applied to soil at rates necessary for control of fungi and nematodes, there is a period of sterilization or partial sterilization of the fumigated zone. This reaction in itself releases nutrients from the killed biomass. The surviving or reinvading flora use the newly released nutrients and fumigant by-products for growth. As much as 10 kg N/ha is released from soil by fumigation (41). There is also an inhibition of the nitrification process, and depending on the season of the year, this may lead to increased or decreased plant responses. The production of ethylene after EDB fumigation may also contribute to examples of increased plant growth beyond the control of soil pests. MIT may react with amino acids in killed microorganisms and decaying plant material to form glucosides which are slowly disappearing phytotoxic residues.

LIVING AND DEAD ORGANIC CONSTITUENTS   Another major biological factor contributing to the activity of fumigants in soil are the live and dead plant debris remaining in the soil. It is well known that reliable fumigation of the soil is difficult to obtain when organic matter is abundant. Goring (24) demonstrated that it took from two to ten times as much EDB and three times as much 1,3-D to kill root-knot nematodes when residues of alfalfa, cotton, peat, wheat, barley, tobacco, and sugar beet crops were added to the soil. McKenry et al (55) have reported that it takes at least eight times more 1,3-D to kill root-knot nematodes within grape and fig roots than to kill the same nematode species in the soil; this suggests that large quantities of the chemical are absorbed by the roots before reaching the nematode. Raski et al (67, 68) suggested massive applications of 1,3-D and MB to kill root-knot nematode and grape fan leaf virus-infected roots to 4 m in the soil. Baines et al (8) have suggested similar problems with the control of citrus nematode in replant situations.

Material Balance

McKenry & Thomason (51) were the first to attempt a material balance for EDB and 1,3-D fumigations. Most of the toxicant was accounted for, but that which was not accounted for principally was represented as toxicant irreversibly adsorbed or as toxicant lost during sampling. The material not accounted for represented less than 10% of the 1,3-D and 10–40% of the EDB. After three days at 15°C, approximately 50% of the 1,3-D and about 40% of the EDB was adsorbed in the soil-particle phase; 30% and 25% respectively, in the soil-water phase; and about 15% of the 1,3-D was hydrolyzed, whereas 20% of the EDB was calcualted to have remained in the liquid state. By ten days, approximately 70% of the 1,3-D was hydrolyzed, and about 40% of the EDB was unaccounted for.

ACTIVITY OF FUMIGANTS IN ORGANISMS

To be effective, a fumigant must remain in contact with the target organisms for sufficient time and in sufficient concentration to kill. Only recently have scientists attempted to measure and obtain accurate and reliable dosage-response data for soil fumigants (49, 51–54, 60) based on chemical analysis as well as biological assay. This sophistication provides a strong base for developing predictable soil fumigations and thus obtaining maximum control of the target pest with a minimum of chemical. There is a strong case for measuring the effect of the toxicant as a function of both time (T) and concentration (C). One of the major difficulties in measuring CT for fumigants is that as they disperse through the soil, the sums of CT products decrease with increase from the point of chemical injection. Not only are there considerable differences between organisms in their response to CT, but there are differences between stages of the same organism (24, 76). Recently, McKenry & Thomason (51) compared the concentration of 1,3-D required to kill some selected organisms in soil at temperatures in excess of 15°C. It took 7–8 times more toxicant to kill all stages of the root-knot nematode within 1.25 cm diameter roots as larvae in soil, 2 times as much to kill *Xiphinena index* all stages, 16 times as much to kill oak root fungus–infected citrus roots 3 cm in diameter, and only 0.7 times as much to kill red fishing worms. Unfortunately, most of the fumigation data reported in the literature cannot be transformed and adequately compared because of deficiencies in data on chemical concentration at the site of the target pest and because of the lack of accurate mortality data.

Response of Soil Fungal Pathogens

Many investigators have attempted to formulate equations to show that the effect of a toxicant is a function of CT, and this was elaborated recently by Munnecke, Bricker & Kolbezen (60). They determined the inherent toxicity of MB to 10 soilborne pathogenic fungi. The exposure required to kill 90% (LD_{90}) of the propagules at a constant concentration was obtained by plotting standard dosage response curves (probit of response versus log time of exposure). When such LD_{90} values for a series of concentrations were plotted using log C versus log T as coordinates, the resulting plot was linear and the positions and slopes of the curves were indicative of the response to MB. Using these curves, it was possible to obtain CT values predicted to be sufficient to kill 90% of the population of a given fungus.

The slopes and positions of the curves differed with each of the fungi. For example, sclerotia of *Sclerotinia rolfsii* and *Whetzelinia sclerotiorum* were much more resistant to MB when treated with low concentrations of the gas for long periods than when treated with high concentrations for short periods of time. With *S. rolfsii* it took 1335 CT units (μl/liter)·h·10^{-3} to obtain LD$_{90}$ when exposed to 5000 μl/liter, but only 225 CT units when treated with 30,000 μl/liter. Fortunately, not all of the fungi were so variable in their response to MB; with *P. cinnamomi* in infected avocado rootlets, the CT values were approximately 150 when exposed to 5000 μl/liter and 110 when treated with 30,000 μl/liter. The difficulties in comparing the relative response of the fungi because of the variations in the slopes were partially adjusted by calculations based on a 27-hr time interval occurring between 3 and 30 hr from the start of fumigation. The average value was calculated for each fungus; these values are useful in predicting the lethal dose in a fumigation as well as in providing a basis of comparison for the susceptibility to MB of the various fungi. These relationships are presented in Table 4, which was adapted from Figure 3 in reference (60).

The members of the Phycomycetes tested (*Phytophthora cinnamomi, P. citrophthora, P. parasitica,* and *Pythium ultimum*) were the most sensitive to MB. *Fusarium oxysporum, Sclerotium rolfsii, Verticillium albo-atrum,* and *Whetzelinia sclerotiorum* were much more resistant to MB. Such relationships have been known (61) for a long time from practical experiences; that is, *Pythium* spp., *Rhizoctonia* spp., and the Phycomycetes in general, are relatively easy to control with MB, whereas *Fusarium* spp. and sclerotial-forming fungi are much more difficult to control. In fact, it is because of this resistance to MB that mixtures of CP and MB were formulated, since

Table 4 Average concentration per hour of methyl bromide required between 3 and 30 hr to kill 90% of the propagules of 10 fungi[a]

Fungus and propagules exposed	Average conc/hr required for LD$_{90}$ between 3 and 30 hr (10^{-3})
Phytophthora cinnamomi (chlamydospores)	8.0
P. cinnamomi (roots)	11.2
P. cinnamomi (mycelium)	11.7
P. citrophthora (mycelium)	12.1
P. parasitica (roots)	12.3
Pythium ultimum (mycelium)	13.3
Phytophthora parasitica (mycelium)	13.6
Armillaria mellea (mycelium)	17.1
Sclerotium rolfsii (mycelium)	17.4
Rhizoctonia solani (mycelium)	17.9
Whetzelinia sclerotiorum (mycelium)	22.2
S. rolfsii (sclerotia)	23.0
Verticillium albo-atrum (mycelium)	24.5
Fusarium oxysporum (mycelium)	26.1
W. sclerotiorum (sclerotia)	27.6
V. albo-atrum (infested stems)	28.6

[a] Adapted from Figure 3, Munnecke et al (60).

the fungi are more susceptible to CP, as well as to mixtures of the two gases (78) than to MB alone.

CT values are useful in evaluating field fumigations. Actual concentrations attained by a field fumigation may be measured accurately, and a plot of the concentration attained with time after application may be made, thus giving a plot of total dosage attained in the field for a particular fumigation. From this, the total CT attained may be calculated by integrating the concentration-time plot.

Unfortunately, the direct equating of such laboratory results to field applications is more difficult that it would first appear. First, a propagule of a fungus in the field is exposed to continuously varying concentrations of a fumigant. Usually the concentration around the particle rises rapidly from zero to a maximum and then falls off to zero at a slower rate (40). Second, the pattern of distribution in a soil profile varies with the depth of the soil and the method of application of the gas; there are few positions in a soil that are exposed to the same CT products. Usually, the upper 1 m shows the distribution pattern described above. As the depth increases, the curves of concentration attained versus time after application are apt to be more symmetrical with the rate of increase in concentration being approximately equal to the rate of decrease after maximum concentration is attained. Usually the maximum concentration attained is much lower than those attained at a position near the surface. Third, some fungi are more susceptible to high than to low concentrations of MB. Fourth, field conditions are impossible to duplicate in the laboratory. It is significant however, that *P. cinnamomi* responded similarly to MB fumigation in laboratory experiments whether the fungus was exposed as infestations of avocado rootlets, as chlamydospores, or as mycelial plugs.

In spite of the vagaries of CT measurements and evaluations as outlined above, it is possible to step integrate the CT values obtained in a field fumigation. By comparing total CT values attained in the field with the CT values shown by experiments to be necessary to kill 90% of the propagules of a fungus, it is possible to estimate the effectiveness of a soil fumigation.

Another complication in attempting to directly compare killing effectiveness with CT values is that it is possible that secondary biological control mechanisms may enhance a field fumigation. This has been shown by experiments with control of *Armillaria mellea*. Ohr & Munnecke (66) confirmed earlier reports by Bliss (13) and Garrett (20) and unified the relationships that exist when roots infected with *A. mellea* are fumigated with MB or carbon disulfide or treated with steam-air mixtures at sublethal treatments and held in nonsterile field soil for various times thereafter (65). They showed that correlations exist between sublethal treatments, time in contact with the nonsterile soil after fumigation, presence of *Trichoderma* spp., and death of the *A. mellea*. The *A. mellea* response may be a special case, but it is possible that other systems may operate in a similar fashion. (In unpublished work, D. Munnecke and J. Bricker have been unable to show a similar response operating in a system involving *P. cinnamomi* and avocado roots.)

Wilhelm et al (78) have shown that mixtures of CP and MB give better results in the field than when either is used alone for control of *Verticillium* sp. Field experience of farmers has confirmed this; better results usually are obtained with the mixtures than with either of the compounds alone. The mechanisms responsible for this interaction remain to be worked out. It is known that MB penetrates tissues and galls more easily than CP and also that CP is a more effective fungicide than MB. How the two compounds interact to make a more effective fumigant remains an intriguing unanswered question.

Response of Mycorrhizal Fungi to Fumigants

Often plants, especially if started directly from seeds, growing in fumigated soil are stunted and off-color. Frequently, these disorders can be overcome by large additions of phosphate or by inoculation with mycorrhizal fungi (21). The decreased growth response (DGR) is directly correlated with the absence of mycorrhizal fungi on the roots of affected plants. These direct effects of MB on mycorrhizal fungi have been reported by Menge et al (56). For example, *Glomus fasciculatus* and *Glomus constrictus* were both more sensitive to MB than all of the soilborne plant pathogenic fungi previously tested. Both species were approximately twice as sensitive to MB as *P. cinnamomi* and *P. parasitica,* about four times more sensitive to MB than *V. albo-atrum,* and about nine times more sensitive to MB than *S. rolfsii.* Menge et al concluded that it is unlikely that MB dosage could be reduced sufficiently to allow survival of mycorrhizal fungi without at the same time allowing survival of pathogenic fungi.

Response of Nematodes

Nematodes, in general, are much more sensitive to fumigants than are plant pathogenic fungi and bacteria (9, 10, 27, 48). For this reason, more fumigants are useful for control of nematodes than for control of fungi and bacteria; all of the compounds listed in Table 1 are nematicidal, but only MB, CP, and MIT are also fungicidal at the usual field doses. Historically, this difference in relative susceptibility to fumigants has been used as a major criterion for proving that nematodes were important pests on crop plants.

The mode of action of soil fumigants in nematodes has already been reviewed (17), and their resistance to alkyl halides has been discussed (75). The concept of dose as a cumulative product of concentration and time is applicable to nematodes except that long exposures are required to kill nematodes with EDB and DBCP. These chemicals reach equilibrium rather rapidly, but irreversible intoxification leading to death takes considerably longer than with the other fumigants (76).

Seinhorst (72) was one of the first to review the nematicide literature and suggest a linear relationship between log dosage of 1,3-D injected into the soil and probit mortality of plant-parasitic nematodes. In most cases, he found that probit mortality increased by about half a unit per doubling of the dosage of the fumigant. McKenry & Thomason (51) made an extensive study of organism-dosage-response to MB, EDB, and 1,3-D. They found that MB was 1.4 times as toxic to root-knot nematode juveniles and 6–7 times as toxic to *A. mellea* as 1,3-D. MB was approximately 6.1 times as toxic to root-knot nematode juveniles as EDB. They also demonstrated that the *cis*-1,3-D was more toxic to nematodes than the *trans*-1,3-D. Further comparisons of the CT in the soil-vapor phase and the soil-water phase for several nematodes exposed to *cis*-1,3-D and EDB are given in Table 5. Van Gundy et al (76) compared the toxicity of MB to *M. incognita* larvae with *X. index* larvae and adults, and with a *Dorylaimus* sp. The CT for a LD_{95} of MB in the air was 18,000, 15,000, 21,600, and 19,200 μl/liter per hr respectively.

Plant Growth Responses

Plants usually grow better in fumigated as compared to nonfumigated soils (3). In an elegant review paper Wilhelm & Nelson (77) have discussed this increased growth response (IGR); thus only brief mention of the phenomenon is made here. A number of factors are responsible for IGR: elimination of obvious plant pathogens; elimination of less-than-obvious pathogens, such as *P. ultimum;* elimination of weeds and soil insects; and alteration of nitrogen content in the soil. Fumigated soils are uniformly higher in NH_4-N content, since the nitrifying bacteria are easily eliminated by fu-

Table 5 Dosage-response data for various nematodes and toxicants at the $LD_{99.99\%}$ level at $25°$ C[a]

| Nematode | $\mu g\ ml^{-1}$ day | | | |
| | *cis*-1,3-D | | EDB | |
	soil vapor	soil water	soil vapor	soil water
Aphelenchus avenae	2.3	39	29.5	1,000
Meloidogyne javanica	1.3	23	1.1	39
Eucephalobus sp.	1.7	29		
Heterodera schachtii				
juveniles	1.4	24		
white cyst	2.6	44		
brown cyst	6.6	112	23	1,150

[a] From McKenry & Thomason (51).

migation. Some ammonifying organisms survive fumigation so that organic $-N$ is converted to NH_4-N which is less easily leached from soil than NO_3-N. This reaction may work to a disadvantage in cool damp soils causing ammonia toxicity to some crops.

Occasionally, plants growing in fumigated soils grow poorly, and may even die. This usually is due either to the elimination of mycorrhizae, or to the presence of toxic residues of the fumigant. Mycorrhizal fungi are very susceptible to MB and they usually are eliminated from fumigated soils. The problem is acute when susceptible plants are started as seeds in freshly fumigated (or steamed) soil. This has been shown to occur when citrus seeds are planted in sandy soils fumigated previously with MB (37). The seedlings fail to become established or they remain chlorotic, and their growth is stunted. The situation can be alleviated by addition of very large amounts of phosphate fertilizer, or better yet, by inoculation with propagules of *G. fasciculatus,* a mycorrhizal fungus (70, 71). If citrus plants are transplanted into fumigated soil, increased growth may occur (58), presumably because the mycorrhizal fungi had a chance to become established on the roots prior to transplanting. This is an exciting area of current research, and much is being done to establish mycorrhizal fungi in fumigated soils as a means of increasing crop production.

A number of crops have been reported as being adversely affected when planted in soil treated with MB, presumably due to bromide residues (22, 35, 79). For many years in California, bromide-containing compounds such as MB or EDB were not recommended for use on soil to be planted with crops such as carnations, onions, and sugar beets because of this toxicity. Recently, it was found that the toxic nature of fumigated soil could be alleviated by several leachings with water before planting carnation cuttings (12, 45, 46). CP is not injurious to crops, providing sufficient aeration ensues before planting.

CURRENT CONCEPTS IN THE APPLICATION OF FUMIGANTS

Current application strategies are generally directed at four different field situations: (*a*) loose thoroughly mixed potting soil in pots or containers, (*b*) shallow rooted annual crops, (*c*) shallow rooted perennial nursery and specialty crops, and (*d*) deep rooted tree and vine crops. Each has its own specifications for control of soil pests. For example, potting soil usually is high in organic amendments and sorption of the fumigant will be high; however, moisture and porosity can be maintained near optimum levels. In shallow rooted annual crops, the objective is to achieve sufficient pest and pathogen control to achieve improved plant growth based on most economic return for cost of chemical used. The shallow rooted perennial nursery and specialty crops usually return high cash values, and fumigations are aimed at a high degree of weed, pest, and pathogen control. The deep rooted tree and vine crops are difficult to fumigate and achieve control of pests and pathogens. In situations where replanting is attempted there are often residual roots to 4 m which may harbor viruses, endoparasitic nematodes, or root pathogens. These roots die very slowly and often remain alive for five years. In addition to the depth of soil required to fumigate these residual roots, they are 8–16 times more resistant to fumigants than the pest or pathogen residing in them. Thus, massive doses of soil fumigants are often injected deeply in the soil profile. In many instances, an additional shallow injection is needed to kill organisms in the surface layers.

There are a variety of practical methods that can be used by pest control advisors to assess the degree of success achieved by soil fumigation. McKenry & Naylor (50) have described some of these methods, their shortcomings, and degree of sophistication.

Movement in Disturbed Versus Undisturbed Soil

In single point applications of MB in a field, usually the gas will diffuse farther in a uniform undisturbed soil than in a disturbed or cultivated soil. This behavior is useful in situations occurring when a single tree or small number of trees become diseased, and it is desired to replant the area. This may be accomplished by injecting the chemical in rows spaced in a staggered pattern. As a general rule, using an undisturbed loamy or sandy soil that is optimum for fumigation, an injection of 0.68 kg of MB will provide fungicidal concentration in an area 3.65 m in diameter. Gas movement in such a case is usually better than if the whole area is worked up and then injected.

The depth at which fumigants are applied affects gaseous distribution in undisturbed soils. Kolbezen et al (40) compared diffusion patterns of MB following application at two levels. Application in holes 0.6 m deep resulted in higher CT at the 0.15 m and 0.93 m levels, while the 1.5 m deep treatment resulted in higher CT at the 2.75 m depth. The CT at 1.83 m were almost identical. In practice, the optimal positioning is to apply MB between 0.75–0.95 m.

Most field applications involve disturbing the soil to various depths prior to fumigating. For annual crops, soil is prepared to seed-bed condition. That is, roots and stems of the previous crop are removed, chopped by cultivating equipment, and allowed to decay before the large clods of soil and compaction layers are broken by cultivation. As discussed above, the soil should be neither too wet nor too dry.

For perennial crops very deep penetration is desired, so "seedbed condition" is usually too moist because the soil at the lower depths will be too wet. In California, where *Armillaria mellea* (40), *Tylenchulus semipene-*

trans (10), *Meloidogyne* spp. (68), and grape fan leaf virus (67) are preplant problems on deeply buried roots, the soil is dried during the hot summer and early fall months and fumigated as late as possible before the winter rains begin. Sometimes sudan grass, safflower, or barley is planted to speed the drying process. In such cases the soil in the upper 0.5 m may be powdery dry and if the pathogen being attacked is present in that area, very poor control will result. This is not the case with *A. mellea* if the large roots and stems of infected trees are removed from the field prior to fumigating. The fungus cannot live saprophytically in the soil, and infestations of small root pieces left behind in the gleaning operations are unlikely to initiate infections. Then, deep (0.75 m) injections of MB may be used. If the target pest is nematodes in perennial crops, it is imperative both that the deeper soil be dry, and that the surface soil be moistened for optimum fumigations with 1,3-D or MB. If tractor-drawn machines are to be used, the soil must be worked prior to fumigating. For deep application where the chisels are set 1.5 m apart and the gas is injected at 0.93 m deep, the soil must be ripped previously to a depth of 0.9 m.

Recently, single tree sites have been prepared with the use of a back-hoe machine. Large roots and trunk of a diseased tree are removed and the remaining soil in the hole is removed to a depth of 1.2–1.5 m. One or more cans (0.68 kg) of MB, or 500 cc 1,3-D is placed in the hole and the hole immediately covered with soil by the machine.

THE USE OF COVERS OR SEALS   Polyethylene covers are typically 25 μm thick when used in machine applications. Thicker covers are used for bulk soils or spot fumigations. The films are quite permeable to the passage of MB and CP and several techniques have been devised to alleviate this condition. Initially, MB was formulated as a thixotropic gel which was designed (*a*) to slow the release of MB after it was applied to the soil and (*b*) to avoid the need for a cover. In some cases, it appeared that a more economical use of MB was obtained in this way, but usually the increase in cost, difficulty in applying, and erratic results (S. D. Van Gundy, unpublished) have negated any widespread usage.

The easiest way to increase gas retention is to use polyethylene sheeting of greater thickness. Grimm & Alexander (28) compared the effectiveness of MB in a sandy soil when enclosed by tarps of differing thickness. *Phytophthora* sp. was killed to 1.2 m deep in soil covered with a 100 μm tarp, but about half of the samples survived the same dose (49 g per m^2) applied under a 25 μm tarp. Using a mixture of 67% MB and 33% CP (49 g per m^2 total) under a 100 μm tarp, they got almost the same results as with a dose of 98 g under a 25 μm tarp.

In a commercial fumigation (58), plots were treated with a mixture of MB and CP (ratio of 2:1) at the rate of 336 kg/ha applied by shanks set 15 cm deep and 30 cm apart. Plots were covered with polyethylene tarps 25, 100, or 150 μm thick. The soil atmospheres were sampled and the CTs attained as shown in Table 6. The CT values increased significantly with increase in tarp thickness. Similar results were reported from other experiments.

Abdalla et al (1) reported that placement of MB in soil at 0.76–0.81 m deep without using polyethylene cover resulted in sufficient concentrations to kill nematodes as deep a 2.44 m. Unfortunately, their paper does not include data, with one exception, as to the concentration attained in the soil profile near the surface. In one experiment, they reported considerable concentration of MB above the 0.3 m level, but the data were obscured by rain. In three of their experiments where a cover also was used, significantly higher concentations of gas were attained at various positions in the soil. Others (62) have shown that covering the sites of injection with a polyethylene cover will enhance the fumigation considerably, even though the gas is released deeply in the soil. In one experiment of this type, in which 0.68 kg MB was injected 1 m deep, CT values at the 30 cm deep level in a citrus orchard soil were increased 2.36 times by tarping and 1.29 times at the 90 cm level.

Other materials much less permeable to MB have been used to confine MB, but to date, none are available commercially. The most promising are the so-called high density polyethylene films.

If an impermeable cheap film were available for use, much better fumigations could be made using less fumigant and presumably avoiding safety and residue problems.

Table 6 Effect of tarp thickness on sums of CT (μl per liter × hr) values of MB attained in the first 120 hr after a field was fumigated[a]

Polyethylene cover thickness (μm)	Depth in soil (m)	Sum of CT (thousands)	Increase (–fold)		
			100–vs– 25 μm	150–vs– 25 μm	150–vs– 100 μm
25	0.15	83			
	0.30	37			
	0.90	10			
	1.50	5			
100	0.15	145	1.8		
	0.30	61	1.7		
	0.90	14	1.5		
	1.50	5	1.1		
150	0.15	259		3.1	1.8
	0.30	96		2.6	1.6
	0.90	32		3.3	2.2
	1.50	10		2.1	2.0

[a] From Munnecke et al (62).

EXAMPLES OF FUMIGANT MOVEMENT IN THE FIELD   Field A (Figure 1) and Field B (Figure 2) of sandy loam soil were covered with polyethylene tarps 37 μm thick. A portion of Field A was ripped in three directions to a depth of 0.7 m and disked prior to fumigating with approximately 850 kg MB per ha using two chisels set 0.75 m deep and 1.65 m apart. Another portion of the orchard soil was not cultivated prior to fumigation, using the multiple injection method at 508 kg per ha. Holes 3 cm in diameter were made in the soil to 0.9 m, tubes were inserted to the bottom and sealed with soil and water, and MB was applied as a gas using 0.68 kg MB per hole. The holes were placed in rows in a staggered pattern on 3.6 m centers. In Field B, a typical strawberry preplant fumigation, the soil was cultivated into seed-bed condition and MB at the rate of 470 kg per ha was applied with chisels set at 0.2 m deep, 0.3 m apart.

With the shallow injection method, Figure 2, high concentrations of MB were attained in the soil profile at 0.3 m and above. Concentrations at 0.6 and 0.9 m depths were insignificant. Consequently, this type of fumiga-

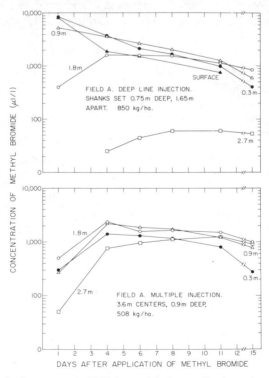

Figure 1   Distribution patterns of MB in a sandy loam soil: *above,* deep line injection by machine; *below*, multiple single-point injection. (Unpublished data of D. E. Munnecke and M. J. Kolbezen.)

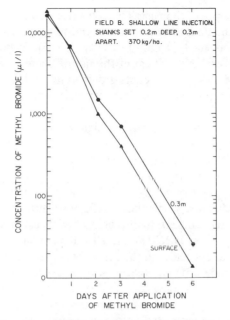

Figure 2 Distribution pattern of MB in a sandy loam soil: shallow line injection by machine. Concentrations attained at 0.6 m and 0.9 m were insignificant. (Unpublished data of D. E. Munnecke and M. J. Kolbezen.)

tion is satisfactory only for use with annual or short-lived crops, and is not recommended for land to be planted to perennial crops.

In contrast, when the line injection was made deeper, better penetration of the soil by MB resulted (Figure 1). Even so, the gas did not penetrate to the 2.7 m level with significant fungitoxic concentration. This method is satisfactory for use on an extensive scale on land to be planted to perennial crops. It is expensive, and the return may not justify the cost.

The multiple injection of undisturbed soil (Figure 1) provides the best distribution of MB throughout the whole soil profile. High concentrations of MB were obtained to 2.7 m deep. This method works well for areas where small areas need to be fumigated before replanting with another tree or vine. Efficacy can be increased by applying a small amount (~350 kg/ha) of MB as a gas beneath the tarp as a surface application.

THE FUTURE

It is difficult to determine what the future holds for soil fumigation. In many parts of the world, agriculture is becoming more intensive and mechanization and monoculture is more common than under the old, but excellent, crop rotation schemes of the past. The collective demands of intensive crop production have made it very profitable (and in some cases imperative) to fumigate soils. Conversely, trends of the last decade indicate that societies are becoming increasingly critical of the use of chemicals in food production. Bromine residues in crop plants grown in methyl bromide–treated soil have become an important issue, and very stringent regulations have been established.

The effects of dispersal of volatile fumigants in the atmosphere have not been studied adequately.

In many cases, the soil residue problems associated with soil fumigants may be greater than with compounds developed as foliage fungicides, for example. Development of foliage fungicides is already difficult for that reason. Also, fumigants, particularly MB, pose serious health hazards to operators and other workers in the field as well as to animals. There is serious question whether DBCP and EDB will be manufactured and distributed as agricultural pesticides in the future because of their health hazards. As more use is made of these toxicants, it is probable that carelessness in their use will ensue, and some serious illness or even fatalities may occur, as with the handling and use of other dangerous commodities such as flammable liquids and gases, radioactive materials, etc. Such accidents may be disproportionately emphasized in the public's mind, however, and the result could be that such products could be banned from use.

The costs of use of soil fumigants have risen greatly in recent years, and this will be another limiting factor in their use in the future.

The negative factors regarding usage of soil fumigants may be counterbalanced by the certainty that continuing higher production must be obtained from agricultural lands if man is to avoid the consequences of overpopulation. Soil fumigation is one way to make agriculture more productive. Also, it should be noted that thousands of applications of soil fumigants have been made without harm to operators or consumers of products grown on fumigated land. The increase in knowledge of fumigant behavior combined with the practical usage of such knowledge has led to an increased efficacy in usage and commensurate lowering of health hazards.

It appears to us that increasingly greater areas of land will be fumigated in the future. The advances in research probably will be made mostly by empirical testing of a few fumigants such as MB, CP, and 1,3-D in field situations to achieve more efficient usage and better timing of applications. Perhaps no new principles will be developed, but fundamental knowledge presently known will be exploited more fully.

Literature Cited

1. Abdalla, N., Raski, D. J., Lear, B., Schmitt, R. V. 1974. Distribution of methyl bromide in soils treated for nematode control in replant vineyards. *Pest. Sci.* 5:259–69
2. Altman, J. 1969. Effect of chlorinated C_3 hydrocarbons on amino acid production by indigenous soil bacteria. *Phytopathology* 59:762–66
3. Altman, J. 1970. Increased and decreased plant growth response resulting from soil fumigation. In *Root Diseases and Soil-Borne Pathogens*, ed. T. A. Toussoun, R. V. Bega, P. E. Nelson, pp. 216–21. Berkeley: Univ. Calif. Press. 252 pp.
4. Altman, J., Fitzgerald, B. J. 1960. Late fall application of fumigants for the control of sugar beet nematodes, certain soil fungi, and weeds. *Plant Dis. Reptr.* 44:868–71
5. Ashley, M. G., Leigh, B. L., Lloyd, L. S. 1963. The action of metham-sodium in soil. II. Factors affecting the removal of methyl isothiocyanate residues. *J. Sci. Food Agric.* 14:153–61
6. Audus, L. J., ed. 1964. *The Physiology and Biochemistry of Herbicides.* London: Academic. 555 pp.
7. Audus, L. J., ed. 1976. *Herbicides, Physiology, Biochemistry, Ecology,* Vol. 2. London: Academic. 564 pp. 2nd ed.
8. Baines, R. C., Foote, F. J., Stolzy, L. H., Small, R. II., Garber, M. J. 1959. Factors influencing control of the citrus nematode in the field with D-D. *Hilgardia* 29:359–81
9. Baines, R. C., Klotz, L. J., DeWolfe, T. A. 1977. Some biocidal properties of 1,3-D and its degradation product. *Phytopathology* 67:936–40
10. Baines, R. C., Klotz, L. J., DeWolfe, T. A., Small, R. H., Turner, G. O. 1966. Nematocidal and fungicidal properties of some soil fumigants. *Phytopathology* 56:691–98
11. Belser, N. O., Castro, C. E. 1971. Biodehalogenation—the metabolism of the nematocides *cis*– and *trans*-3-chloroallyl alcohol by a bacterium isolated from soil. *J. Agric. Food Chem.* 19:23–26
12. Besemer, S. T., McCain, A. H. 1973. Carnation fusarium wilt. Control with soil fumigation and fungicides. *Calif. Agric.* (9):4–5
13. Bliss, D. E. 1951. The destruction of *Armillaria mellea* in citrus soils. *Phytopathology* 41:665–83
14. Castro, C. E., Bartnicki, E. W. 1968. Biodehalogenation. Epoxidation of halohydrins, epoxide opening, and transhalogenation by a *Flavobacterium* sp. *Biochemistry* 7:3213–18
15. Castro, C. E., Belser, N. O. 1966. Hydrolysis of *cis*– and *trans*-1,3-dichloropropene in wet soil. *J. Agric. Food Chem.* 14:69–70
16. Castro, C. E., Belser, N. O. 1968. Biodehalogenation. Reductive dehalogenation of the biocides ethylene dibromide, 1,2-dibromo-3-chloropropane, and 2,3-dibromobutane in soil. *Environ. Sci. Technol.* 2:779–83
17. Castro, C. E., Thomason, I. J. 1971. Mode of action of nematicides. In *Plant Parasitic Nematodes,* Vol. 2, ed. B. M. Zuckerman, W. F. Mai, R. A. Rohde, pp. 289–96. New York: Academic. 347 pp.
18. Crosby, D. G. 1976. Nonbiological degradation of herbicides in the soil. See Ref. 7, pp. 65–97
19. Gandy, D. G., Chanter, D. O. 1976. Some effects of time, temperature of treatment, and fumigant concentration on the fungicidal properties of methyl bromide. *Ann. Appl. Biol.* 82:279–90
20. Garrett, S. D. 1958. Inoculum potential as a factor limiting lethal action by *Trichoderma viride* Fr. on *Armillaria mellea* (Fr.) Quel. *Trans. Br. Mycol. Soc.* 41:157–64
21. Gerdemann, J. W. 1968. Vesicular-arbuscular mycorrhiza and plant growth. *Ann. Rev. Phytopathol.* 6:397–418
22. Gollop, Z. 1974. The problems of bromide residues after soil fumigation. *Agric. Environ.* 1:317–20
23. Good, J. M., Rankin, H. W. 1964. Evaluation of soil fumigants for control of nematodes, weeds, and soil fungi. *Plant Dis. Reptr.* 48:194–99
24. Goring, C. A. I. 1957. Factors influencing diffusion and nematode control by soil fumigants. *ACD Inf. Bull. 110* (Nov. 1957). Midland, Mich.: Dow Chemical Co. 53 pp.
25. Goring, C. A. I. 1962. Theory and principles of soil fumigation. *Adv. Pest Control Res.* 5:47–84
26. Goring, C. A. I. 1967. Physical aspects of soil in relation to the action of soil fungicides. *Ann. Rev. Phytopathol.* 5:285–318
27. Goring, C. A. I. 1972. Fumigants, fungicides, and nematicides. In *Organic Chemicals in the Soil Environment,* Vol. 2, ed. C. A. I. Goring, J. W. Hamaker, pp. 569–632. New York: Dekker. 968 pp.

28. Grimm, G. R., Alexander, A. F. 1971. Fumigation of *Phytophthora* in sandy soil by surface application of methyl bromide and methyl bromide-chloropicrin. *Plant Dis. Reptr.* 55:929–31

29. Grossbard, E. 1976. Effects on the soil microflora. See Ref. 7, pp. 99–148

30. Hartley, G. S. 1964. Physical factors and action through the soil. See Ref. 6, pp. 111–61

31. Hartley, G. S. 1976. Physical behaviour in the soil. See Ref. 7, pp. 1–28

32. Hodges, L. R., Lear, B. 1973. Efficacy of 1,2-dibromo-3-chloropropane for control of root knot nematode as influenced by concentration, exposure time, and rate of degradation. *J. Nematol.* 5:249–53

33. Johnson, D. E., Lear, B. 1969. The effect of temperature on the dispersion of 1,2-dibromo-3-chloropropane in soil. *J. Nematol.* 1:116–21

34. Kaufman, D. D., Kearney, P. C. 1976. Microbial transformations in the soil. See Ref. 7, pp. 29–64

35. Kempton, R. J., Maw, G. A. 1974. Soil fumigation with methyl bromide: The phytotoxicity of inorganic bromide to carnation plants. *Ann. Appl. Biol.* 76:217–29

36. Kenaga, E. E. 1961. Time, temperature and dosage relationships of several insecticidal fumigants. *J. Econ. Entomol.* 54:537–42

37. Kleinschmidt, G. D., Gerdemann, J. W. 1972. Stunting of citrus seedlings in fumigated nursery soils related to the absence of endomycorrhizae. *Phytopathology* 62:1447–53

38. Kolbezen, M. J., Abu-El-Haj, F. J. 1972. Fumigation with methyl bromide. I. Apparatus for controlled concentration, continuous flow laboratory procedures. *Pestic. Sci.* 3:67–71

39. Kolbezen, M. J., Abu-El-Haj, F. J. 1972. Fumigation with methyl bromide. II. Equipment and methods for sampling and analyzing deep field soil atmospheres. *Pestic. Sci.* 3:73–80

40. Kolbezen, M. J., Munnecke, D. E., Wilbur, W. D., Stolzy, L. H., Abu-El-Haj, F. J., Szuszkiewicz, T. E. 1974. Factors that affect deep penetration of field soils by methyl bromide. *Hilgardia* 42:465–92

41. Lebbink, G., Kolenbrander, G. J. 1974. Quantitative effect of fumigation with 1,3-dichloropropene mixtures and with metham sodium on the soil nitrogen status. *Agric. Environ.* 1:283–92

42. Leistra, M. 1970. Distribution of 1,3-dichloropropene over the phases in soil. *J. Agric. Food Chem.* 18:1124–26

43. Leistra, M. 1971. Diffusion of 1,3-dichloropropene from a plane source in soil. *Pestic. Sci.* 2:75–79

44. Leistra, M., Smelt, J. H., Nollen, H. M. 1974. Concentration-time relationships for methyl isothiocyanate in soil after injection of metham-sodium. *Pestic. Sci.* 5:409–17

45. Malkomes, H. P. 1971. Untersuchung verschiedener Einflusse auf die Zersetzung von Methylbromid (TERABOL) bie Begasung von Boden und Pflanzsubstraten. *Z. Pflanzenkr. Pflanzenschutz* 78:464–76

46. Malkomes, H. P. 1972. Der Einfluss von Bodenbegasungen mit Methylbromid (TERABOL) auf gartnerische Kulturpflanzen. I. Bromidaufnahme und Bromidtoleranz bei zierpflanzen *Z. Pflanzenkr. Pflanzenschutz* 79:274–90

47. Marks, C. F., Thomason, I. J., Castro, C. E. 1968. Dynamics of the permeation of nematodes by water, nematocides, and other substances. *Exp. Parasitol.* 22:321–37

48. McKenry, M. V. 1978. Selection of preplant fumigation. *Calif. Agric.* 32(1):15–16

49. McKenry, M. V., Hesse, C. O. 1978. Preplant fumigation of planting sites. *Calif. Agric.* 32(1):14–15

50. McKenry, M. V., Naylor, P. 1978. Practical methods of evaluating soil fumigation. *Calif. Agric.* 32(1):17

51. McKenry, M. V., Thomason, I. J. 1974. 1,3-Dichloropropene and 1,2-dibromoethane compounds: I. Movement and fate as affected by various conditions in several soils. II. Organism-dosage-response studies in the laboratory with several nematode species. *Hilgardia* 42:393–438

52. McKenry, M. V., Thomason, I. J. 1976. Dosage values obtained following preplant fumigation for perennials. I. 1,3-Dichloropropene nematicides in eleven field situations. *Pestic. Sci.* 7:521–34

53. McKenry, M. V., Thomason, I. J. 1976. Dosage values obtained following preplant fumigation for perennials. II. Using special methods of applying methyl bromide and 1,3-dichloropropene nematicides. *Pestic. Sci.* 7:535–44

54. McKenry, M. V., Thomason, I. J., Johnson, D. E., Neja, R., Swanson, F. 1978. The movement and toxicity of preplant soil fumigants for nematode control. Preplant fumigations with 1,3-D nematicides. *Calif. Agric.* 32(1):12–13

55. McKenry, M. V., Thomason, I. J., Naylor, P. 1977. Dosage-response of root-knot nematode-infected grape roots to cis-1,3-dichloropropene. *Phytopathology* 67:709–11

56. Menge, J. A., Munnecke, D. E., Johnson, E. L. V., Carnes, D. W. 1978. Dosage response of the vesicular-arbuscular mycorrhizal fungi *Glomus fasciculatus* and *G. constrictus* to methyl bromide. *Phytopathology* 68:1368–72

57. Munnecke, D. E. 1972. Factors affecting the efficacy of fungicides in soil. *Ann. Rev. Phytopathol.* 10:375–98

58. Munnecke, D. E. 1978. Methyl bromide for control of soilborne fungal pathogens. *Proc. Int. Soc. Citricult. 1977* III. In press

59. Munnecke, D. E., Bricker, J. L. 1978. Effect of temperature on response of *Pythium ultimum* to methyl bromide. *Plant Dis. Reptr.* 62:628–29

60. Munnecke, D. E., Bricker, J. L., Kolbezen, M. J. 1978. Comparative toxicity of gaseous methyl bromide to ten soilborne phytopathogenic fungi. *Phytopathology* 68:1210–16

61. Munnecke, D. E., Ferguson, J. 1953. Methyl bromide for nursery soil fumigation. *Phytopathology* 43:375–77

62. Munnecke, D. E., Kolbezen, J. J., Wilbur, W. D. 1977. Types and thickness of plastic films in relation to methyl bromide fumigation. *Proc. Int. Agric. Plastics Congr. 1977, San Diego, Calif.*, pp. 482–87

63. Munnecke, D. E., Ludwig, R. A., Sampson, R. E. 1959. The fungicidal activity of methyl bromide. *Can. J. Bot.* 37:51–58

64. Munnecke, D. E., Moore, B. J., Abu-El-Haj, F. 1971. Soil moisture effects on control of *Pythium ultimum* or *Rhizoctonia solani* with methyl bromide. *Phytopathology* 61:194–97

65. Ohr, H. D., Munnecke, D. E., Bricker, J. L. 1973. The interaction of *Armillaria mellea* and *Trichoderma* spp. as modified by methyl bromide. *Phytopathology* 63:965–73

66. Ohr, H. D., Munnecke, D. E. 1974. Effects of methyl bromide on antibiotic production by *Armillaria mellea*. *Trans. Br. Mycol. Soc.* 62:65–72

67. Raski, D. J., Hewitt, W. B., Schmitt, R. V. 1971. Controlling fan leaf virus-dagger nematode disease complex in vineyards by soil fumigation. *Calif. Agric.* 25(4):11–14

68. Raski, D. J., Jones, N. O., Kissler, J. J., Luvisi, D. A. 1976. Soil fumigation: One way to cleanse nematode-infested vineyard lands. *Calif. Agric.* 30(1):4–7

69. Rodriguez-Kabana, R., Backman, P. A., Curl, E. A. 1977. Control of seed and soilborne plant diseases. In *Antifungal Compounds*, Vol. I, ed. M. R. Siegel, H. D. Sisler, pp. 117–61. New York: Dekker. 600 pp.

70. Ross, J. P. 1971. Effect of phosphate fertilization on yield of mycorrhizal and nonmycorrhizal soybeans. *Phytopathology* 61:1400–3

71. Ross, J. P., Harper, J. A. 1970. Effect of endogone mycorrhiza on soybean yields. *Phytopathology* 60:1552–56

72. Seinhorst, J. W. 1973. Dosage of nematicidal fumigants and mortality of nematodes. *Neth. J. Plant Pathol.* 79:180–88

73. Smelt, J. H., Leistra, M. 1974. Conversion of metham-sodium to methyl isothiocyanate and basic data on the behaviour of methyl isothiocyanate in soil. *Pestic. Sci.* 5:401–7

74. Van Dijk, H. 1974. Degradation of 1,3-dichloropropenes in the soil. *Agro-Ecosystems* 1:193–204

75. Van Gundy, S. D., McKenry, M. V. 1977. Action of nematicides. In *Plant Disease, An Advanced Treatise. How Disease Is Managed*, Vol. I., ed. J. G. Horsfall, E. B. Cowling, pp. 263–83. New York: Academic. 465 pp.

76. Van Gundy, S. D., Munnecke, D. E., Bricker, J., Minteer, R. 1972. Response of *Meloidogyne incognita, Xiphinema index*, and *Dorylaimus* sp. to methyl bromide fumigation. *Phytopathology* 62:191–92

77. Wilhelm, S., Nelson, P. E. 1970. A concept of rootlet health of strawberries in pathogen-free soil achieved by fumigation. See Ref. 3, pp. 208–15

78. Wilhelm, S., Storkan, R. C., Wilhelm, J. M. 1974. Preplant soil fumigation with methyl bromide-chloropicrin mixtures for control of soil-borne diseases of strawberries—a summary of fifteen years of development. *Agric. Environ.* 1:227–36

79. Williamson, C. E. 1954. Soil-borne disease control with chemicals. *Down Earth* 10(3):6–8

Ann. Rev. Phytopathol. 1979. 17:431–60
Copyright © 1979 by Annual Reviews Inc. All rights reserved

WATER RELATIONS
OF WATER MOLDS

♦3718

J. M. Duniway

Department of Plant Pathology, University of California,
Davis, California 95616

INTRODUCTION

As a plant pathologist my definition of the water molds necessarily includes zoosporic fungi ranging all the way from truly aquatic, nonmycelial forms to completely terrestrial forms with well-developed mycelium (92). In other words, rather than confining the water molds to those Chytridiomycetes, Hypochytridiomycetes, and Oomycetes that occur in fresh water, terrestrial and marine members of these classes are included. In fact, most of what we know about the water relations of water molds comes from research on their salt tolerance or parasitism of terrestrial plants. Notable in this regard are *Phytophthora infestans* and the downy mildews, for rudimentary knowledge of their water relations has facilitated disease forecasting and control [e.g. see references in (21, 84)]. The aerial environment oscillates between extremes of water status and, fortunately, the water requirements of *P. infestans* and the downy mildews are satisfied only part of the time. Most of the substrates on which zoosporic fungi grow have a water status that is very much higher and more constant than the humidities of ambient air would suggest, and for even closely related plant pathogens that inhabit soil, such as *Phytophthora* and *Pythium* species, the determination of permissive conditions for epidemic increase will be more difficult. Nevertheless, the literature and considerable field experience indicate that very wet soil conditions enhance many *Phytophthora* and *Pythium* diseases (e.g. 17, 34, 36, 45, 93, 99, 117), and through more research their water requirements may become sufficiently known for practical use in disease control and irrigation management.

Even the most terrestrial examples of zoosporic fungi have not lost all aquatic affinities, and the response of zoosporic fungi to small changes in

431

0006-4286/79/0901-0431$01.00

water status, of a magnitude that occurs routinely in estuaries, the soil, and other terrestrial habitats, is the main subject of this review. While there are several recent reviews of water relations research on other fungi (17, 34–36), there are no inclusive reviews on the water relations of the water molds. In fact, there is not even a completely coherent body of research on which to base a review, and we are only beginning to learn the water relations of these fungi. Insofar as possible, the direct influences of water on recognizable stages in the life cycles of zoosporic fungi are reviewed, with an emphasis on plant pathogens.

COMPONENTS OF WATER POTENTIAL

The status of water in fungi and the substrates on which they grow can best be described using the concept of water potential (17, 34, 88). Water potential is the chemical potential of water per unit volume and has the dimensions of pressure (88). The usual units are bars and millibars (mb), and pure water at atmospheric pressure is defined arbitrarily as having a water potential of zero. At any given temperature, the potential of liquid water is a logarithmic function of its activity (a_w) and, therefore, its equilibrium relative humidity (RH). For example, at a water potential of –15 bars, a value at which crop plants frequently wilt, the equilibrium RH of air is 98.9%; at a water potential of –100 bars, which is too dry for the growth of most but not all fungi (34), the RH is 92.8%. Useful tables relating water potential to humidity and for interconversions of units can be found in the volume edited by Slavik (88).

The major forces acting on water can be expressed as components of water potential in the equation

$$\Psi = \psi_s + \psi_m + \psi_p$$

where Ψ is the total water potential and ψ_s, ψ_m, and ψ_p are the solute, matric, and pressure components, respectively (34, 88). Solute potential, ψ_s, represents the influence of dissolved substances on water and is equivalent to but opposite in sign from osmotic pressure. The ψ_s of sea water is approximately –24 bars and for purposes of this review, most expressions of salinity, concentration, or osmolality have been converted to ψ_s values in bars (36, 88). Matric potential, ψ_m, represents the reduction in Ψ due to adsorption and capillary forces (34, 36, 88). Older expressions of soil moisture suction, sometimes given as centimeters of water, or pF values (34), and negative pressure terms used in soil hydrology, usually refer to the negative ψ_m component of soil Ψ. As it is used here the pressure potential, ψ_p, represents the hydrostatic pressure component of Ψ, such as the positive ψ_p identified with the turgor pressure of plant cells (88).

Matric forces are of primary importance in soil and, as in any capillary system, there are definite relationships between the size of pores that can retain liquid water and ψ_m (17, 33–35, 88). The structure of most soils, however, is sufficiently heterogeneous that relationships between water content and ψ_m, especially at high ψ_m values, are not consistent, and unless parallel ψ_m measurements are presented for the same soil, expressions of soil water content (e.g. percentage by weight, percent moisture holding capacity) which prevail in literature on soil fungi are of limited value. Although field capacity frequently represents the water content of soil drained to –0.3 bar ψ_m, its definition is not universally the same. Precise measurements and manipulations of low Ψ values in soil are actually hard to achieve, and it is rarely possible to accurately partition low Ψ values of soil into their ψ_m and ψ_s components. Liquid water and vapor equilibration between points in relatively dry soil systems can be very slow, and vapor exchange is always sensitive to temperature gradients. All too frequently, published works on soil fungi do not indicate the precision or care with which substrates or other experimental points in soil were equilibrated to the Ψ values reported. Whenever possible, it is probably wise to actually measure all pertinent Ψ values, even when soil moisture is adjusted by vapor exchange with solutions of known ψ_s or by the addition of water according to an established relationship between soil water content and ψ_m.

Most fungal cells are surrounded by a semipermeable membrane and somewhat rigid wall and, so far as we know, their internal water relations resemble those of higher plant cells (14). Some of the strongest evidence for this resemblance comes from behavior of hyphal tips and isolated protoplasts in osmotica (14, 79). Of course, the most notable exceptions among fungal cells are zoospores and the plasmodia of slime molds that lack cell walls. Walled cells usually have a significantly positive turgor pressure or ψ_p which is an essential element of growth. For example, the ψ_p of growing hyphae has been estimated to be between +1 and +15 bars in a variety of fungi (2, 36), including *Aphanomyces euteiches* (47) and germinating zoospore cysts of *Phytophthora palmivora* (76). The magnitudes of these pressures are realized when one compares them to most auto tires which are inflated to about 2 bars pressure. The protoplasm of fungi, while highly structured, is probably sufficiently fluid to propagate ψ_p more or less evenly in a given cell or coenocytic hypha, and to a large degree localized adjustments in wall rigidity regulate both the growth and morphology of fungi (14).

The relative contribution of ψ_s and ψ_m to intracellular Ψ values is by no means clear. Nevertheless, ψ_s is probably by far and away the most negative component of Ψ in fungal cells. For one thing, at the high water contents usually attributed to living cytoplasm, most substances do not have a significantly negative ψ_m value (114) and the effects of dissolved solutes on ψ_s are more likely to be important. Contractile filaments may still create small internal ψ_p differences and contribute to cytoplasmic streaming. Likewise, vacuoles and other membrane-bound organelles may generate small internal ψ_p values and swell. In coenocytic hyphae, for example, increased vacuolization is thought to cause cytoplasmic streaming and contribute to apical growth (14, 37). Research on plasmodia (63) and amoeba (55), however, indicates that contractile filaments and membranes can only generate or retain pressures less than about 60 mb. Therefore, intracellular pressure gradients even in coenocytic hyphae are probably small when compared to the net turgor pressures retained by walls. The oldest regions of normally coenocytic mycelia frequently contain cross walls (14, 37), and it is conceivable that the cross walls help to maintain turgor in the more active apical hyphae.

It is generally assumed that the Ψ value of a fungus immersed in its substrate or buried in soil will equal the Ψ value of the substrate (e.g. 17, 34, 36). In most instances, this assumption is probably correct, but it is also possible that water uptake associated with growth may depress the Ψ value of fungi somewhat below that of the substrate. We actually know very little about the paths of water uptake and translocation that operate in hyphae (14), and while most of the water taken up in hyphal growth may enter through paths of low resistance at or near growing tips, the capacity of many coenocytic fungi to grow aerial mycelium suggests that water uptake by the more remote and older regions of hyphae can also support growth. Fungi on surfaces exposed to the air, of course, will be at some Ψ value intermediate between those of the air and the substrate. A consideration which is less obvious is that the response of a fungus to Ψ may depend on the nature of the substrate and the components of Ψ that predominate. For example, in solutions ψ_s predominates and cell surfaces are in complete contact with liquid water at all Ψ values. In contrast, as saturated soil dries from a Ψ of nearly zero, ψ_m usually predominates and much of the water in soil is replaced by air. Therefore, in soil not only does Ψ change, but the entire geometry of water and air in the system changes so that there are huge changes in the paths of gas and solute diffusion to and from microorganisms (17, 34). Conceivably, those fungi which invade other plants may encounter all components of Ψ. Not only do most living plants have a generally negative Ψ value, but any intercellular growth that distorts host cells and all intracellular growth [e.g. haustoria, ref. (10)] must counteract a positive ψ_p component. As host cells die, membrane semipermeability and ψ_p are lost, but the cytoplasm that remains will have a negative ψ_s value. Finally, as the water is lost from dead tissues a significantly negative ψ_m probably develops.

INFLUENCE OF WATER POTENTIAL ON GROWTH

Growth of Chytrids and Chytrid-Like Fungi

The chytrids and closely related fungi are among the most ubiquitous fungi in soils and natural waters, including habitats with a wide range of salinity (12, 51, 54, 92). Of interest to plant pathologists are those chytrids which are known to be vectors of soilborne viruses or virus-like diseases of plants (e.g. 113) and those that are active as mycoparasites (92). Recent research indicates that chytrids and chytrid-like fungi are vigorous parasites of oospores of plant pathogenic *Pythium, Phytophthora,* and *Aphanomyces* species in soil and their parasitism of oospores is most prevalent in flooded soils and is largely confined to soils wetter than field capacity (5, 89).

In nonmycelial forms where there is little or no development of a vegetative thallus, measurements of growth usually include increases in population due to sporulation. However, in the chytrid-like fungus *Althornia crouchii* (3) and various members of the Saprolegniaceae and Pythiaceae (12, 39, 41, 48, 103), formation of sporangia and release of zoospores are known to be more sensitive to changing ψ_s than is growth in size or by hyphal extension, and in many experiments on growth by mostly holocarpic forms in osmotica, the comparatively stringent limitations of ψ_s on zoosporic reproduction may have prevailed. The only study to examine quantitatively the influence of water status on a terrestrial chytrid, *Olpidium brassicae,* found that sporangia were produced in lettuce roots at all Ψ values where the host maintained turgor, but the release and movement of zoospores in soil required ψ_m values higher than −60 mb (113).

Species can be found among the nonmycelial zoosporic fungi which respond to changes in salinity in almost every conceivable way. A minority of the species for which data exist are confined in growth to either fresh or sea water (9, 12, 40, 51, 54), but the last group includes a variety of common marine forms such as the Thraustochytriaceae, of which a few are stenohaline (54) but also have a definite requirement for sodium which is not strictly osmotic (32, 51). A majority of the species tested, however, are more plastic and can grow in fresh-water media as well as in full-strength sea water or some moderate dilution thereof (9, 12, 40, 51, 54). One of the more interesting studies to examine the effects of salinity on the growth of lower zoosporic fungi is the one by Booth (9), some features of which are reviewed elsewhere (12, 54). Of the 60 isolates Booth tested from a number of species and genera, including many nonmycelial forms, only two, *Blastocladiella britannica* and one of eight isolates of *Rhizophydium sphaerotheca,* were so intolerant of sea salt that growth was confined to ψ_s values between 0 and −3 bars. All of the remaining isolates were capable of some growth at ψ_s values down to at least −10 bars. More important, Booth found that isolates from habitats of different salinity displayed different interactions with salinity, even if the isolates were of the same morphological species. There was sometimes more similarity in response to salinity among different species from the same site than among isolates of the same species from different sites. There is also strong evidence for ecotypic responses to salinity among isolates of *Pythium* species (40, 48).

Growth of Mycelial Fungi

SOLUTE POTENTIAL EFFECTS ON MYCELIAL GROWTH   Interests in both plant pathology and ecology have stimulated considerable research on the effects of osmotica on mycelial growth by several diverse fungi (34–36, 48, 51, 53, 54). The Saprolegniaceae are relatively common in fresh water but are poorly represented in waters with any degree of salinity (40, 51), and considerable experimentation by Harrison & Jones (40) suggests that their growth, at least when measured as increase in colony area on agar, is relatively sensitive to decreases in ψ_s due to sea salt. For example, they found *Aphanomyces stellatus* to grow well only at $\psi_s \geqslant -5$ bars and growth by *Leptolegnia caudata* decreased rapidly to 0 when they decreased ψ_s from –6 to –9 bars (40). Growth by various *Achlya, Dictyuchus, Isoachlya, Saprolegnia,* and *Protoachlya* species was somewhat less affected by sea salt, but major declines in growth were still evident at –7 to –9 bars ψ_s and their lower ψ_s limits for measurable growth in colony area were between –12 and –14 bars (40). When growth by some of the same species was measured as increase in dry weight, however, several species maintained at least half of their optimum growth rate at –16 bars ψ_s (39, 40). Furthermore, growth by *Saprolegnia parasitica* was somewhat more sensitive to sea salt when grown on hemp seeds than when grown in broth (39). In a separate study, TeStrake (103) found that colony extension by *Achlya diffusa, Dictyuchus monosporus,* and *Saprolegnia diclina* also decreased markedly as the ψ_s of cornmeal agar was decreased from –7 to –15 bars, but some of the data she presented showed that *D. monosporus* made measurable growth at ψ_s values as low as –50 bars. More important, TeStrake noted that the salinity tolerance shown by the vegetative growth on cornmeal agar far exceeded the salinity tolerance of these fungi under natural conditions. Evidently, not only do various measures of growth complicate judgments of salinity tolerance, but the level of nutrition and lack of competition in most laboratory experiments may be unrealistic.

    In contrast to the Saprolegniaceae, the Pythiaceae are well represented in brackish and marine environments (40, 48, 51), and while their salinity tolerance varies with the methods and isolates used (48), their growth is generally less restricted by sea salt. *Pythium* species from marine habitats, for example, grow well on agar media amended with sea salt to a variety of ψ_s values ranging from nearly zero to –26 bars (26, 48, 58). Even though effects of sea salt on growth by various *Pythium* species from nonmarine habitats sometimes differ, the response of *Pythium proliferum* reported by Harrison & Jones (40) is not atypical. Optimum colony growth on agar media occurred at $\psi_s = -5$ bars and growth declined in a somewhat linear fashion with further decreases in ψ_s to about one fifth of the optimum rate in full-strength sea water at $\psi_s \simeq -24$ bars. The growth of *Pythium ultimum* isolates from soil on agar media is affected similarly by NaCl (97). Perhaps more surprising to plant pathologists is the salinity tolerance of *Phytoph-thora* species. For example, growth of *Phytophthora cactorum* on agar is not greatly affected by sea salt at ψ_s values down to –18 bars and it maintained nearly half of its optimum growth rate in full-strength sea water (40). *Phytophthora* species have been isolated from the sea for some time (26, 48), and many of the marine isolates grow well on cornmeal agar made with distilled water or with salt water at ψ_s values down to nearly –30 bars. Similar responses to changes in the NaCl concentration of agar media have even been reported for *Phytophthora cinnamomi* (98). Although sea salt is sometimes found to enhance the growth of marine Oomycetes (40, 58), few appear to be truly stenohaline or require low ψ_s values for growth (3).

Interests in the behavior of plant pathogens have extended research on the effects of changing ψ_s on mycelial growth to osmotica other than sea salt. In one of the more comprehensive studies of radial growth on agar media, Sommers et al (91) found that three *Phytophthora* spp., i.e. *P. cinnamomi, P. parasitica,* and *P. megasperma,* were all capable of growth at ψ_s values ranging from nearly zero down to about –30 or –40 bars. While the precise effects of changing ψ_s on growth depended somewhat on the species, nutrients, and osmotica used, the effects of changes in ψ_s due to sucrose, KCl, or a mixture of salts were sufficiently similar that Sommers et al attributed them to water stress. Furthermore, the general trend of the data presented for the effects of KCl on the growth of *P. cinnamomi* by Sommers et al, including an enhancement of growth at low ψ_s by increased nutrition, has been confirmed by others (71, 98, 115). In comparison to rates of colony extension on agar medium, the growth of *P. cinnamomi,* when measured as dry weight increases in liquid media, is sensitive to decreasing ψ_s (98). For example, additions of KCl or NaCl to liquid me-dium reduced growth by one half at $\psi_s \simeq -10$ bars, whereas the same salts halved radial growth on agar medium at $\psi_s \simeq -20$ bars. Furthermore, effects of ψ_s on growth in liquid medium depended greatly on the actual solutes used, and Sterne et al (98) concluded that some solutes, such as $MgSO_4$, are more inhibitory to *P. cinnamomi* than is low ψ_s per se. In a subsequent study, Sterne & McCarver (97) found no difference between the effects of changing ψ_s due to NaCl, $CaCl_2$, or polyethylene glycol (PEG 6000) on the growth of *Pythium ultimum*. More in keeping with the results for *P. cinnamomi,* the growth of *P. ultimum* in liquid media was about twice as sensitive to depressions in ψ_s as was radial growth on agar media. Both measures of growth, however, suggest that *P. ultimum* has the capacity for growth at ψ_s values down to about –30 bars (97). In comparison, the growth of *Phytophthora cactorum* in liquid media containing NaCl was one quarter of the optimum rate at $\psi_s \simeq$ –25 bars and was reduced to zero at $\psi_s \simeq$ –50 bars (30).

Among the handful of studies on the growth of fungi from different taxonomic classes under the same ψ_s conditions are two that included *P. cinnamomi* (1, 115). Briefly, these studies showed that while growth by *P. cinnamomi* on agar media was confined to ψ_s values higher than –40 to –50 bars, *Fusarium moniliforme, Penicillium canesuns* (115), and *Alternaria tenuis* (1) grew well at –40 to –50 ψ_s and had measurable growth at ψ_s values down to at least –100 bars. In a study employing both liquid and agar media, Sterne & McCarver (97) found that growth by *Pythium ultimum* was invariably more sensitive to decreasing ψ_s than was growth by either *Verticillium dahliae* or *Rhizoctonia solani.* In liquid medium at ψ_s = –32 bars, for example, *P. ultimum* barely increased in dry weight whereas *V. dahliae* and *R. solani* maintained specific growth rates approaching one half of their rates under optimum ψ_s conditions. Recent reviews by Cook & Papendick (17) and Griffin (34) indicate that other Ascomycetes and Fungi Imperfecti of concern to plant pathology grow at ψ_s values down to –80 bars or less. This is not the case for all of the nonzoosporic fungi as several species, including *Phycomyces nitens* (34), *Ophiolobus graminis* (18, 34), *Cenococcum graniforme* (74), *Thelephora terrestris* (74), and, in some experiments, *Rhizoctonia solani* (34), are more similar to *P. cinnamomi* and *P. ultimum* (1, 97, 98, 115) in that they are found to grow only at ψ_s values higher than about –50 bars. Also, with the exception of a *Geastrum* sp. (35, 115), growth of those Homobasidiomycetes for which data exist is confined to ψ_s values as high as or higher than the ψ_s values required by pythiaceous fungi (35, 36). In spite of ecotypic differences within any one taxon (9, 106), the experimental work on mycelial growth reviewed here and elsewhere (17, 34–36, 51, 53, 106) suggests that members of the Saprolegniaceae are among the most sensitive fungi to decreasing ψ_s. The Pythiaceae are intermediate in salt or low ψ_s tolerance between the Saprolegniaceae and most of the Zygomycetes (106), Ascomycetes, and Fungi Imperfecti, but are probably no more sensitive to low ψ_s than are many Basidiomycetes (34, 35).

OSMOTIC ADJUSTMENT AND MYCELIAL GROWTH   The physiological mechanisms underlying fungal responses to changes in substrate ψ_s are largely unknown. There may even be major interactions between the underlying mechanisms of osmotic adjustment and various measures of growth (36) and there is presently no clear basis for deciding whether growth measured as colony radius or dry weight provides a better assessment of ψ_s effects. Although changes in the elasticity of cell walls may permit growth at reduced ψ_p values, the predominant means by which plant cells adjust to major changes in external ψ_s involves a regulation of the amount of osmotically active solute within the cell (20, 42). Such cellular control over internal ψ_s is here termed *osmotic adjustment* and can be accomplished by transport through the cell membrane or by biochemical interconversions (20). It is generally thought that osmotic adjustments in animal cells are linked to internal ψ_s or cell volume. The same linkages may occur in fungi, but some elegant experiments on walled plant cells indicate that their osmotic adjustments are more closely linked to the regulation of an internally positive ψ_p (20). In one of the few studies to assess both the growth and ψ_p adjustments by fungi to decreasing ψ_s, Adebayo et al (2) found that mycelia of *Mucor hiemalis* and *Aspergillus wentii* maintained positive ψ_p values at external ψ_s values ranging down to −31 bars.

The combined effects of selective membrane permeability and of solute compatibility or toxicity, both internally and externally at the cell membrane, to a large extent probably account for the different growth responses among fungi to various osmotica. For example, compatibility of a given solute with cell function varies between organisms, and some salt-tolerant bacteria have relatively salt-tolerant enzymes (49). On the other hand, some plant cells, including sugar-tolerant (13) and halotolerant yeasts (38), respond in part to high external concentrations of an incompatible solute by accumulating some more compatible solute by internal biochemical interconversions (20). In addition, cells may accumulate some minor solute components from the external medium when the predominant solute either is incompatible with the cell internally (20) or is excluded from the cell by low permeability. This last mechanism of osmotic adjustment occurs in many marine bacteria and algae that accumulate potassium in preference to sodium (20, 49), and additions of potassium frequently enhance the growth of marine fungi at the concentrations of NaCl found in sea water (3, 32, 51, 53, 54, 58). Although recent work on *Phytophthora cactorum* has begun to address membrane permeability in a water relations context (30), and in spite of a considerable body of literature on other groups of fungi (14), there is relatively little information about membrane permeability and solute transport in the water molds. It is also important to realize that there can be strong selection pressures in nature for tolerance to only one kind of solute rather than to low ψ_s per se, such as can be found among salt- and sugar-tolerant yeasts (13, 52).

The mechanisms available to fungi for osmotic adjustment, with the possible exception of passive uptake of compatible solute, require metabolic energy. In one of the few studies to have measured both respiration and growth responses by fungi to changing ψ_s, Wilson & Griffin (115) found that decreases in the ψ_s of agar media increased respiration per unit growth more in *Phytophthora cinnamomi* than in three other fungi with growth rates less restricted by low ψ_s. Although cause and effect relationships between efficiency of energy utilization and growth responses to changing ψ_s are yet to be established, the few other studies that have begun to probe mechanisms of osmotic adjustment in fungi (2, 36, 38) also suggest that the energy budget for growth at low ψ_s is a major parameter. The multitude of small deviations in culture conditions that can influence tolerance to low ψ_s (e.g. 38) indicate that any environmental factors which are marginal or limiting for growth may reduce the capacity of fungi for effective osmotic adjustments, and the ability of fungi to adjust to low ψ_s or Ψ values in natural substrates may generally be less than that found in pure culture on nutritious media.

MATRIC POTENTIAL EFFECTS ON MYCELIAL GROWTH   Relatively little is known about the effects of changing ψ_m on the growth of zoosporic fungi, but where it has been examined adequately, growth of fungi is generally found to be somewhat more restricted by decreasing ψ_m values in soil than the effects of decreasing ψ_s would suggest (36). While differences between the limiting effects of low ψ_m and ψ_s on the growth of several Fungi Imperfecti are small (17, 18, 34, 36), one of the most complete studies comparing the effects of ψ_m and ψ_s on mycelial growth found that both *Alternaria tenuis* and *Phytophthora cinnamomi* were considerably more sensitive to decreasing ψ_m than to decreasing ψ_s (1). The Ψ range for growth by *P. cinnamomi* in soil amended to various ψ_s values at a constant water content extended down to −40 or −45 bars; but when soil water content was adjusted to give various ψ_m values growth was confined to Ψ values higher than −20 to −35 bars, the exact value depending on soil texture. The Ψ range for growth by *A. tenuis* extended down to at least −130 and −60 bars in comparable solute or water content amended soils, respectively. The soils used by Adebayo & Harris (1) to obtain these results were amended with dried nutrient broth, brought up to the desired water contents, and finally equilibrated to the Ψ values reported by vapor exchange, i.e. isopiestically. However, with this technique there may have been a significantly negative ψ_s component due to water content-nutrient broth interactions in their matric-controlled system, with ψ_m representing some unknown fraction of the Ψ values reported for growth. If so, the growth of *P. cinnamomi* and *A. tenuis* may be even more sensitive to ψ_m than the Ψ data of Adebayo & Harris (1) suggest. Furthermore, responses of *P. cinnamomi* to ψ_m may not be representative of the genus, for an earlier study (91) found that while growth by *P. cinnamomi* was more sensitive to low Ψ values of agar media achieved isopiestically than osmotically, growth by *P. megasperma* and *P. parasitica* was similar in an osmotically and isopiestically controlled agar system. Though there is little doubt that at low water contents agar media have a negative ψ_m component (114), the actual partition of low Ψ values in partially dried agar media into ψ_s and ψ_m components is unknown. One additional consideration is that while mycelia of soilborne *Phytophthora* spp. are not thought to withstand extensive drying (117), mycelia of at least some species can persist in a viable state in soil (23, 75) and culture (11) at Ψ values too dry for growth.

Clearly, there is a need for more research on the relative effects of ψ_m and ψ_s on the growth of zoosporic fungi, particularly with reference to growth in natural substrates. Considerations of solute diffusion to and from hyphae, and the availability of adequate solute for osmotic adjustment suggest that low ψ_m values may frequently be more limiting to growth than equivalent low ψ_s values (2), but there is presently too little precise data on fungi for further speculation about mechanisms by which low ψ_m values limit growth.

A study by Kouyeas (64), while not directed at differences between ψ_s and ψ_m, is still central because it examined moisture effects on the growth of many diverse fungi. Briefly, Kouyeas found that plant tissues used as bait in soil as it was allowed to dry were colonized heavily by bacteria only while the soil remained very wet. *Pythium* spp. were more active at slightly lower water contents, with optimum activity at about -0.4 to -1.0 bar ψ_m. *Pythium* spp. on baits declined gradually with further drying until water contents were below ψ_m values of -1.4 to -3.8 bars where members of other genera such as *Penicillium, Aspergillus, Fusarium, Mucor,* and *Trichoderma* predominated. In an experiment on hyphal growth from agar plugs at various humidities, Kouyeas found that minimum Ψ values for growth were approximately 0 to -5 bars for species in the Saprolegniaceae, -13 to -41 bars for various *Pythium* spp., -26 bars for *Phytophthora parasitica,* and -44 to -236 bars for a variety of Zygomycetes, Ascomycetes, and Fungi Imperfecti. Evidently, with the possible exception of some Basidiomycetes (35), growth of mycelium by zoosporic fungi is generally more sensitive to water stress from any source than is growth of nonzoosporic fungi.

Although the present review is concerned mainly with direct effects of Ψ on water molds, the Ψ range over which fungi can grow in nature is frequently narrowed by the effects of Ψ on the behavior of competitive or antagonistic microorganisms (17, 34). Saprophytic growth by *Pythium* and *Phytophthora* species in moist soil, for example, is generally limited by competition from other microorganisms (93, 117), and bacteria may be responsible for hyphal lysis of *Phytophthora cactorum* in relatively wet soil (90). Foremost among the indirect effects of changing soil Ψ on growth of fungi are those due to changes in gas and solute diffusion associated with changes in soil water content (34). Such interactions with soil water content are best illustrated by the behavior of *Pythium ultimum* in the ψ_m range from 0 to about −1 bar. Briefly, while growth is somewhat better in drained pore spaces, *P. ultimum* grows well in water-filled pores (33) and evidently tolerates conditions of poor gas exchange that occur in saturated soil (33, 34). High soil water contents also increase the diffusion of nutrients from seed into the surrounding soil, and thereby increase damping off, and presumably growth, by *P. ultimum* (59). As has been noted previously (33, 34, 93), tolerance to conditions of poor gas exchange, and enhanced diffusion of nutrients, may give some competitive advantage and account more for the occurrence of *Pythium* diseases in relatively wet soils than does any positive requirement for high Ψ.

INFLUENCE OF WATER POTENTIAL ON SPORULATION AND GERMINATION

Formation of Zoosporangia

The effects of water status on asexual reproduction by the water molds are of central importance to plant pathology, for while sporulation by fungi is governed by many physiological and physical factors, it can probably be stated unequivocally that water status is one of the most determining factors in the formation and behavior of sporangia and zoospores. Furthermore, in the process of forming zoosporangia or releasing zoospores, plant pathogens leave a somewhat buffered environment within the host and face the elements of climate in the aerial or soil environment more directly. In spite of much research on sporulation by plant pathogens (e.g. 84) and in comparison to what has been demonstrated quantitatively about the water relations of mycelial growth, there is relatively little precise data on the water relations of sporulation by fungi. Nevertheless, one can find evidence that while optium Ψ values for sporulation are sometimes high, asexual sporulation by several Zygomycetes and Fungi Imperfecti occurs throughout most of the Ψ range permitting mycelial growth (36, 51, 53). On the other hand, zoosporangium and zoospore formation are the first processes to be visibly impaired as increasing levels of salinity begin to limit growth by various Oomycetes (12, 39, 41, 48, 51, 103). For example, while maintaining growth at the same or lower ψ_s values in dilutions of sea water, sporangium formation by freshwater and terrestrial *Saprolegnia* spp. is inhibited by ψ_s values less than −2 to −4 bars (12, 39, 41). The lowermost ψ_s values at which some *Pythium* species form sporangia is also on the order of −5 bars, but more typically *Pythium* and even *Phytophthora* spp. form sporangia at ψ_s values considerably lower than −5 bars, the exact values depending on the species or isolates used (48, 51). In the case of *Pythium monospermum, Phytophthora parasitica,* and *P. cinnamomi* isolates from inland soils, sporangia are formed only in dilutions of sea water or other salt solutions at ψ_s values higher than −8 to −16 bars (48, 71, 96, 110). Of course, asexual sporulation by truly marine members of these genera occurs abundantly in full strength sea water at $\psi_s \simeq -24$ bars (26, 48, 51), and the Thraustochytriaceae from marine environments actually require sea salt for normal zoospore formation (32, 54).

AERIAL FORMATION OF SPORANGIA   Research on the water requirements for sporangium formation under terrestrial conditions has been directed mainly at the downy mildews and those *Phytophthora* species which attack aerial plant parts. The importance of water to their sporulation and epidemiology has been reviewed recently (84), and only a few aspects of their sporulation are discussed here. Briefly, the formation of sporangia on aerial host surfaces by members of the Peronosporales is generally confined to ambient humidities higher than 95 to 97% RH, with sporulation increasing as the humidity more nearly approaches saturation (21, 27, 80, 84). There are important exceptions, however, such as *Phytophthora palmivora* which is reported to form numerous sporangia on cocoa pods over the entire 70 to 90% RH range without forming any at 100% RH (83). There are also reports that the presence of abundant liquid water on plant surfaces inhibits sporulation by several downy mildews and *Phytophthora infestans* (27, 84). Evidently, optimum conditions for sporangium formation aerially by members of the Peronosporales may represent a compromise between the requirement for water at very high Ψ values and a requirement to grow aerially or for unimpeded gas exchange with the air. Unfortunately, the substrates on which sporulation was observed were not at equilibrium with the ambient humidities, and it is impossible to infer the exact Ψ requirements for sporangium formation by these aerial forms from the existing data on their humidity or wetness requirements. Nevertheless, existing data on humidity and wetness effects on sporangium formation aerially meet most of our present epidemiological needs (84), and more precisely determined Ψ requirements for sporangium formation in soil are probably of greater value to epidemiological analysis.

FORMATION OF SPORANGIA IN SOIL   There is surprisingly little quantitative data describing the effects of soil moisture on sporulation by *Pythium* and *Phytophthora* species. There are observations of zoosporangium formation by *P. ultimum* in soil over the ψ_m range extending from about −30 mb to −10 bars (6, 95), but in the absence of data on other *Pythium* spp. there is no way of knowing whether the sporulation of *P. ultimum* is at all representative of the genus. Soil moisture effects on sporulation by *Phytophthora* species have received some attention, and an isolate of *P. cryptogea* [formally identified as *P. drechsleri* (24)] has been shown to require soil wetter than −4 bars Ψ for sporangial formation, with soil drained to less than −20 mb ψ_m being more nearly optimal than saturated soil (22, 23). The literature contains little data for comparison of lower Ψ limits on sporangial formation in soil, but while less complete data suggest that some other species may require higher Ψ values (81, 101), a lower limit of −4 bars does agree with data published by Reeves (82) for sporangium formation by *P. cinnamomi*. In any case, the lower Ψ limit for sporangial formation in soil is several bars higher than the lowermost ψ_s values at which soilborne *Phytophthora* species form sporangia in osmotica (48, 71, 96).

Reports of sporangial formation by *Phytophthora* species in soil at ψ_m values approaching zero or saturation, where aeration is a major parameter, are more numerous and diverse. For example, *P. cactorum* is variously reported to form maximum numbers of sporangia only in completely saturated soil ($\psi_m = 0$) or only in somewhat drained soil at −0.1 to −0.3 bar ψ_m (29, 75, 90), with some differences due to the depth of the fungus in soil. An isolate of *P. parasitica* from tomato behaves in the same way as does the isolate of *P. cryptogea* from safflower (22, 23) and does not form any sporangia in saturated soil but forms maximum numbers of sporangia over the entire ψ_m range of −0.025 to −0.3 bar (50). Pfender et al (81) report optimum sporangial formation by *P. megasperma* in soil at saturation, one fourth of optimum formation at −0.1 bar ψ_m, and no formation at −0.6 or −2.8 bars ψ_m. Interestingly, sporangial formation by isolates of *P. megasperma* and *P. cambivora* from cherry which also appears to require saturated soil is, nevertheless, greatly enhanced by artificial aeration of the soil solution (101). Furthermore, once sporangial formation was initiated at $\psi_m = 0$, *P. cambivora* continued to produce sporangia for several days after ψ_m was decreased to −50 mb (101), a ψ_m that was evidently too dry for their initiation. There are also several reports that sporangia of *Phytophthora* spp. form more abundantly on mycelia at the surface of soil extracts or solutions than when submersed (e.g. 22), and additional diversity among *Phytophthora* species in sporangium formation aerially or on substrates (8) or at various O_2 or CO_2 concentrations (77) suggests that species will respond differently to small ψ_m deficits. Furthermore, the manner in which soil is wetted to saturation will affect the degree of aeration, and the time course and abundance of sporangium formation on roots are major parameters in epidemic development that have not been examined adequately. The amount of reported diversity that is inherent among *Phytophthora* species, or even isolates of one species, and the amount that can be attributed to experimental conditions remain to be determined. It is obvious from this brief review that we do not have universally applicable data to describe soil water effects on sporangium formation.

Persistence of Zoosporangia

The importance of moisture or humidity in sporangial viability has been recognized ever since DeBary's time and there is far more information on the behavior of sporangia in aerial environments than can be reviewed here. It is noteworthy, however, that while sporangia of the Peronosporales are usually considered to be more sensitive to desiccation than are many other spore forms, the sporangia of *Peronospora tabacina* can survive extreme desiccation (86) and the sporangia of several downy mildews and of *Phytophthora infestans* are thought to persist for at least a few hours at intermediate ambient humidities (16, 21, 80, 84). Sporangial tolerance of low humidity may also be conditioned by prior wetting (16) or by their attachment to sporangiophores (e.g. 21, 27), and sporangia may require high humidities for maturation (21, 84). Unfortunately, only a small fraction of the data on aerial sporangia of the Peronosporales was obtained under sufficiently equilibrated or controlled humidity conditions for meaningful conversions to equivalent Ψ values, and a recent study by Warren & Colhoun (111) illustrates one of the complexities. Warren & Colhoun examined glass slides bearing sporangia of *P. infestans* at various humidities for the presence or absence of liquid water and found that at all humidities tested sporangia died as soon as adhering water films dried away. Attachment or deposition on substrates of high water status and the relatively high humidities at plant surfaces probably account more for sporangial viability at unsaturated ambient humidities than is generally recognized.

There are innumerable accounts of persistence by zoosporic fungi in soil, but only a few have considered the effects of soil water status on sporangial viability. One of the chief exceptions is the work on *P. ultimum* which found sporangia to persist in a viable condition for months over a wide range of soil moisture levels (95). Although there are no firm data on other *Pythium* spp., the lobate sporangia of some species are sensitive to drying in tissue (25) and are thought to be less persistent in soil (93). Much of the literature on soilborne *Phytophthora* spp. also gives an impression that their sporangia are ephemeral. There is limited evidence, however, that those of *P. palmivora* and *P. infestans* may persist in a viable state for weeks or even months under what appear to be optimum conditions of intermediate soil moisture (107, 116). The most complete studies of soil moisture effects on sporangial persistence by a *Phytophthora* species have been done on *P. cactorum*. Sneh & McIntosh (90) found sporangia of *P. cactorum* to persist for 12 weeks at −5 to −0.2 bars ψ_m and 10°C; at 24°C persistence was somewhat less, especially in more moist soils. Sporangial viability in *P. cactorum* is lost rapidly in both air-dried and saturated soils (29, 90), and a less quantitative study found its optimum and minimum moisture levels for sporangial persistence in soil to be 75 and 25% moisture holding capacity, respectively (75). Studies on *P. cactorum* also suggest that as sporangia are exposed to progressively drier conditions they lose their capacity for indirect germination before they lose their capacity for direct germination (11, 75). Short-term experiments on *P. cryptogea* and *P. megasperma* showed their sporangia to be surprisingly tolerant to moderate drying in soil (68). For example, when sporangia were dried to various Ψ values between 0 and −85 bars for 2 days, those of *P. cryptogea* were unaffected by drying to −10 bars, only 50% died when dried to −45 bars, and viability was completely lost at −85 bars. Sporangia of *P. megasperma* were about twice as sensitive to drying with complete loss of viability at −40 bars (68). Evidently sporangia of both *P. cryptogea* and *P. megasperma* can survive, at least for short periods, all levels of drying they are likely to encounter in the root zones of crop plants.

Germination of Zoosporangia

Wetness and temperature are, without doubt, two of the main climatic factors governing the germination of zoosporangia (e.g. 8, 92), and even in the more complex examples where sporangia germinate directly as well as indirectly by releasing zoospores, both modes of germination are generally thought to have similar wetness requirements (8, 21). There are indications, however, that indirect germination by sporangia of *Phytophthora palmivora* on plant surfaces may actually have a greater requirement for contact with liquid water than does direct germination by growth of germ tubes (105). Experiments on *Phytophthora* species in osmotica, while complicated somewhat by nutrient effects (56), also suggest that there is a shift from indirect to direct germination by sporangia as ψ_s decreases and that indirect germination requires higher ψ_s values than does direct germination (11, 56, 109). Furthermore, when sporangia of *P. parasitica* are placed in solutions favoring direct over indirect germination, they reverse the early stages of zoosporogenesis and lose their competence for indirect germination (44). More definitive studies on the relative effects of ψ_s on direct and indirect germination of sporangia, including ψ_s effects on mycelial growth for comparison, would be of great interest.

Gradations in the germination behavior of sporangia with water status need to be examined more thoroughly in soil, for it is important to know whether infection can occur by means of direct germination at moisture levels less than those usually associated with the release of zoospores. Experiments on *Pythium* spp. are interesting in this regard because periodic saturation of soil increased root infection by *P. vexans,* a species that germinates indirectly, but only increased infection slightly by *P. irregulare,* a species that germinates directly (7). More specifically, sporangia of *P. ultimum* are known to germinate directly in soil at ψ_m values between −50 and −100 mb (95), and enhanced diffusion of nutrients from roots and tolerance of poor aeration in soil may account for some of their germination behavior in wet soils (34, 93). Even less is known about the direct germination by sporangia of *Phytophthora* spp. in soil, and their germination by germ tubes as opposed to zoospores may be modified by soil bacteria (72) and aeration (108, 116), as well as water status. Nevertheless, sporangia of *P. palmivora* have been observed to germinate directly to infect nearby roots in soil (62), and in saturated natural soil sporangia of *P. cactorum* grew germ tubes as long as 380 μm (29). Although there are observations of direct germination by sporangia of *Phytophthora* spp. at ψ_m values between −50 and −300 mb ψ_m (22, 23, 81, 90, 101), there are few data indicating the lowermost ψ_m values at which direct germination is possible in soil. For example, sporangia of *P. cactorum* germinated directly in soil to form secondary sporangia at 25% but not at 10% moisture holding capacity (75), and when stimulated by glucose, a large percentage grew germ tubes at ψ_m values of −0.16 and −0.3 bar (90). Germ tubes of *P. cactorum* that grew within the rhizosphere of apple seedlings at ψ_m = −0.3 bar, however, were terminated by new sporangia (90). Additional clues on germination behavior come from the work on sporangium formation by *P. cryptogea* in which direct germination was observed at Ψ > −3 bars but not at the lower values where sporangia were formed (23).

WATER RELATIONS OF ZOOSPORE RELEASE   Recent experiments on the effects of soil ψ_m provide the best evidence that contact with liquid water is essential for zoospore release. For example, indirect germination by sporangia of *Phytophthora cryptogea* was stimulated within 24 hr only by ψ_m values wetter than −50 mb and was maximal only at ψ_m = 0 (24). Less complete experiments on *Phytophthora megasperma* (81) and *Aphanomyces euteiches* (46) suggest that their release of zoospores is also confined to ψ_m values wetter than −50 mb. In the most detailed study on zoospore release in soil to date, both *P. megasperma* and *P. cryptogea* were found to require ψ_m values greater than −10 mb for 1 hr before zoospore release is stimulated (67). Furthermore, when soil at −150 mb was wetted to induce release, release was related more to soil ψ_m than to soil water content, and intermittent wetting of soil to $\psi_m = 0$ was not detrimental to release, the total release being relative to the total time at $\psi_m = 0$ (67, 68). Results are similar for *P. cambivora, P. drechsleri,* another isolate of *P. megasperma* (101), and with slight modifications, for *Olpidium brassicae* (113). Further experimentation in soil is still very much needed, however, because our knowledge of ψ_m effects is not sufficient to predict zoospore release under many natural conditions. For example, sporangia of *Phytophthora* spp. have been observed to release zoospores into soil in long-term experiments at ψ_m values drier than −25 mb, sometimes at values as dry as −300 mb (22, 81, 82). Slight drying of *P. cryptogea* sporangia may enhance their subsequent release of zoospores (68), and the various kinds of upward shifts in ψ_m that can induce release need to be determined. A need to examine soil texture also arises because most soils are finer than those already tested (67) and release in finer soils may occur at slightly lower ψ_m values (24, 113). There is also a definite need to examine soil ψ_m effects on zoospore release by other forms, such as the *Pythium* species that apparently require contact with liquid water to form discharge tubes and vesicles (25).

The few studies that have compared the effects of ψ_m and ψ_s on zoospore release have invariably found ψ_m to be the more determining component of Ψ. In one study abundant zoospore release by *P. megasperma* and *P. cryptogea* occurred in a number of salt and sucrose solutions at ψ_s values as low as −3 to −5 bars and in PEG 300 solutions at ψ_s values as low as −9 bars (67). These ψ_s values contrast with the −25 mb ψ_m limit on zoospore discharge by the same sporangia (67) and therefore represent a 120- to 360-fold difference in response to the ψ_m and ψ_s components of Ψ. On the other hand, zoospore discharge by sporangia of *P. megasperma* and *P. cryptogea* was confined to ψ_s values greater than −1.3 bars in solutions of PEG 6000 (67). Hoch & Mitchell (46, 47) also found that higher molecular weight PEG prevented zoospore formation in *A. euteiches* at much higher ψ_s values than did other solutes, and that ψ_s limitations on zoospore formation were generally several orders of magnitude less than those of ψ_m. Likewise, zoospores of *O. brassicae* were released into solutions at ψ_s values down to −5.5 bars but were not released into sand unless it was essentially saturated, i. e. $\psi_m \geqslant -10$ mb (113).

It has been hypothesized that sporangia may take up certain solutes from surrounding solutions and thereby discharge zoospores at lower ψ_s values than the limitations of ψ_m would suggest (67). This hypothesis is supported by the relatively restrictive effects of high molecular weight PEG (47, 67), for these compounds are probably more completely excluded from sporangia than are the other solutes which have been tested for ψ_s effects on zoospore discharge. Furthermore, the differential effects of specific ions on zoospore discharge result to a large extent from differential solute uptake. This was demonstrated most clearly by Gisi et al (30) in their studies on zoospore discharge and plasmolysis by sporangia of *Phytophthora cactorum*. With the exception of a more specific inhibition of discharge by Mg^{2+}, they found the effects of various ions on zoospore discharge to parallel their relative permeability and to follow the lyotropic series. The ψ_s values at which salt solutions limited zoospore discharge by *P. cactorum* were invariably several orders of magnitude less negative than the ψ_s values causing plasmolysis (30), but the behavior of *Phytophthora capsici* in solutions described by Katsura (56) indicates that the highest sucrose concentration permitting zoospore release and concentrations of sucrose causing plasmolysis of sporangia have more similar ψ_s values, i.e. −8 to −9 bars. Katsura also observed zoospore release by *P. capsici* in $NaNO_3$, KNO_3, and urea solutions at ψ_s values approaching −40 bars. The lowest ψ_s values at which specific solutes are reported to allow zoospore release by *P. cactorum, P. megasperma,* and *P. cryptogea* can also be found to differ by up to a few orders of magnitude (30, 67, 109), and Gisi et al (30) found that the germination behavior of *P. cactorum* sporangia could be conditioned by their age and prior exposure to solutes. Perhaps when such effects are considered along with the effects of specific solutes and inherent biological differences the diversity of the ψ_s effects on zoospore release might be explained. The effects of ψ_s on zoospore release by forms from marine or other saline habitats are usually less restrictive than for related terrestrial or freshwater forms (12, 39, 48, 51), but the ψ_s requirements for zoospore release are almost universally more narrow than the ψ_s range which permits growth or sporangial formation (3, 12, 30, 47, 48, 51).

Zoospore release appears to have the most exacting Ψ requirements, especially in terms of its sensitivity to a few millibars ψ_m, of any stage in the life cycles of those fungi for which water relations data are published, including germ tube initiation by spores (34–36). The common link of zoospore release to high ψ_m among members of such diverse genera as *Phytophthora* (67), *Aphanomyces* (46), and *Olpidium* (113) suggests that some common mechanisms are involved. Zoospores may exit sporangia singly under their own locomotion, but in many forms, and especially in those with discharge vesicles (8, 112), they appear to be forcibly expelled from sporangia. The water relations question which can be addressed briefly here is by what force are zoospores expelled from sporangia and how is it affected by Ψ? Fuller (28) argues strongly for a pressure buildup within sporangia prior to zoospore release and suggests, among other things, that contractile proteins may function in zoospore discharge. While there is no direct evidence for their involvement, contractile proteins only generate pressures on the order of those which could be countered by the small

ψ_m deficits that prevent zoospore release, and contractile protein activity in slime molds is linked to external ion concentrations (63) in a manner not unlike zoospore release (30). A more plausible explanation might be that osmotic pressures are generated, and Gisi et al (31) have presented good evidence that the discharge vesicle of *P. cactorum* behaves as an osmotic chamber. The spores themselves may not contribute greatly to any pressure within the sporangium or vesicle (31), for if their turgor or ψ_p became significant they might burst upon release into strongly hypotonic media. In fact, zoospore release even bears some resemblance to the release of fungal protoplasts into hypertonic media (79) and zoospore differentiation is not a prerequisite for sporangial discharge (e.g. 110). Interestingly, *A. euteiches* apparently has a lower turgor pressure or ψ_p during zoosporogenesis than during growth (47).

There is still much to be resolved about the role of osmotica, for if zoospore discharge is dependent simply on osmotic adjustments, it is surprising that discharge is limited by extremely small ψ_m deficits, particularly since discharge can operate in solutions of significantly negative ψ_s. Conceivably, though, the effects of ψ_m may be related to the geometry of water contact with sporangia and their rate of water uptake. One possible mechanism which has recently received attention because of ψ_m effects is that of a swelling gel matrix which may generate at least some of the pressures necessary for zoospore discharge (67). Gel matrices lose water very quickly when ψ_m is decreased (114), and when wetted require a ψ_m of nearly 0 for complete hydration to yield any swelling pressure. Such a mechanism could explain the great sensitivity of zoospore discharge to slight ψ_m deficits, and seems particularly attractive for those parts of sporangia which may not be surrounded by semipermeable membranes (26, 47, 66). Other plausible mechanisms have been proposed, such as the surface tension model put forth by Webster & Dennis (112) to account for the final discharge of cytoplasm into the sporangial vesicle of *Pythium middletonii*. A large volume of ultrastructural work has been done on zoosporogenesis, and a complete explanation of zoospore discharge will require more studies of the type conducted by Gisi et al (31) and Webster & Dennis (112) which effectively consider both the histological and physiological aspects of zoospore release.

Water Relations of Zoospores

ZOOSPORE MOTILITY   The biology of fungal zoospores has been reviewed (28, 45, 56, 57), and there is universal agreement that the activities of motile zoospores are confined to liquid water. Aside from the requirement of water as a medium for motility, however, very little seems to be known about the water relations of fungal zoospores. Motile zoospores have no wall and, therefore, probably do not have a significant ψ_p component, and as is the case for other kinds of nonwalled cells (42), they may have very precise mechanisms for osmotic regulation. Although zoospores of *P. cactorum* have been observed to burst on transfer to water or strongly hypertonic solutions (11) and contractile vacuoles are occasionally described in fungal zoospores (e.g. 28, 43, 57), there has been no completely systematic treatment of their responses to ψ_s. Zoospores would be interesting subjects for such study, especially when one considers their importance in plant pathology, the possible interactions between their salt relations and tactic behavior (57, 60), and the range of salinity levels at which various kinds of zoospores function in nature.

Active movement by zoospores of plant pathogenic fungi to infection sites has been demonstrated many times (45), and there is no doubt that zoospores are effectively dispersed, both actively and passively, in surface waters (e.g. 104). Their capacity for movement through soils, however, is less certain. The first experiments dealing with soil constraints on zoospore dispersal found little downward washing of *Phytophthora infestans* to infect tubers (116). As one might expect from filtration theory, experiments on other *Phytophthora* species and on spores other than zoospores have also found that all but very coarse soils rapidly retain spores moving downward with water flow (24, 34, 85). Surface charge interactions with soil (60) and adhesion to surfaces on encystment (87) may further reduce zoospore movement through soil by water flow. The reciprocal question about the extent of active zoospore movement upward through soil to reach surface water where their movement is less obstructed may be equally important (24). Zoospore behavior may enhance such an upward movement because *Phytophthora* zoospores are reported to have both negative geotaxis (78) and positive rheotaxis (56). Interestingly, the most recent work in this sphere indicates that zoospores of *P. megasperma* can move as far as 2 to 6 cm upward through saturated soils, the exact distance depending on soil texture (81).

The most critical questions about zoospore motility have to do with their active movement to host roots (45). Griffin's perceptive reviews (34, 36) point out that soil structure and water status determine the size distribution of water-filled pores in soil and, therefore, the extent of active zoospore movement in soil. In fact, one review of *Pythium* diseases (93) suggests that soil constraints on zoospore movement are sufficiently great that zoospore movement may not be a major factor in infection. Some zoospore movement in soil, however, has been demonstrated in a variety of experiments [see citations in (24)], and we are beginning to learn the extents to which soil texture and water status limit active zoospore movement. Briefly, work on

Phytophthora cryptogea (24) shows that zoospores can readily swim 25–35 mm in standing surface water or through a coarse-textured soil mix at $\psi_m \geq -1$ mb. Active movement in the soil mix was reduced at $\psi_m = -10$ mb and was barely detected at $\psi_m = -50$ mb. Active movement in finer loam soils was limited to a distance of 5 mm at $\psi_m = -1$ mb and was not detected in two of three soils at $\psi_m = -10$ mb. While these results suggest that textural constraints on zoospore movement are great, if zoospore movement through soils has any role in disease, they also demonstrate the levels of moisture which are critical.

More realistic experiments on the contribution of zoospore movement to infection are needed. Soil pores on the order of 300 μm diam drain water at -10 mb ψ_m (24, 34). Therefore the ψ_m requirements for active movement by *P. cryptogea* zoospores suggest that there must not only be continuous water-filled channels that can accommodate zoospores (diam < 60 μm with flagella extended), but probably the larger and less tortuous the water-filled pores the more suitable they will be for active zoospore movement (24). In fact, occurrence of very large soil pores suitable for swimming zoospores will depend on a number of parameters that are not the usual criteria for soil texture. Among these are aggregate structure, compaction, and the prevalence of channels left by dead roots and burrowing animals. The many factors that affect the structure of soil macropores make it difficult to predict zoospore behavior in natural soils from the results obtained in sieved and reconstituted soils (24), and more realistic experiments on zoospore movement in undisturbed soils are needed.

There are biological parameters in zoospore motility that should be given more attention. In *Phytophthora* spp., contact with surfaces reduces the duration of zoospore motility by stimulating encystment and there appear to be major differences in duration of motility by zoospores of *P. cryptogea* and *P. megasperma* in soil (69). Furthermore, zoospores of *Olpidium brassicae* do not encyst and swim farther in finer soils and at slightly lower ψ_m values (113) than do zoospores of *P. cryptogea* (24), the better swimmer of the two *Phytophthora* spp. that have been compared in soil (69). Evidently, species that differ in swimming behavior and motile period might have significantly different capacities for movement through soil. Furthermore, while research on *P. palmivora* suggests that motile zoospores function more effectively as inoculum in natural soils than do encysted spores (61), in most instances we have no firm estimates of the extents to which active zoospore movement enhances root infection over that which could occur by other means such as the growth of germ tubes.

ZOOSPORE PERSISTENCE   Zoospores are released from sporangia in the absence of external nutrients, and soil constraints on zoospore motility suggest that many zoospores encyst in the soil before they reach a suitable spot for growth. Although zoospores of *Phytophthora parasitica* and structures derived from them persist in surface waters (104), the fate of zoospores that fail to reach a substrate or host root in soil and their capacity to function subsequently as inoculum are largely unknown. In spite of impressions that zoospores are ephemeral, Turner (107) reported that soils infested with zoospores of *P. palmivora* remained infective for 6–18 months when stored at 50% moisture holding capacity, and in saturated or air-dried soils, infectivity by *P. palmivora* was lost in 10 to 32 days. One of the few quantitative studies on zoospore persistence in soil was done on *P. cactorum* by McIntosh (73). Among other things, he found that survival was favored in dry (–2 and –5 bars ψ_m) relative to wet soil (–0.1 and –0.2 bar ψ_m), but as in previous studies, zoospores persisted only a few weeks. A more recent German study (75) reported zoospores of *P. cactorum* to persist for 92 days in soil at cooler temperatures and 25% moisture holding capacity, but persistence was less in wetter soils and at warmer temperatures. Recent applications of quantitative techniques to assay both soil Ψ and zoospore populations (70) found that while encysted zoospores of *P. megasperma* persisted in natural soil for only a few days, regardless of Ψ value, those of *P. cinnamomi* persisted for several weeks in soil at –5 or –15 bars Ψ and in the field for more than 10 weeks at –20 bars soil Ψ and 20–28°C. Furthermore, cysts of *P. cinnamomi* persisted in their original form (70). Clearly, more experimental work is needed before we will truly understand potentials for zoospore persistence.

ZOOSPORE GERMINATION   Most of the few clues that can be found in the literature about the water requirements for zoospores to germinate come from experiments in solutions. Even in the most conducive osmotica tested, germ tube growth by zoospores of a freshwater *Aphanomyces* sp. was confined to ψ_s values higher than –5 bars (102), and salinity effects on cyst germination and mycelial growth by *Saprolegnia parasitica* are similar (39). Zoospore cysts of *Phytophthora parasitica* and *P. cactorum* can grow germ tubes in sucrose solutions at ψ_s values down to –8 or –20 bars (11, 43), and at ψ_s values approaching zero, cysts of *P. parasitica* functioned as microsporangia (43) or formed microsporangia on germ tubes to release zoospores (104). With the exception of *P. infestans*, for which germ tubes of zoospore cysts were observed to grow at 25% moisture holding capacity (116), only *P. cactorum* has been used to study soil moisture effects on zoospore germination. Its zoospores germinated quickly in saturated soil (29) and infection of host tissues by cysts occurred only in soils wetter than 50% moisture holding capacity, with optimum infection at 100% moisture holding capacity (75). It would be interesting to know how closely the water requirements for the various modes of germination by zoospores parallel those for sporangial germination.

Water Relations of Chlamydospores and Oospores

CHLAMYDOSPORES   There has been considerable interest in the biology and function of *Phytophthora* chlamydospores in soil, but their interactions with soil water, while examined to some extent, are not clearly known. For example, chlamydospore formation by *P. cactorum* is reported to have water requirements that parallel those for sporangial formation (29) while formation of chlamydospores by *P. cinnamomi* is reported at Ψ values considerably drier than those that permit formation of sporangia (71, 82). Chlamydospores have generally been thought to assist the survival of those *Phytophthora* spp. which form them, and limited experimentation leaves little doubt that soil water is a parameter in their survival (15). The germination of chlamydospores has received more attention, and their mode of germination by continued growth of germ tubes or by germsporangia is influenced to some extent by water status, nutrition, and the presence or absence of host roots (e.g. 100). Germinating chlamydospores of *P. parasitica* appear to have the same water requirements for sporangial formation as do mycelia (50), but chlamydospores of other *Phytophthora* spp. can grow germ tubes at low Ψ values, even at −98 bars ψ_s in the instance of *P. drechsleri* (19, 71). As was noted earlier for sporangia, and as has been recently demonstrated for chlamydospores of *P. cinnamomi* (100), the growth of germ tubes to infect roots may depend on interactions between soil ψ_m and the diffusion of nutrients from roots. Furthermore, infection of roots by chlamydospores of *P. cinnamomi* in soil is influenced by ψ_m values within the 0 to −0.25 bar range much more than by ψ_s values in the −0.4 to −2.0 bar range, a range which represents the ψ_s values of saline soils in California avocado groves (99).

SEXUAL REPRODUCTION   In those few terrestrial and freshwater Ascomycetes for which data are published, sexual reproduction is confined to higher ψ_s values than is either growth or asexual sporulation (36, 53). Similarly, sexual reproduction by at least some terrestrial and freshwater Oomycetes appears to be confined to much higher ψ_s values than is mycelial growth (12, 41, 48, 51). The existing literature on soilborne *Pythium* and *Phytophthora* species, however, appears to be contradictory, for oospore formation is reported to occur at Ψ values only approaching zero (29, 75) and at appreciably negative Ψ values (6, 71, 90). Recent and rather exciting work has shown the probable importance of soil water status to oospore persistence and germination. In relatively wet soils parasitism of oospores can be extensive (5, 89), and conversion of *Pythium ultimum* oospores to a state from which they can germinate readily is enhanced by incubation in nonsterile soils at Ψ values higher than -7 bars (65). Very wet soil conditions are generally the worst for oospore survival, and while oospores can evidently withstand some air drying, intermediate levels of soil moisture are better (4, 75, 90). Oospores of *P. cactorum* can evidently germinate to form sporangia in soil at -0.2 to -5.0 bars ψ_s (90), and while oospores of *Pythium aphanidermatum* have the capacity to germinate directly in soil at Ψ values as low as -15 bars (94), those of *Pythium hydnosporum* evidently require some greater contact with liquid water to germinate (4). Obviously, there is a need for more systematic studies of Ψ effects on oospore formation, persistence, and germination.

CONCLUDING REMARKS

Insofar as possible, the relatively direct effects of water status on various stages in the life cycles of zoosporic fungi have been reviewed. There is much additional literature which pertains to the water relations of zoosporic fungi, but most of it either does not deal with recognizable stages in their life cycles or does not contain adequate data on water status. Even though a relatively small number of studies have focused on the water relations of zoosporic fungi, our knowledge of their water relations is fast approaching our level of knowledge about the water relations of other major groups of fungi. Certainly in comparison to many Zygomycetes, Fungi Imperfecti, and Ascomycetes, we are reassured that zoosporic fungi are truly water molds; that is, their water potential requirements for zoospore release and movement are among the highest that have been found. The zoospore, however, is not their only link to water, for a number of zoosporic fungi are also known to have relatively precise water requirements for growth and to form spores other than zoospores. Broader generalizations are probably risky, for ecotypic variation in organismal water relations can occur within any taxon, including morphological species of fungi. Furthermore, most of our knowledge about the water relations of fungi is descriptive of their water requirements, and a more mechanistic or physiological approach will be required to predict fungal responses to the complex set of parameters usually presented in nature.

For the plant pathologist, even the most complete knowledge of a pathogen's water relations represents only one facet of disease epidemiology. While the water requirements of the pathogen establish some of the climatic constraints on disease development, in the final analysis pathogen behavior and disease epidemiology represent an integration of pathogen physiology with a number of other biological and physical parameters. For example, the water status at which plant pathogens are active may be determined by the activities of competitive or antagonistic microorganisms or by host responses to the environment. Nevertheless, the more narrow the water requirements of a pathogen the more likely they are to influence disease development and, as we have seen, zoosporic fungi have some of the most stringent water requirements known.

Knowledge of the limitations that pathogen water relations place on late blight and downy mildew epidemiology has sometimes facilitated control, and this may also become possible for soilborne diseases such as phytophthora root rot. Knowledge of soil moisture effects on root disease, of course, has the most potential for use in greenhouses and irrigated agriculture where soil moisture is one of the most manipulated components of the physical environment. However, such knowledge also has applications in less controlled circumstances. For example, roughly estimated moisture requirements of *Phytophthora cinnamomi* are among the parameters for regulating industry access to jarrah forests in Western Australia (S. R. Shea, personal communication). Precise knowledge of the water requirements of soilborne *Phytophthora* spp. may also enhance their detection in soil (50) and has universal application in selecting conditions for pathogenicity tests to determine host range or find genetic resistance. Potentially, such knowledge could also aid the judicious use of chemicals that may become available for control. However, despite the many published associations between phytophthora root rots and saturated soils [e.g. see citations in (24, 67, 99)], the environmental factors influencing phytophthora root rots are not sufficiently defined to model the soil environment for permissive characters or, for that matter, for general use in irrigation and disease management. Some of the same deficiencies exist for all diseases incited by zoosporic fungi, and hopefully, plant pathologists and mycologists alike will take a more quantitative interest in those aspects of the water relations of zoosporic fungi that pertain to the control of plant disease.

Literature Cited

1. Adebayo, A. A., Harris, R. F. 1971. Fungal growth responses to osmotic as compared to matric water potential. *Soil Sci. Soc. Am. Proc.* 35:465–69
2. Adebayo, A. A., Harris, R. F., Gardner, W. R. 1971. Turgor pressure of fungal mycelia. *Trans. Br. Mycol. Soc.* 57:145–51
3. Alderman, D. J., Jones, E. B. G. 1971. Physiological requirements of two marine phycomycetes, *Althornia crouchii* and *Ostracoblabe implexa*. *Trans. Br. Mycol. Soc.* 57:213–25
4. Al Hassan, K. K., Fergus, C. L. 1973. The effects of nutrients and environment on germination and longevity of

oospores of *Pythium hydnosporum. Mycopathol. Mycol. Appl.* 51:283–97
5. Ayers, W. A., Lumsden, R. D. 1977. Mycoparasitism of oospores of *Pythium* and *Aphanomyces* species by *Hyphochytrium catenoides. Can. J. Microbiol.* 23:38–44
6. Bainbridge, A. 1970. Sporulation by *Pythium ultimum* at various soil moisture tensions. *Trans. Br. Mycol. Soc.* 55:485–88
7. Biesbrock, J. A., Hendrix, F. F. Jr. 1970. Influence of soil water and temperature on root necrosis of peach caused by *Pythium* spp. *Phytopathology* 60:880–82

8. Blackwell, E. M., Waterhouse, G. M. 1931. Spores and spore germination in the genus *Phytophthora*. *Trans. Br. Mycol. Soc.* 15:294–310

9. Booth, T. 1971. Ecotypic responses of chytrid and chytridiaceous species to various salinity and temperature combinations. *Can. J. Bot.* 49:1757–67

10. Bracker, C. E., Littlefield, L. J. 1973. Structural concepts of host-pathogen interfaces. In *Fungal Pathogenicity and the Plant's Response*, ed. R. J. W. Byrde, C. V. Cutting, pp. 159–317. London & New York: Academic. 499 pp.

11. Braun, H., Kröber, H. 1958. Untersuchungen über die durch *Phytophthora cactorum* (Leb. u. Cohn) Schroet. hervorgerufene Kragenfäule des Apfels. *Phytopathol. Z.* 32:35–94

12. Bremer, G. B. 1976. The ecology of marine lower fungi. In *Recent Advances in Aquatic Mycology*, ed. E. B. G. Jones, pp. 313–33. London: Elek Sci. 749 pp.

13. Brown, A. D. 1974. Microbial water relations: Features of the intracellular composition of sugar-tolerant yeasts. *J. Bacteriol.* 118:769–77

14. Burnett, J. H. 1968. *Fundamentals of Mycology*. New York: St. Martin's. 546 pp.

15. Chee, K. H. 1973. Production, germination and survival of chlamydospores of *Phytophthora palmivora* from *Hevea brasiliensis*. *Trans. Br. Mycol. Soc.* 61:21–26

16. Cohen, Y., Perl, M., Rotem, J., Eyal, H., Cohen, J. 1974. Ultrastructural and physiological changes in sporangia of *Pseudoperonospora cubensis* and *Phytophthora infestans* exposed to water stress. *Can. J. Bot.* 52:447–50

17. Cook, R. J., Papendick, R. I. 1972. Influence of water potential of soils and plants on root disease. *Ann. Rev. Phytopathol.* 10:349–74

18. Cook, R. J., Papendick, R. I., Griffin, D. M. 1972. Growth of two root-rot fungi as affected by osmotic and matric water potentials. *Soil Sci. Soc. Am. Proc.* 36:78–82

19. Cother, E. J., Griffin, D. M. 1974. Chlamydospore germination in *Phytophthora drechsleri*. *Trans. Br. Mycol. Soc.* 63:273–79

20. Cram, W. J. 1976. Negative feedback regulation of transport in cells. The maintenance of turgor, volume, and nutrient supply. In *Transport in Plants, II, Part A, Cells, Encyclopedia of Plant Physiology, New Series*, ed. U. Lüttge, M. G. Pitman, pp. 284–316. Berlin: Springer. 400 pp.

21. DeWeille, G. A. 1964. Forecasting crop infection by the potato blight fungus. *K. Ned. Meterologisch Inst. Meded. Verh.* 82. 144 pp.

22. Duniway, J. M. 1975. Formation of sporangia by *Phytophthora drechsleri* in soil at high matric potentials. *Can. J. Bot.* 53:1270–75

23. Duniway, J. M. 1975. Limiting influence of low water potential on the formation of sporangia by *Phytophthora drechsleri* in soil. *Phytopathology* 65:1089–93

24. Duniway, J. M. 1976. Movement of zoospores of *Phytophthora cryptogea* in soils of various textures and matric potentials. *Phytopathology* 66:877–82

25. Dow, R. L., Lumsden, R. D. 1975. Histopathology of infection of bean with *Pythium myriotylum* compared with infection with other *Pythium* species. *Can. J. Bot.* 53:1786–95

26. Fell, J. W., Master, I. M. 1975. Phycomycetes (*Phytophthora* spp. nov. and *Pythium* sp. nov.) associated with degrading mangrove (*Rhizophora mangle*) leaves. *Can. J. Bot.* 53:2908–22

27. Fried, P. M., Stuteville, D. L. 1977. *Peronospora trifoliorum* sporangium development and effects of humidity and light on discharge and germination. *Phytopathology* 67:890–94

28. Fuller, M. S. 1977. The zoospore, hallmark of the aquatic fungi. *Mycologia* 69:1–20

29. Gisi, U. 1975. Investigations on the soil phase of *Phytophthora cactorum* (Leb. et Cohn) Schroet. with fluorescent optical direct observation. *Z. Pflanzenkr. Pflanzenschutz* 82:355–77 (In German)

30. Gisi, U., Oertli, J. J., Schwinn, F. J. 1977. Water and salt relations of sporangia of *Phytophthora cactorum* (Leb. et Cohn) Schroet. in vitro. *Phytopathol. Z.* 89:261–84 (In German)

31. Gisi, U., Schwinn, F. J., Oertli, J. J. 1979. Dynamics of indirect germination in *Phytophthora cactorum* sporangia. *Trans. Br. Mycol. Soc.* 72: In press

32. Goldstein, S. 1973. Zoosporic marine fungi (Thraustochytriaceae and Dermocystidiaceae). *Ann. Rev. Microbiol.* 27:13–26

33. Griffin, D. M. 1963. Soil physical factors and the ecology of fungi. II. Behaviour of *Pythium ultimum* at small soil water suctions. *Trans. Br. Mycol. Soc.* 46:368–72

34. Griffin, D. M. 1972. *Ecology of Soil Fungi*. Syracuse: Syracuse Univ. Press. 193 pp.

35. Griffin, D. M. 1977. Water potential and wood-decay fungi. *Ann. Rev. Phytopathol.* 15:319–29
36. Griffin, D. M. 1978. Effect of soil moisture on survival and spread of pathogens. In *Water Deficits and Plant Growth*, ed. T. T. Kozlowski, 5:175–97. New York: Academic. 323 pp.
37. Grove, S. N., Bracker, C. E., Morré, D. J. 1970. An ultrastructural basis for hyphal tip growth in *Pythium ultimum*. *Am. J. Bot.* 57:245–66
38. Gustafsson, L., Norkrans, B. 1976. On the mechanism of salt tolerance. Production of glycerol and heat during growth of *Debaryomyces hansenii. Arch. Microbiol.* 110:177–83
39. Harrison, J. L., Jones, E. B. G. 1971. Salinity tolerance of *Saprolegnia parasitica* Coker. *Mycopathol. Mycol. Appl.* 43:297–307
40. Harrison, J. L., Jones, E. B. G. 1974. Patterns of salinity tolerance displayed by the lower fungi. *Veroeff. Inst. Meeresforsch. Bremerhaven Suppl.* 5: 197–220
41. Harrison, J. L., Jones, E. B. G. 1975. The effect of salinity on sexual and asexual sporulation of members of the Saprolegniaceae. *Trans. Br. Mycol. Soc.* 65:389–94
42. Hellebust, J. A. 1976. Osmoregulation. *Ann. Rev. Plant Physiol.* 27:485–505
43. Hemmes, D. E., Hohl, H. R. 1971. Ultrastructural aspects of encystation and cyst-germination in *Phytophthora parasitica. J. Cell Sci.* 9:175–91
44. Hemmes, D. E., Hohl, H. R. 1975. Ultrastructural changes and a second mode of flagellar degeneration during ageing of *Phytophthora palmivora* sporangia. *J. Cell Sci.* 19:563–77
45. Hickman, C. J. 1970. Biology of *Phytophthora* zoospores. *Phytopathology* 60:1128–35
46. Hoch, H. C., Mitchell, J. E. 1970. The effects of water potential on zoospore production in *Aphanomyces euteiches. Phytopathology* 60:1296 (Abstr.)
47. Hoch, H. C., Mitchell, J. E. 1973. The effects of osmotic water potentials on *Aphanomyces euteiches* during zoosporogenesis. *Can. J. Bot.* 51:413–20
48. Höhnk, W. 1953. Studien zur Brack- und Seewassermykologie. III. (Oomycetes: Zweiter Teil.) *Veroeff. Inst. Meeresforsch. Bremerhaven* 2:52–108
49. Ingram, M. 1957. Micro-organisms resisting high concentrations of sugars or salts. In *Microbial Ecology*, ed. R. E. O. Williams, C. C. Spicer. *Symp. Soc. Gen. Microbiol.* 7:90–133
50. Ioannou, N., Grogan, R. G. 1977. The influence of soil matric potential on the production of sporangia by *Phytophthora parasitica* in relation to its isolation from soil by baiting techniques. *Proc. Am. Phytopathol. Soc.* 4:173
51. Johnson, T. W. Jr., Sparrow, F. K. Jr. 1961. *Fungi in Oceans and Estuaries.* Weinheim: Cramer. 668 pp.
52. Joho, M. 1975. Growth behavior of a marine yeast, *Rhodotorula glutinis* var. *salinaria.* I. Effect of high concentrations of salts. *Trans. Mycol. Soc. Jpn.* 16:383–89
53. Jones, E. B. G., Byrne, P. J. 1976. Physiology of the higher marine fungi. See Ref. 12, pp. 135–75
54. Jones, E. B. G., Harrison, J. L. 1976. Physiology of marine Phycomycetes. See Ref. 12, pp. 261–78
55. Kamiya, N. 1964. The motive force of endoplasmic streaming in the ameba. In *Primitive Motile Systems in Cell Biology*, ed. R. D. Allen, N. Kamiya, pp. 257–74. New York: Academic. 642 pp.
56. Katsura, K. 1971. Some ecological studies on zoospore of *Phytophthora capsici* Leonian. *Rev. Plant Prot. Res.* 4:58–70
57. Katsura, K., Miyata, Y. 1971. Swimming behavior of *Phytophthora capsici* zoospores. In *Morphological and Biochemical Events in Plant-Parasite Interaction*, ed. S. Akai, S. Ouchi, pp. 107–28. Tokyo: Phytopathol. Soc. Jpn. 415 pp.
58. Kazama, F. Y., Fuller, M. S. 1973. Mineral nutrition of *Pythium marinum*, a marine facultative parasite. *Can. J. Bot.* 51:693–99
59. Kerr, A. 1964. The influence of soil moisture on infection of peas by *Pythium ultimum. Aust. J. Biol. Sci.* 17:676–85
60. Khew, K. L., Zentmyer, G. A. 1974. Electrotactic response of zoospores of seven species of *Phytophthora. Phytopathology* 64:500–7
61. Kliejunas, J. T., Ko, W. H. 1974. Effect of motility of *Phytophthora palmivora* zoospores on disease severity in papaya seedlings and substrate colonization in soil. *Phytopathology* 64:426–28
62. Ko, W. H. 1971. Direct observation of fungal activities on soil. *Phytopathology* 61:437–38
63. Kobatake, Y., Ueda, T. 1977. Electrochemical processes in stimulus-response relationship in slime mold *Physarum polycephalum.* In *Electrical Phenomena at the Biological Membrane Level*, ed.

E. Roux, pp. 11–23. Amsterdam: El-
sevier. 565 pp.
64. Kouyeas, V. 1964. An approach to the
study of moisture relations of soil fungi.
Plant Soil 20:351–63
65. Lumsden, R. D., Ayers, W. A. 1975.
Influence of soil environment on the
germinability of constitutively dormant
oospores of *Pythium ultimum. Phytopathology* 65:1101–7
66. Lunney, C. Z., Bland, C. E. 1976. An
ultrastructural study of zoosporogenesis in *Pythium proliferum. Protoplasma*
88:85–100
67. MacDonald, J. D., Duniway, J. M.
1978. Influence of the matric and osmotic components of water potential on
zoospore discharge in *Phytophthora.
Phytopathology* 68:751–57
68. MacDonald, J. D., Duniway, J. M.
1978. Temperature and water stress
effects on sporangium viability and
zoospore discharge in *Phytophthora
cryptogea* and *P. megasperma. Phytopathology* 68:1449–55
69. MacDonald, J. D., Duniway, J. M.
1978. Influence of soil texture and temperature on the motility of *Phytophthora cryptogea* and *P. megasperma*
zoospores. *Phytopathology* 68:1627–30
70. MacDonald, J. D., Duniway, J. M.
1979. Use of fluorescent antibodies
to study the survival of *Phytophthora
megasperma* and *P. cinnamomi* zoospores in soil. *Phytopathology.* 69: In
press
71. Malajczuk, N. 1975. *Interactions between Phytophthora cinnamomi Rands
and roots of Eucalyptus calophylla R.
Br. and Eucalyptus marginata Donn. ex.
Sm.* PhD thesis. Univ. Western Australia, Perth. 371 pp.
72. Marx, D. H., Bryan, W. C. 1969. Effect
of soil bacteria on the mode of infection
of pine roots by *Phytophthora cinnamomi. Phytopathology* 59:614–19
73. McIntosh, D. L. 1972. Effects of soil
water suction, soil temperature, carbon
and nitrogen amendments, and host
rootlets on survival in soil of zoospores
of *Phytophthora cactorum. Can. J. Bot.*
50:269–72
74. Mexal, J., Reid, C. P. P. 1973. The
growth of selected mycorrhizal fungi in
response to induced water stress. *Can.
J. Bot.* 51:1579–88
75. Meyer, D., Schönbeck, F. 1975. Investigations on the development of *Phytophthora cactorum* (Leb. & Cohn.) Schroet.
in soil. *Z Pflanzenkr. Pflanzenschutz*
82:337–54 (In German)

76. Meyer, R., Parish, R. W., Hohl, H. R.
1976. Hyphal tip growth in *Phytophthora:* gradient distribution and ul-
trahistochemistry of enzymes. *Arch.
Microbiol.* 110:215–24
77. Mitchell, D. J., Zentmyer, G. A. 1971.
Effects of oxygen and carbon dioxide
tensions on sporangium and oospore
formation by *Phytophthora* spp. *Phytopathology* 61:807–12
78. Palzer, C. 1975. Negative geotaxis by
Phytophthora cinnamomi zoospores.
Phytophthora Newsl. 3:9–11
79. Peberdy, J. F. 1972. Protoplasts from
fungi. *Sci. Prog. Oxford* 60:73–86
80. Pegg, G. F., Mence, M. J. 1970. The
biology of *Peronospora viciae* on pea:
Laboratory experiments on the effects
of temperature, relative humidity and
light on the production, germination
and infectivity of sporangia. *Ann. Appl.
Biol.* 66:417–28
81. Pfender, W. F., Hine, R. B., Stanghellini, M. E. 1977. Production of sporangia and release of zoospores by *Phytophthora megasperma* in soil. *Phytopathology* 67:657–63
82. Reeves, R. J. 1975. Behaviour of *Phytophthora cinnamomi* Rands in different soils and water regimes. *Soil Biol.
Biochem.* 7:19–24
83. Rocha, H. M., Machado, A. D. 1973.
The effect of light, temperature and relative humidity on sporulation of *Phytophthora palmivora* (Butl.) Butl. in
cacao pods. *Rev. Theobroma* 3:22–25
(In Portuguese)
84. Rotem, J., Cohen, Y., Bashi, E. 1978.
Host and environmental influences on
sporulation in vivo. *Ann. Rev. Phytopathol.* 16:83–101
85. Ruddick, S. M., Williams, S. T. 1972.
Studies on the ecology of Actinomycetes in soil. V. Some factors influencing the dispersal and adsorption
of spores in soil. *Soil Biol. Biochem.*
4:93–103
86. Shepherd, C. J., Simpson, P., Smith, A.
1971. Effects of variation in water potential on the viability and behaviour
of conidia of *Peronospora tabacina*
Adam. *Aust. J. Biol. Sci.* 24:219–29
87. Sing, V. O., Bartnicki-Garcia, S. 1975.
Adhesion of *Phytophthora palmivora*
zoospores: Electron microscopy of cell
attachment and cyst wall fibril formation. *J. Cell Sci.* 18:123–32
88. Slavik, B. ed. 1974. *Methods of Studying
Plant Water Relations.* New York:
Springer. 449 pp.
89. Sneh, B., Humble, S. J., Lockwood, J.
L. 1977. Parasitism of oospores of *Phy-*

tophthora megasperma var. sojae, P. cactorum, Pythium sp., and Aphanomyces euteiches in soil by Oomycetes, Chytridiomycetes, Hyphomycetes, Actinomycetes, and bacteria. Phytopathology 67:622–28

90. Sneh, B., McIntosh, D. L. 1974. Studies on the behavior and survival of Phytophthora cactorum in soil. Can. J. Bot. 52:795–802

91. Sommers, L. E., Harris, R. F., Dalton, F. N., Gardner, W. R. 1970. Water potential relations of three root-infecting Phytophthora species. Phytopathology 60:932–34

92. Sparrow, F. K. Jr. 1960. Aquatic Phycomycetes. Ann Arbor: Univ. Michigan Press. 1187 pp. 2nd ed.

93. Stanghellini, M. E. 1974. Spore germination, growth and survival of Pythium in soil. Proc. Am. Phytopathol. Soc. 1:211–14

94. Stanghellini, M. E., Burr, T. J. 1973. Effect of soil water potential on disease incidence and oospore germination of Pythium aphanidermatum. Phytopathology 63:1496–98

95. Stanghellini, M. E., Hancock, J. G. 1971. The sporangium of Pythium ultimum as a survival structure in soil. Phytopathology 61:157–64

96. Sterne, R. E. 1976. The relationship of soil and plant water status to phytophthora root rot of avocado. PhD thesis. Univ. Calif., Riverside. 145 pp.

97. Sterne, R. E., McCarver, T. H. 1979. Osmotic effects on radial growth rate and specific growth rate of three soil fungi. Can. J. Microbiol. In press

98. Sterne, R. E., Zentmyer, G. A., Bingham, F. T. 1976. The effect of osmotic potential and specific ions on growth of Phytophthora cinnamomi. Phytopathology 66:1398–1402

99. Sterne, R. E., Zentmyer, G. A., Kaufmann, M. R. 1977. The effect of matric and osmotic potential of soil on Phytophthora root disease of Persea indica. Phytopathology 67:1491–94

100. Sterne, R. E., Zentmyer, G. A., Kaufmann, M. R. 1977. The influence of matric potential, soil texture, and soil amendment on root disease caused by Phytophthora cinnamomi. Phytopathology 67:1495–1500

101. Sugar, D. 1977. The development of sporangia of Phytophthora cambivora, P. megasperma, and P. drechsleri and severity of root and crown rot in Prunus mahaleb as influenced by soil matric potential. MS thesis. Univ. California, Davis. 56 pp.

102. Svensson, E., Unestam, T. 1975. Differential induction of zoospore encystment and germination in Aphanomyces astaci, Oomycetes. Physiol. Plant 35:210–16

103. TeStrake, D. 1959. Estuarine distribution and saline tolerance of some Saprolegniaceae. Phyton Int. J. Exp. Bot. 12:147–52

104. Thomson, S. V., Allen, R. M. 1976. Mechanisms of survival of zoospores of Phytophthora parasitica in irrigation water. Phytopathology 66:1198–1202

105. Thorold, C. A. 1955. Observations on black-pod disease (Phytophthora palmivora) of cacao in Nigeria. Trans. Br. Mycol. Soc. 38:435–52

106. Tresner, H. D., Hayes, J. A. 1971. Sodium chloride tolerance of terrestrial fungi. Appl. Microbiol. 22:210–13

107. Turner, P. D. 1965. Behavior of Phytophthora palmivora in soil. Plant Dis. Reptr. 49:135–37

108. Uppal, B. N. 1926. Relation of oxygen to spore germination in some species of the Peronosporales. Phytopathology 16:285–92

109. Von Stille, B. 1965. Das Keimverhalten der Sporangien von Phytophthora infestans in Abhängigkeit von Temperatur- und Hydraturbedingungen. Z. Pflanzenkr. Pflanzenschutz 72(4):193–200

110. Vujîcić, R., Colhoun, J. 1966. Asexual reproduction in Phytophthora erythroseptica. Trans. Br. Mycol. Soc. 49: 245–54

111. Warren, R. C., Colhoun, J. 1975. Viability of sporangia of Phytophthora infestans in relation to drying. Trans. Br. Mycol. Soc. 64:73–78

112. Webster, J., Dennis, C. 1967. The mechanism of sporangial discharge in Pythium middletonii. New Phytol. 66: 307–13

113. Westerlund, F. V., Campbell, R. N., Grogan, R. G., Duniway, J. M. 1978. Soil factors affecting the reproduction and survival of Olpidium brassicae and its transmission of big vein agent to lettuce. Phytopathology 68:927–35

114. Wiebe, H. H. 1966. Matric potential of several plant tissues and biocolloids. Plant Physiol. 41:1439–42

115. Wilson, J. M., Griffin, D. M. 1975. Respiration and radial growth of soil fungi at two osmotic potentials. Soil Biol. Biochem. 7:269–74

116. Zan, K. 1962. Activity of Phytophthora infestans in soil in relation to tuber infection. Trans. Br. Mycol. Soc. 45: 205–21

117. Zentmyer, G. A., Erwin, D. C. 1970. Development and reproduction of Phytophthora. Phytopathology 60:1120–27

Ann. Rev. Phytopathol. 1979. 17:461–84
Copyright © 1979 by Annual Reviews Inc. All rights reserved

SMALL PATHOGENIC RNA
IN PLANTS—THE VIROIDS

♦3719

J. S. Semancik

Department of Plant Pathology and Cell Interaction Group,
University of California, Riverside, California 92521

INTRODUCTION

Over the past several years, the low molecular weight plant pathogenic RNA, the viroids, originally proposed as the causal agent of potato spindle tuber disease (12) and visualized in extracts from the exocortis diseases (76), have been reviewed extensively (7, 15, 16, 66). Most of these treatments have focused on evidence for the existence as well as physical characterization of small pathogenic RNAs as a unique class of infectious molecules. Studies of the physical and structural features of the viroids recently have culminated in the elucidation of the primary sequence of nucleotides in the potato spindle tuber viroid (PSTV) (24). This feat of structural chemical analysis was significant from several perspectives. It provided the most precise molecular sizing to which various earlier observations of the structure and conformation could be related. It also suggested many physical-chemical approaches for investigating the postulated unique conformation of this minimal infectious molecule. It did not, however, provide any dramatic insights into the biological activity, or especially the pathogenic activity, of this unique class of molecules.

As with the historical development of plant virology, our understanding of the physical-chemical properties of the infecting viroid RNA has proceeded significantly faster and with greater comprehension than our appreciation of the biological interaction of the viroid with host cells. We have only the barest understanding of the possible mechanisms involved in viroid replication and pathogenesis. The viroid model does not lend itself to a correlation with virus nucleic acid operating at a minimal level. This interpretation is promoted by the unique structural features of the viroid and its

461

0066-4286/79/0901-0461$01.00

lack of messenger RNA activity. Thus we may anticipate that the biological and pathogenic activities of viroid RNA may present a view equally unique as are the physical and chemical characteristics of this important class of plant pathogenic molecules.

This paper contains a discussion of some of the biological properties of viroid RNA with special emphasis on host cell interactions. I hope this discussion will help identify observations that are relevant to both the process of replication and the process of pathogenesis. Recent advances in the physical description of viroid RNA are presented here principally for this purpose and also to resolve certain alternative hypotheses that have been proposed to account for the biological properties of the viroid.

PHYSICAL PROPERTIES

Molecular Size and Conformation

During the earliest efforts to characterize viroid RNA, estimates of the molecular weight of potato spindle tuber viroid (PSTV) and citrus exocortis viroid (CEV) ranged from 50,000–60,000 (12) to 110,000–125,000 daltons (76). This apparent discrepancy was explained by a study of the effect of polyacrylamide gel porosity on the relative migration of CEV-RNA (70). In gel concentrations less than 8–10%, the viroid migrated as a species with the apparent molecular weight of about 100,000, whereas in gel concentrations greater than 8–10%, the estimated molecular weight was about 50,000. These estimates were based on the distinct slopes of relative mobility evident in the two ranges of gel concentration. Explanation for the disparate results based on the segregation of specific dimeric forms by variation in gel concentration (15) discounted the possible importance of the unique conformation of viroid RNA. No evidence of dimers was observed after the viroid was visualized in polyacrylamide gel electrophoresis (PAGE). The broad range of molecular weight estimates presented for PSTV (12) resulted from estimates of molecular size based on the distribution of infectivity as determined in a systemic bioassay and not from direct measurements of a discrete nucleic acid species.

The effect of gel porosity on the relative migration of CEV is dramatically illustrated by a comparison with a small RNA (~9S) present in extracts of healthy *Gynura aurantiaca*. In a 5% gel CEV is the slowest migrating of the major small RNA species. In a 10% gel the 9S component comigrates with CEV, while in a 15% gel the 9S component migrates more slowly than the viroid (70). A similar phenomenon has been observed for PSTV (47) suggesting a phenotypic conformation for viroids or viroid-like molecules. Both CEV and PSTV demonstrate properties characteristic of double-stranded (ds) and single-stranded (ss) molecules. CEV elutes from methylated albumin and Cf-11 cellulose as a ds molecule and is resistant to inactivation by diethyl pyrocarbonate and yet is susceptible to RNase and formaldehyde inactivation (70, 74, 75). These dichotomous properties could provide an explanation for the divergence in migration in the gel series. Migration in gels less than 10% may be influenced principally by the ss regions of the molecule while migration in gels greater than 10% may reflect ds regions. Gel porosities in the region of 10% may constitute a transition zone between the characteristic steep slope of migration by a ss molecule and the more gentle slope by ds molecules (30).

This property provides further indication that, in addition to the composite ss-ds feature of the structure, the potential interactions with host molecules are similarly expanded.

The absence of any messenger RNA capabilities for CEV and PSTV has stimulated a broader range of considerations for the possible interaction of viroid-RNA with the host genome. Specific interactions of hairpin helical RNA in the priming of DNA synthesis (60) may be pertinent to the phenotypic structure of viroid RNA. The viroid RNA may also function in an indirect capacity not as the "activator RNA" as suggested in the model of Britten & Davidson (4), but as the "agent" which is recognized by the sensor gene resulting in the appearance of the "activator RNA." Structurally similar regions characterized by palindromic or inverted repeated sequences resulting in intrastrand base pairing may be important in the organization of the eucaryotic chromosome. These regions conceivably

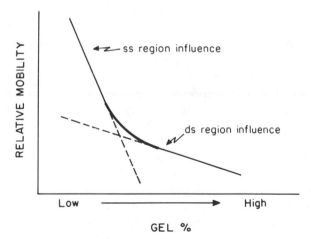

Figure 1 Theoretical description for the effect of single-stranded (ss) and double-stranded (ds) regions of viroid RNA on the relative mobility in electrophoresis in various concentration of polyacrylamide gel electrophoresis.

could function as physically distinct centers for binding of regulator RNA. Small nuclear RNAs have even been implicated in the programming of chromosomal information (22). These extrapolations are presented to introduce alternative mechanisms for the function of a nontranslated small RNA with the implicit importance of structural and conformational properties in this interaction between the viroid and its host.

The sedimentation studies of Säanger et al (64) have confirmed the molecular size of CEV (119,000 daltons), PSTV (127,000 daltons), and cucumber pale fruit viroid (CPFV) (110,000 daltons). These values clearly support the estimates made in the 5% PAGE system (76) under nondenaturing conditions. This electrophoretic system remains valuable as the primary medium for rapid identification (48) since the viroid RNA is easily identified as the slowest migrating low molecular weight species. Purification of preparative quantities of viroid RNA can be also obtained by repeated electrophoresis (47) or coupled with electrophoresis in denaturing gels containing urea or formamide (56, 61).

Secondary Structure and Conformation

Structural stability of the viroid RNA could be predicted from the high specific infectivity and resistance to thermal inactivation of even partially purified preparations. Models for the secondary structure of CEV (64, 70), CPF (31), and PSTV (25) derived from the highly cooperative thermal denaturation profiles and NMR spectroscopy suggest that the viroid is a highly based-paired molecule rich in G-C bonds ($\sim$70%). The intrinsic structural stability of the viroid RNA is greater than that of tRNA and approaches that of dsRNA when in the presence of Mg^{2+} (71). This is not to suggest a structural identity with either tRNA or dsRNA, because the viroid is a totally unique form of RNA. The conformational intricacies will undoubtedly be the subject of further analysis as a minimal form of replicating biologically active molecule.

The closest correlation to a structurally similar replicating nucleic acid in terms of conformation is that of the MDV-1 variant of the RNA bacteriophage Qβ. This aberrant form of Qβ RNA contains significant regions of intrastrand antiparallel nucleotide sequences (46). The viroid RNA differs structurally since the 218 nucleotides of the MDV-1 RNA demonstrates branching while the viroid RNA appears to conform to a "rodlet" structure under nondenaturing conditions. It is very inviting to enlarge our perspective of viroid RNA function by considering the biological activity of MDV-1 RNA. The MDV-1 sequence was derived from rapid passage of Qβ RNA in a Qβ replicase system. The 218 nucleotides that survived this simulated in vitro selection system are presumably enriched for sequences indispensable for recognition of the RNA replicase. Since the MDV-1 sequence is inadequate to translate the entire replicase molecule, a structural dependence on phenotypic recognition reaction for the replicase must constitute the primary biological function of the MDV-1 sequence. By analogy, the structural complexity of the viroid RNA may also be essential for replication by a host-specified RNA-dependent RNA polymerase system. As a corollary, the structural phenotypes of the viroid may also be a significant factor in the process of pathogenesis. The expression of host genetic information may be influenced by the initiation action of viroid RNA hairpin and helical sequences.

Circular and Linear Forms of Viroid-RNA

As early as 1970, fully two years before the positive identification of a viroid RNA, experiments on the inactivation of CEV (74) and PSTV (11) showed that viroid infectivity was partially resistant to phosphodiesterases. From this it was speculated that circular structures were involved. Direct evidence for the existence of circular, single-stranded viroid RNA was reported by McClements' electron microscopy studies of purified PSTV (44, 45). Along with the circular structures, more predominant linear forms of presumably the same viroid RNA were evident. Analysis of 3' and 5' end of a more highly purified cucumber pale fruit viroid indicated the absence of free termini (64) suggesting a covalently closed circular molecule. These studies were not supported by analysis of the specific infectivity of the structural forms, however. The circular form retained a high specific infectivity in the presence of only about 1% contamination with linear structures.

Separation of linear and circular forms was attempted by electrophoresis in denaturing gel containing formamide and urea (56). The linear forms of PSTV were estimated by electron microscopy to be contaminated with only 0.2% circular forms; nevertheless, the specific infectivity of both forms was about equal, suggesting that both forms of PSTV were infectious. Since the unfractionated preparations contained only about 18% circular molecules, if the circular molecules only were infectious, separation must be accomplished in the presence of the large mass of noninfectious linear forms. In addition, the specific activity of the circular forms must also be considered. From the data reported by Owens et al (56), the infection presumably associated with the linear form does not correlate directly with the stained band in PAGE. Since the distribution of linear and circular forms of PSTV was not made from the area of the gel displaying maximum infectivity, the conclusion that the linear molecules are infectious is not unequivocal. The contention that only the circular forms of PSTV are infectious is further supported by similar studies by Morris (47) of the infectivity of forms separated by successive cycles of electrophoresis including denaturing conditions. Infectivity was associated almost exclusively with a slowly migrating, presumably circular form of PSTV. However, positive identification of the molecular structures of the two migrating species was not corroborated by electron microscopy.

Even though an element of uncertainty emerges from these independent studies, the idea of the circular form as the infectious species is consistent and therefore must be accepted as the probable native form of the viroid RNA. The association of infectivity with the linear structures may be influenced by variations in the methods of extraction, and by separation of the structural forms of the viroid RNA resulting in differential cross-contamination of the structural species.

Little consideration and study have been directed to the importance and occurrence of these forms in vivo. Preparations with less than 1% linear form can be achieved (65) although it is not certain whether this value represents an accurate estimate of in vivo concentration. Nevertheless, in the process of replication of the viroid sequence, the synthesis of a complete linear form of the pathogenic RNA must precede ligation to a circular form. Whether these linear structures possess the biological activity of viroid RNA remains to be resolved.

Primary Sequence of Viroid RNA

Initial efforts to determine the sequence of nucleotides in viroid RNA utilized two-dimensional fingerprints of RNase T_1, and pancreatic RNase digests of [125]I-(10), and ^{32}P-labeled (18) viroid RNA. These methods indicated a degree of sequence complexity compatible with a 110,000 to 125,000 daltons molecule. The distinct oligonucleotide patterns demonstrated the unique nucleotide sequences in PSTV, CEV, and chrysanthemum stunt viroid (CSV). This characterization was especially significant since separation of viroids such as PSTV and CEV is very difficult or even impossible with biological tests.

Transmission experiments by Van Dorst & Peters (77) with CPFV suggested that the viroid RNA sequence might be influenced by the host environment resulting in species adaptation and the resultant apparent selective loss of infectivity. Analyses of CEV and PSTV both in *Gynura* and tomato indicated that the fingerprint patterns of both viroids were constant regardless of the host species infected (8). The inability of isolates of CPFV to produce disease with passage to different cucumber varieties could be attributed to inadequate threshold concentrations in the crude inoculum.

Conclusive evidence for the definition of the physical size and structure of a viroid RNA was provided by the contribution of Gross et al (24) in establishing the primary sequence of PSTV. Utilizing (5'-^{32}P)-labeled RNase T_1 and A fragments, a structure of 359 ribonucleotides in a covalently closed ring structure emerged for PSTV-RNA. The nucleotide se-

quence can accommodate a model with the high degree (~80%) of intrastrand base pairing that has been predicted for viroid RNA (31, 70, 71).

Even though the recognition sites for the synthesis of a viroid RNA
template as well as the presumed ligating enzyme are contained within the
sequence, little direct information can be deciphered about the replication
and pathogenesis of viroid RNA. The discovery of nucleotide sequences
coincident with other viroids and selection of specific mutants will help
clarify the significance of discrete regions of the RNA in the recognition
reactions. Since no common AUG nucleotide initiation triplets are contained in the sequence, it appears that the viroid RNA cannot be translated
in vitro (29). An RNA sequence complementary to PSTV-RNA can be
predicted from the primary sequence; but since no AUG triplets are found,
it appears that neither the viroid nor its putative complement may function
in a messenger RNA capacity. This adds to the evidence for the action of
the viroid RNA as a regulating molecule distinct from viral nucleic acids.
This absence of any direct evidence for translation of viroid or viroid
complementary sequences marks a point of distinction between viral and
viroid nucleic acid infection—the former but not the latter is accompanied
by the introduction of new genetic information.

A question related to the previous discussion of the circular and linear
forms of viroid RNA is as follows: does the linear form constitute a precursor or degradation product? If the latter condition is true, then does the
reduction in biological activity parallel any single scission in the polynucleo-
tide chain? Since the complete nucleotide sequence does not appear to be
necessary for the synthesis of protein intermediates for replication and
pathogenesis, perhaps retention of the structural integrity of the viroid
RNA molecule even in the presence of single diester cleavage is possible and
a critical precondition for infection by viroids. The phenotypic properties
common to all viroids would therefore be essential to the expression of
biological activity in viroid infection.

VIROID REPLICATION

The description of the physical structure of viroid RNA presents the formidable task of synthesizing a unique molecule, a single-s¹ ʳanded, circular
RNA of about 350 nucleotides in a plant system. This challenge illustrates
the limited level of our understanding of the viroid replication process with
progeny synthesized even in the absence of demonstrable disease symptoms
(53). The existence of symptomless hosts can be taken as evidence that
synthesis of viroid RNA can occur in the absence of dramatic physiological
disturbances. In fact, it may be valuable in future studies to investigate the
process of viroid replication in a tissue lacking the pronounced stunting

effect characteristic of viroid infection. In this manner, it may be possible to interpret more clearly the various characteristics of viroid RNA and viroid-complementary sequences with replication separated from pathogenesis. Nevertheless, the limited genetic potential of the viroid RNA reinforces the implicit host-dependent nature of viroid replication.

Viroid Template

A critical component in any nucleic acid replication scheme is the resolution of a template or intermediate. Theoretically, progeny viroid could be generated via an RNA or DNA template synthesized either pre- or postinfection. A scheme involving an RNA intermediate synthesized after infection would parallel the process of RNA virus infection. Consideration of a DNA template produced postinfection would require the activity of a RNA-directed DNA polymerase, an enzyme not yet recognized in plant systems. This process might parallel replication of oncogenic viruses. An alternative mechanism for involvement of a DNA intermediate centers on the activation or retrieval of preexisting viroid sequences from the host genome. This process assumes that viroid sequences constitute a normal component of the host genome or have been inserted in the genome via the reverse transcriptase activity in some previous encounter between the host and viroid.

Detection of viroid complementary sequences initially was accomplished in DNA-rich preparations by hybridization with ^{125}I-CEV (67). Complementary CEV (cCEV) sequences were localized in nuclear-rich fractions from infected, but not healthy, *Gynura* and tomato leading to the suggested occurrence of complementary DNA. Even though partially purified preparation of DNA retained complementary sequences, however, DNase could not completely destroy the CEV complementary sequences introducing the possible existence of an RNA variety trapped in the DNA matrix (67).

When nucleic acid preparations from CEV-infected *Gynura* were partitioned with 2M LiCl, the sequences could be identified in RNA fractions (23). Because the viroid RNA displays a high degree of self-complementing sequences, it was necessary to verify that the RNase-resistant sequences were not generated from hybridization of the ^{125}I-CEV probe with unlabeled CEV. Detection of the viroid complementary RNA (cRNA) was specific for the ^{125}I-CEV probe, and could be chased with unlabeled CEV. Sensitivity of the complementary sequences to RNase confirm the presence of an RNA complement to the citrus exocortis viroid only following infection. A similar cRNA form of PSTV has recently been confirmed (M. Zaitlin et al, unpublished). The replication of viroid RNA via an RNA template must therefore assume a prominent role in any consideration of viroid synthesis.

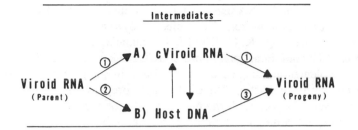

Figure 2 Schemes for replication of viroid RNA. 1. RNA-directed RNA polymerase, 2. "activation" of viroid complementary DNA, 3. DNA-directed RNA polymerase.

An alternative mechanism presented by Hadidi et al (28) centers on the existence of DNA sequences complementary to ^{125}I-PSTV. The reported hybridization of PSTV with cellular DNA was not exclusively correlated with viroid infection since DNA from healthy hosts as well as nonhost demonstrated similar levels of the complementary sequences. DNA from bean, which is not a host of PSTV, contained sequences complementary to a larger portion of the PSTV molecule than DNA from *Gynura,* a host of PSTV. The complementary sequences from bean also formed more credible RNA-DNA hybrids than from *Gynura* as reflected in the optical properties of T_m and thermal transition. Therefore, interpretation of this molecular hybridization data must be viewed with caution.

It is unfortunate that the report (28) did not include a challenge to the ^{125}I-PSTV probe in the form of a totally heterologous DNA, such as from an animal source. Under experimental conditions such as the 336-hr hybridization, it is essential that the probe does not contain traces of host RNA or that ^{125}I is not released from the probe and reacts with cytidine residues in the DNA. The stability of the ^{125}I-label must also be verified. The use of foreign ^{125}I-probes would indicate the extent of nonspecific hybridization to the various DNAs. A further, more theoretical, inconsistency centers on the interpretation that the viroid may function as a regulatory RNA, and yet PSTV hybridizes to infrequent or even single-copy DNA sequences. Thus, by definition, the viroid class of regulatory RNA must be distinctly different from the class of RNA regulator proposed by Britten & Davidson (4) which is related to repetitive DNA sequences. No evidence of RNA sequences complementary to PSTV was detected in these studies; however, more recently, the cRNA form of PSTV suggested by M. Zaitlin et al (unpublished) has been confirmed by A. Hadidi (T. O. Diener, personal communication).

In an attempt to reconcile the evidence for the existence of RNA and DNA sequences complementary to viroid RNA, further analysis of the iodinated probes and DNA extracts have been made. The 2M LiCl-soluble DNA-rich preparations from CEV-infected *Gynura* can be freed of contaminating RNA by (L. K. Grill and J. S. Semancik, unpublished) sedimentation in CsCl containing guanidinium-HCl. No evidence for the extent of viroid complementary sequences can be found in the purified DNA when compared with the starting 2M LiCl-soluble preparations or DNA from healthy tissue. Coupled with the previous data for the existence of cRNA in CEV-infected tissues, these data suggest that only viroid-complementary RNA exists in this system. The precise nature of these RNA sequences remain uncertain. The presence of viroid-complementary RNA has been confirmed in both 2M LiCl soluble (DNA-, tRNA-rich) and 2M LiCl insoluble (rRNA-rich) (22) preparation. Therefore, both a ssRNA sequence, the properties of which may be influenced by the presence of DNA, as well as a dsRNA analogous to the RF and RI structures for virus replication must be considered as possible models.

The authenticity of the DNA sequences complementary to PSTV has been challenged by competition hybridization studies (M. Zaitlin et al, unpublished). Addition of unlabeled tomato ribosomal RNA to the hybridization reaction of ^{125}I-PSTV and tomato DNA reduced the extent of hybridization significantly (fivefold). If PSTV was purified through a urea-containing gel, the competition effect was not observed, demonstrating the problem of contaminating host RNA sequences in the iodinated viroid probe. Neither the percentage of PSTV hybridization nor the presence of viroid-complementary sequences in DNA from healthy extracts as reported by Hadidi et al (28) could be confirmed. Therefore, the existence of viroid-complementary DNA sequences has not been confirmed for even the PSTV system. The possibility that low level hybridization to DNA may exist should still be entertained since the activity of viroid RNA or a genomic regulator may require this direct interaction. This viroid RNA : host-DNA reaction, however, may not reflect complete homology with the entire viroid sequence, but either a partial homology or even some unique reaction of affinity to the host DNA.

Actinomycin D and Viroid Replication

Additional data for the contention that a DNA template is involved in the replication of PSTV center on the inhibition of incorporation of ^{14}C uracil into PSTV when either leaf strips (17) or tomato nuclei (78) was tested with Actinomycin D. In both systems, the incorporation of ^{14}C into PSTV was very low (2 to 3 times over background). Addition of 30–50 μg/ml of Actinomycin D effectively totally reduced incorporation into ribosomal RNA as well as PSTV. Assuming the classical mode of action of Actinomycin D in inhibiting the synthesis of DNA-directed RNA synthesis, these results were interpreted as suggesting that PSTV was replicated via a DNA template.

Before these data can be taken as unequivocal, additional factors must be considered. Actinomycin D has been demonstrated (6, 39) to inhibit the synthesis of viral RNA through a RNA-directed RNA polymerase. Furthermore, the total host dependence of the viroid synthesis might be influenced by the high concentration of the inhibitor which may constitute a near lethal dosage. Finally, if Actinomycin D was effective in the classical mode of action, the most cautious interpretation would suggest an essential DNA-dependent step in viroid synthesis which might be any host-mediated function and not necessarily the function of a DNA template for viroid replication.

The Putative Viroid Replicase System

Even though no experimental data elucidating the role of any in vivo enzymatic system in the synthesis of viroid RNA are available, the template models can be used to predict possible systems. The cRNA intermediate might result from the activity of an RNA-directed RNA polymerase on a parent viroid. An enzyme with this specificity has become accepted as a constitutive host enzyme in many plant species (32, 38, 63). Invoking a cDNA intermediate would require the activity of a RNA-directed DNA polymerase (reverse transcriptase) not described for plant systems. Healthy plant genomes containing homologous sequences might have required these integrated sequences by some previous encounter with viroid RNA. The host-specified, DNA-directed RNA polymerase would then complete synthesis of linear form progeny. Alternatively, a minor enzyme such as RNA polymerase III (59) may accept the viroid or complementary sequences. Most interesting, if the circular form of viroid RNA is the native structure, the activity of an unprecedented RNA ligating enzyme in plants must be predicted for conversion of the linear, single-stranded viroid RNA to the circular form. The existence of "splicing enzymes" of mRNA in eucaryotic systems (3) makes the prediction more tenable.

Viroid RNA can function in a number of in vitro RNA or DNA polymerase systems (21, 54, 55); however, the viroid template molecules were probably linear forms and the products did not approach full size viroid with any degree of efficiency.

Locus of Viroid Synthesis

Evidence for the subcellular locus of viroid synthesis can be deduced principally from the detections of the viroid by bioassay or gel electrophoresis, and the viroid complementary RNA (cViroid) by molecular hybridization.

Recent developments in the synthesis of copy DNA probes (54) may provide a valuable tool for elucidating the early stages of viroid infection by molecular hybridization with small amounts of viroid RNA. This system should increase the sensitivity of detection above the level of bioassay or direct gel analysis. These probes must reflect a random sequence of the intact viroid if detection of the intact biologically active viroid molecule is to be achieved.

The primary role of the nucleus in the replication of viroid RNA has been suggested by the detection of infectivity associated with nuclear-rich preparations (13, 72) as well as the incorporation of ^{14}C uracil into PSTV with nuclei (78). The exclusive association of PSTV with nuclei (13) could not, however, be confirmed with extracts from CEV-infected tissue in which the integrity of the endomembranes was maintained (72). A significant percentage (60%) of the total recoverable viroid infectivity was associated with subcellular components sedimenting more rapidly than free ribosomes. Since neither CEV nor PSTV could be detected in organelle-rich (i.e. chloroplast, mitochondria) preparations, a viroid-membrane complex might be postulated. Whether this structure constitutes a replicating complex or simply the accumulation of progeny viroid has not been determined. CEV-RNA was not detected by bioassay of postribosomal soluble cytoplasmic preparations even when treated under conditions that assure the recovery of exogenously added CEV to healthy extracts.

The occurrence of viroid RNA in nuclear and membrane preparations can be contrasted with the detection of viroid complementary RNA sequence (23) in nuclear and cytoplasmic fractions. If the cCEV sequences can be considered a putative viroid template, the cytoplasmic and nuclear loci might suggest a bipartite synthetic scheme in which the parent CEV-RNA is first copied in the cytoplasm by the well characterized soluble RNA-dependent RNA polymerase, then mobilized to the nucleus where it functions as the template for progeny viroid. The final synthetic step involving ligation to a circular structure is then anticipated as a nuclear function. Association of the progeny viroid with components of the endomembrane system provides an introduction to the process of viroid pathogenesis.

VIROID PATHOGENESIS AND HOST RESPONSES

The perception of viroid pathogenesis is even more vague than present understanding of viroid replication. Perhaps the interaction of the viroid RNA with the host cell represents as unique a biological process as the physical nature of the viroid itself. The interaction of a minimal infectious RNA with the host cell may, nevertheless, describe functions common to other pathogenic agents especially viruses.

Severe stunting, epinasty, and leaf rugosity are the most common symptoms of viroid-induced diseases. This series of common host responses is illustrated by the production of spindle tubers on potatoes inoculated with citrus exocortis viroid (69). Separation of viroids by biological reactions thus remain difficult. In addition, viroid diseases can be described as a persistent infection in which the titer remains constant and high over an extended growth curve in vegetatively propagated plants (75). This stable relationship and intimate association with host cell activity appears analogous to a form of cellular transformation.

An attempt to outline selected aspects of viroid pathogenesis will be made through a discussion of subjects concerned with (a) variation within viroids and its relationship to host symptomatology, (b) host responses to viroid infection, and (c) cell cycle and viroid infection. The distribution of biologically active viroid and the complementary RNA sequences discussed in connection with viroid replication may also be relevant to pathogenesis. Nevertheless, since replication of viroid RNA can occur in the absence of symptoms, observations of the diseased tissue may reflect host-specified secondary responses to the viroid infection.

Viroid Strains and "Cross Protection"

Field isolates of CEV and PSTV (19) display a gradient of symptom expression ranging from severe to almost symptomless conditions. Viroids that induce specific types of symptoms appear to be stable in successive transfers and have been assumed to constitute viroid strains. These isolates may have been derived from near limit dilution inoculations, but it must be simply assumed that the distinct host responses reflect separated strains and not a mixture. Considering the limited size of the viroid genome, the quantitative and qualitative differences in nucleotide sequence between mild and severe strains define a part of the host-viroid interaction. Oligonucleotide mapping of PSTV isolates suggests that the host response is affected by differences of 2–10 nucleotide residues (9).

The relationship between these apparent stable viroid strains and the "cross protection" phenomenon has been reported by Fernow (19). When tomato was inoculated with a mild strain of PSTV prior to challenge inoculation with a severe strain, protection was partial and temporary. Expanding these studies to include four viroids, "interferences in symptom expression" were observed and interpreted as cross protection (52). In many cases the apparent "cross protection" effect as judged by the decrease in severity of symptoms was simulated by substituting buffer in the primary inoculation; this demonstrates that the age of the host plant is a significant factor in symptom response. To minimize the effect of a seven-day delay in challenge inoculation, it would be instructive to doubly inoculate with a variable ratio of the mild and severe strains. The additional parameter of "days to appearance of symptoms" suggests a significant effect of delay of symptoms; however, only three classes of 18-day intervals are reported from a 75-day recording period. Finally, when doubly infected plants were bioassayed, both viroid strains were detected.

These uncertainties present severe questions as to whether the effect on viroid symptom expression can be interpreted as a demonstration of "cross protection." The CEV isolate used in these studies (52) when inoculated to tomatoes either at the cotelydon stage or at various times later in plant development will produce the full range of symptom expression from severe to symptomless (68; J. S. Semancik, unpublished). Also, the mild, moderate, and severe isolates of CEV identified in *Etrog* citron demonstrate no cross protection (E. C. Calavan, unpublished), even though they are highly stable variants. This phenomenon of development of symptom expression may be related to the discussion of initiation of viroid infection and the "receptive" host cells to be discussed later.

Host Response to Viroid Infection

Analysis of the distribution of viroid RNA by bioassay describes the subcellular locus of the biologically active molecule. The survival of viroid RNA through such a procedure when exogenously added to extracts from healthy tissue demonstrated that CEV was rendered resistant to degradation in tissue sap presumably by associating with rapidly sedimenting cellular constituents and postribosomal supernatant fractions (72). Accumulation of both PSTV (13) and CEV (72) in the cell nucleus is well established. Isolated chloroplasts, mitochondria, and ribosomes do not show evidence of viroid infectivity. When extraction is performed under conditions preserving the integrity of membrane constituents, however, a significant portion (60%) of the recovered viroid activity was associated with membranes (72). Contamination by broken nuclei could not account for this level infectivity. The data point not only to a nuclear, but also to a membrane association of viroid RNA. These loci must be considered first, simply as sites of accumulation of progeny viroid, and second, as possible descriptions of regions of synthetic or pathogenic expression. Nevertheless, the absence of viroid RNA in soluble (postribosomal) fractions from infected tissues, as is the case of control experiments of CEV added to healthy sap, implies a functional association. The continuity of the endomembrane system with the nuclear envelope is consistent with a possible integration in the processes of synthesis and pathogenic expression.

The developmental aberration resulting in the severe stunting symptom has not been associated with a distinct cytopathic lesion. Observations of the sites of viroid accumulation indicate no changes in nuclear structure; however, the manifestation of vesicular and tubular proliferation is evident in the paramural region (73). The plasmalemmosomes, or endocytic invaginations from the plasma membrane are not viroid-specified cytopathic structures (40, 41). Nevertheless, the increased frequency of these structures may reflect a developmental lesion introduced as a result of viroid infection.

Symptom expression in *Gynura aurantiaca* following CEV infection is also accompanied by the accumulations of a low molecular weight protein, CEV-P_1 (5). The 14,000 dalton protein can be detected in preparations from lysed protoplasts as well as postribosomal supernatant preparations. No specific in vitro or in vivo biological property can yet be attributed to the CEV P_1 protein (20). The direct correlation with symptom expression and the influence of the host species on molecular weight (V. Conejero, unpublished) suggests a host-specified product.

The effect of viroid infection resembles a hormonal malfunction resulting in abnormal cell development. A preliminary investigation of endogenous plant growth substances indicates that while no change occurred in amounts of abscisic acid and indoleacetic acid, a significant decrease was detected in the amounts of gibberellins in CEV-infected *Gynura* (62). The reduction in gibberellic acid (GA) concentration observed at 30 and 60 days after inoculation correlated well with the severe stunting host response. Since GA "can regulate nuclear events and modify metabolism of some or all RNA species" (34), viroid replication and/or pathogenesis may be accompanied via this class hormonal intermediary.

Included in this section are data describing a wide range of host responses induced by viroid infection. Whether the nuclear and membrane accumulation of viroids is related to the cytopathic structures and altered protein patterns as an intermediate response in what could ultimately represent a hormonal imbalance remains unclear.

Cell Cycle and Viroid Infection

As the final topic in this section on viroid pathogenesis, I would like to present a hypothesis for the synchrony of viroid infection with cell differentiation and/or mitotic activity. These thoughts are offered as gross speculation. Nevertheless, some of the observations support an intimate association between CEV and the host cell that might be anticipated for this class of highly successful minimal infection molecules.

The "half-leaf system" of CEV in *Gynura* (73) clearly demonstrates that it is possible to obtain viroid-containing, symptom-expressing tissue on the same leaf as viroid-free, asymptomatic tissue. If the site of inoculation is parallel with the position of petiole attachment to the stem, the initiating viroid inoculum can become established in only one half-leaf. An interpretation of this reaction is that the inoculum is carried in the vascular system more directly to the half-leaf proximal to the inoculation site. The distal

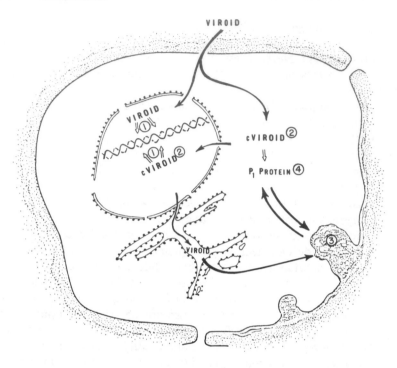

Figure 3   A perspective of viroid pathogenesis. 1. Association of viroid or viroid complementary RNA with host nucleus (chromatin), 2. synthesis of viroid complementary RNA, 3. cytopathic response in endomembrane system, 4. alteration in host-specified products (low molecular weight proteins, endogenous hormone levels).

half-leaf has evaded infection by passing through the "receptive" phase. All more mature leaves similarly remain resistant to viroid infection. The target cells susceptible to viroid infection have been suggested as being linked to cellular development (5, 73). All leaves less differentiated than the "half-leaf" become uniformly infected as do all axial shoots that develop later. The infected/resistant half-leaf resulting from this critical temporal progression can be thought of as bracketed between younger susceptible and older resistant tissue. Following this initiation phase, CEV is persistant and all subsequent apical tissue maintains a stable titer of viroid RNA over a period of years (70, 71). Maximum titer of CEV occurs in mitotically active tissues with no evidence of a recovery phenomenon to healthy conditions as observed with many viral infections.

Since CEV-infected tissue can be freed of the viroid by meristem tip culture (51), the apparent coordination with cell division does not coincide with mitosis, but must involve invasion at some later developmental stage.

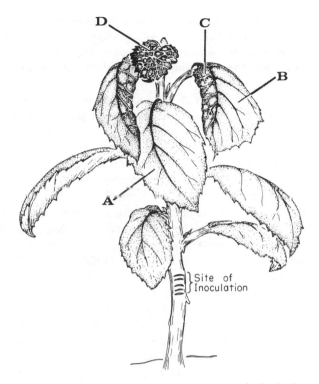

Figure 4 Diagram for the initiation of citrus exocortis viroid infection in *Gynura aurantiaca* —the "half-leaf" model. *A.* Symptomless mature leaf, *B.* symptomless half leaf, *C.* "infected" half leaf, *D.* systemically "infected" leaf [from (73)].

The number of cells in these conditions would be influenced by the age of the plant and perhaps be reflected in the severity of symptoms.

An additional experimental procedure that reinforces this relationship of viroid infection to cell division utilizes double infection with the crown gall organism, *Agrobacterium tumefaciens* and CEV (68). Stem tissue of tomato, which is a very poor source of CEV, when stimulated by oncogenic activity rapidly proliferates cells that provide an excellent source of CEV. This response can be interpreted as defining a homology of cellular environment between the transformed cells and cells developmentally receptive to viroid replication. The yield of viroid RNA from this neoplastic tissue is significantly greater than from apical tissue.

The observation by Hadidi (27) that ^{32}P, when applied to tomato plants at two weeks after inoculation, was incorporated into PSTV, but not when administered at three weeks after inoculation, might be treated as incompatible data. To the contrary, however, viroid infection normally takes 10–14 days to become established, and in the case of severe infection by PSTV, by three weeks postinoculation, the growth of the plant may have essentially stopped or at least mitotic activity decreased to the level where detection of synthesis of a low concentration component like PSTV would be very difficult.

Protoplasts isolated from viroid-infected tissue (49) as well as inoculated in vitro (50) have been reported to incorporate (^{3}H) uridine. It was conceded that the low rate of viroid synthesis may result from the low capacity of differentiated cells. Viroid synthesis could not be detected when protoplasts from two varieties of tomato were inoculated with three viroids that sustain good viroid titers in plants. Incorporation into CPFV, but not CEV and PSTV, was observed in protoplasts from the variety Hilda, which is a symptomless host of all these viroids. Acceptance of this evidence for viroid synthesis in protoplasts would be strengthened by detection of labeled precursors into viroid characterized by slab gel electrophoresis and direct autoradiography. Nevertheless, the data do not refute the possiblity that the bulk of viroid synthesis does not occur in mitotically active tissue, but that a "low-rate" of viroid replication occurs in other cells.

The analogy may be drawn to DNA synthesis in which the major activity occurs prior to cell division but maintenance activity is detected through the cell cycle. This replication optimum may be controlled or influenced by the appearance of a critical host component that oscillates during the cell cycle. Implicit in this scheme for the temporal susceptibility to viroid infection is the alteration in some structural feature of the cell as induced by developmental progression. Since the viroid infection centers on a nuclear locus, the differential replication or composition of DNA during development may expose a viroid-related transitory event. Pearson et al (58) reported major differences in satellite DNA from meristematic tissue as opposed to mature, differentiated tissues. Stimulation of satellite DNA synthesis has been reported following stress conditions such as wounding and infection (26) as well as application of gibberellic acid (35). It is interesting to note that plant species in the family Rutaceae, susceptible to CEV, and Curcurbitaceae, susceptible to CPFV, are especially high in satellite DNA content (33). The primitive stage of our comprehension of viroid pathogenesis encourages me to present the possible correlation among such factors as chromosomal structures, satellite DNA, GA, and viroid infection. Verification, much less elaboration of mechanistic pathways, must survive the challenge of experimental data. This focusing on alterations of the host genome regulation is stimulated by the common symptom expression of most viroid infections that simulate a permanent impairment in the endogenous hormone system. Since the viroid and the viroid complementary RNA sequences of nucleotides have been detected in nuclear preparations and in association with DNA, both molecules either directly or indirectly might mediate control of the regulatory mechanism.

Theoretical schemes have been proposed for the function of small RNAs as regulators of host genome expression characterized in association with middle repetitive DNA (4). Support for this mechanism can be drawn from the description of small nuclear RNAs (snRNAs) (79). Data on the functional significance of snRNAs are lacking; however, a class of small stable nuclear (ssn) RNAs have been found associated with the chromatin of mouse L cells (1). These ssnRNAs demonstrate quantitative and qualitative differences dependent upon growth condition of cells, organ of origin, and tumorigenesis.

The behavior of snRNAs during mitosis has been studied in *Amoeba proteus* (22). The pattern of snRNA association with mitotic chromosomes suggests a role in the "programming" of the genome. The mobilization of snRNAs away from condensed chromosomes (metaphase) and in association with anaphase chromosomes suggests a differential affinity. Since only a small amount of snRNAs remain associated with the interphase chromatin, a positive role in gene expression has been conjectured.

The viroid RNA could qualify as a class of snRNA both in terms of its locus and metabolic stability. The distinctive nature of this "pathogenic" class of snRNA, the viroids, then centers on the ability to induce severe stunting, an aberrant host mechanism interpreted as a disease syndrome. Yet as it has been emphasized here already, viroid synthesis may proceed in the absence of any symptom expression. Therefore, it would appear that the sketchy observation and the intuition presented for the CEV system are compatible with the outline of the occurrence and possible metabolic activity of the snRNAs. The special advantages inherent in a pathological condition for the characterization and identification of a biologically active molecule such as the viroid may comprise a valuable tool in continued investigation of molecule counterparts in normal cellular metabolism.

EXTRAPOLATION TO OTHER BIOLOGICAL SYSTEMS

Implicit in the previous discussion of the relationship of viroids to other small nuclear RNAs is the functioning of similar molecules in the regulation of normal cellular genome expression. If the activity of these molecules is integrated with developmental events, the concentration and half-life of these regulators may present formidable obstacles to detection, much less characterization.

A technically more feasible approach centers on the investigation of other disease conditions in plants and animals. With the hypothesis that some lesion in developmental, perhaps even hormonal, control is involved, attention is directed to cellular transformation in plants by *Agrobacterium tumefaciens*. Beljanski (2) reported that viroid-like RNA from bacterium may initiate tumorigenesis in *Datura*. Investigation of primary tumors (68) and cell suspensions of transformed tobacco cells (J. S. Semancik, unpublished) could not confirm the presence of viroid RNA associated with transformed cells. Nevertheless, the concept that the new characteristics expressed by neoplastic or transformed cells might be derived from some gene normally functioning during an earlier stage of development is compatible with the view of viroid pathogenesis expressed here. The inability to detect viroid RNA may be influenced by concentration or survival parameters of intermediary small RNA components functioning in a manner analogous to viroids.

Low molecular weight RNAs of the same order of magnitude as viroid RNA accompany plant virus infection (36, 65). These RNA species, such as the satellite of tobacco ringspot virus (S-TRSV) and cucumber mosaic virus RNA 5, differ from viroid RNA in being dependent for replication upon viral RNA synthesis, and therefore, have been termed *satellite* or *defective RNA*. Since the RNA sequence retains recognition for the viral polymerases, interference with viral replication is observed (65). This property, along with the absence of a detectable mRNA function (57), resembles viroid RNA properties. Although the satellite RNAs alone do not elicit a host response, when combined with virus infection they may produce a dramatic increase in disease severity (37). These anomalous RNA components suggest that the expression of plant viral infection may be mediated, at least in part, by viroid-like RNAs.

A major extrapolation to animal systems may not be extreme as might appear at first hand. Are the viroids a unique form of pathogen limited to plant systems, or, as in the development of the science of virology, are these molecules to become the classical progenitors of nucleic acid species in other biological systems?

The estimated target size and stability characteristics of the scrapie agent (SA), the cause of a neurological degeneracy in animals, has stimulated interest in the theoretical (14) as well as experimental demonstration (43) of the similarity of the SA to plant viroids. Recent reports indicate that the SA can be distinguished from viroid RNA by the demonstration of a DNA component critical to scrapie infectivity (42). The SA does, however, migrate in polyacrylamide gel electrophoresis as a low molecular weight species. Therefore, the animal disease model may expand the perspective of viroid-like molecules by including the first DNA viroid. These properties of the SA have been deduced from bioassay data and would be strengthened by resolution of a scrapie-specified nucleic acid species.

CONCLUSION

The unique structure and conformation of the small pathogenic RNAs, the viroids, stimulate equally novel considerations for viroid synthesis and pathogenesis. The apparent absence of a messenger RNA function coupled with the association of viroid and complementary viroid RNA sequences with the nucleus reinforces the role of either molecule as a regulator of host genome expression. Host response to viroid infection appears to be influenced by developmental factors, and perhaps vulnerability of selected species to aberration in a control function. Viroid diseases in which one component in the cell-pathogen encounter has been reduced to a minimal form have already expanded the concept of a pathogenic nucleic acid and promise continued insight into the biological activity of nucleic acids.

ACKNOWLEDGMENTS

The author wishes to acknowledge the various colleagues listed in the cited papers for continued interest and efforts in the development of the CEV system. Special thanks to L. K. Grill for discussion and review of this manuscript. Supported in part by NSF Grant No. PCM 7813622 and the John Simon Guggenheim Memorial Foundation.

Literature Cited

1. Apirion, D., Berek, I., Gill, B. S., Podder, U.S. 1977. Small stable RNA molecules in the nucleus: Possible mediators in gene expression. In *Molecular Approaches to Eukaryotic Genetic Systems,* pp. 431–39, ed. G. Wilcox, J. Abelson, C. F. Fox. New York: Academic
2. Beljanski, M., Aaron-da-Cunha, M. I., Beljanski, P., Manigault, P., Bourgarel, P. 1974. Isolation of the tumor-inducing RNA from oncogenic and nononcogenic *Agrobacterium tumefaciens. Proc. Natl. Acad. Sci. USA* 71:1585–89
3. Berget, S. N., Moore, C., Sharp, P. A. 1977. Spliced segments at the 5' terminus of adenovirus 2 late mRNA. *Proc. Natl. Acad. Sci. USA* 74:3171–75
4. Britten, R. J., Davidson, E. H. 1969. Gene regulation for higher cells: A theory. *Science* 165:349–57
5. Conejero, V., Semancik, J. S. 1977. Exocortis viroid: Alteration in the proteins of *Gynura aurantiaca* accompanying viroid infection. *Virology* 77:221–32
6. Dawson, W. O., Schlegel, D. E. 1976. The sequence of inhibition of tobacco mosaic virus synthesis by actinomycin D, 2-thiouracil, and cycloheximide in a synchronous infection. *Phytopathology* 66:177–81
7. Dickson, E. 1978. Viroids: Infectious RNA in plants. In *Nucleic Acids in Plants,* ed. T. C. Hall, J. W. Davies. *Uniscience Series.* Gainesville: CRC Press. In press
8. Dickson, E., Diener, T. O., Robertson, H. D. 1978. Potato spindle tuber and citrus exocortis viroids undergo no major sequence changes during replication in two different hosts. *Proc. Natl. Acad. Sci. USA* 75:951–54
9. Dickson, E., Niblett, C. L., Robertson, H. D., Horst, R. K., Zaitlin, M. 1978. Mild and severe strains of potato spindle tuber viroid have only minor differences in their nucleotide sequences. *Nature* 277:60–62
10. Dickson, E., Prensky, W., Robertson, H. D. 1975. Comparative studies of two viroids: Analysis of potato spindle tuber and citrus exocortis by RNA fingerprinting and polyacrylamide-gel electrophoresis. *Virology* 68:309–16
11. Diener, T. O. 1970. Isolation of exonuclease-resistant ribonucleic acid from healthy and potato spindle tuber virus-

infected tomato leaves. *Phytopathology* 60:1014 (Abstr)

12. Diener, T. O. 1971. Potato spindle tuber "virus." IV. A replicating, low molecular weight RNA. *Virology* 45:411–28

13. Diener, T. O. 1971. Potato spindle tuber virus: A plant virus with properties of a free nucleic acid. III. Subcellular location of PSTV-RNA and the question of whether virions exist in extracts or in situ. *Virology* 43:75–89

14. Diener, T. O. 1972. Is the scrapie agent a viroid? *Nature New Biol.* 235:218

15. Diener, T. O. 1974. Viroids: The smallest known agents of infectious disease. *Ann. Rev. Microbiol.* 28:23–39

16. Diener, T. O., Hadidi, A. 1977. Viroids. In *Comprehensive Virology*, Vol. 2, ed. H. Frankel-Conrat, R. R. Wagner, pp. 285–337. New York: Plenum

17. Diener, T. O., Smith, D. R. 1975. Potato spindle tuber viroid. XIII. Inhibition of replication by actinomycin D. *Virology* 63:421–27

18. Domdey, H., Jank, P., Sänger, H. L., Gross, H. J. 1978. Studies on the primary and secondary structure of potato spindle tuber viroid: Products of digestion with ribonuclease A and ribonuclease T, and modification with bisulfite. *Nucleic Acid Res.* 5:1221–34

19. Fernow, K. H. 1967. Tomato as a test plant for detecting mild strains of potato spindle tuber virus. *Phytopathology* 57:1347–52

20. Flores, R., Chroboczek, J., Semancik, J. S. 1978. Some properties of the CEV-P_1 protein from citrus exocortis viroid-infected *Gynura aurantiaca* D. C. *Physiol. Plant Pathol.* 13:193–201

21. Geelen, J. L. M. C., Weathers, L. G., Semancik, J. S. 1976. Properties of RNA polymerases of healthy and citrus exocortis viroid-infected *Gynura aurantiaca* D. C. *Virology* 69:539–46

22. Goldstein, L. 1976. Role for small nuclear RNAs in "programming" chromosomal information? *Nature* 261:519–21

23. Grill, L. K., Semancik, J. S. 1978. RNA sequences complementary to citrus exocortis viroid in nucleic acid preparations from infected *Gynura aurantiaca*. *Proc. Natl. Acad. Sci. USA* 75:896–900

24. Gross, H. J., Domdey, H., Lossaw, C., Jank, P., Raba, M., Alberty, H., Sänger, H. L. 1978. Nucleotide sequence and secondary structure of potato spindle tuber viroid. *Nature* 273:203–8

25. Gross, H. J., Domdey, H., Sänger, H. L. 1977. Size and secondary structure of potato spindle tuber viroid. *Virology* 76:477–84

26. Guille, E., Grisvard, J. 1971. Modifications of genetic information in crown-gall tissue cultures. *Biochem. Biophys. Res. Commun.* 44:1402–9

27. Hadidi, A., Diener, T. O. 1977. De novo synthesis of potato spindle tuber viroid as measured by incorporation of ^{32}P. *Virology* 78:99–107

28. Hadidi, A., Jones, D. M., Gillespie, D. H., Wong-Staal, F., Diener, T. O. 1976. Hybridization of potato spindle tuber viroid to cellular DNA of normal plants. *Proc. Natl. Acad. Sci. USA* 73:2453–57

29. Hall, T. C., Wepprich, R. K., Davies, J. W., Weathers, L. G., Semancik, J. S. 1974. Functional distinction between ribonucleic acids from citrus exocortis viroid and plant viruses: Cell-free translation and aminoacylation reactions. *Virology* 61:486–92

30. Harley, E. H., White, J. S., Rees, K. R. 1973. The identification of different structural classes of nucleic acids by electrophoresis in polyacrylamide gels of different concentrations. *Biochem. Biophys. Acta* 299:253–63

31. Henco, K., Riesner, D., Sänger, H. L. 1977. Conformation of viroids. *Nucleic Acid Res.* 4:177–94

32. Ikegami, M., Fraenkel-Conrat, H. 1978. RNA dependent RNA polymerase of tobacco plants. *Proc. Natl. Acad. Sci. USA* 75:2122–24

33. Ingle, J., Pearson, G. G., Sinclair, J. 1973. Species distribution and properties of nuclear satellite DNA in higher plants. *Nature New Biol.* 242:193–97

34. Jacobsen, J. V. 1977. Regulation of ribonucleic acid metabolism by plant hormones. *Ann. Rev. Plant Physiol.* 28:537–64

35. Kadouri, A., Atsman, D., Edelman, M. 1975. Satellite-rich DNA in cucumber: Hormonal enhancement of synthesis and subcellular identifications. *Proc. Natl. Acad. Sci. USA* 72:2260–64

36. Kaper, J. M., Tousignant, M. E., Lot, H. 1976. A low molecular weight replicating RNA associated with a divided genome plant virus: Defective or satellite RNA? *Biochem. Biophys. Res. Commun.* 72:1237–43

37. Kaper, J. M., Waterworth, H. E. 1977. Cucumber mosaic virus associated RNA 5: Causal agent for tomato necrosis. *Science* 196:429–31

38. LeRoy, C., Stussi-Garaud, C., Hirth, L. 1977. RNA dependent RNA polymerase in uninfected and in alfalfa mosaic

virus-infected tobacco plants. *Virology* 82:48–62

39. Lockhart, B. E. L., Semancik, J. S. 1969. Differential effect of actinomycin D on plant virus multiplication. *Virology* 39:362–65

40. Mahlberg, P., Olson, K., Walkinshaw, C. 1971. Origin and development of plasma membrane derived invaginations in *Vinca rosea* L. *Am. J. Bot.* 58:407–16

41. Marchant, R., Robards, A. W. 1968. Membrane systems associated with the plasmalemma of plant cells. *Ann. Bot.* 32:457–71

42. Marsh, R. F., Malone, T. G., Semancik, J. S., Lancaster, W. D., Hanson, R. P. 1978. Scrapie agent: Evidence for an essential DNA component. *Nature* 275: 146–47

43. Marsh, R. F., Semancik, J. S., Mcdappa, K. C., Hanson, R. P., Rueckert, R. R. 1974. Scrapie and transmissible mink encephalopathy: Search for infectious nucleic acid. *J. Virol.* 13: 993–96

44. McClements, W. L. 1975. *Electron microscopy of RNA: Examination of viroids and a method for mapping single-stranded RNA.* PhD thesis. Univ. Wisconsin, Madison

45. McClements, W. L., Kaesberg, P. 1977. Size and secondary structure of potato spindle tuber viroid. *Virology* 76: 477–84

46. Mills, D. R., Kramer, F. R., Spiegelman, S. 1973. Complete nucleotide sequence of a replicating RNA molecule. *Science* 180:916–27

47. Morris, T. J. 1979. Evidence for a single infectious species of potato spindle tuber viroid. *Intervirology.* In press

48. Morris, T. J., Smith, E. M. 1977. Potato spindle tuber disease: Procedures for the detection of viroid RNA and certification of disase-free potato tubers. *Phytopathology* 67:145–55

49. Mühlbach, H. P., Camacho-Henriquez, A., Sänger, H. L. 1977. Isolation and properties of protoplasts from leaves of healthy and viroid-infected tomato plants. *Plant Sci. Lett.* 8:183–89

50. Mühlbach, H. P., Sänger, H. L. 1977. Multiplication of cucumber pale fruit viroid in inoculated tomato leaf protoplasts. *J. Gen. Virol.* 35:377–86

51. Navarro, L., Roistacher, C. N., Murashige, T. 1975. Improvement of shoot-tip grafting in vitro for virus-free citrus. *J. Am. Soc. Hortic. Sci.* 100:471–79

52. Niblett, C. L., Dickson, E., Fernow, K. H., Horst, R. K., Zaitlin, M. 1978.

53. O'Brien, M., Raymer, W. B. 1964. Symptomless hosts for the potato spindle tuber virus. *Phytopathology* 54: 1045–47

54. Owens, R. A. 1978. *In vitro* synthesis and characterization of DNA complementary to potato spindle tuber viroid. *Virology* 89:380–87

55. Owens, R. A., Diener, T. O. 1977. Synthesis of RNA complementary of potato spindle tuber viroid using Qβ replicase. *Virology* 79:109–20

56. Owens, R. A., Erbe, E., Hadidi, A., Steere, R. L., Diener, T. O. 1977. Separation and infectivity of circular and linear forms of potato spindle tuber viroid. *Proc. Natl. Acad. Sci. USA* 743:3859–63

57. Owens, R. A., Kaper, J. M. 1977. Cucumber mosaic virus-associated RNA 5. II. *In vitro* translation in a wheat germ protein-synthesis system. *Virology* 80:196–203

58. Pearson, G. G., Timmis, J. N., Ingle, J. 1974. The differential replication of DNA during plant development. *Chromosoma* 45:281–94

59. Price, R., Penman, S. 1972. A distinct RNA polymerase activity synthesizing 5.5S, 5S, and 4S RNA in nuclei from Adenovirus 2-infected HeLa cells. *J. Mol. Biol.* 70:435–50

60. Ravetch, J. V., Horiachi, K., Zinder, N. D. 1977. Nucleotide sequences near the origin of bacteriophage F₁. *Proc. Natl. Acad. Sci. USA* 74:4219–22

61. Reijnders, L., Sloof, P., Sival, J., Borst, P. 1973. Gel electrophoresis of RNA under denaturing conditions. *Biochim. Biophys. Acta* 324:320–33

62. Rodriguez, J. L., Garcia-Martinez, J. L., Flores, R. 1979. The relationship between plant growth substances content and infection of *Gynura aurantiaca* D. C. by citrus exocortis viroid. *Physiol. Plant Pathol.* 13:355–63

63. Romaine, C. P., Zaitlin, M. 1978. RNA dependent RNA polymerases in uninfected and tobacco mosaic virus-infected tobacco leaves: viroid-induced stimulation of a host polymerase activity. *Virology* 86:241–53

64. Sänger, H. L., Klotz, G., Riesner, D., Gross, H. J., Kleinschmidt, A. L. 1976. Viroids are single-stranded covalently closed circular RNA molecules existing as highly base-paired rod-like structures. *Proc. Natl. Acad. Sci. USA* 73:3852–56

65. Schneider, J. R. 1971. Characteristics of

a satellite-like virus of tobacco ringspot virus. *Virology* 45:108–22

66. Semancik, J. S. 1976. Structure and replication of plant viroids. In *Animal Virology*, Vol. IV, ed. D. Baltimore, A. S. Huang, and C. F. Fox, pp. 529–45. New York: Academic. 824 pp.

67. Semancik, J. S., Geelen, J. L. M. C. 1975. Detection of DNA complementary to pathogenic viroid RNA in exocortis disease. *Nature* 256:753–56

68. Semancik, J. S., Grill, L. K., Civerolo, E. L. 1978. Accumulation of viroid RNA in tumor cells after double infection by *Agrobacterium tumefaciens* and citrus exocortis viroid. *Phytopathology* 68:1288–92

69. Semancik, J. S., Magnuson, D. S., Weathers, L. G. 1973. Potato spindle tuber disease produced by pathogenic RNA from citrus exocortis disease: Evidence for identity of the causal agents. *Virology* 52:314–17

70. Semancik, J. S., Morris, T. J., Weathers, L. G. 1973. Structure and conformation of low molecular weight pathogenic RNA from exocortis disease. *Virology* 53:488–56

71. Semancik, J. S., Morris, T. J., Weathers, L. G., Rordorf, B. F., Kearns, D. R. 1974. Physical properties of the minimal infectious RNA from exocortis disease. *Virology* 63:160–67

72. Semancik, J. S., Tsuruda, D., Zaner, L., Geelen, J. L. M. C., Weathers, L. G. 1976. Exocortis disease: Subcellular distribution of pathogenic (viroid) RNA. *Virology* 69:669–76

73. Semancik, J. S., Vanderwoude, W. J. 1976. Exocortis viroid: Cytopathic effects at the plasma membrane in association with pathogenic RNA. *Virology* 69:719–26

74. Semancik, J. S., Weathers, L. G. 1970. Properties of the infectious forms of exocortis virus of citrus. *Phytopathology* 60:732–36

75. Semancik, J. S., Weathers, L. G. 1971. Exocortis virus: An infectious free nucleic acid plant virus with unusual properties. *Virology* 47:456–66

76. Semancik, J. S., Weathers, L. G. 1972. Exocortis disease: Evidence for a new species of "infectious" low molecular weight RNA in plants. *Nature New Biol.* 237:242–44

77. Van Dorst, H. J. M., Peters, D. 1974. Some biological observations on pale fruit, a viroid-incited disease of cucumber. *Neth. J. Plant Pathol.* 80:85–96

78. Takahashi, T., Diener, T. O. 1975. Potato spindle tuber viroid. XIV. Replication in nuclei isolated from infected leaves. *Virology* 64:106–14

79. Weinberg, R. A. 1973. Nuclear RNA metabolism. *Ann. Rev. Biochem.* 42:329–54

Ann. Rev. Phytopathol. 1979. 17:485–502
Copyright © 1979 by Annual Reviews Inc. All rights reserved

RELATION OF SMALL SOIL ♦3720
FAUNA TO PLANT DISEASE

Marvin K. Beute and D. Michael Benson

Department of Plant Pathology, North Carolina State University, Raleigh,
North Carolina 27650

INTRODUCTION

This review focuses attention on an area of plant pathology which is currently receiving increased interest, namely, associations of soilborne plant pathogens with small animals that increase severity or incidence of disease. This review is limited mainly to soil fauna described by Kevan as microfauna (>100 μm) or meiofauna (>1 cm) (54). Although plant parasitic nematodes certainly are included in this category, they have been omitted from this review because their importance in disease interaction is adequately recognized (85). For similar reasons transmission of viruses by soilborne vectors are not considered in detail (19). Although we have attempted to review all pertinent publications, we have emphasized current research whenever possible.

Insects and plants, including microorganisms, have been closely associated throughout their evolutionary history (55). During their evolution many types of insect-plant-microbe associations with varying degrees of complexity have arisen. Several early plant pathologists recognized the importance of soilborne fauna (other than nematodes) to plant disease (64), but interest in this area, as reflected in numbers of publications, has increased strikingly in the last decade. The reasons for renewed awareness of the possible role of soilborne fauna in plant disease are varied. Many studies were probably initiated because observations in the field differed from predictions based on laboratory or greenhouse tests. Most plant pathologists who have attempted to relate basic laboratory and/or greenhouse data to field performance have experienced the frustration of partial or total failure at one time or another. The reasons for lack of correlation between laboratory, greenhouse, and field data are generally not obvious. Organisms not normally considered to be part of a disease interaction, e.g. soil microar-

485

thropods or insects, may play a major but unrecognized role in infection of subterranean plant tissues.

Although interest in plant disease complexes has developed rapidly during the last two decades and much progress has been achieved in understanding their etiology, few plant pathologists have worked cooperatively with soil zoologists. Recent ecological studies indicate that close associations between soil fauna and microflora are frequent and involve a great diversity of both fauna and microflora (18, 21, 54, 55). Most animal phyla that are not purely marine are represented in the soil. Diversity and abundance of animal life in soil is much greater than generally recognized (54). Numerous publications involving interrelations between soil microflora occur throughout literature (15, 28, 54, 64, 74). Most of this research, however, has been conducted in forest soils and generally was related to soil fertility or decomposition of organic residues (54).

The dominant soil animals are either nematodes (most abundant in terms of individuals) or arthropods (most numerous in terms of species) (54). For example, over 834,500 arthropods have been recorded per square meter from cultivated heathland (32). The most abundant soil arthropods are the mites (Arachnida, Acarina), and the springtails (Collembola) (32). The soil mites constitute one of the most successful, ubiquitous, and yet neglected animal groups. Mites outnumber all other soil arthropods combined by more than three to one (32). Some of the most abundant species of mites belong to the Acariformes, including the pale, soft-bodied Acaridiae and the dark, hard-bodied Oribatei. Both of these groups feed principally on dead organic matter, moribund vegetation, or fungi. Springtails (Collembola) constitute the bulk of the remaining arthropod fauna in most soils. A few species are reasonably well known but the biology of the majority has been little studied. Some springtails have a fairly general diet of decaying leaves, whereas others rarely eat anything but fungi. Although much less numerous, small insects and their larvae are usually more visible than microarthropods. They may also play a significant role in transmission, initiation, or development of plant diseases. Most research on soilborne pathogens involving soil fauna (other than nematodes) has focused on the role of insects in disease.

GENERAL RELATION OF SMALL FAUNA TO PLANT DISEASES

Before considering the role of small fauna in the development of diseases caused by soilborne pathogens, it is appropriate to reflect briefly on certain classic examples of general fauna-pathogen associations. Nearly 90 years ago Waite demonstrated that fireblight of pear is transmitted by bees and thereby established that insects are significant as vectors of plant pathogens (103). Leach summarized very well the early studies in this area in his book on insect transmission of plant diseases (64). In the preface of the book he emphasized the importance and promise of the subject as a field of research.

The relation of dipterous insects to bacterial soft rot of various plants was described in the mid-1920s (19, 64). Bacterial wilt of cucurbits, transmitted by the striped cucumber beetles (*Acalymma vittata*) (97), and bacterial wilt of corn, transmitted by the corn flea beetle (*Chaetocnema pulicaria*), are classic examples of insect-bacterial associations in plant pathology.

No group of plant pathogens is more closely associated with insects than viruses and mycoplasma-like organisms (19). The first vectors associated with a plant virus disease were leafhoppers. They were first reported by Takata in 1895 and Takakomi in 1901 to cause rice dwarf disease in Japan, but later were shown to transmit the disease-causing virus (38). Today about 400 species of insects are known to transmit over 200 different viruses. The green peach aphid *Myzus persicae,* which transmits over 60 viruses, is probably the most important. At least 10 viruses are transmitted by leaf and bud mites (Eriophyidae) or spider mites (Tetranychoidea). Gastropoda (slugs and snails) are reported to transmit 6 viruses. All of these vectors penetrate unwounded plant cells during feeding and acquire virus from infected plants, which can be transmitted to healthy plants they may feed on subsequently.

The appearance of the destructive and spectacular Dutch elm disease in Europe and America and the proof of its dependence upon bark beetles reemphasized to both entomologists and plant pathologists the importance of insects as vectors of plant pathogens (64). The relationship between elm bark beetles (*Scolytus multistriatus*) and Dutch elm disease is similar in many respects to that between the bark beetles of coniferous trees and the blue stain disease of conifers (36). More recent examples of insects which are consistently associated with diseases of forest trees have also been described (37, 47). Although Leach (64) and Carter (19) reported several examples of associations between fauna and diseases of herbacious plants, general interest in this area of plant pathology has developed slowly. One recent, notable contribution was Laemmlen's description in 1972 of the interdependence of a mite, *Siteroptes reniformis,* and a fungus, *Nigrospora oryzae,* in the nigrospora lint rot of cotton (59). He reported that a mutualistic form of symbiosis exists between the mite and fungus. The presence of the fungus is important for normal growth and reproduction of the mite, and the mite aids the fungus in dissemination, inoculation, and early growth.

Programs of biological control of weed species have resulted in practical application of knowledge of at least two insect-pathogen associations. *Ceu-*

torhynchus litura, a European weevil, was released in Ontario, Canada in 1967 as a potential agent for biological control of Canada thistle (74, 82). Thistle shoots decreased to 4% of their former density near the center of the release site. It was suggested that the spread of thistle rust (*Puccinia punctiformis*) by the weevil contributed significantly to the decline of the thistle density. Successful biological control of the prickly pear cactus in Australia by release of the insect, *Cactoblastis cactorum,* is thought to have resulted from pathogen-vector relationships in which associated fungi and bacterial soft-rot organisms extended the damage caused by the insect and completed the eradication (109).

RELATION OF SMALL FAUNA TO SOILBORNE PATHOGENS

Evidence for the Role of Soil Fauna in Disease

ASSOCIATION OF FAUNA WITH DISEASE   The numerous associations observed by forest pathologists between bark beetles and pathogenic fungi (2, 9, 20, 35, 39, 44, 46, 61, 77, 93, 96, 98, 99, 105) on elm, fir, larch, and pine species have led to the discovery of the importance of these insects in survival, dissemination, and inoculation of these hosts with pathogenic fungi. The impetus provided by these studies probably did much to stimulate research on interactions between fauna and soilborne pathogens of herbaceous plants. One of the most intensively studied insect-fungus relationships is the association of clover root curculio (*Sitona* sp.) and other insect larvae with the decline of various legumes (11, 27, 40, 45, 51, 56, 57, 63, 65–68, 70, 83, 90, 111). Although it has been stated categorically that the curculio is by itself a primary factor in decline of legumes, evidence implicates the curculio-fungus association as a major cause of death of the plants. Root-worm infestations have been associated with fusarium root rot of corn and pythium (*Pythium myriotylum*) pod rot of peanut (79, 80, 84). A similar association between pythium pod rot of peanut and soilborne mites (Acarina) and springtails (Collembola) stimulated investigation of the role of soilborne mites in peanut pod rot (12, 13, 91, 92). Similarly, the observation that *Collembola* spp. are commonly found in Alabama field soil led to a study of the effect of interactions of Collembola and cotton rhizosphere microflora on incidence and severity of disease on cotton seedlings (106–108).

Bulb mites (*Rhizoglyphus* sp.) have been found in high populations on corms and bulbs of numerous flowers (8, 31, 36, 41, 52). The role of these mites is still under investigation, but they appear to contribute significantly to rot of corms, bulbs, and roots of flower crops (8, 41). Phytophthora root rot of avocado in central western Brazil is reported to be associated with the attack of roots by wood borers (26), but the relationship between the insects and pathogen and possible strategies for control of this disease have not been studied.

EFFICACY OF INSECTICIDES IN DISEASE CONTROL   Many interesting biological interactions are discovered by accident; for example, the use of selected pesticides to elucidate the role of a suspected arthropod in a disease complex may indicate instead that previously unsuspected organisms are major contributors to the problem. A good example of this is the research on the mite-pythium association in peanut pod rot (12, 13). The fumigant nematicide D-D (1,3-dichloropropene + 1,2-dichloropropane) and the nematicide-insecticide carbofuran were used to control nematodes in a North Carolina peanut field having a history of both nematode damage (*Meloidogyne hapla*) and pythium pod rot. Effective control of nematodes with D-D did not reduce pod rot incidence. Less effective control of nematodes with the nematicide-insecticide, however, reduced the incidence of pod rot. Control of pod rotting was accompanied by a reduction or eradication of soilborne mites and springtails. In further field and greenhouse tests, other acaricides reduced (or eradicated) soilborne mites with a concomitant reduction in pod rotting (92).

A somewhat similar situation was reported for decline of red clover in North Carolina (51). Repeated applications of the fungicide benomyl, benomyl plus carbofuran, or carbofuran alone in field plots resulted in increased growth and survival of plants only when carbofuran was applied, even though it was concluded that *Fusarium* spp. actually were causing the death of plants. These results support a previous report from Pennsylvania that drenching field plots of red clover with the fungicide, methyl arsinoxide, had no effect on plant survival, but drenching with either heptachlor or lindane insecticides resulted in increased yield and survival of plants (111). Caging plots on methyl bromide fumigated soil to prevent reinfestation by insects was equal to use of insecticide drenches in promoting root growth and survival of plants. Similar results have been reported for ladino clover (90) and alfalfa (27). Experiments in Oregon using soil insecticides also indicated that an association between clover root curculio injury to roots of alsike clover and the incidence of pathogenic fungi was involved in clover decline (65).

There are numerous other examples of the use of insecticides to provide evidence of an insect-pathogen association in a disease complex. Application of insecticides to dryland pinto beans in Colorado decreased root rot severity and increased yields (53). Rhizoctonia root infection was suppressed on cabbage in Poland by application of insecticides to soil (71), and black rot of sugar beet in Bratislav was reduced with a combined application of insecticides and fungicides but not with fungicides alone (94). Control of the northern corn root worm (*Diabrotica longicornis*) with insecticides in Minnesota decreased incidence of fusarium infection and lodging, and increased corn yield (80). Enhancement of fungicidal control of seed rot and seedling diseases of rice by an insecticidal seed treatment was reported in Louisiana (89). Although mechanisms of action have not been elucidated, use of soil insecticides are reported to decrease bacterial and fungal infections of potatoes and chrysanthemum (17, 23).

Mechanism of Disease Interaction

WOUNDING AS MEANS OF ENTRY BY PATHOGEN   Soil fauna may be involved in disease complexes through several mechanisms, including injury which enhances entry of pathogen into plant tissue (19). As early as 1929 it was reported that potato scab caused by *Actinomyces scabies* commonly occurred in tissues injured by the larvae of the potato flea beetle (*Epitrix cucumeris*) (64). In other situations, lesions caused by the flea beetle were invaded by *Rhizoctonia solani,* a fungus that usually does not infect tubers. *Erwinia carotovora* causes blackleg of potatoes but insects may be critical in this disease as they breach the wound barrier of seed tubers (62).

It was reported in Holland in 1907 that blackleg of cabbage caused by *Phoma lingam* occurred only when roots were injured by the cabbage maggot (*Hylemya brassicae*) (87). Further investigation indicated that maggot injury was not always necessary but was a significant contributory factor to disease (64). Rhizoctonia infection of cabbage roots may be enhanced by insect injuries, and at low soil moistures *R. solani* preferentially colonizes moist microhabitats in injured roots (71). Infection of young shoots of cassava in Zaire by *Xanthomonas manihotis* is reported to occur through insect punctures (72). As mentioned earlier, phytophthora root rot of avocado in Brazil was associated with attack by wood borers (26). In Puerto Rico about 97% of the *Fusarium oxysporum* f. sp. *vanillae* root infections in vanilla were associated with mechanical damage caused by insects, nematodes, or other agents (1).

Field-grown peanut pods injured by the feeding of southern corn rootworm larvae (*Diabrotica undecimpunctata howardi*) were more susceptible to fungal colonization than noninjured pods (84). When both rootworm larvae and *Pythium myriotylum* were present, incidence of pod rotting was almost twice that observed when only the fungus was present at the same inoculum density. Feeding sites were thought to provide entrance into peanut pods for many fungi, including *P. myriotylum.* Four species of

Fusarium were regularly isolated from rootworm-infected roots of corn in Minnesota (79, 80). Surprisingly, two of the *Fusarium* spp. grew only on tissues damaged by the rootworm. As the incidence of roots infected by rootworm (*Diabrotica longicornis*) increased in the field, the incidence of roots infected by *Fusarium* spp. increased also. Control of the rootworm reduced incidence of fusarium-infected roots. Studies in Oregon referred to in the preceding section demonstrated a positive relationship between injury to alsike clover roots by the clover root curculio and the incidence and severity of vascular decay caused by pathogenic fungi (65). Similarly, incidence of *R. solani* root infection in lupine was increased in Egypt by insect injury (30). Deterioration of alfalfa roots by fungi was also enhanced by insect injury, but alfalfa plants appeared to be very hardy and tolerant to root rot even with insect damage (27).

Gladiolus corm and root rot in California was greatly enhanced by injury to gladioli associated with the root mite, *Rhizoglyphus rhizophagus* (8). Within six or eight weeks, most of a vigorous root system was destroyed in soil infested with both mites and fungi. Most of the fungi isolated from injury-induced rot, however, had no or very limited pathogenicity when introduced into mechanically wounded corm tissue. The smoulder of narcissus disease caused by *Sclerotinia narcissicola* was also enhanced by injury from the bulb scale mite (*Steneotarsonemus laticeps*) (41). Similarly, a high incidence of rhizopus decay of sweet potato seed pieces was associated with injury by the tobacco wireworm (*Conoderus vespertinus*) in Virginia (42). Rhizopus root rot of mature sugar beets in Arizona resulted from wounding by lepidopterous larvae (95).

Injury of tree roots by small soil fauna may play an important role in initiating rotting in various tree species. Wounds caused by root weevils provide courts for entry of root rotting and staining fungi in white spruce and balsam fir but not in Norway spruce, Sitka spruce, red pine, or Scots pine (105). Attack of the black turpentine beetle (*Dendroctonus terebrans*) on roots of slash pine and resulting decay of these roots may be responsible for the death of many trees that have a relatively small number of beetle attacks on trunks (93). Damage by the poplar borer (*Saperda calcarata*) in the junction of the root and stem of Balsam poplar in Canada caused little direct tree mortality (29), but diseases that can kill poplar were frequently associated with borer injury. In Indonesia and Malaysia, a Lepidoptera oil-palm root miner facilitated entry of the fungus, *Ganoderma,* causing severe damage to the roots of oil-palm in plantations (24).

SMALL FAUNA PREDISPOSING PLANTS TO INFECTION AND DECAY
As indicated previously, providing a court of entry for pathogenic microorganisms often occurs when soil fauna are involved in a disease complex.

Nevertheless, the role of the soil fauna frequently may be more than simply an agent of injury (88). Several authors have reported apparent physiological changes in host susceptibility resulting from insect feeding that are similar to changes noted in certain nematode-fungus interactions (85). The gladiolus-root mite association, discussed previously, results in pathogenesis by several fungi which are normally not pathogens of gladiolus corms or roots (8). Feeding by the larvae of the fungus gnat, *Bradysia* sp., causes damage of economic importance to clover and alfalfa both in the greenhouse and field and predisposes plants to attack by root pathogens (69). As previously indicated, rootworm damage to corn also facilitated infection by *Fusarium* spp., which normally do not attack corn roots (80). The role of clover root borers (*Hylastinus obscurus*) and root feeding weevils in the clover decline problem may be a combination of transmission of pathogen, wounding for entry, and physiological predisposition. Latent infection by *Fusarium* spp. is common in red clover but rotting generally does not occur unless the plants are stressed (51). This observation is further supported by a report that fusarium root rot and plant death were increased in alfalfa, and red and white clover when plants were stressed by the feeding of pea aphids (*Acyrthosiphon pisum*) and potato leafhoppers (*Empoasca fabae*) on the foliage (67). Similarly, sugar maple trees defoliated in June or July in Connecticut by insects had more armillaria root rot and higher mortality than nondefoliated trees (104). Early defoliation was associated with a chemical change in the outer wood of maple roots that is favorable for the growth of *A. mellea*. Severe defoliation of western larch in Idaho by foliar feeding insects weakens trees and predisposes them to root disease and wood borers which are common killers of weakened trees (101).

ROOT DISEASES PREDISPOSING PLANTS TO FAUNAL ATTACK There are several examples of root diseases that result in increased attack of diseased plants by insects (20, 25, 35, 43, 66, 73, 81). *Fomes annosus* root decay was implicated in California as an important agent in predisposing white fir to infestation by bark beetles (35). *Armallaria mellea* root rot predisposed grand fir to bark beetles in Idaho (44, 73, 81) and white fir to bark beetles in California (20). *Fomes annosus* and *Verticicladiella wagenerii* infection of ponderosa pine roots in California predisposed pine trees to the western (*Dendroctonus brevicomis*) and mountain (*Dendroctonus ponderosae*) pine beetles (43, 96). It was concluded that trees with root rot, moisture stress, or other conditions that result in a subnormal physiological state are prone to bark beetles and other insect infestations. Researchers in Malaya concluded that in unsuitable soil, the tap roots of pine trees become infected with decay fungi which predispose them to termite attack, resulting in death of the trees (99). Clover root borers (*Hylastinus obscurus*) can discern between diseased and healthy red clover roots (66). In tests using root pieces or aqueous leachates from root pieces, adult borers were preferentially attracted to diseased roots and their leachates as compared with healthy roots, leachates from healthy roots, or bacteria or fungi isolated from diseased roots on potato-dextrose agar.

PATHOGEN TRANSMISSION   Other factors being equal, the probability of a root becoming infected by a soilborne pathogen is generally assumed to be a function of propagule density and rooting intensity of the plant (15). Investigations of soil fauna-pathogen interactions, however, suggest that certain soilborne pathogens (and resultant root infections) may be influenced by specific vector-plant relationships. Collembola migrate to the rhizosphere and root surfaces of cotton seedlings in slowly drying soil (106, 107). The animals readily contribute to inoculum build-up by transporting *Trichoderma viride, Fusarium oxysporum* f. sp. *vasinfectum,* and other sporulating fungi and bacteria to the rhizoplane of cotton. *Rhizoctonia solani* and *Pythium* spp. apparently were not transported by Collembola but these fungi did provide favorable habitats for growth and survival of the animals in soil culture. *Pythium myriotylum* may be transmitted to healthy peanut pods by soilborne mites (91, 92). Caloglyphus mites isolated from decaying peanut pods exhibited a pronounced attraction to *P. myriotylum* in food preference tests. Both fungal fragments and oospores of *P. myriotylum* remained viable after passing through the alimentary canal. Mites were observed to exhibit a definite klinotaxic response in laboratory tests. After total decay and desiccation of rotted peanut pods, mites are thought to migrate to adjacent healthy pods and enhance disease development by introducing propagules of *P. myriotylum* to the pod surface or interior. Addition of soilborne mites to field soil infested with *P. myriotylum* increased incidence of peanut pod rot (92).

Large numbers of Astigmatid mites associated with peanut pods in South Africa were usually heavily contaminated with fungus spores (3). Under certain circumstances mites penetrated peanut pods and fed on the kernels, disseminating fungus spores in the process. Although the role of the bulb mite (*Rhizoglyphus echinopus*) in transmitting *Verticillium albo-atrum* propagules to cotton was not established, it produced fecal pellets containing viable conidia and fragments of microsclerotia after feeding on fungal cultures and was commonly found in decomposing cotton debris in California (86). The bulb mite, however, may transmit *Pseudomonas marginata* in gladiolus, causing bacterial scab disease (36). *Siteroptes* spp. of mites are vectors of the cereal pathogens *Fusarium* and *Nigrospora* in Manitoba, Canada (58). Larvae of the clover cuculio carried fusaria along with other fungi (57), and *Drosophila* spp. were occasional vectors of sporangia of

Phytophthora parasitica on papaya in Hawaii (49). Adult flies of the seed corn maggot (*Hylemya platura*), during ovipositing, spread blackleg of potato caused by *Erwinia atroseptica* (60).

Fomes annosus is disseminated in pine forests by various beetles in Europe and North America (46, 76, 77). Black stain root disease of pine caused by *Verticicladiella wagenerii* can spread from tree to tree by root grafts but beetles are implicated in long-distance spread (39, 61). The beetle, *Hylastes macer,* is thought to be the major vector of this fungus while attacking small roots of weakened trees.

SOIL FAUNA EFFECTS ON THE INOCULUM DENSITY-DISEASE CURVE

As illustrated by numerous examples cited in previous sections, soil fauna may have a profound effect on disease development (8, 84, 92) which may be reflected in the inoculum density-disease curve. Activities of soil fauna may affect inoculum density-disease curves by a direct effect on inoculum potential and/or an indirect effect on disease potential. These effects may result in a change in position (quantitative effect) or a change in slope (qualitative effect) of the inoculum density-disease curve (48). The term *inoculum potential* as used in this review is a resultant of inoculum density as affected by environmental factors (4, 5). Disease potential is a resultant of host genetic susceptibility as affected by disease proneness (environmental effects) (4, 5).

Root disease models proposed by Baker et al assume that inoculum is randomly distributed in field soils that are cultivated (7). In soils where small fauna are abundant, inoculum could be accumulated at the infection court by animals that had fed on pathogen propagules or carried propagules externally. Soil moisture gradients that developed in the root-soil continuum (106) would influence soil fauna aggregation in the vicinity of roots and other subterranean plant tissues (86, 92, 107). Inoculum potential increases as inoculum accumulates in the infection court, and hence, more infections are observed. In terms of the inoculum density-disease curve (5–7, 48), one effect of the accumulation of fungal propagules in the infection court would be quantitative; that is, the position of the inoculum density-disease curve would be displaced to the left of its position in the absence of the fauna effect (Figure 1A).

Another possible effect of the accumulation of inoculum in the infection court by soil fauna would be the opportunity for inoculum units to "pool" energy for infection. Thus a synergistic effect might develop if increases in inoculum result in more than a proportional increase in the number of infections (6). In terms of the root disease model proposed by Baker et al

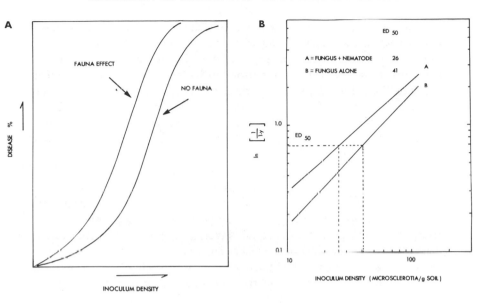

Figure 1   A. Effect of soil fauna on change in position (quantitative effect) of the inoculum density-disease curve for a hypothetical root disease due to accumulation of inoculum in the infection court or an increase in the number of infection sites (wounding). *B.* Transformed data of Conroy et al (22) for *Verticillium albo-atrum* and the nematode *Pratylenchus penetrans* on tomato, demonstrating a shift in position (quantitative effect) but not in slope of the inoculum density-disease curve.

(7), synergism is evident when transformed inoculum density-disease curves have a slope value greater than 1.0 (log-log basis), when inoculum is non-motile and the infection court is fixed (rhizosphere effect). However, synergistic effects apparently are rare in nature for soilborne fungi, having been demonstrated only for *Rhizoctonia solani* (10).

Soil fauna also may predispose plant roots to infection by a pathogen by altering disease potential. Two mechanisms may operate here. Wounding of plant roots by the soil fauna may increase the number of susceptible sites (greater disease potential) that the pathogen can infect (Figure 1A). This mechanism has been demonstrated on tomato for *Verticillium albo-atrum* and the nematode *Pratylenchus penetrans* using a split root technique (22).[1] Here, inoculum becomes more efficient because the host is predisposed to infection, which results in a shift of the inoculum density-disease curve to the left (a quantitative effect). This shift is demonstrated by trans-

[1]While this article excludes nematodes in general, no specific examples suitable for graphic analysis could be found for other soil fauna. The authors hope this review will stimulate interest in the involvement of other soil fauna in this area.

forming (log-log) the data of Conroy et al (22) (Figure 1B). In the presence of *V. albo-atrum* alone, ED_{50} (inoculum density causing 50% infection) for tomato was 41 microsclerotia/gram of soil, but in combination with *P. penetrans*, ED_{50} dropped to 26 microsclerotia/gram, a quantitative effect. Under field conditions it would be difficult to separate quantitative effects due to accumulation of inoculum in the infection court from an increase in the number of susceptible sites on the root because they both result in a similar effect, i.e. a shifting to the left of the inoculum density-disease curve unless synergism is involved.

Wounding of plant roots by soil fauna also may result in a second effect on disease potential, thus altering the inoculum density-disease curve. The wounding process may alter host response (48, 85) such that the mechanism of disease induction is changed. A change in host susceptibility so that susceptible sites are more easily invaded increases disease potential. Van der plank (102) conceptualizes the number of susceptible sites and the ease with which they are invaded as the "two paths to susceptibility." As the number of susceptible sites increases, but the ease with which they are infected is low, there is less curvature to the inoculum density-disease curve. On an arithmetic basis, thus, the inoculum density-disease curve is nearly straight. When the number of susceptible sites is low, but the ease with which they are infected is high, the inoculum density-disease curve bends to the right (102). Horsfall & Dimond (48) maintain that a change in the disease induction mechanism is reflected by a change in slope (on a log-probit basis) of the inoculum density-disease curve (a qualitative effect). At present, we know of no quantitative data useful for testing the hypothesis that soil fauna (excluding nematodes) alter the disease induction mechanism. A change in slope has been demonstrated for diseases due to differences in pH effects, tissue age effects, and species susceptibility (48). In another study of soil-borne diseases where hosts of varying susceptibilities were analyzed, slopes of the inoculum density-disease curves were parallel but the position was shifted (5). More work is needed to quantify the effect of soil fauna on disease development. Of particular importance is the question of whether activities of soil fauna cause simply quantitative differences in disease development or are there qualitative and synergistic effects as well.

CONCLUSION

Soil animals may influence many aspects of plant pathogen behavior. Soil fauna are important in dispersal of soil microbes including pathogens in the soil (16, 28, 50, 101) and, in certain situations, survival of pathogens over lengthy periods of time (19, 64). Complex interactions between soil fauna, pathogens, and plant hosts probably occur at a much higher frequency in nature than now recognized. The relationship between a plant pathogen and soil fauna could be rather simple; for example, wounding of root tissue provides a court of entry for resident microbes. In other situations the role of soil fauna in disease may be extremely complex, depending on the particular relationship existing between the pathogen, soil animal, and plant species.

Various insects and mites, including those found in soil, are able to induce species-specific phytotoxemias in host plants (75). Resultant changes in attractiveness and/or susceptibility of hosts may significantly influence disease incidence and severity if pathogen inoculum is present. Similar effects occur with certain plant parasitic nematodes, frequently resulting in disease caused by weak pathogens or microbes, not normally considered pathogens on these hosts (85). Predisposition effects also operate in reverse order; that is, root decay of tree hosts is known to increase the attractiveness of hosts to insect attack. Although there has been extensive interest in the role of nematodes in disease complexes, studies involving plant pests should be expanded to include a broader spectrum of soil animals.

Insects, mites, and other small fauna obviously are vectors and "inoculators" of many more species of plant pathogens than is now recognized (33, 34, 110). Norris (75) suggested that many small fauna are symbiotic hosts for pathogenic microbes that spend part of their life cycle as endosymbiotes and the rest of their lives as ectosymbiotes of the animal and as pathogens of plants. Certain insects representing this group have special ectodermal structures prominent in various developmental forms (larva, nymph, adult) which protect and nourish (microbes may increase in number) pathogens while disseminating them among host plants. However, the result of an association wherein inoculum propagules are simply carried as contaminants on external projections of the animal may be equally as important as are the more complex relationships.

The effect of both simple and complex relationships between soil fauna and pathogenic microbes on induction of root disease is entirely relevant to current efforts in simulating disease epidemics. Whether the role of fauna is to create ports of entry, accumulate inoculum at infection sites, or alter disease proneness of the host (Figure 2), studies on these associations can provide valuable insights into the ecology of soilborne pathogens in natural soil environments. Soil animals of diverse types undoubtedly play a vital role in population equilibria of both microfauna and microflora of the soil. Although not considered in this review, feeding by soil animals on fungal spores and mycelia as well as other soil animals is part of a dynamic situation characteristic for each soil habitat (14). For example, some mites feed principally on fungi while others are predators of these mites. Collembola feed on fungi, mites, and soil nematodes. Certain nematodes also feed

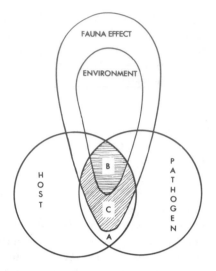

Figure 2 Theoretical increase in plant disease resulting from the role of soil fauna in one or more of the following aspects: 1. vector and inoculator of pathogen, 2. increased disease proneness of host, and 3. expanded environment through creation of microhabitats, etc. *A* is the zone of potential disease. *B* is the zone of disease as influenced by the soil environment (including soil microorganisms). *C* is the zone of disease as increased by the soil fauna effect.

on fungi while other nematodes parasitize mites, Collembola, and insects. Recent work indicates that protozoan fauna, e.g. giant vampyrellid soil amaebae, may play a significant role in reducing spore populations of certain pathogenic fungi in soil (78). Extensive work is needed to evaluate the significance of these multifaceted interactions in the biology of soilborne plant pathogens. It is hoped that this review will stimulate new and continued efforts in cooperative studies on soilborne pathogens by plant pathologists, entomologists, and soil zoologists.

Literature Cited

1. Alconero, R. 1968. Infection and development *Fusarium oxysporum* f. sp. *vanillae* in vanilla roots. *Phytopathology* 58:1281–83
2. Alma, P. J., Van Boven, P. J. 1976. Insect invasion and survival of Douglas fir stumps in New Zealand. *NZ J. Sci.* 5:306–12
3. Aucamp, J. L. 1969. The role of mite vectors in the development of aflatoxin in groundnuts. *J. Stored Prod. Res.* 5:245–49
4. Baker, R. 1965. The dynamics of inoculum. In *Ecology of Soil-Borne Plant Pathogens*, ed. K. F. Baker, W. C. Snyder, pp. 395–419. Berkeley: Univ. Calif. Press
5. Baker, R. 1971. Analyses involving inoculum density of soil-borne plant pathogens in epidemiology. *Phytopathology* 61:1280–92
6. Baker, R. 1978. Inoculum potential. In *Plant Disease: An Advanced Treatise*, Vol. 2, ed. J. G. Horsfall, E. B. Cowling, pp. 137–154. New York: Academic
7. Baker, R., Maurer, C. L., Maurer, R. A. 1967. Ecology of plant pathogens in soil. VIII. Mathematical models and

inoculum density. *Phytopathology* 57: 662–66

8. Bald, J. G., Jefferson, R. N. 1952. Injury to gladioli associated with the root mite, *Rhizoglyphus rhizophagus* (BKS). *Plant Dis. Reptr.* 36:435–37

9. Basham, J. T., Hudak, J., Lachance, D., Magasi, L. P., Stillwell, M. A. 1976. Balsam fir death and deterioration in eastern Canada following girdling. *Can. J. For. Res.* 6:406–14

10. Benson, D. M., Baker, R. 1974. Epidemiology of *Rhizoctonia solani* preemergence damping-off of radish: Inoculum potential and disease potential interaction. *Phytopathology* 64: 957–62

11. Bertran, M., Jourdheuil, P. 1968. Ecological bases for control of clover seed weevil in purple cover. *Ann. Ephiphyt.* 19:335–65

12. Beute, M. K. 1974. A quantitative technique for the extraction of soil-inhabiting mites (Acarina) and springtails (Collembola) associated with pod rot of peanut. *Phytopathology* 64:571–72

13. Beute, M. K. 1974. Evidence for the role of soilborne mites in peanut pod rot disease using a new extraction technique. *J. Am. Peanut Res. Educ. Assoc.* 6:61

14. Boosalis, M. G., Mankau, R. 1965. Parasitism and predation of soil microorganisms. See Ref. 4, pp. 374–91

15. Bowen, G. D., Rovira, A. D. 1976. Microbial colonization of plant roots. *Ann. Rev. Phytopathol.* 14:121–44

16. Broadbent, L. 1960. Dispersal of inoculum by insects and other animals, including man. In *Plant Pathology. An Advanced Treatise*, Vol. 3, ed. J. G. Horsfall, A. E. Dimond, pp. 97–135. New York: Academic

17. Busch, L. V., Rowberry, R. G. 1972. Potato seed piece treatment in Ontario. *Am. Potato J.* 49:7–11

18. Butcher, J. W., Snider, R., Snider, R. J. 1971. Bioecology of edaphic Collembola and Acarina. *Ann. Rev. Entomol.* 16:249–88

19. Carter, W. 1973. *Insects in Relation to Plant Disease*. New York: Wiley. 759 pp.

20. Cobb, F. W. Jr., Parmeter, J. R. Jr., Wood, D. L., Stark, R. W. 1974. Root pathogens as agents predisposing ponderosa pine and white fir to bark beetles. *4th Int. Conf. Fomes annosus, Proc.*, pp. 8–15

21. Coleman, D., McGinnis, J. T. 1970. Quantification of fungus-small arthropod food chains in soil. *Oikos* 21: 134–37

22. Conroy, J. J., Green, R. J. Jr., Ferris, J. M. 1972. Interaction of *Verticillium albo-atrum* and the root lesion nematode, *Pratylenchus penetrans*, in tomato roots at controlled inoculum densities. *Phytopathology* 62:362–66

23. Cruz, C., Streu, H. T., Snee, R. D. 1970. Growth response of greenhouse chrysanthemum to root drenches of diazinon and Demeton. *J. Econ. Entomol.* 63:1446–51

24. DeChenon, R. D. 1975. Presence in Indonesia and Malaysia of a Lepidoptera oil palm root miner, *Sufetula sunidesalis* Walker, and its relationship to attacks by *Ganoderma. Oleugeneux* 30. 449–56

25. Dengler, K. 1975. Control of *Phaenops cyanea*. *Z. Angew. Entomol.* 78:5–9

26. Dianese, J. C., Freire, I. D. L., Takatsu, A., Bokan, H. A. 1976. *Phytophthora cinnamomi* root rot of avocado in the Brazilian capital. *Fitopatologia Bras.* 1:91–98

27. Dickason, E. A., Leach, C. M., Gross, A. E. 1968. Clover root curculio injury and vascular decay of alfalfa roots. *J. Econ. Entomol.* 61:1163–68

28. Dobbs, C. G., Hinson, W. H. 1960. Some observations on fungal spores in soil. In *The Ecology of Soil Fungi*, ed. D. Parkinson, J. W. Waid, pp. 33–42. Liverpool: Liverpool Univ. Press

29. Drouin, J. A., Wong, H. R. 1975. Biology, damage and chemical control of the Poplar borer (*Saperda calcarata*) in the junction of the root and stem of Balsam Poplar in Western Canada. *Can. J. For. Res.* 5:433–39

30. El-Rafei, M. E. E., Hamed Badr, M. 1970. Studies on root rot disease of Lupine. *Agric. Res. Rev.* 48:100–5

31. Engelhard, A. W. 1969. Bulb mites associated with diseases of gladiola and other crops in Florida. *Phytopathology* 59:1025

32. Evans, G. O. 1955. Identification of terrestrial mites. In *Soil Zoology*, ed. D. K. M. Kevan, pp. 55–61. New York: Academic

33. Fennell, D. I., Kwolek, W. F., Lillehoj, E. B., Adams, G. A., Bothast, R. J., Zuber, M. S., Calvert, O. H., Guthrie, W. D., Bockholt, A. J., Manwiller, A., Jellum, M. D. 1977. *Aspergillus flavus* presence in silks and insects from developing and mature corn ears. *Cereal Chem.* 54:770–78

34. Fennell, D. I., Lillehoj, E. B., Kwolek, W. F. 1975. *Aspergillus flavus* and other

fungi associated with insect-damaged field corn. *Cereal Chem.* 52:314–21

35. Ferrell, G. T., Smith, R. S. Jr. 1976. Indicators of *Fomes annosus* root decay and bark beetle susceptibility in sapling white fir. *For. Sci.* 22:365–69

36. Forsberg, J. L. 1959. Relationship of the bulb mite *Rhizoglyphus echinopus* to bacterial scab of gladiolus. *Phytopathology* 49:538

37. Frederick, D. J., Sloan, N. F., Skowron, W. S. Jr. 1976. Potential insect transmission of *Scleroderris lagerbergii* by scolytid beetles. *Plant Dis. Reptr.* 60:411–13

38. Gibbs, A., Harrison, B. *Plant Virology.* New York: Wiley. 292 pp.

39. Goheen, D. J., Cobb, F. W. Jr. 1978. Occurrence of *Verticicladiella wagenerii* and its perfect state, *Ceratocystis wageneri* sp. nov., in insect galleries. *Phytopathology* 68:1192–95

40. Graham, J. H., Newton, R. C. 1960. Relationship between the clover root curculio and incidence of fusarium root rot in Ladino white clover. *Plant Dis. Reptr.* 44: 534–35

41. Gray, E. G., Shaw, M. W., Shiel, R. S. 1975. The role of mites in the transmission of smoulder of narcissus. *Plant Pathol.* 24:105–7

42. Graves, B., Savage, C. P. Jr. 1975. Some aspects of sweet potato seed piece decay control. *Hortic. Sci.* 10:148

43. Helms, J. A., Cobb, F. W. Jr., Whitney, H. S. 1971. Effect of infection by *Verticicladiella wagenerii* on the physiology of *Pinus ponderosa. Phytopathology* 61: 920–25

44. Hertert, H. D., Miller, D. L., Partridge, A. D. 1975. Interaction of bark beetles (Coleoptera:Scolytidae) and root rot pathogens in grand fir in northern Idaho USA. *Can. Entomol.* 107: 899–904

45. Hill, R. R. Jr., Murray, J. J., Zeiders, K. E. 1971. Relationships between clover root curculio injury and severity of bacterial wilt in alfalfa. *Crop Sci.* 11:306–7

46. Himes, W. E., Skelly, J. M. 1972. An association of the black turpentine beetle, *Dendroctonus terebrans*, and *Fomes annosus* in loblolly pine. *Phytopathology* 62:670

47. Hinds, T. E. 1972. Insect transmission of Ceratocystis species associated with aspen cankers. *Phytopathology* 62: 221–25

48. Horsfall, J. G., Dimond, A. E. 1963. A perspective on inoculum potential. *J. Indian Bot. Soc.* 42A:46–57

49. Hunter, J. E., Buddenhagen, I. W. 1969. Field biology and control of *Phytophthora parasitica* on papaya (*Carica papaya*) in Hawaii. *Ann. Appl. Biol.* 63:53–60

50. Hutchison, S. A., Kamel, M. 1956. The effect of earthworms on the dispersal of soil fungi. *J. Soil Sci.* 7:213–18

51. James, J. R. 1978. *Role of diseases and insects in the decline of Ladino clover.* MS thesis. NC State Univ., Raleigh. 67 pp.

52. Jefferson, R. N., Bald, J. G., Morishta, R. S., Close, D. H. 1956. Effect of vapam on Rhizoglyphus mites and gladiolus soil diseases. *J. Econ. Entomol.* 49:584–89

53. Keenan, J. G., Moore, H. D., Oshima, N., Jenkins, L. E. 1974. Effect of bean root rot on dryland pinto bean production in southwestern Colorado. *Plant Dis. Reptr.* 58:890–92

54. Kevan, D. K. M. 1965. The soil fauna. See Ref. 4, pp. 33–51

55. Kevan, P. G., Chaloner, W. G., Savile, D. B. O. 1975. Interrelationships of early terrestrial arthropods and plants. *Palaeontology* 18:391–417

56. Kilpatrick, R. A. 1961. Fungi associated with larvae of Sitona spp. *Phytopathology* 51:640–41

57. Kilpatrick, R. A., Dunn, G. M. 1961. Fungi and insects associated with deterioration of white clover tap roots. *Crop Sci.* 1:147–49

58. Kines, K. A., Sinha, R. N. 1975. A laboratory study of fauna and flora in an agricultural soil in Manitoba Canada. *Manit. Entomol.* 7:59–66

59. Laemmlen, F. F., Hall, D. H. 1973. Interdependence of a mite, *Siteroptes reniformis*, and a fungus *Nigrospora oryzae*, in the Nigrospora lint rot of cotton. *Phytopathology* 63:308–15

60. Landis, B. J., Onsager, J. A., Fox, L., Foiles, L. L. 1971. Chemical control of the seed-corn maggot, *Hylemya platura* (Meigen), and seed-piece decay in potato seed pieces. *Am. Potato J.* 48:374–80

61. Landis, T. D., Helburg, L. B. 1976. Black stain root disease of Pinyon pine in Colorado. *Plant Dis. Reptr.* 60: 713–17

62. Lapwood, D. H. 1976. Field observations on blackleg in England. *Eur. Plant Prot. Organ.* 6:237–39

63. Lau, N. E., Filmer, R. S. 1959. Injury of clover root curculios to red clover in New Jersey. *J. Econ. Entomol.* 52: 1155–56

64. Leach, J. G. 1940. *Insect Transmission of Plant Diseases*. New York: McGraw-Hill. 615 pp.
65. Leach, C. M., Dickason, E. A., Gross, A. E. 1961. Effects of insecticides on insects and pathogenic fungi associated with alsike clover roots. *J. Econ. Entomol.* 54:543–56
66. Leath, K. T., Byers, R. A. 1973. Attractiveness of diseased red clover roots to the clover root borer. *Phytopathology* 63:423–31
67. Leath, K. T., Byers, R. A. 1977. Interactions of fusarium root rot with pea aphid and potato leafhopper feeding on forage legumes. *Phytopathology* 67: 226–29
68. Leath, K. T., Lukezic, F. L., Crittenden, H. W., Elliot, E. S., Halisky, P. M., Howard, F. L., Ostazeski, S. A. 1970. The fusarium root rot complex of selected forage legumes in the Northeast. *Pa. Agric. Exp. Stn. Bull.* 777. 67 pp.
69. Leath, K. T., Newton, R. C. 1969. Interaction of a fungus gnat, *Bradysia* sp. (Sciaridae) with *Fusarium* spp. on alfalfa and red cover. *Phytopathology* 59:257–58
70. Leath, K. T., Zeiders, K. E., Myers, R. A. 1978. Increased yield and persistence of red clover after a soil drench application of benomyl. *Agron. J.* 65:1008–9
71. Maciejowska, Z. 1967. Results of experiments on control of root diseases of cabbage on a muck soil in Reguly. *Pr. Nauk. Inst. Ochr. Rosl. Warszawa* 9:41–49
72. Maraite, H., Meyer, J. A. 1975. *Xanthomonas manihotis* (Arthaud-Berthet) Starr, causal agent of bacterial wilt, blight and leaf spots of cassova in Zaire. *PANS* 21:27–37
73. Miller, D. L., Partridge, A. D. 1974. Root rot indicators in grand fir. *Plant Dis. Reptr.* 58:275–76
74. Mills, J. T., Alley, B. P. 1973. Interactions between biotic components in soils and their modification by management practices in Canada: A Review. *Can. J. Plant Sci.* 53:425–41
75. Norris, D. M. 1969. How insects induce disease. In *Plant Diseases: An Advanced Treatise,* Vol. 4, ed. J. G. Horsfall, E. B. Cowling, New York: Academic. In press
76. Nuroteva, M., Laine, L. 1968. Über die möglichkeiten der insekten als überträger des wurzelschwamma (*Fomes annosus* (Fr.) Cooke). *Ann. Entomol. Fenn.* 34:113–35
77. Nuroteva, M., Laine, L. 1972. Lebensfahige diasporen des wurzelschwamms (*Fomes annosus* (Fr.) Cooke) in den exkrementen von *Hylobius abietis* L. (Col., curculionidae). *Ann. Entomol. Fenn.* 38:119–21
78. Old, K. M., Darbyshire, J. F. 1978. Soil fungi as food for giant amoebae. *Soil Biol. Biochem.* 10:93–100
79. Palmer, L. T., Kommedahl, T. 1968. The fusarium root rot complex of *Zea mays* as affected by *Diabrotica longicornis,* the northern corn rootworm. *Phytopathology* 58:1062
80. Palmer, L. T., Kommedahl, T. 1969. Root-infecting *Fusarium* species in relation to rootworm infestations in corn. *Phytopathology* 59:1613–17
81. Partridge, A. D., Miller, D. L. 1972. Bark beetles and root rots related in Idaho conifers. *Plant Dis. Reptr.* 56:498–500
82. Peschken, D. P., Beecher, R. W. 1973. *Ceutorhynchus litura* (Coleoptera: Curculionidae): Biology and first releases for biological control of the weed Canada thistle (*Cirsium arvense*) in Ontario, Canada. *Can. Entomol.* 105: 1489–94
83. Pesho, G. R. 1975. Clover root curculio-estimates of larval injury to alfalfa tap roots. *J. Econ. Entomol.* 68:61–65
84. Porter, D. M., Smith, J. C. 1974. Fungal colonization of peanut fruit as related to southern corn rootworm injury. *Phytopathology* 64:249–51
85. Powell, N. T. 1971. Interactions between nematodes and fungi in disease complexes. *Ann. Rev. Phytopathol.* 9:253–74
86. Price, D. W. 1976. Passage of *Verticillium albo-atrum* propagules through the alimentary canal of the bulb mite. *Phytopathology* 66:46–50
87. Quanjer, H. M. 1907. Neue kohlkronkheiten in Nord-Holland (Drehherzkrankheit, Fallsucht und Krebs). *Z. Pflanzenkr.* 17:258–67
88. Rawlings, W. A. 1955. *Rhizoglyphus solani,* a pest of onions. *J. Econ. Entomol.* 48:334
89. Rush, M. C., Gifford, J. R. 1972. Enhancement of fungicidal control of seed rot and seedling diseases of rice by an insecticidal seed treatment. *Plant Dis. Reptr.* 56:154–57
90. Schillinger, J. A. Jr., Leffel, R. C. 1964. Persistence of ladino clover, *Trifolium repens* L. *Agron. J.* 56:11–14
91. Shew, H. D. 1977. *Evidence for the involvement of soilborne mites in pythium pod rot of peanut*. MS thesis. NC State Univ., Raleigh, NC. 42 pp.

92. Shew, H. D., Beute, M. K. 1978. Evidence for the involvement of soilborne mites in pythium pod rot of peanut. *Phytopathology* 69:204–7

93. Smith, R. H., Kowa, R. J. 1968. Attack of the black turpentine beetle on roots of slash pine. *J. Econ. Entomol.* 61: 1430–32

94. Smrz, J., Gaher, S. 1973. The effectiveness of various seed disinfectants and their combinations on pelleted monogerm sugar beet seed. *Ochr. Rostl.* 9:15–21

95. Stanghellini, M. E., Kronland, W. C. 1977. Root rot of mature sugar beets by *Rhizopus arrhizus. Plant Dis. Reptr.* 61:255–56

96. Stark, R. W., Cobb, F. W. Jr. 1969. Smog injury, root diseases, and bark beetle damage in Ponderosa pine. *Calif. Agric.* 23:13–15

97. Sutherland, D. W. S. 1978. *Common Names of Insects and Related Organisms Spec. Publ. 78–1.* College Park, Md.: Entomol. Soc. Am. 132 pp.

98. Talbot, P. H. B. 1977. The Sirex woodwasps—Amylostereum-pinus association. *Ann. Rev. Phytopathol.* 15:41–54

99. Thapa, R. S., Soon, S. P. 1971. Termite damage in plantation hoop pine, *Araucaria-cunninghamii* D. Don, in Sabah and its control. *Malays. For.* 34:47–52

100. Thornton, M. L. 1970. Transport of soil-dwelling aquatic Phycomycetes by earthworms. *Trans. Br. Mycol. Soc.* 55:391–97

101. Tunnock, S., Denton, R. E., Carlson, C. E., Janssen, W. 1969. *Larch Gasebearer and Other Factors Involved with Deterioration of Western Larch Stands in Northern Idaho. US For. Serv. Res. Pap. INT–68.* 10 pp.

102. Van der Plank, J. E. 1975. *Principles of Plant Infection.* New York: Academic. 216 pp.

103. Waite, M. B. 1891. Results from recent investigations in pear blight. *Bot. Gaz.* 16:259

104. Wargo, P. M. 1971. Enhanced growth of *Armillaria mellea* on extracts from roots of defoliated sugar maple trees. *Phytopathology* 61:915

105. Warren, G. L., Singh, P. 1970. Hylobius weevils and armillaria root rot in a coniferous plantation in Newfoundland. *Can. Dep. Fish. For. Res. Notes* 26:55

106. Wiggins, E. A. 1976. *Interactions of the Collembola and microflora of cotton rhizosphere in relation to seedling disease potential.* MS thesis. Auburn Univ., Auburn, Alabama. 149 pp.

107. Wiggins, E. A., Curl, E. A. 1979. Interactions and microflora of cotton rhizosphere and potential relation to root pathogen association. *Phytopathology* 69:244–49

108. Wiggins, E. A., Curl, E. A., Harper, J. D. 1978. Effects of soil fertility and cotton rhizosphere on populations of Collembola. *Pedobiologia* 18:In press

109. Wilson, C. L. 1969. Use of plant pathogens in weed control. *Ann. Rev. Phytopathol.* 7:411–34

110. Windels, C. E., Windels, M. B., Kommedahl, T. 1976. Association of *Fusarium* species with picnic beetles on corn ears. *Phytopathology* 66:328–31

111. Zeiders, K. E., Newton, R. C., Graham, J. H. 1969. Influence of caging and pesticide drenches on yield and persistence of red clover, *Trifolium pratense* L. *Agron. J.* 61:952–53

AUTHOR INDEX

503

Maramorosch, K., 38, 40, 41,
43, 44, 46, 48, 50
Marchant, R., 475
Marchoux, G., 41, 46, 47
Marfey, P., 110
Marie, R., 82
Marinos, N. G., 108, 110
Markham, P. G., 39-42, 44, 49
Marks, C. F., 408, 409
Marley, S. J., 345
Marschall, K. J., 47
Marschner, H., 97, 99-106,
108-10, 113
Marsh, R. F., 480
Marshall, M. I., 59
Martens, J. W., 79, 80, 395
Martin, C. R., 350
Martin, D., 107, 109, 113,
114
Martin, G. C., 290
Martin, M. M., 40, 44
Martin, N. E., 270
Martinez, A. L., 44
Marwitz, R., 38, 47, 48
Marx, D. H., 447
Maskell, E. J., 99
Mason, T. G., 99
Massenot, M., 372, 376
Master, I. M., 437, 443, 450
Mathys, M. L., 172
Matsumoto, T., 125
Matthee, F. N., 127
Mattoo, A. K., 112
Matveenko, A. N., 63
Maurer, C. L., 494, 495
Maurer, R. A., 494, 495
Maurer, S. G., 352
Mavrodineau, S., 291
Maw, G. A., 420
Mazliak, P., 110
Mazzucchi, U., 127, 129, 139,
165
McCain, A. H., 420
McCann, J., 175
McCardell, B. A., 169
McCarver, T. H., 437, 438
McClellan, W. D., 9
McClements, W. L., 465
McClure, W. F., 348
McCoy, R. E., 41
McElroy, F. D., 289
McGee, D. C., 346
McGinnis, J. T., 486
McGinnis, R. C., 380
McGinty, R. J., 348
McGlohon, N. E., 285
McGregor, 65
McIntosh, A. H., 41, 43, 48
McIntosh, D. L., 442, 444, 446,
447, 453-55
McIntyre, J. L., 50
McKenry, M. V., 405, 407-9,
411, 413-15, 418, 419, 421
McKenzie, B. A., 344, 351, 357

McKenzie, R. I. H., 79, 80,
395
McKinney, D. E., 302, 307
McKinney, R. M., 124, 139
McMillian, W. W., 359
McNeilan, R. A., 173
McNew, G. L., 127, 134
McWhorter, F. P., 258
Meagher, J. W., 285
Medappa, K. C., 480
Meddins, B. M., 39, 41, 42, 44,
49
Medeiros, A. G., 126
Mehta, Y. R., 211
Melouk, H. A., 85
Mence, M. J., 443, 445
Mendonça, A. de V. F., 47
Mendonca, M. M., 129
Menge, J. A., 418
Mengel, K., 100, 113
Merlo, D. J., 170, 171, 182-84,
197
Meronuck, R. A., 350-53
Message, B., 187
Messenger, P. S., 303
Messens, E., 191
Meuser, R. U., 106
Mexal, J., 438
Meyer, D., 441, 444, 446, 447,
453-55
Meyer, J. A., 490
Meyer, R., 433
Michael, G., 104, 110
Michel, M., 197
Micke, A., 75, 77, 85
Middleton, R. B., 185, 193
Milani, V. J., 191
Milbrath, D. G., 258
Millard, W. A., 231
Millaway, R. M., 97, 113
Miller, D. L., 488, 492
Miller, J. D., 395
Miller, L. I., 284
Miller, P. M., 291
Miller, P. R., 230
Miller, P. W., 125
Miller, T. A., 153
Millikan, C. R., 102
Millis, W. A., 21
Mills, D., 189
Mills, D. R., 464
Mills, J. T., 346, 486, 488
Milne, D. L., 287
Milward, Z., 355
Minteer, R., 415, 419
Minteer, R. J., 291
Miranowski, J., 152, 155
Mircetich, S. M., 39-41
Mirocha, C. J., 346
Mitchell, D. J., 303, 304, 444
Mitchell, J. E., 433, 447-50
Mišiga, S., 45
Mix, G. P., 97, 100, 102-5
Miyata, Y., 450, 451

Moberg, R., 331
Mode, C. J., 368, 394
Moës, A. J., 76
Mohr, H. E., 345, 346, 359
Moll, J. N., 40, 44, 50
Mollenhauer, H. H., 40, 41, 49
Moller, W. J., 41, 164-66, 172
Monroe, R. L., 50
Montoya, A. L., 170, 171, 173,
182-84, 197
Moody, M. D., 134
Moore, B. J., 410
Moore, C., 471
Moore, H. D., 489
MOORE, L. W., 163-79;
164-68, 170-75, 184, 269
Moran, E. J. Jr., 353, 355
Moreno-Martinez, E., 358
Morey, D. D., 395
Morey, R. V., 350-53
Morgan, G. T., 293
Morishta, R. S., 488
Moritsugu, M., 97
Morré, D. J., 112, 434
Morris, T. J., 462-65, 467, 476
Mortensen, J. A., 49
Morton, D. J., 127, 130, 131,
134
Morvan, G., 49
Moseley, M., 39, 49
Moseman, J. G., 371, 372,
385-87, 389, 390, 394-96
Mossin, J., 152
Mostafa, M. A. E., 102
Mostafawi, M., 103, 104, 114
Mota, M., 47
Mountain, W. B., 288
Mowat, D. N., 353, 355
Mugniery, D., 288
Mühlbach, H. P., 478
Muir, W. E., 348, 351
Müller, H. M., 38, 42, 44, 48
Müller, K. O., 230
MUNNECKE, D. E., 405-29;
169, 405, 409-12, 415-23
Murant, A. F., 292
Murashige, T., 476
Murphy, H. C., 379, 380, 395
Murray, E. G. D., 123
Murray, J. J., 488
Murray, M. J., 85
Murty, B. R., 81, 86
Mushin, R., 126, 134
Musil, M., 45
Myers, R. A., 488

N

Nadler, K. D., 112
Naegele, J. A., 253
Naghski, J., 126, 130
Nain, K., 126
Nakai, H., 78, 81, 87
Nakajima, K., 85
Nakamura, K., 172

SUBJECT INDEX

A

Abscisic acid, 475
 ion uptake, 99
Acalymma vittala
 disease transmission, 487
Acetic acid, 355
Acetobacter roseum
 summer bunch rot, 8
Acetobacter sp., 7–8
Achlya diffusa
 solute potential effect, 436
Achlya sp.
 cytokinins
  calcium uptake
   stimulation, 106
Acholeplasma laidlawii
 membrane surface, 39
 protein patterns, 49
Acholeplasma sp.
 saprophytic species
  multiplication, 44
 transmission, 43
Acridine orange, 189, 192–93
Actinomycin D
 viral RNA synthesis
  inhibition, 471
 viroid replication, 470–71
Acyrthosiphon pisum, 492
cAdenosine monophosphate
 production
  glucose inhibition, 172
  tumor inhibition, 173
Adenosine triphosphatase
 calcium effect, 111
Aeration
 nematode populations,
  289–90
Aeschynomene
 see Indian jointvetch;
  Northern jointvetch
Aflatoxin, 343
 ammonia destruction, 355–56
 production inhibition, 356
 test, 348
Agaricus bisporus
 methyl bromide sensitivity,
  409
Agerarina
 see Pamakani weed
Agrobacterium
 radiobacter, 183
  agricin 84, 169
  animal toxicity, 174
  inoculum stability, 175
  serological differentiation,
   126
  strain 84 attributes as
   biocontrol agent,
   174–76

 see also Biocontrol of
  crown gall
 rhizogenes
  root proliferation transfer,
   183
  serological differentiation,
   126
 rubi
  serological identification,
   126
 tumefaciens
  agrocin 84 sensitivity,
   165–66
  biocontrol, 269
  citrus exocortis viroid
   interaction, 477
  genetic exchange, 168
  oncogenicity, 182
  oncogenicity transfer, 183
  oncogenicity transformation, 191
  plasmid transfer, 190
  plasmid and virulence,
   183, 198
  serological differentiation,
   125–26
  Ti plasmid, 170–72
  transposon, 197
  viroid-like RNA, 480
  see also Biocontrol of
   crown gall
Agrobacterium spp.
 identification, 138
 Ouchterlony double diffusion
  tests, 127
Agrocin 84
 Agrobacterium strains
  sensitivity, 165–66
  affecting factors, 170–73
  glucose effect, 166
 mode of action, 169
  alternative explanations,
   173
  physiochemical nature,
   169–70
 sensitive Agrobacterium
  strains
  phenotypic characteristics,
   171
Air pollution
 see Lichens as air pollution
  indicators
Alectoria implexa
 air pollution indicator,
  331–32
 fluorine effect, 332
Alfalfa (Medicago)
 soil depth
  Xiphinema population,
   282–83

 soil pH
  Pratylenchus growth rates,
   294
Alfalfa dwarf, 42, 46, 49
Allium
 see Onion; Rust fungi-host
  coevolution
Almond (Prunus)
 crown gall biological control
  breakdown, 168
Almond leaf scorch, 42, 46, 49
Alternaria
 solani
  control, 317
  methyl bromide sensitivity
   and moisture, 410
 tenuis
  matric potential effect,
   440–41
  solute potential and
   growth, 438
  tobacco curing loss, 244
Alternaria spp., 359
 grain drying, 353
Althornia crouchii
 water solute potential and
  growth, 435
American Phytopathological
 Society, 19
 formation, 16
Ames test, 175
Ammonia
 aflatoxin destruction, 355–56
 grain storage, 356
 toxicity
  fumigation, 420
Ammonium
 effect on calcium uptake, 99
Ammonium butyrate, 355
Amo 1618, 106
Amoeba proteus
 small nuclear RNAs, 479
α-Amylase
 calcium requirement, 112
Anaptychia ciliaris
 air pollution indicator, 330,
  333
 distribution maps, 337
Anguina
 agrostis
  osmotic pressure tolerance,
   291
 tritici
  desiccation tolerance, 289,
   291
  osmotic pressure tolerance,
   291
  restriction by moisture,
   289

CUMULATIVE INDEXES

CONTRIBUTING AUTHORS, VOLUMES 13–17

541

CHAPTER TITLES, VOLUMES 1–17

Please list on the order blank on the reverse side the volumes you wish to order and whether you wish a standing order (the latest volume shipped to you automatically upon publication each year). Volumes not yet published will be shipped in month and year indicated. Out of print volumes subject to special order.

PRICE CHANGE NOTICE

All volumes of the Annual Reviews published on or after July 1, 1980 will increase in price by $3.00 per copy (USA) and $3.50 per copy (elsewhere). This price change is effective regardless of when the order is placed. Volumes published prior to July 1, 1980 will not be affected by the price increase.

NEW SERIES.... Volume 1 to be published May 1980

Annual Review of PUBLIC HEALTH $17.00 (USA), $17.50 (elsewhere) per copy

SPECIAL PUBLICATIONS

ANNUAL REVIEW REPRINTS: CELL MEMBRANES, 1975-1977 (published 1978)
A collection of articles reprinted from recent Annual Review series.
    ISBN 0 8243 2601 X   Soft cover: $12.00 (USA), $12.50 (elsewhere) per copy

THE EXCITEMENT AND FASCINATION OF SCIENCE (published 1965)
A collection of autobiographical and philosophical articles by leading scientists.
    ISBN 0-8243-1601-0   Clothbound: $6.50 (USA), $7.00 (elsewhere) per copy

THE EXCITEMENT AND FASCINATION OF SCIENCE, VOLUME 2:
Reflections by Eminent Scientists (published 1978)
    ISBN 0-8243-2601-6   Hard cover: $12.00 (USA), $12.50 (elsewhere) per copy
    ISBN 0-8243-2602-4   Soft cover: $10.00 (USA), $10.50 (elsewhere) per copy

THE HISTORY OF ENTOMOLOGY (published 1973)
A special supplement to the Annual Review of Entomology series.
    ISBN 0-8243-2101-7   Clothbound: $10.00 (USA), $10.50 (elsewhere) per copy

ANNUAL REVIEW SERIES

Annual Review of ANTHROPOLOGY   ISSN 0084-6570
    Vols. 1-7 (1972-78) now available
    Vol. 8 available Oct. 1979 $17.00 (USA), $17.50 (elsewhere) per copy

Annual Review of ASTRONOMY AND ASTROPHYSICS   ISSN 0066-4146
    Vols. 1-16 (1963-78) now available
    Vol. 17 available Sept. 1979 $17.00 (USA), $17.50 (elsewhere) per copy

Annual Review of BIOCHEMISTRY   ISSN 0066-4154
    Vols. 28-48 (1959-79) now available $18.00 (USA), $18.50 (elsewhere) per copy

Annual Review of BIOPHYSICS AND BIOENGINEERING   ISSN 0084-6589
    Vols. 1-8 (1972-79) now available $17.00 (USA), $17.50 (elsewhere) per copy

Annual Review of EARTH AND PLANETARY SCIENCES   ISSN 0084-6597
    Vols. 1-7 (1973-79) now available $17.00 (USA), $17.50 (elsewhere) per copy

Annual Review of ECOLOGY AND SYSTEMATICS   ISSN 0066-4162
    Vols. 1-9 (1970-78) now available
    Vol. 10 available Nov. 1979 $17.00 (USA), $17.50 (elsewhere) per copy

Annual Review of ENERGY   ISSN 0362-1626
    Vols. 1-3 (1976-78) now available
    Vol. 4 available Oct. 1979 $17.00 (USA), $17.50 (elsewhere) per copy

Annual Review of ENTOMOLOGY   ISSN 0066-4170
    Vols. 7-24 (1962-79) now available $17.00 (USA), $17.50 (elsewhere) per copy

Annual Review of FLUID MECHANICS   ISSN 0066-4189
    Vols. 1-11 (1969-79) now available $17.00 (USA), $17.50 (elsewhere) per copy

Annual Review of GENETICS   ISSN 0066-4197
    Vols. 1-12 (1967-78) now available
    Vol. 13 available Dec. 1979 $17.00 (USA), $17.50 (elsewhere) per copy

Annual Review of MATERIALS SCIENCE   ISSN 0084-6600
    Vols. 1-8 (1971-78) now available
    Vol. 9 available Aug. 1979 $17.00 (USA), $17.50 (elsewhere) per copy

(continued on reverse side)

Annual Review of MEDICINE: Selected Topics in the Clinical Sciences
Vols. 1-3, 5-15, 17-30 (1950-52,   ISSN 0066-4219
1954-64, 1966-79) now available $17.00 (USA), $17.50 (elsewhere) per copy
----
Annual Review of MICROBIOLOGY   ISSN 0066-4227
Vols. 15-32 (1961-78) now available
Vol. 33 available Oct. 1979 $17.00 (USA), $17.50 (elsewhere) per copy
----
Annual Review of NEUROSCIENCE   ISSN 0147-006X
Vols. 1-2 (1978-79) now available $17.00 (USA), $17.50 (elsewhere) per copy
----
Annual Review of NUCLEAR AND PARTICLE SCIENCE   ISSN 0066-4243
Vols. 10-28 (1960-78) now available
Vol. 29 available Dec. 1979 $19.50 (USA), $20.00 (elsewhere) per copy
----
Annual Review of PHARMACOLOGY AND TOXICOLOGY   ISSN 0362-1642
Vols. 1-3, 5-19 (1961-63, 1965-79)
now available $17.00 (USA), $17.50 (elsewhere) per copy

Annual Review of PHYSICAL CHEMISTRY   ISSN 0066-426X
Vols. 10-21, 23-29 (1959-70,
1972-78) now available $17.00 (USA), $17.50 (elsewhere) per copy
Vol. 30 available Nov. 1979

Annual Review of PHYSIOLOGY   ISSN 0066-4278
Vols. 18-41 (1956-79) now available $17.00 (USA), $17.50 (elsewhere) per copy
----
Annual Review of PHYTOPATHOLOGY   ISSN 0066-4286
Vols. 1-16 (1963-78) now available
Vol. 17 available Sept. 1979 $17.00 (USA), $17.50 (elsewhere) per copy
----
Annual Review of PLANT PHYSIOLOGY   ISSN 0066-4294
Vols. 10-30 (1959-79) now available $17.00 (USA), $17.50 (elsewhere) per copy
----
Annual Review of PSYCHOLOGY   ISSN 0066-4308
Vols. 4, 5, 8, 10-30 (1953, 1954,
1957, 1959-79) now available $17.00 (USA), $17.50 (elsewhere) per copy
----
Annual Review of SOCIOLOGY   ISSN 0360-0572
Vols. 1-4 (1975-78) now available
Vol. 5 available Aug. 1979 $17.00 (USA), $17.50 (elsewhere) per copy

---

To ANNUAL REVIEWS INC., 4139 El Camino Way, Palo Alto, CA 94306 USA (415-493-4400)

Please enter my order for the following publications:
(If a standing order, indicate which volume you wish order to begin with)

_____, Vol(s). _____ Standing order _____

_____, Vol(s). _____ Standing order _____

_____, Vol(s). _____ Standing order _____

_____, Vol(s). _____ Standing order _____

Amount of remittance enclosed $_____   California residents please add sales tax.
Please bill me _____   Prices subject to change without notice.

SHIP TO (Include institutional purchase order if billing address is different)

Name _____

Address _____

_____

_____ Zip code _____

Signed _____   Date _____

_____ Send free copy of annual Prospectus for current year

_____ Send free brochure listing contents of recent back volumes for Annual Review(s)

of _____